1st International Conference on Geoscience (ICoGes 2018)

Geoscience and Technology for Better Environment

IOP Conference Series: Earth and Environmental Science
Volume 279

Makassar, Indonesia
1 – 2 November 2018

Editor:

Moh. Ivan Azis

ISBN: 978-1-5108-9644-4
ISSN: 1755-1307

Printed from e-media with permission by:

Curran Associates, Inc.
57 Morehouse Lane
Red Hook, NY 12571

proceedings.com

Some format issues inherent in the e-media version may also appear in this print version.

Institute of Physics
Dirac House, Temple Back
Bristol BS1 6BE UK

Phone: 44 1 17 929 7481
Fax: 44 1 17 920 0979

techtracking@iop.org

Additional copies of this publication are available from:

Curran Associates, Inc.
57 Morehouse Lane
Red Hook, NY 12571 USA
Phone: 845-758-0400
Fax: 845-758-2633
Email: curran@proceedings.com
Web: www.proceedings.com

TABLE OF CONTENTS

GEOENGINEERING

GEO-INFORMATICS

GEOLOGY AND SOLID GEOPHYSICS

GEOPHYSICAL METHODS

OCEAN AND ATMOSPHERIC SCIENCES

IOP Conf. Series: Earth and Environmental Science **279** (2019) 011001 doi:10.1088/1755-1315/279/1/011001

Preface

The International Conference on Geoscience (ICoGeS) is organized by the Study Program of Geophysics, Faculty of Mathematics and Natural Sciences, Hasanuddin University, Indonesia in cooperation with Indonesian Association of Geophysicist (HAGI), Indonesian Student Association of Geophysics–V Region (HMGI), Society of Exploration Geophysicists (SEG) Hasanuddin University Student Center, Society of Petroleum Engineer (SPE) Hasanuddin University Student Chapter, European Association of Geoscientist & Engineers (EAGE) Hasanuddin University Student Chapter.

The objectives are to bring together scholars, researchers and experts from diverse areas of backgrounds and applications. A special goal of the conference is to promote an interaction and exchange of knowledge and experiences among scientists and practitioners especially in geoscience theoretical and experimental studies.

This 1st ICoGes was held in Makassar (Capital city of South Sulawesi Province) on 1st – 2nd of November 2018 with the main theme *"Geoscience and Technology for Better Environment"*. We have six keynote speakers coming from Terra Australis Geophysica Pty. Ltd. Australia (Mr. Koya Suto), Hasanuddin University Indonesia (Prof. Dr. Dadang A. Suriamihardja), Bandung Institute of Technology Indonesia (Prof. Satria Bijaksana, PhD), Universiti Malaysia Kelantan Malaysia (Prof. Madya Dr. Aweng A/I Eh Rak), Universiti Malaysia Kelantan Malaysia (Dr. Arham Muchtar A. Bahar), Bandung Institute of Technology Indonesia (Dr. David P. Sahara).

Erfan Syamsuddin (Chairman) and Moh. Ivan Azis (Editor in Chief) *
The 1st International Conference on Geoscience (ICoGes 2018) Publication
Geophysics Study Program
Faculty of Mathematics and Natural Sciences
Hasanuddin University
Tamalanrea, Makassar, 90245, Indonesia
*E-mail: mohivanazis@yahoo.co.id

IOP Conf. Series: Earth and Environmental Science **279** (2019) 011002 doi:10.1088/1755-1315/279/1/011002

The Committees

The 1st International Conference on Geoscience (ICoGes 2018)

Advisory Editorial Board

Dadang Ahmad Suriamihardja	(Hasanuddin University, Indonesia)
Halmar Halide	(Hasanuddin University, Indonesia)
Moh. Ivan Azis	(Hasanuddin University, Indonesia)
Dahlang Tahir	(Hasanuddin University, Indonesia)
Amiruddin	(Hasanuddin University, Indonesia)
Altin Massinai	(Hasanuddin University, Indonesia)
Adi Maulana	(Hasanuddin University, Indonesia)
Purwanto	(Hasanuddin University, Indonesia)
Ardy Arsyad	(Hasanuddin University, Indonesia)
Hendra Pachri	(Hasanuddin University, Indonesia)
Peter Lepong	(Mulawarman University, Indonesia)
Rustam Effendi	(Tadulako University, Indonesia)
Virman	(Cenderawasih University, Indonesia)
Ida Usman	(Haluoleo University, Indonesia)
Idris Mandang	(Mulawarman University, Indonesia)
Rusydi Hasanuddin	(Tadulako University, Indonesia)
Koya Suto	(Terra Australis Geophysica Pty Ltd, Australia)
Arham Muchtar Ahmad Bahar	(Universiti Kelantan Malaysia, Malaysia)
Aweng A/I Eh Rak	(Universiti Kelantan Malaysia, Malaysia)
Satria Bijaksana	(Bandung Institute of Technology, Indonesia)
David P. Sahara	(Bandung Institute of Technology, Indonesia)

Organizing Committee

Erfan Syamsuddin	(Chairman)
Muhammad Fawzy Ismullah	(Secertary)
Maria	(Treasurer)
Sabrianto Aswad	(Secretariat Member)
Lantu	(Secretariat Member)
Hasanuddin	(Secretariat Member)
Bambang Harimei	(Secretariat Member)
Sakka	(Secretariat Member)
Samsu Arif	(Secretariat Member)
Paharuddin	(Secretariat Member)
Muh. Alimuddin Hamzah	(Secretariat Member)
Muhammad Hamzah	(Secretariat Member)
Makhrani	(Secretariat Member)
Syamsuddin	(Secretariat Member)
Nur Hasanah	(Secretariat Member)
Edy Saputra	(Secretariat Member)
Dewi Ayu Swastika	(Secretariat Member)
Muhammad Nasri	(Secretariat Member)
Edi Wahyudi	(Secretariat Member)
Nur Ayu Anas	(Secretariat Member)
Sufridah Hardiyanti	(Secretariat Member)
Muh. Resky Ariansyah	(Secretariat Member)
Aslam	(Secretariat Member)
Muh. Shabran	(Secretariat Member)
Sri Wahyuni	(Secretariat Member)
Fatma Jamil	(Secretariat Member)
Muh. Adimaher Zamhuri	(Secretariat Member)
Diky Prayudi Anggara	(Secretariat Member)
Harmita Lestari	(Secretariat Member)

IOP Conf. Series: Earth and Environmental Science **279** (2019) 011002 doi:10.1088/1755-1315/279/1/011002

Muh. Firmansyah	(Secretariat Member)
Fajriani Samrin	(Secretariat Member)

Advisory Committee

Amiruddin	Dean of Faculty of Mathematics and Natural Sciences
Moh Ivan Azis	Vice-Dean for Academic Affairs
Muhammad Zakir	Vice-Dean for Finance and Human Resources
Andi Ilham Latunra	Vice-Dean for Student Affairs and Alumni Relations

Steering Committee

Koya Suto (Terra Australis Geophysica, Australia)
Satria Bijaksana (Bandung Institute of Technology, Indonesia
Dadang A. Suriamihardja (Hasanuddin University, Indonesia)
Aweng A/I Eh Rak (Universiti Kelantan Malaysia, Malaysia
Arham Muchtar Ahmad Bahar (Universiti Kelantan Malaysia, Malaysia)
David P. Sahara (Bandung Institute of Technology, Indonesia
Moh. Ivan Azis (Hasanuddin University, Indonesia)
Muh. Alimuddin Hamzah (Hasanuddin University, Indonesia)
Ardy Arsyad (Hasanuddin University, Indonesia)

Scientific and Reviewer Committee

Dahlang Tahir (Hasanuddin University)
Isnaeni (Indonesian Institute of Sciences)
Halmar Halide (Hasanuddin University)
Taufik Sutanto (Syarif Hidayatullah State Islamic University)
Paulina Taba (Hasanuddin University)
Misita Anwar (Monash University, Australia)
Utami Dyah Syafitri (Bogor Agricultural University)
Junaidi (Tadulako University)
Sri Astuti Thamrin (Hasanuddin University)
Achmad Effendi (Brawijaya University)
Mawardi Bahri (Hasanuddin University)

IOP Conf. Series: Earth and Environmental Science **279** (2019) 011003 doi:10.1088/1755-1315/279/1/011003

Conference Photographs

The First International Conference on Geoscience (ICoGes 2018)

Sponsor or funding acknowledgements

We would like to thank our main sponsor, who has supported the 1st International Conference on Geoscience (ICoGes 2018).

IOP Conf. Series: Earth and Environmental Science **279** (2019) 011005 doi:10.1088/1755-1315/279/1/011005

IOP Conf. Series: Earth and Environmental Science **279** (2019) 011005 doi:10.1088/1755-1315/279/1/011005

9	The Analysis of Salt Pond Cluster in Arungkeke sub-district Jeneponto Regency *Sakka, Baharuddin, W Samad*
10	Petrogenesis of Kalisongo Tertiary Lava, Girimulyo Kulon Progo, Yogyakarta based on Mineralogy and Geochemical Analysis *Yoga Aribowo, Rinal Khaidar Ali, Abdul Najib*
11	Derivative using Complex Variable Conjugate Approach for Analytic Signal of Magnetic Field Anomaly Due to 2D Finite Prism *Abdul Manan, La Hamimu, Rani Chahyani*
12	Efforts to Improve Community Awareness Toward the Potential of A Great Earthquake Which Threats Jakarta Based on Geographic Information and 3D Simulation Systems in Matraman District, East Jakarta *Zidny Ilma Andromeda, Cattleya Randi, Khairunnisa Nazhifah, Reza Syahputra, Iskandarsyah, Muhammad Rizqy Septyandy*
13	Behavior of Mercury Around the Landfill of Tamangapa Antang, Makassar City, Indonesia *Andi Artiningsih, Hazairin Zubair, A. M. Imran, Sri Widodo*
14	Two-Dimensional Inversion Magnetotelluric Investigation of the Pariangan Geothermal Field, Indonesia *Wahyuni, Sabrianto Aswad, Maria, Wiwid Joni, Jamaluddin*
15	Groundwater Conservation with Hole Infiltration of Biopore Cube *Muhammad Hamzah Syahruddin, Amiruddin, Halmar Halide, Sakka, Makhrani*
16	Analysis of Regional Anomaly on Magnetic Data Using the Upward Continuation Method *Ilapadila, Bambang Harimei, Maria*
17	Analysis of Seismicity Level in Central Sulawesi Period 1997 - 2017

IOP Conf. Series: Earth and Environmental Science **279** (2019) 011005 doi:10.1088/1755-1315/279/1/011005

IOP Conf. Series: Earth and Environmental Science **279** (2019) 011005 doi:10.1088/1755-1315/279/1/011005

	Rahmi Mulyasari, Hari Wiki Utama, Nandi Haerudin
35	Geology and Genesis of the Tanco Isolated Hill on the Kerinci Lake, Jambi *Fiqhy Nabella, Hari Wiki Utama, Yulia Morsa Said*
36	Groundwater Identification at East Sentani Auwena Formation (Tema) Based on Geoelectric *Virman, Muhammad Altin Massinai, Muhammad Fawzy Ismullah Massinai*
37	Detection and Delineation of Shallow Channel Morphology Using 3D Seismic Data in K Field, Malay Basin, Malaysia *Nur Hazira Hamsan, Md Yazid Mansor, Liaw Kim Kiat*
38	Study of Climate Determination Analysis Based on Pallontara/ Papananrang and Rainfall Opportunities in Sidrap District *Amir Yassi, Kaimudin, Abdul Haris Bahrun, Asmiaty Sahur*
39	Distribution of Soft Coral Octocorallia (Alcyonacea) in coastal waters of Gonda, Polewali-Mandar, West Sulawesi *Ayub Wirabuana Putra, Magdalena Litaay, Dody Priosambodo*
40	Estimation Mean by the Bayesian Approach on the Exponentially Weight Moving Average Control Chart *Meylina Siruddin, Erna Tri Hardiani, Sri Astuti Thamrin*
41	Geographical Mapping of Dengeu Fever Incidence 2012 -2016 in Makassar, Indonesia *Sri Astuti Thamrin, Alimun*
42	Heavy Metal Cadmium (Cd) Content in the Medaka Fish Oryzas Javanicus Bleeker *Magdalena Litaay, Andi Iin Fadliah, Irma Adriani, Dahlang Tahir*

	Deniyatno, A Armid, Harisma, E Anshari and E S Hasan
52	Identify the Distribution of Galena using Induced Polarization and Resistivity Methods in central of Lombok, West Nusa Tenggara *Asnur Azis, Muh. Adimaher Zamhuri, Muh. Iqbal Rais, Syamsuddin, Sabrianto Aswad, Obed Patiung, Yanto Sudianto*

IOP Conf. Series: Earth and Environmental Science **279** (2019) 011006 doi:10.1088/1755-1315/279/1/011006

Peer review statement

All papers published in this volume of *IOP Conference Series: Earth and Environmental Science* have been peer reviewed through processes administered by the proceedings Editors. Reviews were conducted by expert referees to the professional and scientific standards expected of a proceedings journal published by IOP Publishing.

Comparative analysis of ANFO and emulsion application on overbreak and underbreak at blasting development activity in underground Deep Mill Level Zone (DMLZ) PT Freeport Indonesia

S Widodo[1], H Anwar[2] and N A Syafitri[3]

[1]Mining Engineering Department, Hasanuddin University, Indonesia
[2,3]Mining Engineering Department, Moslem University of Indonesia, Indonesia

[1]srwd007@yahoo.com

Abstract. Underground Deep Mill Level Zone (DMLZ) mine PT Freeport Indonesia (PTFI) is still in development stage and currently has no production activities yet. In underground DMLZ mine, a trial of emulsion explosive has conducted since March 15th, 2017 where previously used was Ammonium Nitrate Fuel Oil (ANFO) explosive for tunnel blasting. This study purposes to analyse the results and impact of ANFO and emulsion explosive in order to know the advantages and disadvantages of each explosives. In this study, an overbreak and underbreak comparison evaluation was performed by ANFO explosives and emulsion in underground DMLZ mine tunnel blasting. Furthermore, processing data based on explosives specification data, explosives quantity, heading location and time that detonated, solid design and actual tunnel DMLZ of PT. Freeport Indonesia, also cross section data. After all the data were processed and analysed, it was concluded that the emulsion explosive was more effective for use in tunnel blasting. To optimize the effective blasting results it is necessary to redesign new drilling patterns and blasting geometry by used of radial crack method, so that they can be applied to DMLZ tunnels in the future.

1. Introduction

PT. Freeport Indonesia (PTFI) is a mining company that using surface and underground mining systems. PTFI's surface mine called Grasberg is one of the largest open pit in the world, now the reserves of Grasberg will be exhausted in a few years. PTFI's underground mine has five sites: Kucing Liar, Big Gossan (BG), Deep Ore Zone (DOZ), Grasberg Block Cave (GBC), and Deep Mill Level Zone (DMLZ).

Underground mining activities require the best possible technical planning for ore retrieval. One such plan is the establishment of a tunnel (Drift) as access, Conveyor Drift, Intake and Exhaust Drift.

PTFI DMLZ underground mine is still in development stage and has not done production activities yet. One of the development stage activities is the opening of the tunnel to open the access path for employees and mining equipment. The activity is to blast the tunnel by burn cut method.

In the DMLZ underground mine, a trial of emulsion explosives has been conducted since March 15th, 2017 where previously used was ANFO for tunnel blasting. The heading is detonated with ANFO explosives, the perimeter hole using detonating cord is not filled with ANFO so that when mucking the materials of explosion and dumping in the crusher can cause explosion in the crusher. In addition,

ANFO explosives can cause excessive cracking of the tunnel.

ANFO and emulsion explosives has their respective advantages and disadvantages such as ANFO is not water resistant and emulsion is water resistant. ANFO's diameter is 2 mm and emulsion 0.001 mm where ANFO mixing is easy to do anywhere, but emulsions require enhanced sensitivity by adding a sensitizing agent (sensitizer). Since there are several consequences of using ANFO explosives in DMLZ tunnel blasting, the researchers conducted further observations on emulsion use trials as well as the effect of emulsion on overbreak and underbreak, also evaluated comparative effects of the use of ANFO and emulsions.

2. Methodology

There are literature study and data retrieval at this stage with primary data and secondary data. The data required in the final project study is obtained from Orica's direct observation as well as from UG DMLZ Engineering department in underground mine Deep Mill Level Zone PT. Freeport Indonesia.

The primary data that needed for this study are quantity of explosives used, actual tunnel condition, actual perimeter area and circumference, overbreak and underbreak area and circumference, the furthest distance of overbreak and underbreak, and actual tunnel length and volume. And the secondary data are explosive specification, the location and time of heading being detonated, tunnel solid design and actual, also drilling pattern and blasting geometry.

Explosive and rock specification data, quantity of explosives used, the location and time of heading detonated, as well as tunnel solid design and actual DMLZ PTFI be processed and obtained cross section data using Auto CAD. The cross section results will be processed again to analyze the perimeter, overbreak and underbreak, as well as the area and volume of the tunnel. Once everything is analyzed, conclusions and suggestions on effective explosives can be found to use in future DMLZ tunnels.

2.1. Overbreak and Underbreak

Many previous studies have been conducted to examine overbreak and underbreak in mining operation actvities. Overbreak and underbreak have been known as the main causal of hazards and damages to mining industry management, especially in underground mining. Refer to some previous studies, factors that causing overbreak and underbreak in mining industry can be catagorized into two points. Geological and blasting factors were the two major points that play a big role for overbreak and underbreak occurences in tunnel [1], [7].

Blasting geometry are changeable components. In modern underground blasting, smooth blasting and presplitting methods are reasonably organized techniques with advanced final wall blasting methods. Final wall customized explosives and computer base drill operating systems significantly minimize the possible failures on blasting operations. However, geological factors are unchangeable and they have a significant influence on the overbreak and underbreak phenomena. In fact, if the rock is not strong enough to support itself, possibly no blasting techniques can stop the occurrence of overbreak and underbreak. Overbreak is a surplus blasted area of rock beyond the theoretical contour in a blasting activity that can occur in any kind of underground development method. Otherwise, underbreak is rock remaining (a minus blasted area) within a specific blasting perimeter that should have been thrown out by the blast and needeed secondary blasting or excavation. The form of overbreak and underbreak in a tunnel developmnet after blasting are shown in **Figure 1.**

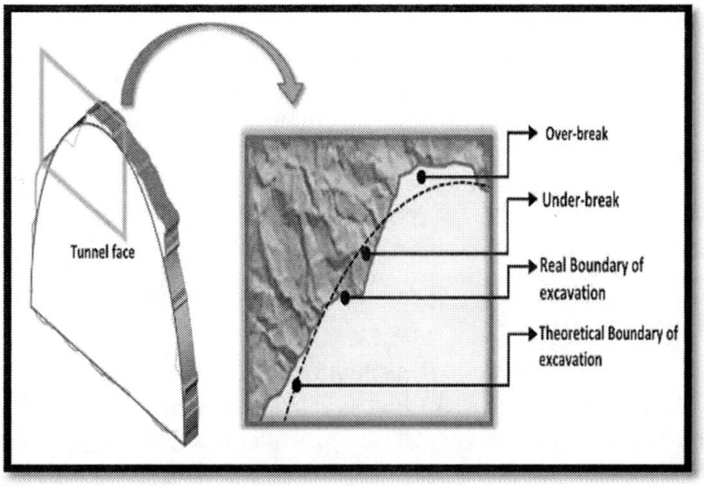

Figure 1. Forms of overbreak and underbreak in the tunnel after blasting activity [8]

2.1.1. Parameters of blasting. Overbreak can be managed by set these parameters because parameters of blasting are able to be changed. The parameters of blasting include blasting geometry, sub-drilling, firing sequences, guide holes, cut design, blasting hole, deviation, explosive characteristics, charge concentrations, powder factors, coupling ratio, blast-induced shock wave, energy levels, and others. Within just a few milliseconds, all blasting parameters affect overbreak in complex mutual correlation. Accurate blasting and drilling design should take priority and play role the selection of other blasting parameters to obtain a smooth (minimum) fracture plane without any wall damage. For example, [6] has been done model and field blasting tests to define the optimal delay time between contour holes in lock blasting by comparing instantaneous and micro-sequential initiation systems. Instantaneous initiation system was shown to be superior in minimizing overbreak. Field tests indicated the maximum radial crack length into the remaining rock mass of instantaneous initiation systems created a 1.3–9.0 times less than the micro-sequential initiation system which had only 1 ms firing delay between contour holes.

2.1.2. Geological Parameters. Geological parameters are unchanging factors and almost of them such as the strength of rock mass, discontinuous characteristics, stress and water conditions, as well the topography of the surrounding area have importance impact on the overbreak and underbreak phenomena. For example, Hagan [5] emphasised the importance of beddings and pre-existing joints on in-situ rock. In accord with his study, fractures in the rock tend to predominate the nature of the blast-induced fracture pattern and it commonly affect the overbreak more than the physical and mechanical characteristics of the rock.

The orientation of discontinuity is one of the main factors affect the overbreak phenomena. According to the study by Hoek and Brown [4], a discontinuity plane having strike parallel to the tunnel axis is considered to have an unfavorable effect on overbreak. In general, less overbreaks and underbreaks are discovered where the strike of the discontinuity is nearly vertically to the tunnel axis and contrary greater when they are nearly parallel. In specific of other oriented, drives with dip are more gainful than drive against dip where the strike of discontinuity is vertically to tunnel axis and fair and very unprofitable for dip with angle of 20 to 45° and 45 to 90°, respectively, when the strike is parallel to tunnel axis [2].

2.2. Radius Crack Propagation

Radius crack propagation is the optimum distance where rocks can be ruptured by explosives, so with the determination of radius crack propagation can be determined from the location of the blast hole and the distance between the blast holes. According to Oucterlony [3], the equation to determine the magnitude of radius crack propagation that can be used is:

$$Q = \text{E eff ANFO x } 10^6 \text{ x REEws} \quad (1)$$

$$\gamma = \sqrt{(1 + D^2/Q)} \quad (2)$$

$$K_{Ic} = \text{Tensile Strength x } \sqrt{\pi \; x \; crack \; length} \quad (3)$$

where Q is explosion energy (J/kg); REEws is relative weight energy of explosive; γ is exponential of adiabatic expansion; D is explosive VOD (m/s); K_{Ic} is cracked strength of rocks (Pa.m$^{0.5}$). Blast hole pressure can be determined as follows:

$$Ph = \gamma^{\gamma} / \, (\gamma + 1)^{(\gamma+1)} \, . \rho_e . D^2 . (f)^{2.2} \quad (4)$$

$$Ph = \gamma^{\gamma} / (\gamma + 1)^{(\gamma+1)} \;\; \rho e \;\; D^2 \left(\Phi e / \Phi h \right)^{2.2} \quad (5)$$

where Ph is blast hole pressure (pa) ; ρe is explosive density (kg/m^3); Φe is charge diameter (m); Φh is blast hole diameter (m). Blast hole pressure with crack using following the equation:

$$P_{h,crack} = 3{,}30 Klc / \sqrt{\Phi_h} \quad (6)$$

where, $P_{h,Crack}$ = blast hole pressure with crack (pa). Rco = radial crack (m) can be determined as follows:

$$Rco = 0{,}5 \, x \, \Phi h \, x \left(P_h / P_{h,crack} \right)^{2/\left[3 \left(D/c \right)^{0.25} - 1 \right]} \quad (7)$$

3. Result and Discussion

3.1. Rock Specification

The physical, mechanical, and dynamic properties of an intact rock are important to know the behavior of rocks against the detonation process of explosives. The study area of diorite rock type has a uniaxial compressive strength are 156.5 MPa, tensile strength are 11.5 MPa, and density are 2.6 gr / cm^3 (see **Table 1**.).

Table 1. Rock mechanics and physical properties [9]

Rock Type	Diorite (Fresh)
Uniaxial Compressive Strength (MPa)	156.5
Tensile Strength (MPa)	11.5
Modulus Young (GPa)	51.66
Poisson Ratio	0.296
Friction Angle (^{0})	38.6
Cohesion (MPa)	6.3
Density (kg/m^3)	2703.5
Velocity (m/s)	5730.24

The blasting process in rock mass is very important to be studied in order to know how far the effect of these rock characteristics on the effectiveness of blasting activity. Based on data from UG

DMLZ Geology PTFI, obtained RQD value between 70% to 100% so it shows good rock quality in Drainage Level has been shown in **Table 2.**, Truck Haulage Level in **Table 3.**, Extraction Level in **Table 4.**, and Undercut Level see **Table 5.**

Table 2. Rock mass data in drainage level [9]

Rock Type	Parameters	Lowest Values	Middle Values	Highest Values
Diorite	Q-System	5.83	8.89	13.33
	RQD (%)	70-90	80-90	90-100
	Joint spacing (m)	1	2	3
	Number of joint sets	One Joint Set	Two Joint Sets	Two Joint Sets Plus Random
	Joint alteration (ja)	3	5	6

Table 3. Rock mass data in Truck Haulage Level [9]

Rock Type	Parameters	Lowest Values	Middle Values	Highest Values
Diorite	Q-System	5.83	8.89	13.33
	RQD (%)	70-90	80-90	90-100
	Joint spacing (m)	1	2	3
	Number of joint sets	One Joint Set	Two Joint Sets	Two Joint Sets Plus Random
	Joint alteration (ja)	3	5	6

Table 4. Rock mass data in Extraction Level [9]

Rock Type	Parameters	Lowest Values	Middle Values	Highest Values
Diorite	Q-System	5.83	10	14.965
	RQD (%)	70-90	80-90	90-100
	Joint spacing (m)	1	2	3
	Number of joint sets	One Joint Set Plus Random	Two Joint Sets	Two Joint Sets Plus Random
	Joint alteration (ja)	3	5	6

Table 5. Rock mass data in Undercut Level [9]

Rock Type	Parameters	Lowest Values	Middle Values	Highest Values
Diorite	Q-System	6.67	10.585	20
	RQD (%)	70-90	80-90	90-100
	Joint spacing (m)	1	2	3
	Number of joint sets	One Joint Set	Two Joint Sets	Two Joint Sets Plus Random
	Joint alteration (ja)	3	5	6

3.2. Overbreak and Underbreak Analysis

From the results data of Drainage, Truck Haulage, Extraction, and Undercut Level that has been averaged, then **Table 6.** showed results data of ANFO and emulsion application for area and circumference of overbreak and underbreak. Afterwards, **Table 7.** showed results data of ANFO and emulsion application for the furthest distance of overbreak and underbreak from design.

Table 6. Results data of ANFO and Emulsion application for area and circumference of overbreak and underbreak average at study area

Number	Location	Explosive	Overbreak Area (m^2)	Underbreak Area (m^2)	Overbreak Circumference (m)	Underbreak Circumference (m)
1	Drainage Level	Emulsion	5.4959	0.9468	35.6423	7.5315
		ANFO	9.0302	0.4324	39.8952	5.1857
2	Truck Haulage Level	Emulsion	8.3949	1.3237	35.4118	8.3723
		ANFO	7.7661	1.0388	34.2499	11.1498
3	Extraction Level	Emulsion	2.0389	0.414	21.4225	8.4789
		ANFO	8.612	0.9720	24.9592	4.0931
4	Undercut Level	Emulsion	1.808567	0.374422	20.04629	7.393733
		ANFO	1.196072	0.994789	16.27197	12.62942

Table 7. Results data of ANFO and Emulsion application for the furthest distance of overbreak and underbreak average from design at study area

Number	Location	Explosive	The Furthest Distance of Overbreak from Design (m)	The Furthest Distance of Underbreak from Design (m)
1	Drainage Level	Emulsion	0.7275	0.35495
		ANFO	0.86665	0.26915
2	Truck Haulage Level	Emulsion	1.2175	1.945
		ANFO	1.23745	0.94665
3	Extraction Level	Emulsion	0.36	0.22025
		ANFO	1.354	0.330238
4	Undercut Level	Emulsion	0.320333	0.154533
		ANFO	0.293	0.235333

Based **Table 6.**, then get ANFO and emulsion application chart at study area for area and circumference of overbreak and underbreak average showing the ANFO and emulsion overbreak and underbreak area and circumference is fluctuated, but ANFO's more than emulsion's [9]. See **Figure 2.**

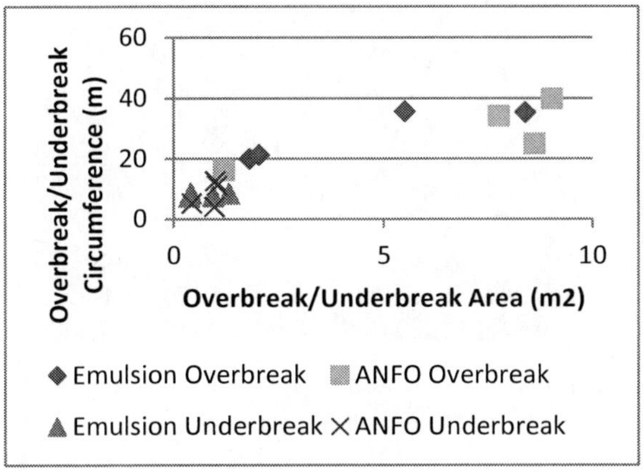

Figure 2. ANFO and Emulsion chart for area and circumference of overbreak and underbreak average at study area

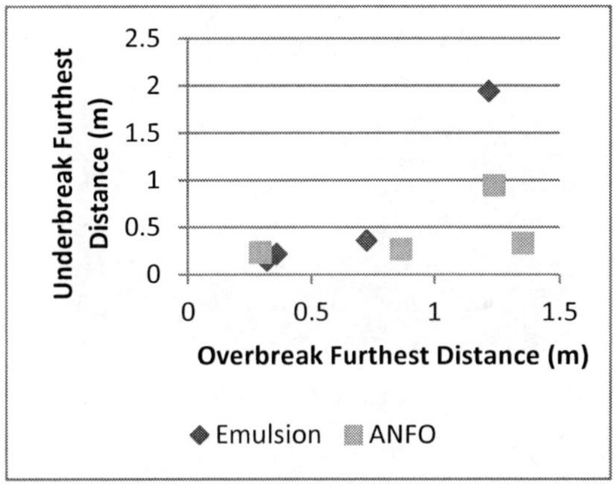

Fig 3. ANFO and Emulsion chart for the furthest distance of overbreak and underbreak from design average at study area

ANFO and emulsion application chart at Undercut Level for the furthest distance of overbreak and underbreak from design based on **Table 7.** has been shown in **Figure 3.** The result of ANFO and emulsion overbreak and underbreak furthest distance from design fluctuated and then ANFO overbreak increased so that more than emulsion's. However, emulsion underbreak more than ANFO's [9].

Review from result that has obtained at Drainage Level, Truck Haulage Level, Extraction Level, and Undercut Level without average all or based in all data, can be concluded that emulsion application produced some undebreak and few of overbreak. However, ANFO application produced more overbreak than emulsion and some underbreak. The result has been shown in **Table 8**. There is one of sample heading that blasted with ANFO at Extraction Level in **Figure. 4.** and blasted with emulsion at Undercut Level in **Figure. 5.**

Table 8. Final results data of ANFO and Emulsion for highest value of overbreak and underbreak

Number	Location	Explosive	Overbreak Area	Underbreak Area	Overbreak Furthest Distance from Design	Underbreak Furthest Distance from Design
1	Drainage Level	Emulsion	-	√	-	√
		ANFO	√	-	√	-
2	Truck Haulage Level	Emulsion	-	√	-	√
		ANFO	√	-	√	-
3	Extraction Level	Emulsion	-	-	-	-
		ANFO	√	√	√	√
4	Undercut Level	Emulsion	√	-	-	-
		ANFO	-	√	√	√

Figure 4. Heading at P#6 DP#12 East Blasted with ANFO

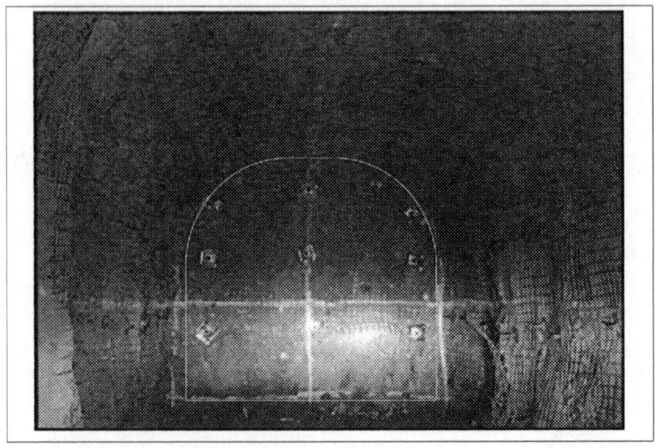

Figure 5. Heading at MAD 2 DD#8 EN Blasted with Emulsion

4. Conclusion

From results that has obtained at Drainage Level, Truck Haulage Level, Extraction Level, and Undercut Level, can be concluded that emulsion application produced some undebreak and few of overbreak. However, ANFO application produced more overbreak than emulsion and some underbreak. The impact of overbreak is excessive usage of ground support and the impact of underbreak is doing blasting again that's mean disadvantageous and also dangerous in geotechnical side. The advantages of ANFO are inexpensive or economic, easy to create, safe handling. However, ANFO also has disadvantages i.e. low density and density can't be adjusted, not water resistance, so using cartridge emulsion at lifter hole, and usually leaving detonating cord at perimeter hole. Afterwards, the advantages of emulsion are water resistance, safe handling, not using detonating cord at perimeter hole, and density can be adjusted depend the type of gasser used. However, the disadvantages of emulsion are expensive, emulsion consumption more than ANFO because have to do QA (Quality Assurance) and QC (Quality Control) before loading at blast hole, and not easy to create. Thereby, the explosives that effective to use in blasting development at underground Deep Mill Level Zone (DMLZ) PT. Freeport Indonesia is emulsion. However, it should be redesign drill pattern and blasting geometry for emulsion. To use emulsion in blasting development UG DMLZ PTFI, be recommended to redesign drill pattern and blasting geometry using radial crack method.

References

[1] Mahtab, M.A., Bostald, D.D., Alldredge, J.R. *and* Shanley, R.J. 1972. *Analysis of Fracture Orientations for Input to Structural Models of Discontinuous Rock*. USBM-RI 7669: USA

[2] Bieniawski, Z.T. 1973. *Engineering Classification of Rock Masses*. Wiley: Trans. South Africa.

[3] Ouchterlony, Finn. 1977. *Analysis of Crack Related to Rock Fragmentation*. Springer-Verlag Wien: Stockholm, Sweden

[4] Hoek E. and Brown, E.T. 1980. *Underground Excavation in Rock*. London

[5] Hagan, T.N. 1992. *Personal Communication, Blasting*. Queensland, Australia.

[6] Rustan, A. 1998. Rock Blasting Terms and Symbols. CRC Press: Rotterdam.

[7] Mandal, S.K., and Singh M.M. 2008. Analysis of Blast Pattern to Evaluate Extent and Causes of Overbreak in Large Cavern. Blasting and Fragmentation 2. Page 1-26. ISSN 1937-6359.

[8] Jang H 2012 *Optimizing Overbreak Prediction Based on Geological Parameters Comparing Multiple Regression Analysis and Artificial Neural Network*. Australia. 161-162.

[9] PT Freeport Indonesia 2017 Underground DMLZ Geo-technical Services and Geology

Acknowledgments

The authors wist to thank PT Freeport Indonesia (PTFI) for the contribution and data collection in Mimika Regency, Papua Province, Indonesia. The authors also gratefully acknowledge the reviews provided by anonymous reviwers of International Conference on Geoscience 2018.

IOP Conf. Series: Earth and Environmental Science **279** (2019) 012002 doi:10.1088/1755-1315/279/1/012002

Mapping of Landslide Susceptibility Using Analytical Hierarchy Process in Sukamaju Area, Tenggarong Seberang, Regency of Kutai Kartanegara

M A Syam, Heryanto, M D Balfas
Mulawarman University, Kuaro Street, Gn. Kelua, Samarinda, 75119, Indonesia

Abstract. A landslides susceptibility mapping produced as a mitigation effort to reduce risk of landslide. Landslide susceptibility mapping aims to identify most susceptible area to landslide. Analytical Hierarchy Process (AHP) is quantitative method based on qualitative assessment to certain phenomena. This method was applied in Sukamaju area, Tenggarong Seberang, regency of Kutai Kartanegara to determine the weights of each data influencing landslise occured. Data were used to map susceptibility of landslide in study area are relief and slope data, detail geological and geomorphological map, topographic wetness index (TWI), and Normalized Difference Vegetation Index (NDVI) data. The data then were analyzed spatially using overlay function in ArcGIS 10.5 to produce susceptible area to landslide. Susceptibility of landslide in the study area consist of four classes, they are low, moderate, high, and very high. Generally, in the study area dominated by moderately susceptible to landslide by percentage of area 61.68%, low susceptibility about 28.28%, high susceptibility about 9.22%, and very high susceptibility about 0.92%. Very high susceptibility area is located in the eastern part of study area, geologically covered by sandstone form Pamaluan formation and denudational morphology.
Keywords: landslide susceptibility, AHP method, spatial analysis

1. Introduction

1.1 Background

Landslide is one of natural disaster occured most in Indonesia. It is the movement of soil or rock mass in short periods and big volume. The movement of this mass can cause damage of infrastructure and people psyhcology. An area can be stated susceptible to lanslide if it has slope more than 25%, plastic and semipermeable lithology or more geological structure, groundwater can increase susceptibility of landslide.

The information about landslide susceptibility can be presented well and clearly if it is made in a map. Determination of landslide susceptible area is commonly complicated because the less of data needed to predict it. Intarawichian dan Dasananda (2010) stated thar the accuracy of landslide susceptibility map depended on scale, the number and quality of the data, and appropriate method selection in analysis and modelling. Guzzetti *et al.* (1999) stated that mapping of landslide susceptibility was very suitable using statistics approach in analyzing historical correlation between factor controlling landslide and distribution of landslide.

The landslide can be influenced by precipitaton, geological, and geotechnical condition, such as relief, slope, lithology, groundwater, mechanical and physical properties of the rock or soil, and even earthquake. In Sukamaju area, if it is assessed from geological condition, it has big potential to landslide. Therefore, mapping of landslide susceptibility very needed in Sukamaju area as preventive effort of landslide risk. This research using geological investigation data, such as detail geological and geomorphological map, and secondary data such as DEM and Landsat 8 OLI. The data analyzed by using Analytical Hierarchy Process (AHP) method to map the landslide susceptible area. This method is selected because very useful on conditions need consideration and complex. It combines logics, knowledge, and experience, then optimized into systematic process. This method used as alternative in analyzing landslide susceptibilty from qualitative data or judgement to quantitative data.

2. Method

Sukamaju area is located in Tenggarong district, Kutai Kartanegara regency, East Kalimantan province. In UTM, study area is lying on 507358.851–518165.202 mE and 9988990.296–9971961.093

mS. Spatial analysis was applied to obtain the landslide susceptibility map. Spatial analysis using overlay function in ArcGIS 10.5. Overlay function can be applied to raster data. Applying this function, each ovariables must have influence weight to occuring of certain phenomena and each variables have attribute data devided to value classes. The influence weights of variables were calculated using Analytical Hierarchy Process (AHP) method.

2.1 Analytical Hierarchy Process (AHP)

This method is solved by developing pairwise comparison matrix (Table 1) by assigning the rank to each factor against other factors. Assigning rank using Table 2 based on degree of preference of factors influencing landslide.

Tabel 1 Matrix of pairwise comparison (Saaty, 2000)

C	A_1	A_2	...	A_n
A_1	a_{11}	a_{12}	...	a_{1n}
A_2	a_{21}	a_{22}	...	a_{2n}
:	:	:	...	:
A_m	a_{m1}	a_{m2}	...	a_{mn}

Table 2 The basic scale of preference between two variables for pairwise comparison matrix in AHP method (Saaty, 1980 in Mandal, 2018)

Importance Rank	Degree of Preference	Explanation
1	Equal importance	Two criteria contribute equally to the objective
3	Moderately importance of one over another	Judgment and experience slightly to moderately favour one criteria over another
5	Strongly importance	Judgment and experience essentially favor one criteria over another
7	Very strongly importance	Experience and judgements is strongly favored over another and its dominance is showed in practice
9	Extremely importance	The evidence favouring a criteria over another is the highest degree probable of an affirmation
2, 4, 6, and 8	Intermediate values between two adjacent judgements	Used to represent compromising between the preferences in weights 1, 3, 5, 7 and 9
Reciprocals	Opposites	Used for inverse comparison

In this model, consistency index, known as the ratio of consistency (CR), is for indicating the probability that the matrix judgments were generated randomly (Saaty, 1977, 1980, 1994 in Mandal, 2018). The consistency ratio is stated valid if it is less or equal to 10%.

$$CR = CI / RI \tag{1}$$

Random Index is the average result of the consistency index that depending on the matrix order in Table 3 given by Saaty (1977) in Mandal (2018) and CI is the consistency index and could be formulated as:

IOP Conf. Series: Earth and Environmental Science **279** (2019) 012002 doi:10.1088/1755-1315/279/1/012002

$$CI = (\lambda_{maks} - n) / (n - 1) \quad (2)$$

Where, λmax is the principal or largest eigenvalue of the matrix and could be calculated easily from the matrix and n is the matrix number (Saaty, 1977 in Mandal, 2018).

Table 3 Random Consistency Index (Saaty, 2000)

n	*1*	*2*	*3*	*4*	*5*	*6*	*7*	*8*	*9*
RI	*0.00*	*0.00*	*0.58*	*0.90*	*1.12*	*1.14*	*1.32*	*1.41*	*1.46*

3. Results and Discussion

3.1 Data

There are five data used in this study. They are lithology and geomorphology map from detail geological mapping, slope and Normalized Topographic Wetness Index (TWI) map produced from Digital Elevation Model (DEM) data with resolution 8 meters, data obtained from Indonesian Geospatial Agency, and Normalized Difference Vegetation Index (NDVI) map produced from Landsat 8 OLI data with resolution 8 meters, data obtained from Earth Explorer United State Geological Survey. All data must be converted to raster data based on the attribute data before analyzed using Spatial Analyst in ArcGIS 10.5.

3.2 Weights Calculation

The weight of raster data calculated using Analytical Hierarchy Process (AHP). The calculation started by assigning importance rank in pairwise comparison matrix, then calculate the value and weights for each data. The last step, consistency index and consistency ratio calculated, the weight is accepted if consistency ratio value is less or equal to 10% or 0.1.

Table 4 Pairwise comparison matrix for landslide susceptibility mapping

DATA	Slope	Lithology	Morphlogy	NTWI	NDVI
Slope	0.455	0.506	0.456	0.370	0.333
Lithology	0.227	0.253	0.304	0.296	0.278
Morphology	0.152	0.127	0.152	0.222	0.222
NTWI	0.091	0.063	0.051	0.074	0.111
NDVI	0.075	0.051	0.038	0.037	0.056
Total	1.000	1.000	1.000	1.000	1.000
Value	2.120	1.358	0.874	0.390	0.257
Weights	42%	27%	18%	8%	5%
CI	***0.0245***				
CR	***0.0219***				

The weight calculated from the matrix shown that slope is the most influencing data, the weight is exactly 42%, lithology 27%, morphology 18%, Normalized-TWI 8%, and Normalized DVI 5%. The value of index consistency is 0.0245, this value used to calculated the consistency ratio. The value of ratio of consistency is 0.0219. less than 0.1, it means the consistency ratio is valid and the weight is accepted to be used in mapping the landslide susceptibilty.

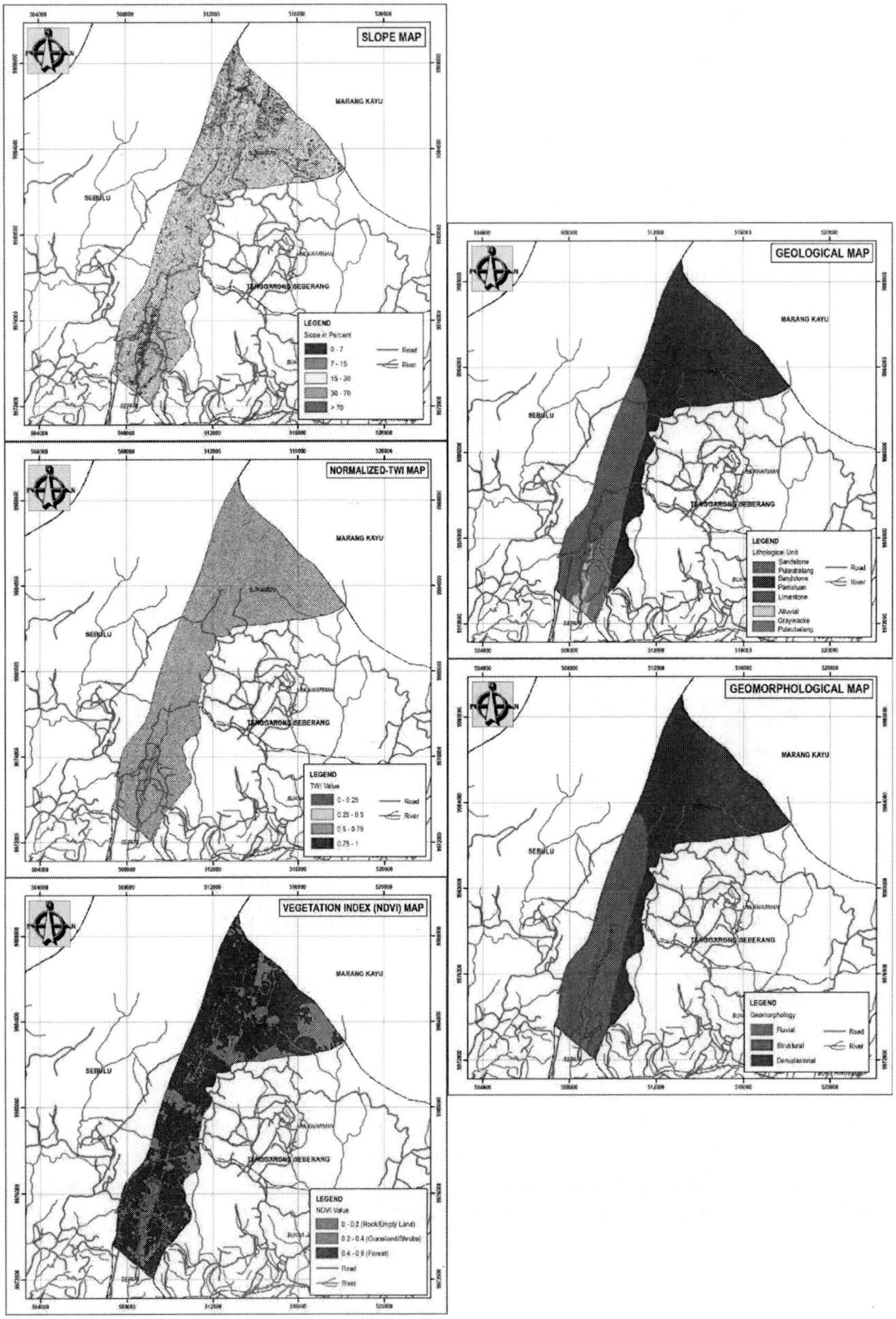

Figure 1 Map used in analysis of landslide susceptibility

3.3 Spatial Analysis

Landslide susceptibility mapping produced by applying spatial analysis. Spatial analysis applied to the map is overlay function. This function combines some the raster data to produce a new raster based on the weight of raster data and attribute value of the rasters. Before applying in overlay function, attribute data must be reclassified to be integer value only based on the influence controlling landslide. The result of spatial analysis presented in Figure 2.

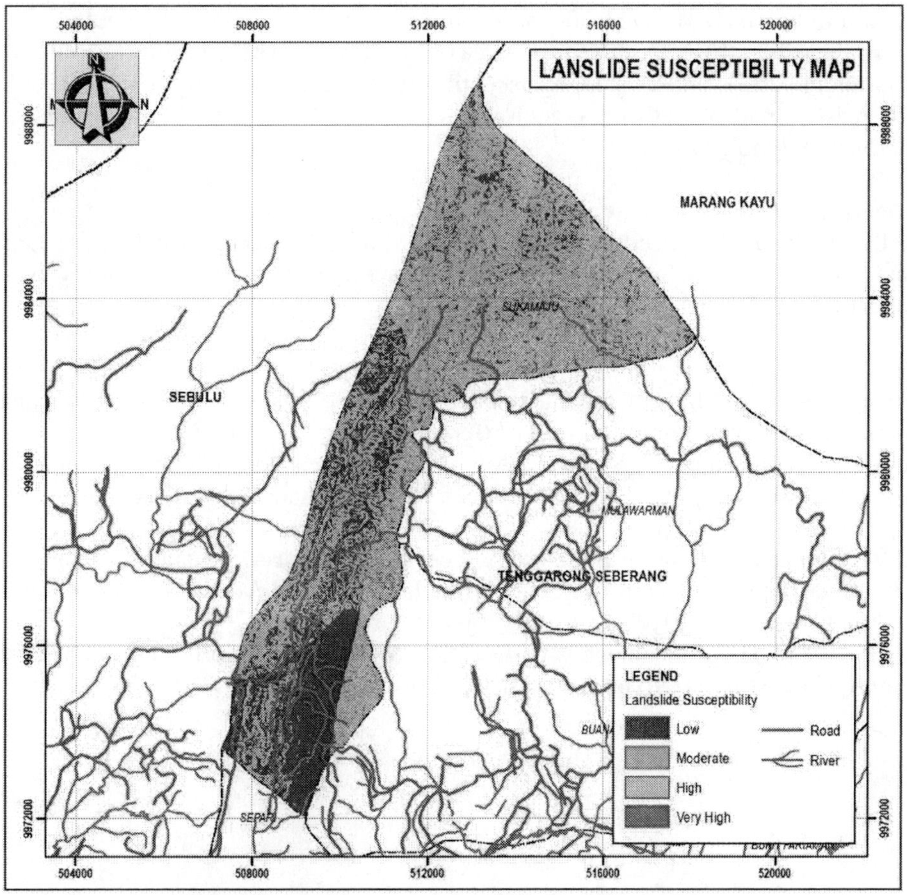

Figure 2 Landslide susceptibility map in study area

Based on Figure 2, susceptibility of landslide in the study area consist of four classes, they are low, moderate, high, and very high. Generally, in the study area dominated by moderately susceptible to landslide by percentage of area 61.68%, low susceptibility about 28.28%, high susceptibility about 9.22%, and very high susceptibility about 0.92% (Table 5). Very high susceptibility area is located in the eastern part of study area, geologically covered by sandstone form Pamaluan formation and denudational morphology.

Table 5 Landslide susceptibility class, area, and percentage

Landslide Susceptibility	Area (km^2)	Percentage (%)
Low	14.34	28.18
Moderate	31.39	61.68
High	4.69	9.22
Very HIgh	0.47	0.92

4. Conclusion

Mapping of landslide susceptibility on this study using five raster data, they are slope, lithology, morphology, Nomalized TWI, and Normalized VDI. Based on calculation using Analytical Hierarchy Process (AHP), the slope is the most influencing data for landslide occuring by weight 42%, then lithology 27%, morphology 18%, Nomalized TWI 8%, and Normalized VDI 5%. The value of ratio consistency of this weight is about 0.0219, less than 0.1, and this shown that the weight is valid and acceptable used in spatial analysis.

Based on the map of landslide susceptibility, the study area dominated by moderately susceptible to landslide by percentage of area 61.68%, low susceptibility about 28.28%, high susceptibility about 9.22%, and very high susceptibility about 0.92%. Very high susceptibility area is located in the eastern part of study area, geologically covered by sandstone form Pamaluan formation and denudational morphology.

Acknowledgement

We would like to thank Engineering Faculty, Mulawarman University for helping in funding this research and village head of Sukamaju for helping facilities to present this study for people and government officials in Sukamaju.

References

[1] Guzzetti, F., Cardinali, M., Carrara, A., Reichenbach, P., 1999, Landslide hazard evaluation: an aid to a sustainable development, *Journal of Geomorphology* 31:181-216.

[2] Intarawichian, N., Dasananda, S., 2010, Analytical hierarchy process for landslide susceptibility mapping in lower Mae Chaem Watershed, Northern Thailand, *Journal of Science and Technology* 17(3):277-292.

[3] Mandal, B., Mandal, S., Analytical hierarchy process (AHP) based landslide susceptibility mapping of Lish river basin of Eastern Darjeeling Himalaya, India, *Advances in Space Research* (2018), doi: https://doi.org/10.1016/j.asr.2018.08.008

[4] Saaty, T.L., 2000, The Fundamentals of Decision Making and Priority Theory with the Analytic Hierarchy Process, Pennsylvia. University of Pittsburgh. Vol 1.

Efforts to Improve Community Awareness Towards the Potential of a Great Earthquake Which Threats Jakarta Based On Geographic Information and 3D Simulation Systems in Matraman District, East Jakarta

Z I Andromeda[2], C Randi[2], K Nazhifah[2], R Syahputra[1], Iskandarsyah[1] and M R Septyandy[1]

[1] Geophysics & Geology lecturer FMIPA, Universitas Indonesia
[2] Geophysics student FMIPA, Universitas Indonesia

syahputra.reza@sci.ui.ac.id

Abstract. Indonesia is located in an area of active tectonic plate. This area suffers from natural disasters. Recorded almost every year disasters, such as earthquakes, tsunamis, floods, and fires, occur in several regions in Indonesia. One of the major city in Indonesia is Jakarta. Jakarta as a capital city with urban area has an important position in terms of economy and government. The issue of earthquakes with the magnitude of 8 SR is one of the major disasters that threatening Jakarta. The effort to alleviate losses caused by disasters, it is necessary to inform the community regarding preparation in the face of disasters that will occur. This preparation is needed especially in densely populated areas as part of disaster mitigation measures like Matraman District in East Jakarta. The initial step in this program is to distribute questionnaires to determine the level of knowledge and information about the disaster. Furthermore, by utilising the geographical information system (GIS), evacuation routes when a disaster occurs can be made to adjust the situation of the area. The evacuation routes produced include the fastest and safest routes in evacuation by considering conditions such as road width, car parking, electricity poles, railroad crossings, and blind alley. This evacuation route will be modelled with 3D simulation in several conditions such as railroad crossings, spill markets, crossroads, and one-way lanes when a disaster occurs. This simulation will be the end result in a program to increase public awareness of potential disasters that threaten their area. This research obtains evacuation routes and 3D simulations that have been made are information media that can facilitate people to get the right knowledge about disaster mitigation. The results of the community awareness raising program on potential disasters can be used as a reference for other regions in the effort to mitigate disasters.

Keyword: Disaster mitigation, Matraman, GIS, 3D simulation

1. Introduction

Indonesia is located at the meeting of four active plates, namely Eurasia, Indo-Australia, Pacific, and Philippine Sea. Subduction zones due to collision of Indo-Australian plates with eurasian plates are located on the coast west of Sumatra, south of Java, and continue to the Nusa Tenggara. And in the

IOP Conf. Series: Earth and Environmental Science **279** (2019) 012004 doi:10.1088/1755-1315/279/1/012004

Eastern Indonesia (Papua and Maluku) collides with Pacific and Philippine Sea Plate. It was in this plate meeting area that earthquakes, tsunamis and landslides often occurred. This is caused by the accumulation of energy that cannot be held back so that it becomes a natural disaster [1]. No wonder Indonesia often experiences. At least according to Meteorological, Climatological, and Geophysical Agency (BMKG) data for 2018, there were five earthquakes with a strength of > 6.0 SR. First on January 23, 2018 an earthquake occurred with epicenter of 43 km southwest of Lebak with magnitude of 6.1 SR (felt up to Jakarta). Secondly on July 29, 2018 an earthquake centered on 28 km northwest of East Lombok with magnitude of 6.4 SR. Then one week later an earthquake which centered 18 km northwest of East Lombok with 7 SR. Then on September 28, 2018 an earthquake occurred again in 26 km North of Donggala with magnitude of 7.4 SR. and the last occurred on October 11, 2018 at 61 km northeast of Situbondo with magnitude of 6 SR. This proves that seismicity in Indonesia is still very active.

Based on seismicity data from PUSGEN (National Earthquake Study Center), in 2018 there were five megathrusts found in western South Sumatra and southern Java (figure 1). Three of them were recorded to have produced earthquakes with magnitude >7.0 SR, namely Mentawai-Pagai Megathrust, Enggano Megathrust, and West-Central Java Megathrust. While the other two, namely Sunda Strait Megathrust and East Java Megathrust, have never recorded an earthquake with magnitude >7.0 SR. This raises fears of a mega earthquake due to the seismic gap in the area. Seismic gap is a term used for tectonic active areas but earthquakes rarely occur for long periods of time. The term seismic gap was first introduced by Fedotov in 1965 and Mogi in 1969 when they mapped earthquake events in the Alaska-Aleutian subduction zone [2].

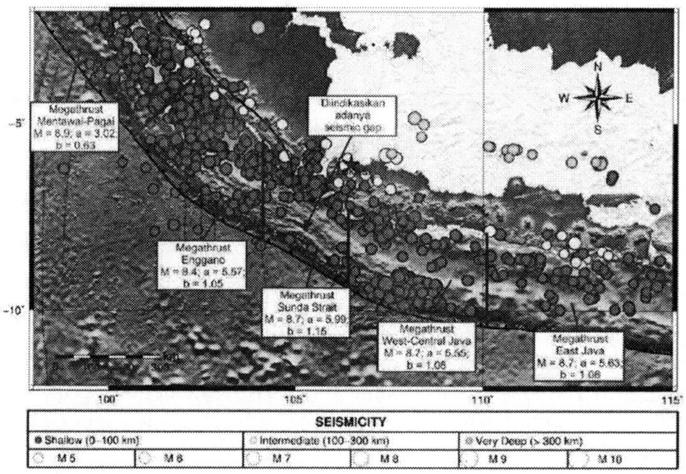

Figure 1. Map of earthquake distribution in the southern region of Sumatra and Java indicated seismic gaps (modification from PUSGEN, 2018).

2. Literature review

There are three main discussions in this paper including area of study, criteria of evacuation points, and network analysis in GIS.

2.1. Area of study

On January 23, 2018, an earthquake with magnitude 6.1 SR shook southwest of Lebak, Banten. The earthquake shock can be felt by the people of Jakarta and surrounding areas. One of them is Matraman District, which is the smallest area in East Jakarta. With an area of 4.88 km^2, 82.68% of the Matraman area is a residential area, where the population of Matraman Subdistrict reaches 193,700 people and its population density is 38,482 people/km^2 [3]. With a very dense population and a very small area, Matraman Subdistrict has a very high potential number of disaster victims if the entire population does not have a sense of care to disaster response. In addition, areas around Matraman often experience disasters, such as floods and fires, so this can be a threat to Matraman Subdistrict which is the region with the highest population density in East Jakarta. Based on the research we did by distributing questionnaires to the community with the aim of knowing the level of community understanding about disaster. And the results show that the level of community understanding of disaster is still quite low. Therefore, it is very important to educate and instill a disaster response attitude for all Matraman people, one way is to conduct sociallization on disaster management efforts.

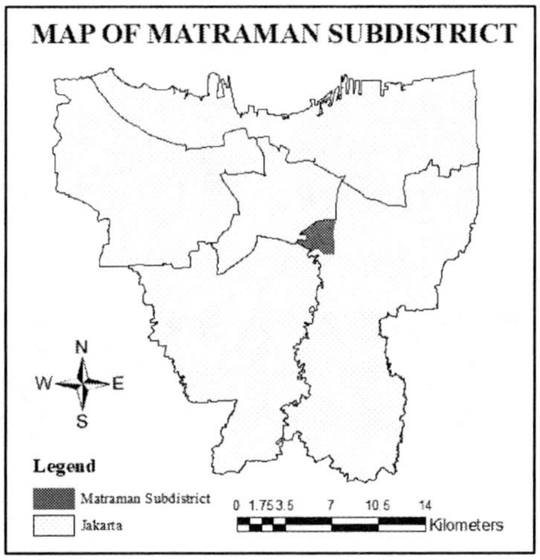

Figure 2. Map of Matraman Subdistrict location.

2.2. Criteria of evacuation points

Based on government regulation PU No. 20/PRT/M/2011 regarding natural disaster hazard zones, the general criteria for determining evacuation points are: [4]

1. The absence of buildings that are at risk of being affected by a disaster is quite high (the building is not earthquake resistant).
2. Not close to sources of disasters (high cliffs, sea, river, bank, volcano, and earthquake fault area).

2.3. Network analysis in GIS

Generally, Network Analysis is a tool that create transportation model to find out the relationship between elements (building, street, and tree) based on traffic situation [5]. There are five analyzes that can be done on ArcGIS software, namely Route, Closest Facility, Service Areas, OD Cost Matrix, and Vehicle Routing Problem Analysis [5].

Making a network analysis map begins with the creation of a basic map that will be included in several components, including: the shp map of the area, road network data, and speed obtained automatically from ArcGIS 10.5. As for knowing the travel time using the formula:

$$Time\ (s) = \frac{Distance\ (m)}{Velocity\ (\frac{m}{s})} \quad (1)$$

By using the Network Analysis method on this GIS, we want to find out the nearest route to get some facilities, in this case the evacuation point in the event of a disaster.

3. Research implementation phase

3.1. Research flowchart

In this paper, the first step we did was to research the knowledge of the Matraman community about disasters, by distributing questionnaires. Then we examine the best road routes that can be passed to the evacuation site, and we also make a simple simulation video about the process of self-evacuation in the event of a disaster. From this result, we conduct socialization about disaster to them as a form of community service. The stages of this research can be seen in figure 3. The next step is to determine areas that often occur in floods and fires, buildings that must be avoided when an earthquake occurs, and analysis of effective and safe evacuation routes to the holding location.

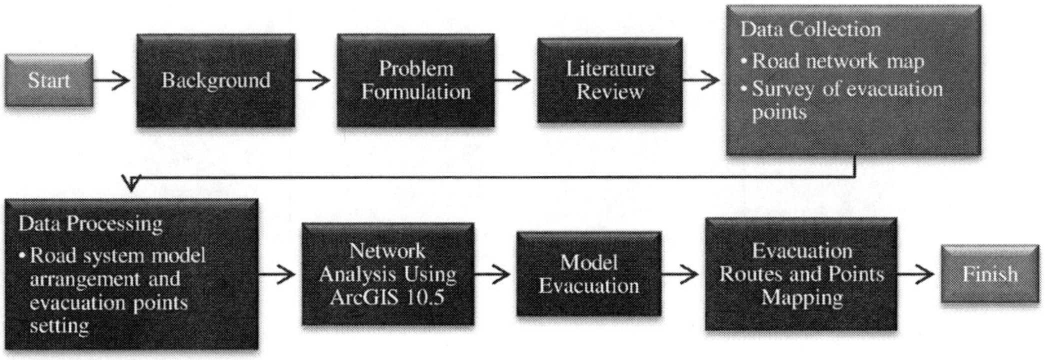

Figure 3. Research flowchart.

3.2. Research methodology

The method used in this paper is the Network Analysis using Closest Facility Analysis. This is based on the aim to find the nearest field or vacant land that is feasible to be used as a muster point. The data used are:

1. Matraman map with shp file type (downloaded from tanahair.indonesia.go.id)
2. The width of the road
3. Road situation
4. Determination of evacuation point

This Matraman map with shp file type is used in processing Network Analysis in ArcGIS 10.5. This road width data is used to estimate how many people and vehicles can pass through the road or alley. Road situation data, such as vehicles parked on the side of the road, traders who sell on the side of the road, one-way roads, and the road situation in the markets, are very important to be used to estimate how long it takes people to reach the evacuation point when a disaster occurs. Determination of evacuation point is very necessary to find out where a safe place can accommodate the community when a big earthquake really shook Jakarta. The research area is a densely populated area and has a narrow environment and many obstacles on the streets, such as there are electric poles in small alleys that are only about 1.5-2.5 m so that it can only be passed by one to two people, and there are still many vehicles parked carelessly on the road which is only about 4-5 m wide so that it often causes congestion.

Figure 4. Traffic condition in Matraman Subdistrict.

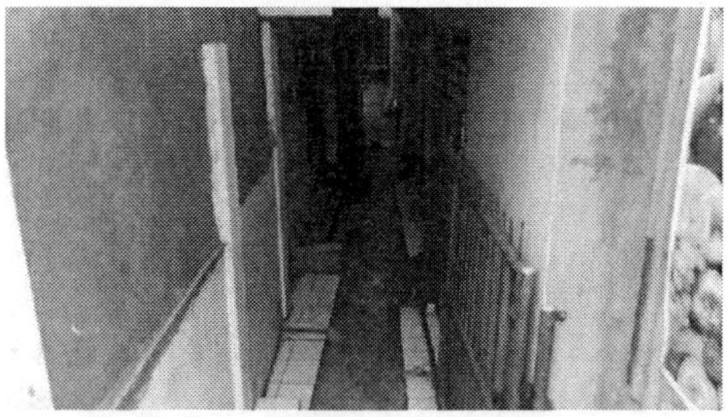

Figure 5. A small alley that can only be passed by one or two people.

4. Discussion and result

4.1. Utan Kayu Selatan Village

Utan Kayu Selatan Village has the largest area in Matraman Subdistrict, which is 1.12 km^2, but has the lowest population density in Matraman Subdistrict, which is 34,699.7 people/km^2 [3]. Traffic conditions in Utan Kayu Selatan are quite smooth and some roads are one-way roads with the aim of minimizing congestion, so that the evacuation process in this area can be carried out smoothly and easily. Based on our survey, we determined that there were four places that could be used as evacuation sites, namely SMPN 97 Jakarta, SMAN 31 Jakarta, SMAN 22 Jakarta, and UPBJJ-UT (figure 6). The time needed by citizens to reach this evacuation point varies depending on speed and distance. For example, by calculating the average pedestrian speed is 10 km/hour (this assumption used in all of sub-district in Matraman), the time needed by residents located at the farthest location with the evacuation point is around 7 minutes 10 seconds, while the fastest time needed by residents located near the evacuation point is around 60 seconds. This data is based on surveys that we have done with normal road conditions and environment.

4.2. Pisangan Baru Village

The area of Pisangan Baru Village is not wider that the three villages above, which is 0.68 km^2. And the total population density in this region is the lowest in Matraman Subdistrict, which is 5,501.36 people/km^2 [3]. Even so, the environmental conditions in this region are very crowded and so do the road conditions because in this area there are many markets. This has caused difficulty in finding evacuation points in Pisangan Baru Village. So based on our survey, we can only find three points that can be used as evacuation poins, namely SMKN 5 Jakarta, SDN Pisangan Baru 13 Pagi, and another in the open field (figure 7). With these conditions, the time needed to get to the evacuation point by people who are far from the evacuation point is around 8 minutes 10 seconds, while the fastest time that can be reached by those closest to the evacuation point is around 1 minute 35 seconds.

4.3. Palmeriam Village

Palmeriam Village has the second smallest area after Kayu Manis Village, which is 0.65 km^2, but the population density is the second highest after Kayu Manis Village, which is 35,965.74 people/km^2 [3]. Based on our survey, we found three points that could be used as evacuation points, namely DISLITBANG TNI AD, SDN Palmeriam 02 Petang, and the Palmeriam market parking lot (figure 8). Road conditions in the DISLITBANG TNI AD and SDN Palmeriam 02 Petang is relatively smooth because it is on a big road, but the road conditions in the Palmeriam market area are very crowded because many traders are on the side of the road and many public transportation stops at the side of the road causing congestion, it is feared that this can hamper the evacuation process. With such dense conditions, the time needed for people who are far from the evacuation point is around 12 minutes, while the fastest time to reach the nearest evacuation point with normal road and environmental conditions is around 60 seconds.

IOP Conf. Series: Earth and Environmental Science **279** (2019) 012004 doi:10.1088/1755-1315/279/1/012004

4.4. Kayu Manis Village

Kayu Manis Village is the smallest area in Matraman Subdistrict but has the highest population density, 0.58 km^2 and 51,902.78 people/km^2 [3]. Based on our survey, there are three points that can be used as evacuation points, namely SMP Al Washliyah, SMA & STMIK Muhammadiyah Jakarta, and the other is in a parking lot around Kayu Manis II Street (figure 9). Road conditions in this area are quite crowded but there are several roads which are one-way roads so that it can be reduce congestion. Usually, congestion that occurs in this area because many vehicles stop on the side of the road and many traders sell on the side of the road. Thus, the time needed to reach the evacuation point by people who are far from the evacuation point is around 5 minutes. And the fastest time that can be reached by people who are close to the evacuation point is around 60 seconds.

4.5. 3D simulation

Simulation is a way to study or predict something that has not happened by imitating or making a system model that is studied and then conducting experiments numerically using a computer [6]. So, simulation is a way to duplicate or describe the characteristics, appearance and characteristics of a real system which is then studied the nature and character of its operations, and the last create conclusions and take decisions based on the suggested simulation. The advantage of this method is that the simulation produced does not replace the real conditions of traffic and provides input if something unexpected happens such as a traffic accident, fallen tree, or other disaster. Whereas 3D visual simulation is a simulation with dimensions of space and depth on objects created [7]. So that the purpose of conveying ideas or information related to this disaster mitigation process can be conveyed well in the form of 3D visual simulations in digital format (figure 10).

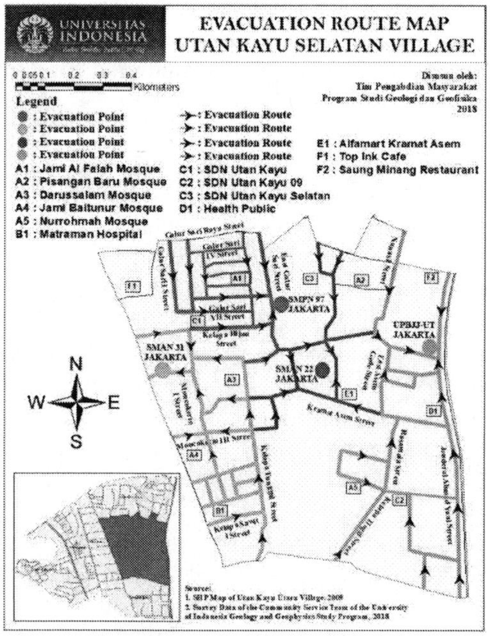

Figure 6. Evacuation route map Utan Kayu Selatan Village.

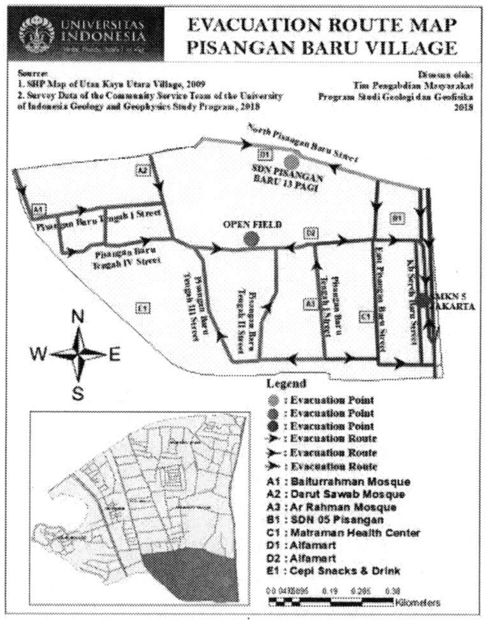

Figure 7. Evacuation route map Pisangan Baru Village.

IOP Conf. Series: Earth and Environmental Science **279** (2019) 012004 doi:10.1088/1755-1315/279/1/012004

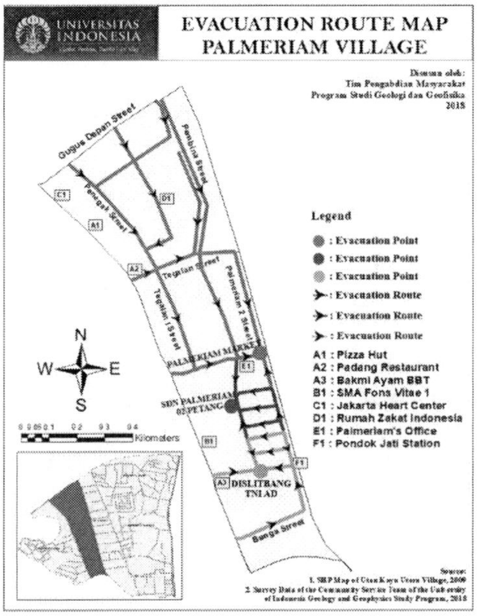

Figure 8. Evacuation route map Palmeriam Village.

Figure 9. Evacuation route map Kayu Manis Village.

Figure 10. 3D simulation of evacuation process to Batalyon Zeni field as evacuation point.

5. Conclusion

Based on our research, there are at least 2 evacuation points in each village in Matraman Subdistrict in the hope of accommodating the entire community when a disaster occurs. The Network Analysis method can easily choose the route that is most effective in determining the evacuation route. Each region has travel time to different evacuation points, depending on the distance between their current position and the evacuation point. The fastest time that can be reached is around 1-3 minutes and the farthest is about 4-12 minutes. Describing the state of a real system with 3D visual simulation can convey ideas or information related to this disaster mitigation process well and easily understood.

IOP Conf. Series: Earth and Environmental Science **279** (2019) 012004 doi:10.1088/1755-1315/279/1/012004

6. References

[1] Yudhicara and Budiono K 2008 *Tsunamigenik Di Selat Sunda: Kajian Terhadap Katalog Tsunami Soloviev* vol 3 (Bandung: Jurnal Geologi Indonesia) p 241–251

[2] Nishenko S P and Sykes L R 1993 *Seismic gap hypothesis: ten years after* (Amsterdam: Journal of Geophysical Research) p 9909-9916

[3] Agus W, Baskoro P Y W and Fitriyani 2017 *Matraman Dalam Angka* ed Patrianto (Jakarta: BPS Kota Administrasi Jakarta Timur) p 3–6

[4] Kementrian Pekerjaan Umum 2011 *Pedoman Penyusunan RDTR dan Peraturan Zonasi Kabupaten/Kota* (Jakarta: Kementrian Pekerjaan Umum)

[5] Fischer M M 2003 *Handbook 5 Transport Geography and Spatial Systems* eds Hensher D, Button K, Haynes K and Stopher P (Austria: Pergamon)

[6] Nasution M 2014 *Pembuatan Simulasi Index Model Splay Tree untuk Pemahaman Struktur Data dengan Menggunakan Metode Binary Search Tree* vol 8 (Pelita Informatika Budi Darma)

[7] Romadhon A P and Murinto 2013 *Media Pembelajar Proses Rendering Objek pada Mata Kuliah Grafika Komputer Berbasis Multimedia* vol 1 (Jurnal Sarjana Teknik Informatika)

Acknowledgments

This research was supported by Community Responsibility Grant University of Indonesia No.4709/UN2.R3.1/HKP.05.00/2018. Special gratitude for SMA Fons Vitae 1 and SMAN 31 Jakarta who have helped in organizing community service events. We would like to acknowledge the BMKG, BPBD DKI Jakarta, Head of Matraman District, and program study of geophysics and geology University of Indonesia for cooperate in support our events.

Developing Agricultural Land Geospatial Information in Supporting Regional Food Resilience

Nurmiaty[1], Sumbangan Baja[2], Samsu Arif[3], Andi Ridwan[4], Rahmad. D.[5], Sukmawati[6]

[1] Department of Estate Crops Cultivation, Pangkep State Polytechnic of Agriculture, South Sulawesi, 90655, Indonesia E-mail: nurmiatyamin1@gmail.com
[2]Department of Soil Science, Faculty of Agriculture, Hasanuddin University, Makassar, Indonesia 90245, email: sbja02@yahoo.com.au, s.baja@unhas.ac.id
[3] Geophysics, Hasanuddin University, Makassar, South Sulawesi, Indonesia Phone +62-411 586016, Fax +62-411 586588 E-mail: samsu_arif@unhas.ac.id
[4] Department of Estate Crops Cultivation, Pangkep State Polytechnic of Agriculture, South Sulawesi, 90655, Indonesia E-mail: ridwanassad@gmail.com
[5] Department of Estate Crops Cultivation, Pangkep State Polytechnic of Agriculture, South Sulawesi, 90655, Indonesia E-mail: rahmadd_rah@yahoo.co.id
[6] Universitas Muhammadiyah Parepare, Parepare, South Sulawesi, Indonesia E-mail: sukmakuuh76@gmail.com

Abstract. One of the major obstacles in increasing the agricultural production yields is the inappropriate management of land use due to the differences in land biophysical characteristics in each region. Therefore, it is necessary to have spatial-based land use management that is appropriate for handling agricultural land. This research aims at structuring geospatial information of agricultural land biophysical characteristics and agricultural land appropriateness (potential and limiter) for food plants development in Pangkep Regency. This research uses the up-to-date approach of multi-scale observation, so the problems becoming the restricting factors of productivity can immediately be overcome. The method used consists of 2 stages, namely stage 1, analysis of agricultural land (biophysical) climate and land characteristics through land survey, laboratory analysis, and land characteristic spatialization; stage 2, analysis of agricultural land appropriateness (potential and limiter) using land appropriateness method of *fuzzy set*/FAO, satellite image information extraction, *ground truth,* and *ground sampling*. The results of the research are in the form of geospatial information on agricultural land biophysical characteristics and agricultural land appropriateness (potential and limiter) for food plants development in Pangkep Regency. With the information, decision makers will have more ease and they integrate into making policies for agricultural land use management in Pangkep Regency that will end up aiming to increase food production.

1. Introduction

The amount of food production from rice commodities in Pangkep Regency is quite large, which is 189,022 tons. Nevertheless, the average productivity is still relatively low, which is 6.3 tons/ha from the potential production of 7-9 tons/ha. The same goes for the amount of corn production, which is 4,711 tons with an average productivity of 6.4/ha from the production potential which can reach 13 tons/ha. The same thing happens to plants in the plantation sector, cocoa production in Pangkep Regency 55.36 tons with productivity averaging 0.17 tons/ha from production potential of 1.5 tons/ha and coffee production 126.24 tons with an average productivity of 0.16.ton/ha of potential production of 1.5-2 tons/ha. This indicates that in general agricultural production in Pangkep Regency is mainly food crops and plantations, far lower than the production potential that can be achieved (Pangkep Regency Agricultural Service, 2017).

The low production and average productivity of agricultural commodity land, especially food crops, which are far below the production potential that can be achieved are triggered by the inaccuracy of land use management in the form of inaccurate space (location), time, and management of agricultural land in Pangkep Regency. The inaccuracy of this space, among others, agricultural commodities are cultivated in areas that are not suitable, for example, food crops are cultivated on sloping areas, which in addition to reducing production, also affect the damage to the environment and spur the occurrence of flooding. Inaccuracies can be inaccuracies in determining planting time. Whereas inaccurate management relates to fertilizer input. Possibly, the input of the fertilizer that is used too generically, using commonly used doses. Meanwhile, each region has different land agro-ecological characteristics (land and climate), so the recommended general dosage is not always suitable for the land.

The accuracy of the management of space, time and management is influenced by the characteristics of land agroecology, socio-economic, and government policies. On the other hand, this information is minimally available and generally is not spatially based (space) and is not integrated between the existing aspects. For this reason, careful research is needed on the land in Pangkep Regency so that land use is in the right space, the right time, and appropriately managed, with effective spatial, time and land design on land with a spatial and integrated approach. The approach with a user-friendly Spatial Decision Support System (SDSS) was built to facilitate policymakers in making decisions for agricultural land management. This study aims to compile geospatial information on the biophysical characteristics of agricultural land and the suitability of agricultural land (potential and limitation) for the development of food crops in Pangkep Regency.

2. Research Method

2.1. Area of Study
This research was conducted in Pangkep Regency, South Sulawesi Province. Pangkep Regency is geographically located between 40 33 ‘20.0 "to 40 57’ 5.6" South Latitude and 1190 22 ‘54.8 "to 1190 48’ 35.8" East Longitude. The area is 145,266 ha which consists of 9 sub-districts, namely: Balocci, Bungoro, Labakkang, Mandalle, Ma’rang, Minasatene, Pangkajene, Segeri, Tondong Tallasa. Pangkep Regency is one of the largest rice producers in South Sulawesi (Figure 1). Variability is quite complex, consisting of beaches, lowlands, and highlands. Bordering Maros and Makassar, the capital of South Sulawesi Province, population pressure has led to greater land use change.

2.2. Analysis Method
This research employed a cutting-edge approach to multi-scale observation so that problems that can be a barrier to productivity can be solved immediately. The method used consists of two stages, namely stage 1, analysis of land and climate (biophysical) characteristics of agricultural land, through land surveys, laboratory analysis, and spatial characteristics of land; stage 2, analysis of agricultural land suitability (potential and limiting), using fuzzy set/FAO land suitability methods, extraction of satellite image information, ground truth, and ground sampling.

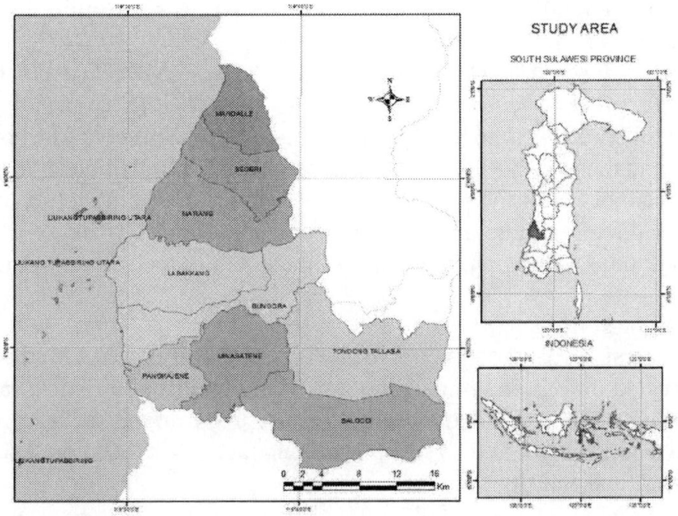

Figure 1. *The Location of Study and South Sulawesi Province*

Database sources consist of: (i) digital topographic maps; (ii) land maps and soil characteristics; (ii) climate data; and (iii) satellite data (SPOT 6 imagery). The digital topographic map of the study area with a scale of 1: 25,000 from the Geospatial Information Agency was used as a reference for mapping. Digital topographic maps in GIS vector format so that it is easier to build databases in the standard GIS vector, which is then converted to raster format.

Each land attribute in each component was assessed from 0 (minimum) to 1.0 (maximum) according to the suitability of corn [1, 2, 3, 4, 5]. Values represented as membership values ranging from 0 to 1.0. Analytical procedures are carried out through the following steps: (i) selecting and designing evaluation criteria, (ii) standardizing data sets, (iii) determining field/climate attribute values, (iv) selecting appropriate scoring functions and parameters [6], (v) converting data formats between software programs (raster to vector or vice versa). The analysis procedure is then followed by three main steps [7].

Step 1, determining individual ratings of land attributes, using equations [8]:

$$MF(xi) = [1/(1 + \{(xi - b)/d\}^2)] \text{ if } 0 < MF(xi) < 1 \quad (1)$$

For the optimum range (Model 2):

$$MF(xi) = 1 \text{ if } (b1 + d1) < xi < (b2 - d2) \quad (2)$$

For the asymmetric left (Model 3):

$$MF(xi) = [1/(1 + \{(xi - b1 - d1)/d1\}^2)] \text{ if } xi < (b1 + d1) \quad (3)$$

For the asymmetric right (Model 4):

$$MF(xi) = [1/(1 + \{(xi - b2 + d2)/d2\}^2)] \text{ if } xi > (b2 - d2) \quad (4)$$

in which MF (xi) = individual membership value for land with attributes x, d = transition zone range (i.e., x at MF = 0.5 or mentioned as crossover point, CP), xi = land attribute i (x) value, and b = the value of the land attribute x at the ideal point [7].

Step 2, the derivation of the rank attribute class of land, through membership integration of land attributes values using the following convex combination:

$$JMF(X) = \sum_{i=1}^{n} \lambda i \, MF(xi) \quad (5)$$

in which $0 < JMF(X) < 1$; $0 < MFxi < 1$; $\lambda 1 + \lambda 2 + \lambda 3 + ... + \lambda n = 1$, and $0 < \lambda i < 1.0$. JMF (X) symbol is a plural membership function of all variables considered in class X, λi is a value factor for land attributes x which in i, MF (xi) are membership values for land attributes xi, and n is the number of land attributes considered.

IOP Conf. Series: Earth and Environmental Science **279** (2019) 012005 doi:10.1088/1755-1315/279/1/012005

Step 3, calculate the overall land suitability index (LSI), using a multiplication function based on cells with cells in the GIS raster, as follows:

LSIP (corn) = JMF (S (p)) x JMF (T (p) x JMF (C (p)) (6)

in which LSIP (corn) is LSI for corn in cell p, JMF (S (p)), JMF (T (p)), and JMF (C (p)) is the JMF value in cell p for soil profile, topography, and climate, respectively. The use of the multiplication function indicates that there is no compensation between land attributes. In other words, land with extreme limitations cannot be compensated, with other characteristics that are very good on land, or vice versa [6, 7]. The multiplication function then produces a land suitability index (LSI) which is represented in the form of a continuous value, ranging from 0 (not suitable) to 1.0 (very suitable) [7].

3. Findings and Discussion

The results of the analysis of land characteristics with the Boolean FAO method and continuous fuzzy set approach are divided into three groups namely 1) climate characteristics: rainfall, temperature, and dry month, 2) soil physical characteristics: slope, texture, rock on the surface, soil depth, drainage soil, erosion hazard, and flood hazard, and 3) soil fertility characteristics: Cation Exchange Capacity (CEC), soil acidity, organic C, and salinity.

3.1. ***Climate***

The climate type of Pangkep Regency based on Oldeman climate classification is climate types B2 and C, while according to the Schmidt-Ferguson classification are climate types A and B (very wet). The average amount of rainfall is 3,346 mm/year with the dry season rainfall distribution pattern falling in June - October every year. The average number of dry months ranges from 3-4 months. The pattern of rainfall distribution is shown in Figure 2.

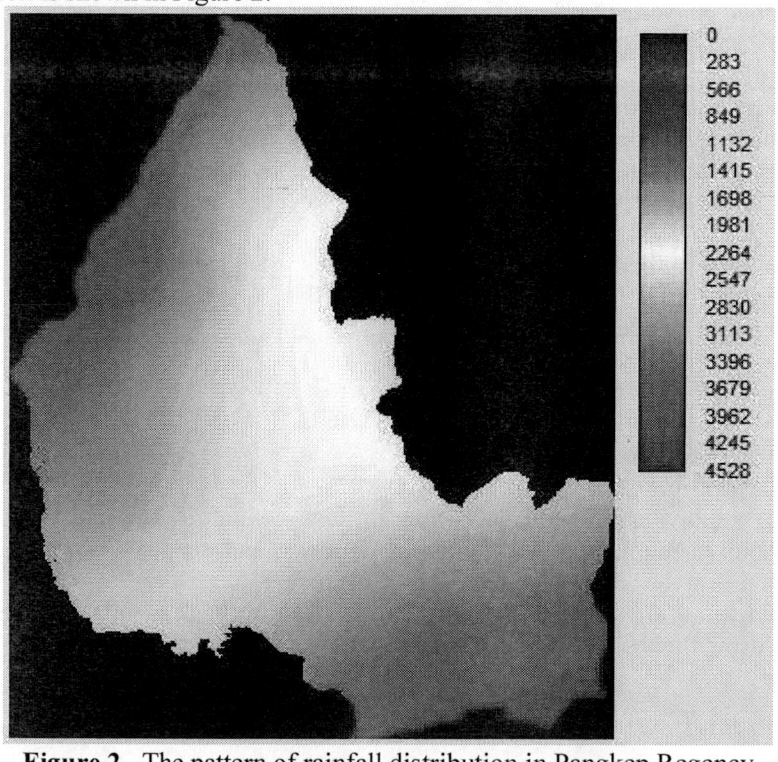

Figure 2. The pattern of rainfall distribution in Pangkep Regency

3.2. Soil

Based on the soil type map (RePPProT, 1988), land in Pangkep Regency consists of 8 types of soil, namely Alluvial Hydromorph (1201.10 ha), Alluvial Gray (4,474.64 ha), Comprehensive Mediterranean

IOP Conf. Series: Earth and Environmental Science **279** (2019) 012005 doi:10.1088/1755-1315/279/1/012005

brown (2,117.89 ha), Yellowish red lithosol (1,475.12 ha), Lithosol (42,494.23 ha), dark brown Mediterranean (8,539.77 ha), red Podsolic (12,448.69 ha), brownish brown Regosol (6,062.67 ha). The type of land that dominates the area of Pangkep Regency is the type of Lithosol soil which covers a wider area compared to other types of soil. However, the characteristics of lithosol soil is a soil with a low fertility level, because it has a very little nutrient in the soil, which only reaches 45 cm

3.3. Characteristics of Slope Land Based on the FAO Boolean Method and the Continuous Fuzzy Set Method

Figure 3a shows the distribution of regions and the extent of each category of slope land characteristics based on the FAO Boolean method. Each polygon category is bounded with a strict and homogeneous attribute. This does not describe the condition of the land in the field, because of the facts on the ground, the slope has no firm boundaries, but rather continuous. While image 3b shows the distribution of the area and the area of each category of slope characteristics based on the continuous fuzzy set method after the reclass. The values of land attributes are converted to continuous values (starting from 0 to 1.0) [9], according to class boundaries determined based on field experience, the results of conventional trial results [10, 11, 12, 13, 14]. Each coordinate has a different slope value, this condition is more representative of the condition of the land slope in a continuous field and has no firm boundaries.

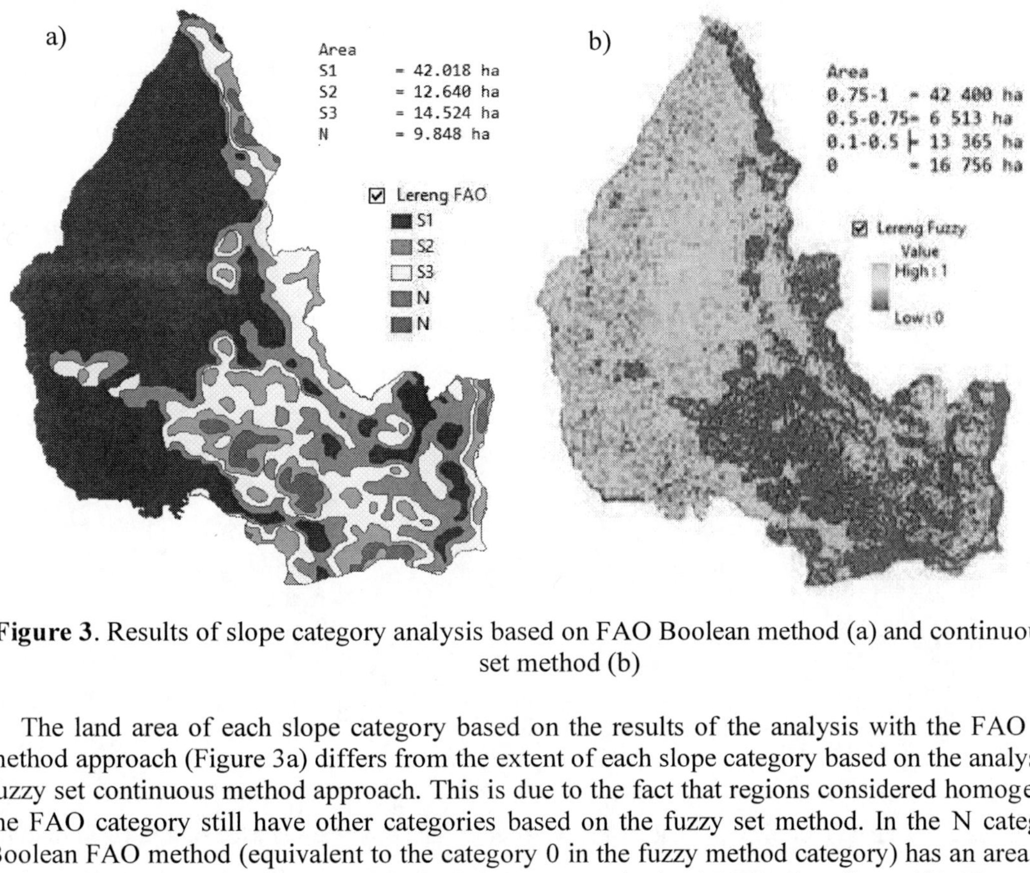

Figure 3. Results of slope category analysis based on FAO Boolean method (a) and continuous fuzzy set method (b)

The land area of each slope category based on the results of the analysis with the FAO Boolean method approach (Figure 3a) differs from the extent of each slope category based on the analysis of the fuzzy set continuous method approach. This is due to the fact that regions considered homogeneous in the FAO category still have other categories based on the fuzzy set method. In the N category, the Boolean FAO method (equivalent to the category 0 in the fuzzy method category) has an area of 9.848 ha (see Figure 3a) and enhancement in the 0 fuzzy set method, 16.756 ha (see figure 3b). Figure 4 shows the results of the slope category analysis on the same land area using the Boolean FAO method (black box) and continuous fuzzy set (inset) method. The facts in the field also indicate heterogeneous conditions, so that categorization with continuous fuzzy set methods is more effective in representing conditions in the field.

IOP Conf. Series: Earth and Environmental Science **279** (2019) 012005 doi:10.1088/1755-1315/279/1/012005

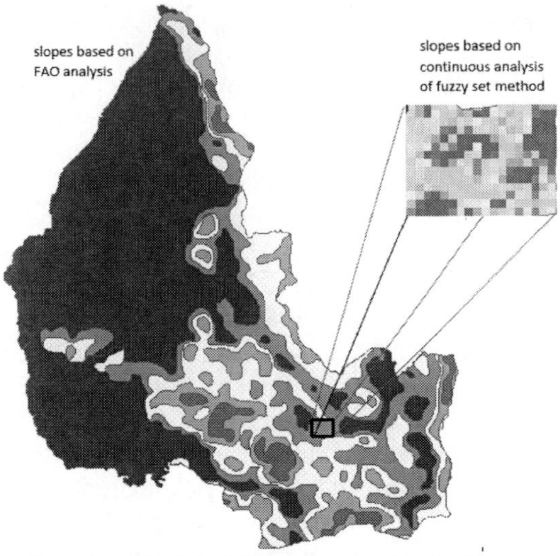

Figure 4. Results of analysis of slope categories in the same land area

3.4. Spatial Distribution of Land Suitability and Land Suitability Index
Land suitability classes for food crops (rice, corn, and soybeans) using the FAO method are shown in Figure 5. While figure 6 displays the Land Suitability Index (LSI) for corn plants. In general, the appearance of the corn crop LSI image indicates that the dominant low LSI value is found in the southeastern part of Pangkep Regency. This area is dominated by limiting factors where there are Bantimurung-Bulusaraung National Parks, karst areas and other ancient sites. Nevertheless, in the region, there are still areas that have high LSI values, which can be identified by the appearance of green in the image. As for the western central region, it is dominated by a higher LSI value; this is because the western region has a relatively sloping topography and is flat. It's just that, getting to the west, getting closer to the coastal area so that it is limited by the limiting factor of salinity.

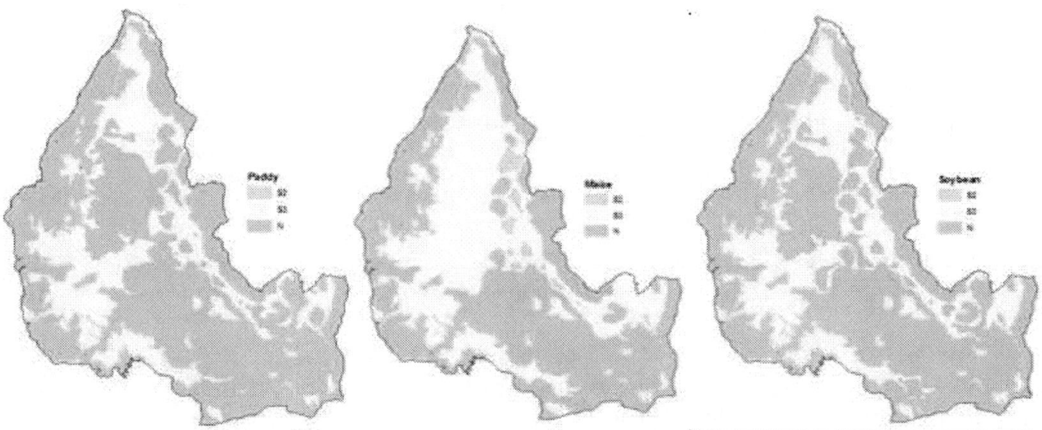

Figure 5. Class of land suitability of food crops (rice, corn, and soybeans) based on the FAO method

IOP Conf. Series: Earth and Environmental Science **279** (2019) 012005 doi:10.1088/1755-1315/279/1/012005

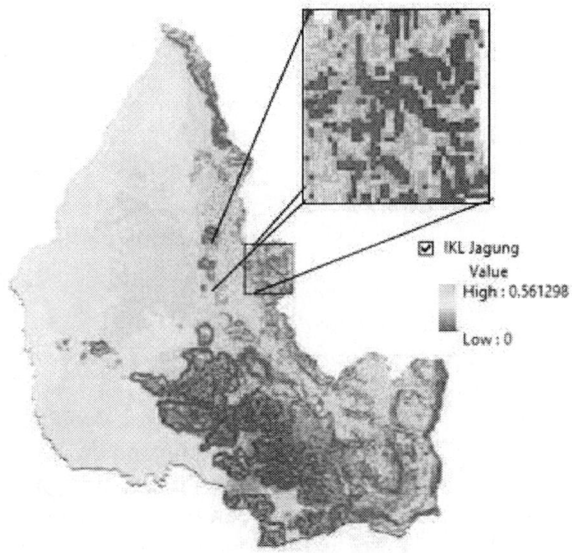

Figure 6. Land suitability index for corn plants based on a continuous fuzzy set method

Land suitability evaluation using the fuzzy set method approach has a range of index values from 0 (very poor) to 1 (very good). The maize land suitability index value which reached 0.561 indicates suitability and production for maize accordingly, although it is limited to regions that have a range of these index values. This value is equivalent to the S2 land suitability class on FAO land suitability evaluation criteria (1976).

3.5. Food Resilience

Detailed spatial information on LSI characteristics and land that better represents land conditions in the field will support efforts to increase agricultural production. The land that has been considered homogeneous is not suitable for the allocation of food crops, with the use of a continuous fuzzy set method, in which land is found to be suitable for the allocation of food crops. With more detailed data and more represent conditions in the field, decision making on agricultural land development will be more effective and efficient, so that higher productivity leads to regional food resilience.

4. Conclusion

The conclusions that can be drawn in this study are as in the following:

1. The characteristics of the land and its potential for land suitability for food crops are spatially distributed based on the magnitude and extent of the land boundaries both physically and chemically. In general, the characteristics of the central part of the land favor the use of agricultural land compared to other parts of the land.
2. The height of the Land Suitability Index (LSI) is dominantly found in the middle segment area. The index value reached 0.561 in corn plants, which is equivalent to the S2 land suitability class on FAO land suitability evaluation criteria (1976). Validating with the FAO category system, in certain regions (especially in high slope areas with low FAO land suitability classes), the standard deviation (SD) value of LSI is relatively high, indicating a high variation in LSI in this area, which means there are regions which actually has high potential (high LSI) even in narrow areas.

5. References

[1] Keshavarzi A, Sarmadian F, Heidari A, Omid M 2010 *Modern Applied Science* **4** 72-81.
[2] Kurtener D, Torbert H A, Krueger E 2008 *Computational Science and Its Applications–ICCSA 2008* 475-490 Springer Berlin Heidelberg.
[3] Mohammadrezaei N, Pazira E, Sokoti R, Ahmadi A 2013 *International Journal of Agronomy and*

Plant Production **4** 3438-3443.
[4] Reshmidevi T V, Eldho T I, Jana R 2009 *Agricultural Systems* **101** 101-109.
[5] Sarmadian F, Keshavarzy A, Rajabpour B, Askari S 2010 Application of MCDM method in the fuzzy modeling of land suitability evaluation *19th World Congress of Soil Science, Soil Solutions for a Changing World* 1 – 6 August Brisbane Australia.
[6] Baja S, Chapman D M, Dragovich D 2001 Fuzzy modeling of environmental suitability index for rural land use systems: an assessment using a GIS *Proc. of the Sixth International Conference on GeoComputation* 24-26 September 2001 The University of Queensland: Brisbane.
[7] Baja S, Chapman D M, Dragovich D 2002 *Environmental Management* **29** 647-661.
[8] Burrough P A 1989 *Journal of Soil Science* **40**(3) 477- 492.
[9] Zadeh L A 1965 *J Information and Control* **8** 338-353.
[10] Baja S 2012 *Metode Analitik Evaluasi Sumber Daya Lahan - Aplikasi GIS, Fuzzy Set, dan MCDM* Identitas Universitas Hasanuddin: Makassar.
[11] Galindo J, Urrutia A, & Piattini M 2005 *Fuzzy Databases*: *Modelling, Design, Implementation* Idea Group Publishing, London the United Kingdom and the United States of America.
[12] Moreno J F S 2007 Applicability of knowledge-based and fuzzy theory-oriented approaches to land suitability for upland rice and rubber, as compared to the farmers' perception, a case study of Lao PDR. *M.S. Thesis*, Consortium partners: University of Southampton (UK), Lund University (Sweden), University of Warsaw (Poland), International Institute for Geo-Information Science and Earth Observation (ITC) (The Netherlands).
[13] Sediyono E, Setiawan A, Kaparang D R 2013 *International Journal of Computer Applications* **84** 26-29.
[14] Sui D Z 1992 *J Computer, Environment, and Urban Systems* **16** 101-115.

IOP Conf. Series: Earth and Environmental Science **279** (2019) 012006 doi:10.1088/1755-1315/279/1/012006

Land Use Conflict with a Particular Reference to Spatial Planning Implementation in South Sulawesi

Sumbangan Baja[1], Dwia A. T. Pulubuhu[2], Risma Neswati[1], Samsu Arif[3], Nurmiaty[4]

[1]Department of Soil Science, Faculty of Agriculture, Hasanuddin University, Makassar, Indonesia 90245, email: sbja02@yahoo.com.au, s.baja@unhas.ac.id
[2]Department of Social Science, Faculty of Social and Political Sciences, Hasanuddin University, Makassar, Indonesia 90245, email: dwiatn@yahoo.com
[1]Department of Soil Science, Hasanuddin University Makassar; email: neswati76@gmail.com
[3]Department of Geophysics, Hasanuddin University Makassar; email: samsu_arif@)unhas.ac.id
[4]Department of Estate Crops Cultivation, Pangkep State Polytechnic of Agriculture, South Sulawesi, Indonesia, phone (+fax) +62-0410-2312720, email: nurmiatyamin@yahoo.co.id

Abstract. Land use changes and conflicts have recently become a major issue in newly developed regions of Indonesia, since the implementation of regional autonomy policies, which put emphasis on decentralization of land use governance. In South Sulawesi, there are two main issues related to land use conflict. First is the invasion on protected forest land in rural areas, and second is the growing spread of *urban development* into rural areas which has created many types of *conflicts* over *land use.* The primary aim of this paper is to assess land use conflict within these two contexts in terms of unsuitable spatial location of existing land use with land use functions in the spatial regulation at two different scales: provincial and regency. At a provincial level, a careful detection was done at some selected area in South Sulawesi, while at regency level, some forms of conflict were assessed in the southern parts of South Sulawesi. The study employed integrated techniques of spatial data analyses and interviews, and the data were analyzed in geographic information systems (GIS). This study indicates that, at a provincial level significant areas of dry land agriculture, estate plantation, mining, and residential were found in the land designated as protected forest, and within GIS data base it can be identified a total areas of 2,075 ha, 677 ha, 1,415 ha, and 193 ha, respectively. At a regency level, conflicts occurred between existing land use and land functions in local spatial planning regulation, particularly at agriculture and urban-suburban areas. Results of interviews and questionnaires by purposive sampling with the community indicate several factors causing land use change include: a) low incomes of local community, b) increasing of price of basic needs, c) attempts to gain more revenue from land use, c) land use existed before spatial regulation being implemented, and d) ecological condition that forces farmers to practice shifting cultivation. The results of this study will give insights into how to design future land use policies both at provincial and regency levels in this region.

Keyword: *agriculture land, land use conflict, GIS, spatial planning, and South Sulawesi.*

IOP Conf. Series: Earth and Environmental Science **279** (2019) 012006 doi:10.1088/1755-1315/279/1/012006

1. Introduction

Since implementation of regional autonomy that concerned about land use management, land use changes and conflicts become urge on developing areas. Increasing population and high level economy, is proportional with land necessary, and will be affect to land use change and conflicts. In South Sulawesi, there are two issues correlate with land use conflicts. The first was invasion on protected forests at rural areas, and the other one is spreading of urban development towards rural areas which has created various conflicts of land use [1]. Traditionally, land use choice depends very much on the criteria used, but in essence it should embrace three different and sometimes conflicting aspects: productivity, workability and the environment [2].

The primary aim of this paper was to assess land use conflict within these two contexts in terms of unsuitable spatial location of existing land use with land use functions in the spatial regulation at two different scales: provincial and regency. At a provincial level, a careful detection was done at some selected area in South Sulawesi, while at regency level, some forms of conflict were assessed in the southern parts of South Sulawesi.

2. Methodology

Study Area

The study area was in district of Jeneponto (Figure 1). South Sulawesi province was located on latitude 1°53'34.0" and 7°30'18.4"S and longitude 119°15'0.0" until 121°50'21.3"W, instead Jeneponto was located on latitude 5°42'9.8"and 5°23'8.9"S, and longitude 119°29'12.1" until 119°56'21.4"W.

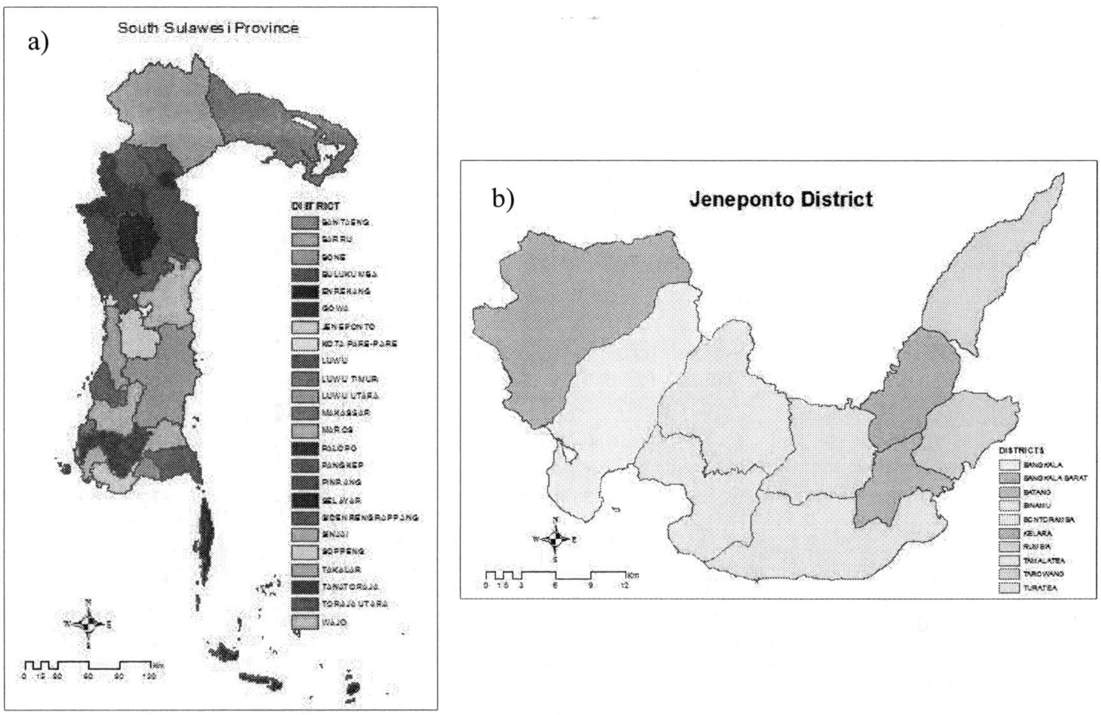

Figure 1. Administrative maps of South Sulawesi (a) and Jeneponto District (b)

South Sulawesi was covering an area of 4,453,049 ha, inclusive 24 districts/city in there, which is the largest city as capital city is Makassar city. According to statistics, population of South Sulawesi was about 8.7 million [3]. Area of Jeneponto was about 80,069 ha, there were 10 sub-districts consists of Bangkala, West Bangkala, Batang, Binamu, Bontoramba, Kelara, Rumbia, Tamalatea, Tarowang, and Turatea. As statistically, population of Jeneponto was proximately 359,787.

Database
This study used basic-spatial data resource, consists of: (i) Indonesia topographic digital map; (ii) Map of forest estate status; (iii) land use map; (iv) climate map; (v) administrative map; (vi) map of land use planning of South Sulawesi and Jeneponto district, and (vii) satellite imagery data (Landsat 8) that has been acquired on 2017. Indonesia topographic digital map of area study was on scale 1: 25.000 from Geospatial Information Agency on 2015, it was used as maps references. Indonesia topographic digital map was in vector so it is easier to build vector basic data based on GIS standard. All data layers were saved using UTM (Universal Transverse Mercator) coordinate system. Another used data is interview results of informants.

Analysis method
The study employed integrated techniques of spatial data analyses and interviews, and the data were analyzed in Geographic Information Systems (GIS). Procedure of analysis that was used shown as following:

1. Land use analyze
   Digital image processing for land suitability analysis was done by extracted the satellite recording of Landsat 8 in 2017 in order to make a land use map. Image extraction is done by using multispectral classification. Multispectral classification is process of digital image processing which is generate thematic map (generally, it is land use map), moreover it usually be an input on spatial model on Geographic Information System (GIS). The first assumption on multispectral classification was the objects could be recognized from one another based on its spectral value. In fact, showing that every objects is given specific pattern of spectral respond. The accuracy of multispectral classification is increase as narrow and as much the channels. As principal, how the multispectral classification algorithm work was marked every kind of objects and it will be seen the differences between one and another, based on features of spectral value and the channels outright. In the other words, multispectral classification algorithm simply has procedure as following:
   a. Determine the representative spectral value for every object by sampling method. Average value each sample will be grip for object identification.
   b. Place the representative value of sample on multidimensional diagram.
   c. Determine tolerance limit as spectral distance from representative value. It means counted pixels vector on outside of this distance will be classed not as object.
   d. Decision making is be like counting a whole of pixels and inserted it into available classes, as long as they are shorter or same with each object's tolerance distance, and group it was non classified since they are not able to classify in any class.
2. Spatial analysis in GIS using software Arc-GIS was an analysis which is done based on space considerations, include the point acquisition/locational accuracy of object or phenomenon with specific study scale. There is the first step should do before the analysis, is equate delineation area for all used parameters. Especially for land use in raster format should be converted become vector, due to the analysis is in vector. After that is coding process for describe every polygon on map layers, and save it in the attribute table. Next step, overlay all the parameter maps to gain land mapping unit, the used method is Union method. The results of land mapping unit have more polygon based on overlaying all parameters. Every polygon was measured its large using PIVOT. The result of PIVOT give illustration of land use conflict, both based on forest estate status or land use planning.
3. Interviews and questioner using purposive sampling
4. Ground truth, to see output of land suitability analysis with ground condition.

All the analyses was carried out all over South Sulawesi Province and continued in Jeneponto as a detailed archetype.

3. Results and discussion
Land use conflicts in provincial level

IOP Conf. Series: Earth and Environmental Science **279** (2019) 012006 doi:10.1088/1755-1315/279/1/012006

Land use is dynamic parameter of environment [4, 5]. Land use conflicts are indeed a widespread phenomenon, and can occur at any time or place between different stakeholders, mainly for different land expectations [6; 7; 8].The changes quickly happen on developing area. Distribution of land use in South Sulawesi is showed at Figure 2. Dominated land use was agriculture namely dry land-shrubs as much as 1,565,894 ha, and followed by secondary dry land forest (765,380 ha), paddy field (605,871 ha), primary dry land forest (587,853 ha) and shrubs (507,798 ha). Overall protected areas in South Sulawesi have been reached 2.083.950 ha or equal with 44.96% of total large of land (Table 1). This value was good based on DAS/DPS and districts/city, it is various (Figure 2b).

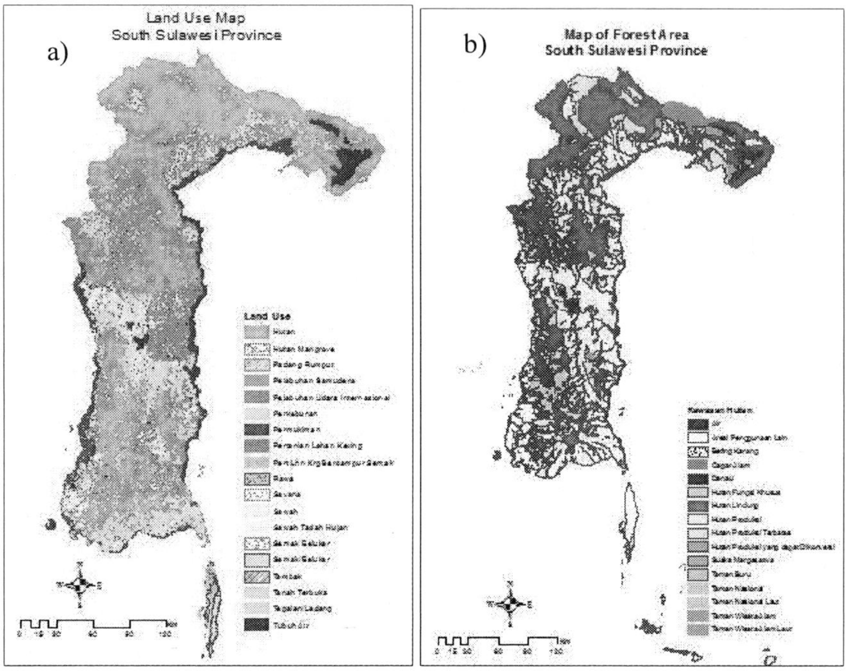

Figure 2. Land Use Map (a) and Forest Estate Status (b) South Sulawesi

Table 1. Wide Area of Protected Areas and Cultivation Area in South Sulawesi

No.	Areal Function	Hectares	Percentage
1.	Cultivation (Other land use areal)	1.909.226	41,19%
2.	Production forest	641.846	13,85%
3.	Protected areal	2.083.950	44,96%
	Total	4.635.022	100,00%

According to spatial analysis (Table 2), there was land use conflict on protected areas. At protected areas, there were some nonprotected areas, such as dry land agriculture as much as 2.075 ha, farm 677 ha. Inside limited-production forest, there were dry land agriculture 1.049 ha, pond 4.500 ha, and open spaces 1.631 ha.

Table 2. Protected areal pattern and large of land use

Spatial patterns	Dry land agriculture	Pond	Open spaces	Plantation	Settlement	Mining
Nature preserve			43.42	10.67	9.49	
Protected forest	2,075.27	17,626.05	2,378.24	677.13	193.98	1,415.10
HP	572.20		173.39	230.57	53.45	
HPK		1,374.48	2.55	98.48	10.54	
HPT	1,049.90	4,500.74	1,631.35	7.43	64.11	
Civil forest	678.51	222.02	15.50	84.27	7.87	
Mangrove	98.25	5,085.70				
National park	194.91		159.18			
Marine national park			10.46			
Nature park	5.62		79.52	52.08		

This study indicates that, at a provincial level significant areas of dry land agriculture, estate plantation, mining, and residential were found in the land designated as protected forest, and within GIS data base it can be identified a total area of 2,075 ha, 677 ha, 1,415 ha, and 193 ha, respectively. At a regency level, conflicts occurred between existing land use and land functions in local spatial planning regulation, particularly at agriculture and urban-suburban areas. Results of interviews and questionnaires by purposive sampling with the community indicate several factors causing land use change include: a) low incomes of local community, b) increasing of price of basic needs, c) attempts to gain more revenue from land use, c) land use existed before spatial regulation being implemented, and d) ecological condition that forces farmers to practice shifting cultivation.

Land Use Conflict on District Level

Land use distributon in Jeneponto was shown on Figure 3a and Table 3. Type of land use was dominated was non irrigated dry land as much as 36,581 ha, and followed by paddy fields (24,556 ha), shrubs (5,932 ha), and forest (5,658.09 ha). Moreover, forest estate status was shown on Figure 4b.

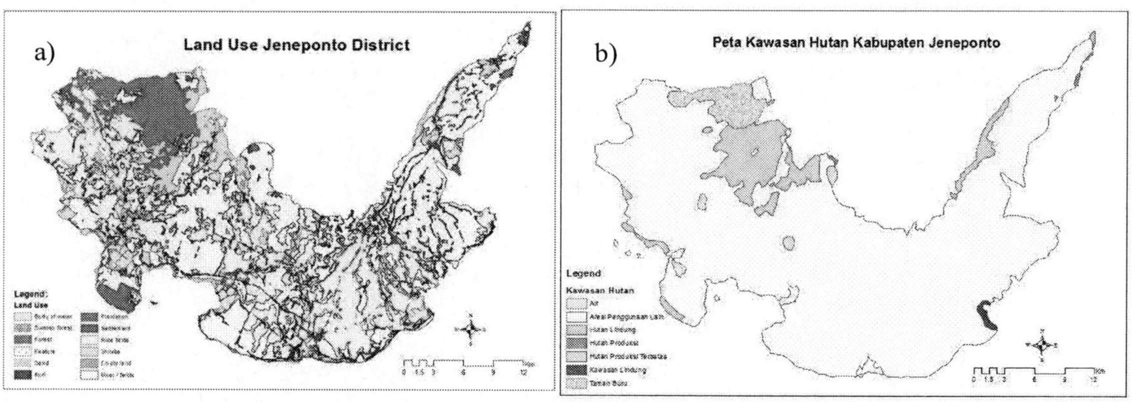

Figure 3. Land use Map (a) and Forest Estate Status (b) Jeneponto District

IOP Conf. Series: Earth and Environmental Science **279** (2019) 012006 doi:10.1088/1755-1315/279/1/012006

Table 3. The large area of land use in Jeneponto

Land Use	Hectares
Swamp forest	38.73
Forest	5,658.09
Savana	1.69
Dune	44.81
Ocean harbor	0.03
Plantation/farm	1,396.72
Settlements	2,602.43
Irrigated-field	24,556.40
Paddy field	343.15
Shrubs	5,932.97
Vacant land	37.40
Dry land agriculture	36,581.40
Water resources	2,381.07

On district level, there was conflict between existing land use and land function on land use planning, especially on agriculture sector and suburb area (Table 4).

Table 4. Protected area pattern and wide area (ha) of land use

Land Use	Spatial Pattern		
	Cultivation areas	Protected areas	Water Resources
Swamp forest	38.73		
Forest	2,210.53	3,447.56	
Savana	1.69		
Dune	40.68	2.68	1.45
Ocean harbor	0.03		
Plantation/farm	1,222.63	174.03	0.06
Settlements	2,552.62	49.80	0.00
Irrigated-field	24,010.73	542.89	2.79
Paddy field	220.20	121.70	1.25
Shrubs	3,605.68	2,297.12	30.06
Vacant land	37.40		
Dry land agriculture	34,426.14	2,150.47	4.76
Water resources	1,968.69	210.58	9.61

Reasons of Land Use Conflicts

According to Baja et al research [1], an adequate wide land use changing in South Sulawesi happened for 2 decades ago was shrubs and farm become paddy fields, and only a dew from settlements become industries. However, the changing of farm, shrub, and paddy field were lately happen, and usually become urban areas, specially settlements and industries. According to this study, the dynamics was really affected by some driving factors, consists of (i) actors, (ii) physics, (iii) economic, (iv) social-culture, and (v) institutional and policy.

Actors was dominant factor on spatial dynamics of land use changing, they are civil, investor, and government. Physics factor as main infrastructure development that change landscape and land use, also the catastrophe. Economic as infrastructure development (road, industry, beach reclamation, paddy field build (by government), plantation development on large scale (like cocoa and oil palm) especially by investors, and opening of mining concession (specifically in East Luwu done by investors). Other factor was social-culture, depending on society habitual, such as villagers, land use on the past was still hold on local wisdom and then it is enriched by science; for forest community, the ignorance about forest ecology aspect cause land using in forest area, even though the scale is not significant in this study; for middle class, most of them leave in urban area, they are change by their attitudes and mindsets significantly. Ecology was main component of indigenous culture and it was also shifted. Natural resources exploitation was become infinite without noticed about environment capability. Figure 4 shown the changing of land use structure in South Sulawesi by Baja [1].

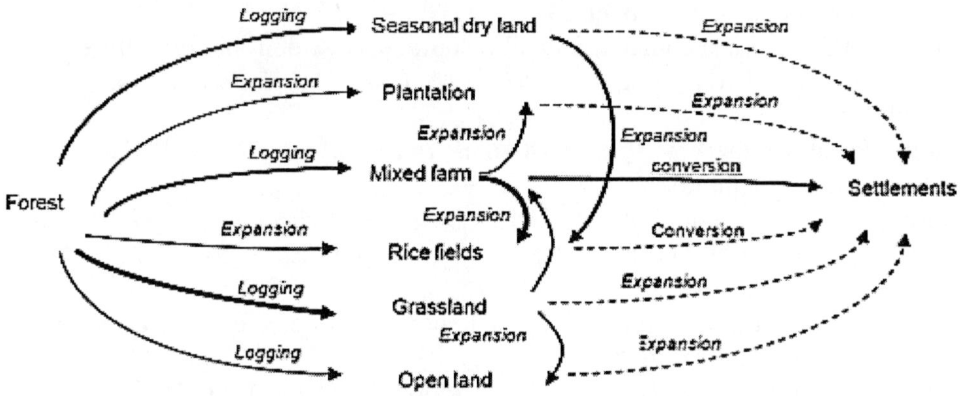

Figure 4. Structure of land use changing in South Sulawesi

Land use problems are mostly created by influential landlords or politicians, because the owners of the land are illiterate to understand their rights, and at times, even they do not know about the economic value of their land. Thus, the process and structure in which use and control of the land is managed forcibly, and decisions are implemented without democratic process, can be termed as land governance failure [9, 10]

This study indicates that, at a provincial level significant areas of dry land agriculture, estate plantation, mining, and residential were found in the land designated as protected forest, and within GIS data base it can be identified a total areas of 2,075 ha, 677 ha, 1,415 ha, and 193 ha, respectively. At a regency level, conflicts occurred between existing land use and land functions in local spatial planning regulation, particularly at agriculture and urban-suburban areas.

Results of interviews and questionnaires by purposive sampling with the community indicate several factors causing land use change include: a) low incomes of local community, b) increasing of price of basic needs, c) attempts to gain more revenue from land use, c) land use existed before spatial regulation being implemented, and d) ecological condition that forces farmers to practice shifting cultivation. The results of this study will give insights into how to design future land use policies both at provincial and regency levels in this region.

4. Conclusion

Conclusion that could be knew from this paper are:

1. This study indicates that, at a provincial level significant areas of dry land agriculture, estate plantation, mining, and residential were found in the land designated as protected forest, and within GIS data base it can be identified a total areas of 2,075 ha, 677 ha, 1,415 ha, and 193 ha, respectively.

2. At a regency level, conflicts occurred between existing land use and land functions in local spatial planning regulation, particularly at agriculture and urban-suburban areas which was driven by a) low incomes of local community, b) increasing of price of basic needs, c) attempts to gain more revenue from land use, c) land use existed before spatial regulation being implemented, and d) ecological condition that forces farmers to practice shifting cultivation.

5. References

[1] Baja S, Amrullah A, Ramli M, Ramlan A 2011 Spatial-based fuzzy classification of land suitability index for agriculture development: A model validation perspective. Pages 435-440 *Proc. of 3rd International Joint Conference on Computational Intelligence-International Conference on Fuzzy Computation Theory & Applications IJCCI-FCTA* Paris, France, 24-26 October 2011.

[2] Baja S 2009 Land use choice and land resource assessment in agriculture *CAB Reviews: Perspectives in Agriculture, Veterinary Science, Nutrition and Natural Resources* **4** No. 015.

[3] Badan Pusat Statistik (BPS) 2018 *Jeneponto dalam Angka* Badan Pusat Statistik Kabupaten Jeneponto.

[4] Camacho C, Barahona A P 2012 *Land use dynamics and the environment* https://www.parisschoolofeconomics.eu/IMG/pdf/draft322012.pdf (Accessed on February 2018).

[5] Eklund L, Degerald M, Brandt M, Prishchepov A V Pilesjö P 2017 *Environ. Res. Lett* **12** 054004. IOP Publishing. https://doi.org/10.1088/1748-9326/aa673a.

[6] Torre A, Melot R, Magsi H, Bossuet L, Cadoret A, Caron A, Darly S, Jeanneaux P, Kirat T, Pham H V, Kolokouris O 2014 *Identifying and Measuring Land-use and Proximity Conflicts: Methods and Identification*. Springer Plus 3:85.

[7] Wehrmann B 2008 *Land Conflicts: A Practical Guide to Dealing with Land Disputes. Handbook* Deutsche GesellschaftfürTechnischeZusammenarbeit (GTZ), Eschborn, Germany.

[8] Magsi H, Torre A, Liu Y, Sheikh M J 2017 *The Pakistan Development Review* **56** 1 Spring 2017 Pp 19-30.

[9] Palmer D, Fricska S, Wehrmann B 2009 *Improved Land Governance* United Nations Human Settlements Programme. Land Tenure Working Paper 11 (accessed: 14 February 2017) ftp://ftp.fao.org/docrep/fao/012/ak999e/ak999e00.pdf.

[10] Borras-Jr S M and Franco J C 2010 *Journal of Agrarian Change* **10**(1) 1–32.

Acknowledgments
The author deeply thankful to The Ministry of Research, Technology, and Higher Education, Republic of Indonesia, through a PSNI research scheme, The Pangkep Agricultural Polytechnic, and Government of South Sulawesi specifically Environmental Management Agency for providing data, and other supporting facilities for this study.

IOP Conf. Series: Earth and Environmental Science **279** (2019) 012007 doi:10.1088/1755-1315/279/1/012007

Numerical solutions to a class of scalar elliptic BVPs for anisotropic quadratically graded media

A. Haddade[1], M. I. Azis[2,*], Z. Djafar[3], St. N. Jabir[4], B. Nurwahyu[2]

[1]Department of Physics, Hasanuddin University, Makassar, Indonesia
[2]Department of Mathematics, Hasanuddin University, Makassar, Indonesia
[3]Department of Mechanical Engineering, Hasanuddin University, Makassar, Indonesia
[4]Department of Electrical Engineering, ATI Polytechnic, Makassar, Indonesia

E-mail: mohivanazis@yahoo.co.id

Abstract. Boundary value problems (BVPs) governed by a class of elliptic equations for anisotropic quadratically graded media are solved using Boundary Element Method (BEM). The variable coefficient governing equation is transformed to a constant coefficient equation which is then transformed to a boundary integral equation. The results show the convergence, consistency, and accuracy of the BEM solutions.

1. Introduction

Several types of constant coefficient equations have been solved using BEM (see for examples [1, 2, 3, 4]. But in general this is not the case for variable coefficient equation. There is some progress in using BEM to solve several types of variable coefficient governing equations (see for examples [5, 6, 7, 8, 9, 10, 11])

The governing equation considered by Salam et. al in [11] takes the form

$$\frac{\partial}{\partial x_i}\left[\lambda_{ij}\left(x_1,x_2\right)\frac{\partial\phi\left(x_1,x_2\right)}{\partial x_j}\right]=0 \tag{1}$$

This paper is intended to extend the work by Salam et. al [11] for problems with governing equation (1) to for 2D boundary value problems governed by another type of (dimensionless) elliptic equation of the form

$$\frac{\partial}{\partial x_i}\left[\lambda_{ij}\left(x_1,x_2\right)\frac{\partial\phi\left(x_1,x_2\right)}{\partial x_j}\right]+\beta\left(x_1,x_2\right)\phi\left(x_1,x_2\right)=0 \tag{2}$$

where the coefficients λ_{ij} depend on x_1 and x_2 and the repeated summation convention (summing from 1 to 2) is employed.

The matrix of coefficients $[\lambda_{ij}]$ is a real symmetric positive definite matrix so that equation (2) is a second order elliptic partial differential equation and may be written explicitly as

$$\frac{\partial}{\partial x_1}\left(\lambda_{11}\frac{\partial\phi}{\partial x_1}\right)+2\frac{\partial}{\partial x_1}\left(\lambda_{12}\frac{\partial\phi}{\partial x_2}\right)+\frac{\partial}{\partial x_2}\left(\lambda_{22}\frac{\partial\phi}{\partial x_2}\right)+\beta\phi=0$$

IOP Conf. Series: Earth and Environmental Science **279** (2019) 012007 doi:10.1088/1755-1315/279/1/012007

Further, the coefficients λ_{ij} and β are required to be twice differentiable functions of the two independent variables x_1 and x_2. The analysis here is specially relevant to an anisotropic medium but it equally applies to isotropic media. For isotropy, the coefficients in (2) take the form $\lambda_{11} = \lambda_{22}$ and $\lambda_{12} = 0$ and use of these equations in the following analysis immediately yields the corresponding results for an isotropic medium.

Steady infiltration problems (when $\beta < 0$, see for examples [12, 13]), acoustic problems (when $\beta > 0$, see for examples [14, 15]), and antiplane strain in elastostatics and plane thermostatic problems (when $\beta = 0$) are the areas for which the governing equation is of the type (2).

The technique of transforming (2) to a constant coefficient equation will be used for obtaining a boundary integral equation for the solution of (2).

2. The boundary value problem

Referred to a Cartesian frame Ox_1x_2 a solution to (2) is sought which is valid in a region Ω in R^2 with boundary $\partial\Omega$ which consists of a finite number of piecewise smooth closed curves. On $\partial\Omega_1$ the dependent variable $\phi(\mathbf{x})$ ($\mathbf{x} = (x_1, x_2)$) is specified and on $\partial\Omega_2$

$$P(\mathbf{x}) = \lambda_{ij} \left(\partial\phi/\partial x_j\right) n_i \tag{3}$$

is specified where $\partial\Omega = \partial\Omega_1 \cup \partial\Omega_2$ and $\mathbf{n} = (n_1, n_2)$ denotes the outward pointing normal to $\partial\Omega$.

3. The boundary integral equation

The boundary integral equation is derived by transforming the variable coefficient equation (2) to a constant coefficient equation. The coefficients λ_{ij} and β are required to take the form

$$\lambda_{ij}(\mathbf{x}) = \overline{\lambda}_{ij} g(\mathbf{x}) \tag{4}$$

$$\beta(\mathbf{x}) = \overline{\beta} g(\mathbf{x}) \tag{5}$$

where the $\overline{\lambda}_{ij}$ and $\overline{\beta}$ are constants and g is a differentiable function of $\mathbf{x}$. Use of (4) and (5) and in (2) yields

$$\overline{\lambda}_{ij} \frac{\partial}{\partial x_i}\left(g \frac{\partial\phi}{\partial x_j}\right) + \overline{\beta} g \phi = 0 \tag{6}$$

Let

$$\phi(\mathbf{x}) = g^{-1/2}(\mathbf{x})\,\psi(\mathbf{x}) \tag{7}$$

so that (6) may be written in the form

$$\overline{\lambda}_{ij} \frac{\partial}{\partial x_i}\left[g \frac{\partial\left(g^{-1/2}\psi\right)}{\partial x_j}\right] + \overline{\beta} g^{1/2}\psi = 0$$

That is

$$\overline{\lambda}_{ij}\left[\left(\frac{1}{4} g^{-3/2} \frac{\partial g}{\partial x_i}\frac{\partial g}{\partial x_j} - \frac{1}{2} g^{-1/2} \frac{\partial^2 g}{\partial x_i \partial x_j}\right)\psi + g^{1/2} \frac{\partial^2 \psi}{\partial x_i \partial x_j}\right] + \overline{\beta} g^{1/2}\psi = 0 \tag{8}$$

Use of the identity

$$\frac{\partial^2 g^{1/2}}{\partial x_i \partial x_j} = -\frac{1}{4} g^{-3/2} \frac{\partial g}{\partial x_i}\frac{\partial g}{\partial x_j} + \frac{1}{2} g^{-1/2} \frac{\partial^2 g}{\partial x_i \partial x_j}$$

permits (8) to be written in the form

$$g^{1/2}\overline{\lambda}_{ij} \frac{\partial^2 \psi}{\partial x_i \partial x_j} - \psi \overline{\lambda}_{ij} \frac{\partial^2 g^{1/2}}{\partial x_i \partial x_j} + \overline{\beta} g^{1/2}\psi = 0 \tag{9}$$

If we further restrict the function $g(\mathbf{x})$ to take the exponential form

$$g(\mathbf{x}) = [A(\alpha_0 + \alpha_1 x_1 + \alpha_2 x_2)]^2 \tag{10}$$

where α_m are constant, then

$$\overline{\lambda}_{ij}\frac{\partial^2 g^{1/2}}{\partial x_i \partial x_j} = 0 \tag{11}$$

Subtitution (11) into (9) implies a constant coefficients equation

$$\overline{\lambda}_{ij}\frac{\partial^2 \psi}{\partial x_i \partial x_j} + \overline{\beta}\psi = 0 \tag{12}$$

Also, substitution of (4) and (7) into (3) gives

$$P = -P_g\psi + P_\psi g^{1/2} \tag{13}$$

where

$$P_g(\mathbf{x}) = \overline{\lambda}_{ij}\frac{\partial g^{1/2}}{\partial x_j}n_i \qquad P_\psi(\mathbf{x}) = \overline{\lambda}_{ij}\frac{\partial \psi}{\partial x_j}n_i$$

A boundary integral equation for the solution of (12) is given in the form

$$\eta(\mathbf{x}_0)\psi(\mathbf{x}_0) = \int_{\partial\Omega}\left[\Gamma(\mathbf{x},\mathbf{x}_0)\psi(\mathbf{x}) - \Phi(\mathbf{x},\mathbf{x}_0)P_\psi(\mathbf{x})\right]ds(\mathbf{x}) \tag{14}$$

where $\mathbf{x}_0 = (a,b)$, $\eta = 0$ if $(a,b) \notin \Omega \cup \partial\Omega$, $\eta = 1$ if $(a,b) \in \Omega$, $\eta = \frac{1}{2}$ if $(a,b) \in \partial\Omega$ and $\partial\Omega$ has a continuously turning tangent at (a,b).

The so called fundamental solution Φ in (14) is any solution of the equation

$$\overline{\lambda}_{ij}\frac{\partial^2 \Phi}{\partial x_i \partial x_j} + \overline{\beta}\Phi = \delta(\mathbf{x} - \mathbf{x}_0)$$

and the Γ is given by

$$\Gamma(\mathbf{x},\mathbf{x}_0) = \overline{\lambda}_{ij}\frac{\partial \Phi(\mathbf{x},\mathbf{x}_0)}{\partial x_j}n_i$$

where δ is the Dirac delta function. Following Azis in [16], for two-dimensional problems Φ and Γ are given by

$$\begin{aligned}
\Phi(\mathbf{x},\mathbf{x}_0) &= \begin{cases} \frac{K}{2\pi}\ln R & \text{if } \overline{\beta} = 0 \\ \frac{\imath K}{4}H_0^{(2)}(\omega R) & \text{if } \overline{\beta} > 0 \\ \frac{-K}{2\pi}K_0(\omega R) & \text{if } \overline{\beta} < 0 \end{cases} \\
\Gamma(\mathbf{x},\mathbf{x}_0) &= \begin{cases} \frac{K}{2\pi}\frac{1}{R}\overline{\lambda}_{ij}\frac{\partial R}{\partial x_j}n_i & \text{if } \overline{\beta} = 0 \\ \frac{-\imath K\omega}{4}H_1^{(2)}(\omega R)\overline{\lambda}_{ij}\frac{\partial R}{\partial x_j}n_i & \text{if } \overline{\beta} > 0 \\ \frac{K\omega}{2\pi}K_1(\omega R)\overline{\lambda}_{ij}\frac{\partial R}{\partial x_j}n_i & \text{if } \overline{\beta} < 0 \end{cases}
\end{aligned} \tag{15}$$

where

$$\begin{aligned}
K &= \ddot{\tau}/\zeta \\
\omega &= \sqrt{|\beta|/\zeta} \\
\zeta &= \left[\overline{\lambda}_{11} + 2\overline{\lambda}_{12}\dot{\tau} + \overline{\lambda}_{22}\left(\dot{\tau}^2 + \ddot{\tau}^2\right)\right]/2 \\
R &= \sqrt{(\dot{x}_1 - \dot{a})^2 + (\dot{x}_2 - \dot{b})^2} \\
\dot{x}_1 &= x_1 + \dot{\tau}x_2 \\
\dot{a} &= a + \dot{\tau}b \\
\dot{x}_2 &= \ddot{\tau}x_2 \\
\dot{b} &= \ddot{\tau}b
\end{aligned}$$

where $\dot{\tau}$ and $\ddot{\tau}$ are respectively the real and the positive imaginary parts of the complex root τ of the quadratic

$$\overline{\lambda}_{11} + 2\overline{\lambda}_{12}\tau + \overline{\lambda}_{22}\tau^2 = 0$$

and $H_0^{(2)}$, $H_1^{(2)}$ denote the Hankel function of second kind and order zero and order one respectively. K_0, K_1 denote the modified Bessel function of order zero and order one respectively, $\imath$ represents the square root of minus one. The derivatives $\partial R/\partial x_j$ needed for the calculation of the Γ in (15) are given by

$$\begin{aligned}
\frac{\partial R}{\partial x_1} &= \frac{1}{R}(\dot{x}_1 - \dot{a}) \\
\frac{\partial R}{\partial x_2} &= \dot{\tau}\left[\frac{1}{R}(\dot{x}_1 - \dot{a})\right] + \ddot{\tau}\left[\frac{1}{R}\left(\dot{x}_2 - \dot{b}\right)\right]
\end{aligned}$$

Use of (7) and (13) in (14) yields

$$\begin{aligned}
\eta(\mathbf{x}_0)\, g^{1/2}(\mathbf{x}_0)\, \phi(\mathbf{x}_0) &= \int_{\partial\Omega} \left\{ \left[g^{1/2}(\mathbf{x})\, \Gamma(\mathbf{x}, \mathbf{x}_0) - P_g(\mathbf{x})\, \Phi(\mathbf{x}, \mathbf{x}_0) \right] \phi(\mathbf{x}) \right. \\
&\quad \left. - \left[g^{-1/2}(\mathbf{x})\, \Phi(\mathbf{x}, \mathbf{x}_0) \right] P(\mathbf{x}) \right\} ds(\mathbf{x}) \qquad (16)
\end{aligned}$$

This equation provides a boundary integral equation for determining ϕ and P at all points of Ω.

4. Numerical examples

In order to show the appropriateness of the BEM and the validity of the analysis used above for deriving the boundary integral equation (16), some particular boundary value problems will be solved. The integrals in equation (16) are evaluated numerically using the Bode's quadrature (see Abramowitz and Stegun [17]).

4.1. Examples with analytical solutions

In order to see the convergence and accuracy of the BEM we will consider some examples of problems with analytical solutions. The parameters for the quadratical inhomogeneity function $g(\mathbf{x})$ are $A = 3, \alpha_0 = 1, \alpha_1 = 0.25, \alpha_2 = 0.75$. Plot of $g(\mathbf{x})$ is shown in Figure 1. The geometry of the region Ω and the boundary conditions are as depicted in Figure 2. The values of the constant coefficients $\overline{\lambda}_{ij}$ for the governing equation (2) are

$$\overline{\lambda}_{11} = 0.75, \overline{\lambda}_{12} = 0.5, \overline{\lambda}_{22} = 1$$

The function $g(\mathbf{x})$ satisfies (10). Therefore equation (12) has to be the corresponding constant coefficient equation for $\psi(\mathbf{x})$ in which $\overline{\beta} > 0$, $\overline{\beta} < 0$ or $\overline{\beta} = 0$. Three possible forms of function $\psi(\mathbf{x})$ satisfying (12) that will be taken are

$$\begin{aligned}\psi(\mathbf{x}) &= B\left[\cos(\gamma_i x_i) + \sin(\gamma_i x_i)\right] \qquad \overline{\beta} = \overline{\lambda}_{ij}\gamma_i\gamma_j \\ \psi(\mathbf{x}) &= B\exp(\gamma_i x_i) \qquad \overline{\beta} = -\overline{\lambda}_{ij}\gamma_i\gamma_j \\ \psi(\mathbf{x}) &= B(\gamma_0 + \gamma_1 x_1 + \gamma_2 x_2) \qquad \overline{\beta} = 0\end{aligned}$$

The parameters for the analytical solutions $\psi(\mathbf{x})$ are taken to be

$$B = 2, \gamma_0 = 1, \gamma_1 = 0.75, \gamma_2 = 0.35$$

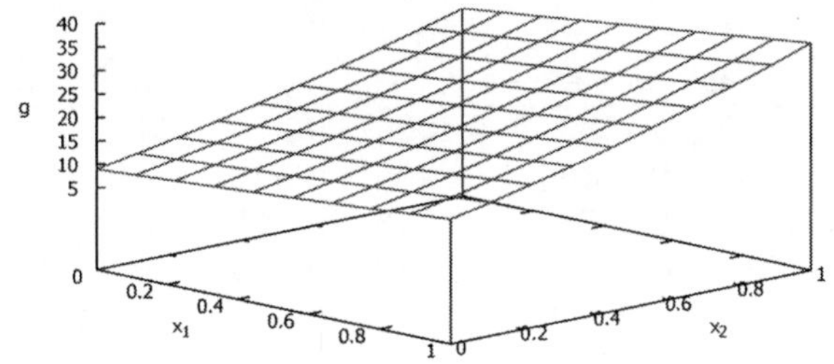

Figure 1. A quadratic inhomogeneity function $g(\mathbf{x}) = \left[3\left(1 + 0.25x_1 + 0.75x_2\right)\right]^2$

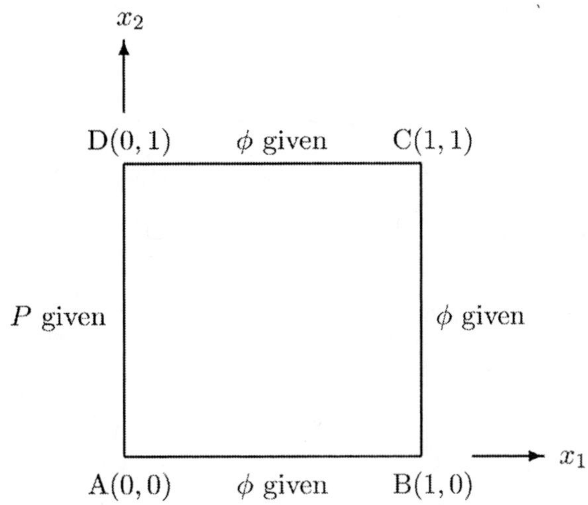

Figure 2. The geometry of all problems in Section 4.1

4.1.1. Problem 4.1.1: Case $\overline{\boldsymbol{\beta}} = \overline{\boldsymbol{\lambda}}_{ij}\gamma_i\gamma_j$ *in equation* (12) We take analytical solutions

$$\begin{aligned}\psi(\mathbf{x}) &= B\left[\cos(\gamma_i x_i) + \sin(\gamma_i x_i)\right] \quad \text{thus } \overline{\beta} = 0.806875 \\ \phi(\mathbf{x}) &= B\left[\cos(\gamma_i x_i) + \sin(\gamma_i x_i)\right] / \left[A\left(\alpha_0 + \alpha_1 x_1 + \alpha_2 x_2\right)\right]\end{aligned}$$

IOP Conf. Series: Earth and Environmental Science **279** (2019) 012007 doi:10.1088/1755-1315/279/1/012007

Table 1 shows the results of the analytical and BEM solutions with 20, 40 and 80 elements of equal length. The BEM solution converges to the analytical solution as the number of elements increases.

Table 1. BEM and analytical solutions for Problem 4.1.1

(x_1, x_2)	ϕ	$\partial\phi/\partial x_1$	$\partial\phi/\partial x_2$	ϕ	$\partial\phi/\partial x_1$	$\partial\phi/\partial x_2$
	BEM 20 elements			BEM 40 elements		
(0.1,0.5)	0.5799	0.1620	-0.1944	0.5797	0.1521	-0.1888
(0.3,0.5)	0.6028	0.0759	-0.2246	0.6026	0.0775	-0.2247
(0.5,0.5)	0.6109	0.0061	-0.2518	0.6109	0.0070	-0.2529
(0.7,0.5)	0.6055	-0.0597	-0.2728	0.6057	-0.0590	-0.2741
(0.9,0.5)	0.5879	-0.1015	-0.2966	0.5876	-0.1207	-0.2878
(0.5,0.1)	0.7297	0.0867	-0.3737	0.7300	0.0660	-0.3545
(0.5,0.3)	0.6651	0.0330	-0.2928	0.6654	0.0338	-0.2945
(0.5,0.7)	0.5637	-0.0164	-0.2223	0.5636	-0.0162	-0.2222
(0.5,0.9)	0.5220	-0.0536	-0.1832	0.5215	-0.0352	-0.1995
	BEM 80 elements			Analytical		
(0.1,0.5)	0.5793	0.1539	-0.1900	0.5792	0.1543	-0.1900
(0.3,0.5)	0.6025	0.0788	-0.2255	0.6025	0.0794	-0.2261
(0.5,0.5)	0.6111	0.0078	-0.2538	0.6112	0.0081	-0.2543
(0.7,0.5)	0.6059	-0.0587	-0.2748	0.6060	-0.0588	-0.2751
(0.9,0.5)	0.5880	-0.1199	-0.2892	0.5880	-0.1204	-0.2890
(0.5,0.1)	0.7306	0.0651	-0.3562	0.7310	0.0638	-0.3560
(0.5,0.3)	0.6658	0.0343	-0.2959	0.6661	0.0341	-0.2965
(0.5,0.7)	0.5636	-0.0152	-0.2229	0.5636	-0.0148	-0.2233
(0.5,0.9)	0.5215	-0.0355	-0.1993	0.5214	-0.0350	-0.1998

4.1.2. Problem 4.1.2: Case $\overline{\beta} = -\overline{\lambda}_{ij}\gamma_i\gamma_j$ *in equation* (12) Analytical solutions are

$$\psi(\mathbf{x}) = B\exp(\gamma_i x_i) \quad \text{thus } \overline{\beta} = -0.806875$$
$$\phi(\mathbf{x}) = B\exp(\gamma_i x_i) / [A(\alpha_0 + \alpha_1 x_1 + \alpha_2 x_2)]$$

The results are shown in Table 2. Again, the BEM solution converges to the analytical solution as the number of elements increases.

Table 2. BEM and analytical solutions for Problem 4.1.2

(x_1, x_2)	ϕ	$\partial\phi/\partial x_1$	$\partial\phi/\partial x_2$	ϕ	$\partial\phi/\partial x_1$	$\partial\phi/\partial x_2$
	BEM 20 elements			BEM 40 elements		
(0.1,0.5)	0.6140	0.3543	-0.1132	0.6127	0.3479	-0.1114
(0.3,0.5)	0.6881	0.3938	-0.1097	0.6869	0.3956	-0.1123
(0.5,0.5)	0.7723	0.4493	-0.1101	0.7713	0.4496	-0.1127
(0.7,0.5)	0.8685	0.5146	-0.1122	0.8673	0.5118	-0.1138
(0.9,0.5)	0.9788	0.5920	-0.1158	0.9768	0.5856	-0.1164
(0.5,0.1)	0.8365	0.4705	-0.2545	0.8363	0.4480	-0.2203
(0.5,0.3)	0.7986	0.4516	-0.1544	0.7983	0.4513	-0.1596
(0.5,0.7)	0.7539	0.4491	-0.0743	0.7525	0.4499	-0.0766
(0.5,0.9)	0.7403	0.4244	-0.0689	0.7402	0.4477	-0.0461
	BEM 80 elements			Analytical		
(0.1,0.5)	0.6121	0.3486	-0.1126	0.6114	0.3494	-0.1136
(0.3,0.5)	0.6864	0.3957	-0.1134	0.6859	0.3962	-0.1147
(0.5,0.5)	0.7708	0.4496	-0.1142	0.7703	0.4494	-0.1156
(0.7,0.5)	0.8667	0.5109	-0.1148	0.8661	0.5099	-0.1159
(0.9,0.5)	0.9758	0.5810	-0.1152	0.9749	0.5788	-0.1158
(0.5,0.1)	0.8366	0.4522	-0.2256	0.8371	0.4534	-0.2302
(0.5,0.3)	0.7982	0.4510	-0.1617	0.7981	0.4508	-0.1640
(0.5,0.7)	0.7518	0.4497	-0.0777	0.7511	0.4495	-0.0785
(0.5,0.9)	0.7393	0.4510	-0.0482	0.7384	0.4513	-0.0492

4.1.3. Problem 4.1.3: Case $\overline{\boldsymbol{\beta}} = \mathbf{0}$ *in equation* (12) Now we choose analytical solutions

$$\begin{aligned} \psi(\mathbf{x}) &= B(\gamma_0 + \gamma_1 x_1 + \gamma_2 x_2) \\ \phi(\mathbf{x}) &= B(\gamma_0 + \gamma_1 x_1 + \gamma_2 x_2) / [A(\alpha_0 + \alpha_1 x_1 + \alpha_2 x_2)] \end{aligned}$$

The results are shown in Table 3. Once again, the BEM solution converges to the analytical solution as the number of elements increases.

Table 3. BEM and analytical solutions for Problem 4.1.3

(x_1, x_2)	ϕ	$\partial\phi/\partial x_1$	$\partial\phi/\partial x_2$	ϕ	$\partial\phi/\partial x_1$	$\partial\phi/\partial x_2$
	BEM 20 elements			BEM 40 elements		
(0.1,0.5)	0.5963	0.2587	-0.1533	0.5957	0.2501	-0.1505
(0.3,0.5)	0.6446	0.2316	-0.1688	0.6440	0.2332	-0.1704
(0.5,0.5)	0.6895	0.2180	-0.1854	0.6892	0.2185	-0.1871
(0.7,0.5)	0.7320	0.2071	-0.2009	0.7316	0.2057	-0.2020
(0.9,0.5)	0.7728	0.2099	-0.2198	0.7715	0.1950	-0.2151
(0.5,0.1)	0.7824	0.2738	-0.3167	0.7825	0.2514	-0.2896
(0.5,0.3)	0.7305	0.2354	-0.2271	0.7307	0.2354	-0.2304
(0.5,0.7)	0.6557	0.2027	-0.1538	0.6551	0.2037	-0.1549
(0.5,0.9)	0.6270	0.1660	-0.1296	0.6267	0.1892	-0.1297
	BEM 80 elements			Analytical		
(0.1,0.5)	0.5955	0.2503	-0.1514	0.5952	0.2509	-0.1522
(0.3,0.5)	0.6438	0.2335	-0.1711	0.6437	0.2338	-0.1720
(0.5,0.5)	0.6890	0.2185	-0.1880	0.6889	0.2185	-0.1889
(0.7,0.5)	0.7314	0.2052	-0.2026	0.7312	0.2046	-0.2033
(0.9,0.5)	0.7712	0.1930	-0.2152	0.7708	0.1921	-0.2155
(0.5,0.1)	0.7829	0.2541	-0.2931	0.7833	0.2535	-0.2951
(0.5,0.3)	0.7308	0.2352	-0.2318	0.7309	0.2350	-0.2332
(0.5,0.7)	0.6548	0.2037	-0.1555	0.6545	0.2039	-0.1561
(0.5,0.9)	0.6263	0.1904	-0.1305	0.6259	0.1908	-0.1312

4.2. Examples without analytical solutions

In this section we will consider some examples of problems without simple analytical solutions. We setup some problems for a homogeneous isotropic material by taking $g(\mathbf{x}) = 9$ and with symmetrical boundary conditions. This function $g(\mathbf{x})$ satisfies equation (10) thus we will take $\psi(\mathbf{x})$ that satisfies (12) to be put into the integral equation (14). The main purpose is to see the consistency of whether the BEM produces symmetrical solutions.

4.2.1. Problem 4.2.1: Case $\overline{\boldsymbol{\beta}} > \mathbf{0}$ in equation (12) For this problem we take $\overline{\beta} = 1.5$ and the symmetrical boundary conditions are as shown in Figure 3. Table 4 shows the results of the BEM solution using 20, 40, 80 and 160 elements of equal length. As expected, the results converge as the number of elements increases and also they are symmetrical about the axes $x_2 = 0.5$.

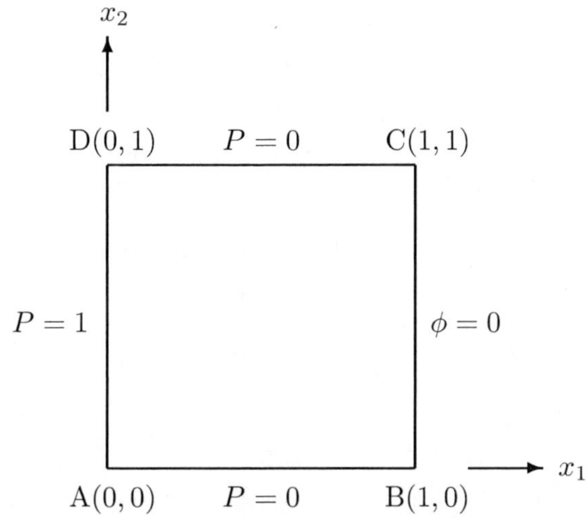

Figure 3. The geometry of Problem 4.2.1 and Problem 4.2.2

Table 4. BEM solution for Problem 4.2.1

(x_1, x_2)	ϕ	$\partial\phi/\partial x_1$	$\partial\phi/\partial x_2$	ϕ	$\partial\phi/\partial x_1$	$\partial\phi/\partial x_2$
	BEM 20 elements			BEM 40 elements		
(0.1,0.5)	0.2402	-0.1497	0.0000	0.2396	-0.1492	0.0000
(0.3,0.5)	0.2032	-0.2179	0.0000	0.2029	-0.2162	0.0000
(0.5,0.5)	0.1539	-0.2721	0.0000	0.1540	-0.2702	0.0000
(0.7,0.5)	0.0955	-0.3094	0.0000	0.0959	-0.3079	0.0000
(0.9,0.5)	0.0315	-0.3278	0.0000	0.0321	-0.3269	-0.0000
(0.5,0.1)	0.1541	-0.2620	-0.0012	0.1541	-0.2713	-0.0004
(0.5,0.3)	0.1540	-0.2724	-0.0002	0.1540	-0.2704	-0.0001
(0.5,0.7)	0.1540	-0.2724	0.0002	0.1540	-0.2704	0.0001
(0.5,0.9)	0.1541	-0.2620	0.0012	0.1541	-0.2713	0.0004
	BEM 80 elements			BEM 160 elements		
(0.1,0.5)	0.2391	-0.1485	-0.0000	0.2389	-0.1482	-0.0000
(0.3,0.5)	0.2026	-0.2152	-0.0000	0.2024	-0.2148	-0.0000
(0.5,0.5)	0.1539	-0.2691	-0.0000	0.1538	-0.2685	-0.0000
(0.7,0.5)	0.0960	-0.3068	-0.0000	0.0960	-0.3062	-0.0000
(0.9,0.5)	0.0324	-0.3261	-0.0000	0.0325	-0.3256	-0.0000
(0.5,0.1)	0.1539	-0.2693	-0.0002	0.1538	-0.2686	-0.0001
(0.5,0.3)	0.1539	-0.2692	-0.0001	0.1538	-0.2686	-0.0000
(0.5,0.7)	0.1539	-0.2692	0.0001	0.1538	-0.2686	0.0000
(0.5,0.9)	0.1539	-0.2693	0.0002	0.1538	-0.2686	0.0001

4.2.2. Problem 4.2.2: Case $\overline{\boldsymbol{\beta}} < \mathbf{0}$ *in equation* (12) We take $\overline{\beta} = -1.5$ and boundary conditions are as shown in Figure 3. Table 5 shows the results of the BEM solution using 20, 40, 80 and 160 elements of equal length. The results converge as the number of elements increases and also they are symmetrical about the axes $x_2 = 0.5$.

4.2.3. Problem 4.2.3: Case $\overline{\boldsymbol{\beta}} = \mathbf{0}$ *in equation* (12) We consider a problem with $\overline{\beta} = 0$ and the boundary conditions are as shown in Figure 3. Table 6 shows the results of the BEM solution

Table 5. BEM solution for Problem 4.2.2

(x_1, x_2)	ϕ	$\partial\phi/\partial x_1$	$\partial\phi/\partial x_2$	ϕ	$\partial\phi/\partial x_1$	$\partial\phi/\partial x_2$
	BEM 20 elements			BEM 40 elements		
(0.1,0.5)	0.0644	-0.0992	-0.0000	0.0652	-0.1000	0.0000
(0.3,0.5)	0.0464	-0.0823	0.0000	0.0470	-0.0831	0.0000
(0.5,0.5)	0.0312	-0.0706	0.0000	0.0316	-0.0713	0.0000
(0.7,0.5)	0.0179	-0.0631	0.0000	0.0182	-0.0638	0.0000
(0.9,0.5)	0.0057	-0.0595	0.0000	0.0059	-0.0601	-0.0000
(0.5,0.1)	0.0311	-0.0679	0.0002	0.0316	-0.0715	-0.0000
(0.5,0.3)	0.0311	-0.0706	0.0000	0.0316	-0.0713	0.0000
(0.5,0.7)	0.0311	-0.0706	-0.0000	0.0316	-0.0713	-0.0000
(0.5,0.9)	0.0311	-0.0679	-0.0002	0.0316	-0.0715	0.0000
	BEM 80 elements			BEM 160 elements		
(0.1,0.5)	0.0655	-0.1003	-0.0000	0.0656	-0.1004	-0.0000
(0.3,0.5)	0.0472	-0.0834	-0.0000	0.0473	-0.0835	-0.0000
(0.5,0.5)	0.0318	-0.0715	-0.0000	0.0319	-0.0716	-0.0000
(0.7,0.5)	0.0183	-0.0640	-0.0000	0.0184	-0.0641	-0.0000
(0.9,0.5)	0.0059	-0.0604	-0.0000	0.0060	-0.0605	-0.0000
(0.5,0.1)	0.0318	-0.0716	-0.0000	0.0319	-0.0717	-0.0000
(0.5,0.3)	0.0318	-0.0715	0.0000	0.0319	-0.0717	0.0000
(0.5,0.7)	0.0318	-0.0715	-0.0000	0.0319	-0.0717	-0.0000
(0.5,0.9)	0.0318	-0.0716	0.0000	0.0319	-0.0717	0.0000

Table 6. BEM solution for Problem 4.2.3

(x_1, x_2)	ϕ	$\partial\phi/\partial x_1$	$\partial\phi/\partial x_2$	ϕ	$\partial\phi/\partial x_1$	$\partial\phi/\partial x_2$
	BEM 20 elements			BEM 40 elements		
(0.1,0.5)	0.0979	-0.1100	-0.0000	0.0992	-0.1108	0.0000
(0.3,0.5)	0.0759	-0.1097	0.0000	0.0770	-0.1106	0.0000
(0.5,0.5)	0.0540	-0.1094	0.0000	0.0549	-0.1105	0.0000
(0.7,0.5)	0.0322	-0.1091	0.0000	0.0328	-0.1104	0.0000
(0.9,0.5)	0.0104	-0.1087	0.0000	0.0108	-0.1102	-0.0000
(0.5,0.1)	0.0540	-0.1053	0.0003	0.0549	-0.1108	0.0001
(0.5,0.3)	0.0540	-0.1095	0.0002	0.0549	-0.1105	0.0001
(0.5,0.7)	0.0540	-0.1095	-0.0002	0.0549	-0.1105	-0.0001
(0.5,0.9)	0.0540	-0.1053	-0.0003	0.0549	-0.1108	-0.0001
	BEM 80 elements			BEM 160 elements		
(0.1,0.5)	0.0996	-0.1110	-0.0000	0.0998	-0.1110	-0.0000
(0.3,0.5)	0.0775	-0.1109	-0.0000	0.0776	-0.1110	-0.0000
(0.5,0.5)	0.0553	-0.1109	0.0000	0.0554	-0.1110	-0.0000
(0.7,0.5)	0.0331	-0.1108	0.0000	0.0332	-0.1110	-0.0000
(0.9,0.5)	0.0110	-0.1108	-0.0000	0.0110	-0.1110	-0.0000
(0.5,0.1)	0.0553	-0.1109	0.0000	0.0554	-0.1110	0.0000
(0.5,0.3)	0.0553	-0.1109	0.0000	0.0554	-0.1110	0.0000
(0.5,0.7)	0.0553	-0.1109	-0.0000	0.0554	-0.1110	-0.0000
(0.5,0.9)	0.0553	-0.1109	-0.0000	0.0554	-0.1110	-0.0000

using 20, 40, 80 and 160 elements of equal length. The results converge as the number of elements increases and also they are symmetrical about the axes $x_2 = 0.5$.

5. Conclusion

The scalar elliptic governing equation (2) is used for modeling physical problems such as steady infiltration problems (when $\beta < 0$), acoustic problems (when $\beta > 0$), and antiplane strain in elastostatics and plane thermostatic problems (when $\beta = 0$). The boundary integral equation (16) was derived from this governing equation (2) and straight from (16) a BEM was then constructed for calculation of numerical solutions to the problems for anisotropic quadratically graded media. The results show the convergence, consistency, and accuracy of the BEM solutions. Together with its ease in implementation, it may be concluded that BEM is a good numerical method for solving such kind of problems.

Acknowledgements

This work was supported by Hasanuddin University and The Ministry of Higher Education of Indonesia.

References

[1] Azis M I, Kasbawati, Haddade A and Thamrin S A 2018 On some examples of pollutant transport problems solved numerically using the boundary element method *Journal of Physics: Conference Series* **979** 1
[2] Azis M I, Asrul L, Khaeruddin and Paharuddin 2018 BEM solutions for unsteady transport problems in anisotropic media *JP Journal of Heat and Mass Transfer* **15** 4
[3] Haddade A, Salam N, Khaeruddin and Azis M I 2017 A Boundary Element Method for 2D Diffusion-Convection Problems in Anisotropic Media *Far East Journal of Mathematical Sciences* **102** 8
[4] Azis M I 2019 Numerical solutions for the Helmholtz boundary value problems of anisotropic media *Journal of Computational Physics* **381** 42
[5] Cheng A H-D 1984 Darcy's Flow with Variable Permeability: A Boundary Integral Solution *Water Resources Research* **20** 980
[6] Clements D L and Azis M I 2000 A Note on a Boundary Element Method for the Numerical Solution of Boundary Value Problems in Isotropic Inhomogeneous Elasticity *Journal of the Chinese Institute of Engineers* **23** 3 261
[7] Azis M I and Clements D L 2014 On some problems concerning deformations of functionally graded anisotropic elastic materials *Far East Journal of Mathematical Sciences*, **87** 2 173
[8] Azis M I, Toaha S, Bahri M and Ilyas N 2018 A boundary element method with analytical integration for deformation of inhomogeneous elastic materials *Journal of Physics: Conference Series* **979** 1
[9] Azis M I and Clements D L 2014 A Boundary Element Method for Transient Heat Conduction Problem of Nonhomogeneous Anisotropic Materials *Far East Journal of Mathematical Sciences*, **89** 1 51
[10] Azis M I and Clements D L 2008 Nonlinear transient heat conduction problems for a class of inhomogeneous anisotropic materials by BEM *Engineering Analysis with Boundary Elements* **32** 1054
[11] Salam N, Haddade A, Clements D L and Azis M I 2017 A boundary element method for a class of elliptic boundary value problems of functionally graded media *Engineering Analysis with Boundary Elements* **84** 186 doi: 10.1016/j.enganabound.2017.08.017
[12] Clements D L and Lobo M 2010 A BEM for time dependent infiltration from an irrigation channel *Engineering Analysis with Boundary Elements* **34** 1100
[13] Solekhudin I and Ang K-C 2012 A DRBEM with a predictor–corrector scheme for steady infiltration from periodic channels with root-water uptake *Engineering Analysis with Boundary Elements* **36** 1199
[14] Barucq H, Bendali A, Fares M, Mattesi V and Tordeux S 2017 A symmetric Trefftz-DG formulation based on a local boundary element method for the solution of the Helmholtz equation *Journal of Computational Physics* **330** 1069
[15] Loeffler C F, Mansur W J, Barcelos H-d-M and Bulcão A 2015 Solving Helmholtz problems with the boundary element method using direct radial basis function interpolation *Engineering Analysis with Boundary Elements* **61** 218
[16] Azis M I 2017 Fundamental solutions to two types of 2D boundary value problems of anisotropic materials *Far East Journal of Mathematical Sciences* **101** 11 2405
[17] Abramowitz M and Stegun I A 1972 *Handbook of mathematical functions: with formulas, graphs and mathematical tables* Dover Publications Washington

IOP Conf. Series: Earth and Environmental Science **279** (2019) 012008 doi:10.1088/1755-1315/279/1/012008

Numerical solutions for BVPs governed by a Helmholtz equation of anisotropic FGM

B. Nurwahyu[1], B. Abdullah[2], A. Massinai[2], M. I. Azis[1,*]
[1]Department of Mathematics, Hasanuddin University, Makassar, Indonesia
[2]Department of Physics, Hasanuddin University, Makassar, Indonesia
E-mail: mohivanazis@yahoo.co.id (*Corresponding author)

Abstract. A Boundary Element Method (BEM) is used for obtaining solutions to anisotropic functionally graded media (FGM) boundary value problems (BVPs) governed by a Helmholtz type equation. A technique of transforming the variable coefficient governing equation to a constant coefficient equation is utilized for deriving a boundary integral equation. Some particular problems are considered to illustrate the application of the BEM. The results show the convergence, consistency, and accuracy of the BEM solutions.

1. Introduction

The BEM has been successfully used for solving many types of problems of homogeneous media. Some works on homogeneous media problems have been recently done by Azis et. al [1, 2] and Haddade et. al [3] in which the authors considered pollutant transport problems governed by 2D diffusion-convection, and also Azis [4] in which numerical solutions for the Helmholtz boundary value problems were obtained.

However, this is generally not the case for FGM. There are two techniques usually used to deal with problems of FGM. The first one uses a technique of deriving a relevant Green function or fundamental solution to the FGM problem. Cheng [5] had applied this technique. The second technique is by transforming the variable coefficient governing equation to a constant coefficient equation. Some progress on using the second technique has been made. For examples, Clements and Azis [6] considered the case for isotropic FGM. For anisotropic FGM some works have been studied by Azis and Clements [7, 8, 9] and Azis et. al [10]. In addition to this, recently Salam et. al [11] have been working on a class of elliptic problems for anisotropic FGM.

This paper discusses derivation of a boundary integral equation for numerically solving 2D boundary value problems governed by the (dimensionless) Helmholtz type equation of the form

$$\frac{\partial}{\partial x_i}\left[\lambda_{ij}\left(x_1,x_2\right)\frac{\partial\phi\left(x_1,x_2\right)}{\partial x_j}\right]+\beta^2\left(x_1,x_2\right)\phi\left(x_1,x_2\right)=0 \tag{1}$$

where the coefficients λ_{ij} and β^2 depend on x_1 and x_2 and the repeated summation convention (summing from 1 to 2) is employed.

A variety of problems of both isotropic and anisotropic inhomogeneous media are usually modeled with equation (1). Acoustic problems (when $\beta^2 > 0$), and antiplane strain in elastostatics and plane thermostatic problems (when $\beta^2 = 0$) are the areas for which the governing equation is of the type (1). Recently some works on the Helmholtz equation for isotropic and/or homogeneous media have been done. For example, Barucq et. al. in [12], Loeffler et. al. in [13] considered Helmholtz equation as the governing equation for isotropic and/or homogeneous media which is a special case of the equation (1).

The technique of transforming (1) to constant coefficient equations will be used for obtaining a boundary integral equation for the solution of (1). It is necessary to place some constraint on the class of coefficients λ_{ij} and β for which the solution obtained is valid. The analysis of this paper is purely formal; the main aim being to construct effective BEM for class of equations which falls within the type (1).

2. The boundary value problem

Referred to a Cartesian frame Ox_1x_2 a solution to (1) is sought which is valid in a region Ω in R^2 with boundary $\partial\Omega$ which consists of a finite number of piecewise smooth closed curves. On $\partial\Omega_1$ the dependent variable $\phi(\mathbf{x})$ ($\mathbf{x} = (x_1, x_2)$) is specified and on $\partial\Omega_2$

$$P(\mathbf{x}) = \lambda_{ij}\left(\partial\phi/\partial x_j\right) n_i \tag{2}$$

is specified where $\partial\Omega = \partial\Omega_1 \cup \partial\Omega_2$ and $\mathbf{n} = (n_1, n_2)$ denotes the outward pointing normal to $\partial\Omega$.

For all points in Ω the matrix of coefficients $[\lambda_{ij}]$ is a real symmetric positive definite matrix so that throughout Ω equation (1) is a second order elliptic partial differential equation. Therefore equation (1) may be written explicitly as

$$\frac{\partial}{\partial x_1}\left(\lambda_{11}\frac{\partial\phi}{\partial x_1}\right) + 2\frac{\partial}{\partial x_1}\left(\lambda_{12}\frac{\partial\phi}{\partial x_2}\right) + \frac{\partial}{\partial x_2}\left(\lambda_{22}\frac{\partial\phi}{\partial x_2}\right) + \beta^2\phi = 0$$

Further, the coefficients λ_{ij} and β are required to be twice differentiable functions of the two independent variables x_1 and x_2.

The method of solution will be to obtain boundary integral equations from which numerical values of the dependent variables ϕ and P may be obtained for all points in Ω. The analysis here is specially relevant to an anisotropic medium but it equally applies to isotropic media. For isotropy, the coefficients in (1) take the form $\lambda_{11} = \lambda_{22}$ and $\lambda_{12} = 0$ and use of these equations in the following analysis immediately yields the corresponding results for an isotropic medium.

3. The boundary integral equation

The boundary integral equation is derived by transforming the variable coefficient equation (1) to a constant coefficient equation. The coefficients λ_{ij} and β are required to take the form

$$\lambda_{ij}(\mathbf{x}) = \overline{\lambda}_{ij}g(\mathbf{x}) \tag{3}$$

$$\beta^2(\mathbf{x}) = \overline{\beta}^2 g(\mathbf{x}) \tag{4}$$

where the $\overline{\lambda}_{ij}$ and $\overline{\beta}$ are constants and g is a differentiable function of $\mathbf{x}$. Use of (3) and (4) and in (1) yields

$$\overline{\lambda}_{ij}\frac{\partial}{\partial x_i}\left(g\frac{\partial\phi}{\partial x_j}\right) + \overline{\beta}^2 g\phi = 0 \tag{5}$$

Let

$$\phi(\mathbf{x}) = g^{-1/2}(\mathbf{x})\,\psi(\mathbf{x}) \tag{6}$$

so that (5) may be written in the form

$$\overline{\lambda}_{ij}\frac{\partial}{\partial x_i}\left[g\frac{\partial\left(g^{-1/2}\psi\right)}{\partial x_j}\right] + \overline{\beta}^2 g^{1/2}\psi = 0$$

That is

$$\overline{\lambda}_{ij}\left[\left(\frac{1}{4}g^{-3/2}\frac{\partial g}{\partial x_i}\frac{\partial g}{\partial x_j}-\frac{1}{2}g^{-1/2}\frac{\partial^2 g}{\partial x_i \partial x_j}\right)\psi+g^{1/2}\frac{\partial^2 \psi}{\partial x_i \partial x_j}\right]+\overline{\beta}^2 g^{1/2}\psi=0 \tag{7}$$

Use of the identity

$$\frac{\partial^2 g^{1/2}}{\partial x_i \partial x_j}=-\frac{1}{4}g^{-3/2}\frac{\partial g}{\partial x_i}\frac{\partial g}{\partial x_j}+\frac{1}{2}g^{-1/2}\frac{\partial^2 g}{\partial x_i \partial x_j}$$

permits (7) to be written in the form

$$g^{1/2}\overline{\lambda}_{ij}\frac{\partial^2 \psi}{\partial x_i \partial x_j}-\psi\overline{\lambda}_{ij}\frac{\partial^2 g^{1/2}}{\partial x_i \partial x_j}+\overline{\beta}^2 g^{1/2}\psi=0$$

It follows that if g is such that

$$\overline{\lambda}_{ij}\frac{\partial^2 g^{1/2}}{\partial x_i \partial x_j}=0 \tag{8}$$

then the transformation (6) carries the variable coefficients equation (5) to the constant coefficients equation

$$\overline{\lambda}_{ij}\frac{\partial^2 \psi}{\partial x_i \partial x_j}+\overline{\beta}^2\psi=0 \tag{9}$$

And if g is such that

$$\overline{\lambda}_{ij}\frac{\partial^2 g^{1/2}}{\partial x_i \partial x_j}-\overline{\beta}^2 g^{1/2}=0 \tag{10}$$

then

$$\overline{\lambda}_{ij}\frac{\partial^2 \psi}{\partial x_i \partial x_j}=0 \tag{11}$$

Also, substitution of (3) and (6) into (2) gives

$$P=-P_g\psi+P_\psi g^{1/2} \tag{12}$$

where

$$P_g(\mathbf{x})=\overline{\lambda}_{ij}\frac{\partial g^{1/2}}{\partial x_j}n_i \qquad P_\psi(\mathbf{x})=\overline{\lambda}_{ij}\frac{\partial \psi}{\partial x_j}n_i$$

A boundary integral equation for the solution of (9) and (11) is given in the form

$$\eta(\mathbf{x}_0)\psi(\mathbf{x}_0)=\int_{\partial\Omega}\left[\Gamma(\mathbf{x},\mathbf{x}_0)\psi(\mathbf{x})-\Phi(\mathbf{x},\mathbf{x}_0)P_\psi(\mathbf{x})\right]ds(\mathbf{x}) \tag{13}$$

where $\mathbf{x}_0=(a,b)$, $\eta=0$ if $(a,b)\notin\Omega\cup\partial\Omega$, $\eta=1$ if $(a,b)\in\Omega$, $\eta=\frac{1}{2}$ if $(a,b)\in\partial\Omega$ and $\partial\Omega$ has a continuously turning tangent at (a,b).

The so called fundamental solution Φ in (13) is any solution of the equation

$$\overline{\lambda}_{ij}\frac{\partial^2 \Phi}{\partial x_i \partial x_j}+\overline{\beta}^2\Phi=\delta(\mathbf{x}-\mathbf{x}_0)$$

and the Γ is given by

$$\Gamma(\mathbf{x},\mathbf{x}_0)=\overline{\lambda}_{ij}\frac{\partial\Phi(\mathbf{x},\mathbf{x}_0)}{\partial x_j}n_i$$

where δ is the Dirac delta function. Following Azis [14], for two-dimensional problems Φ and Γ are given by

$$\begin{aligned}
\Phi(\mathbf{x}, \mathbf{x}_0) &= \begin{cases} \frac{K}{2\pi} \ln R & \text{if } \overline{\beta}^2 = 0 \\ \frac{\imath K}{4} H_0^{(2)}(\omega R) & \text{if } \overline{\beta}^2 > 0 \end{cases} \\
\Gamma(\mathbf{x}, \mathbf{x}_0) &= \begin{cases} \frac{K}{2\pi} \frac{1}{R} \overline{\lambda}_{ij} \frac{\partial R}{\partial x_j} n_i & \text{if } \overline{\beta}^2 = 0 \\ \frac{-\imath K \omega}{4} H_1^{(2)}(\omega R) \overline{\lambda}_{ij} \frac{\partial R}{\partial x_j} n_i & \text{if } \overline{\beta}^2 > 0 \end{cases}
\end{aligned} \tag{14}$$

where

$$\begin{aligned}
K &= \ddot{\tau}/\zeta \\
\omega &= \sqrt{|\overline{\beta}^2/\zeta|} \\
\zeta &= \left[\overline{\lambda}_{11} + 2\overline{\lambda}_{12}\dot{\tau} + \overline{\lambda}_{22}\left(\dot{\tau}^2 + \ddot{\tau}^2\right)\right]/2 \\
R &= \sqrt{(\dot{x}_1 - \dot{a})^2 + (\dot{x}_2 - \dot{b})^2} \\
\dot{x}_1 &= x_1 + \dot{\tau} x_2 \\
\dot{a} &= a + \dot{\tau} b \\
\dot{x}_2 &= \ddot{\tau} x_2 \\
\dot{b} &= \ddot{\tau} b
\end{aligned}$$

where $\dot{\tau}$ and $\ddot{\tau}$ are respectively the real and the positive imaginary parts of the complex root τ of the quadratic

$$\overline{\lambda}_{11} + 2\overline{\lambda}_{12}\tau + \overline{\lambda}_{22}\tau^2 = 0$$

and $H_0^{(2)}$, $H_1^{(2)}$ denote the Hankel function of second kind and order zero and order one respectively and $\imath$ represents the square root of minus one. The derivatives $\partial R/\partial x_j$ needed for the calculation of the Γ in (14) are given by

$$\begin{aligned}
\frac{\partial R}{\partial x_1} &= \frac{1}{R}(\dot{x}_1 - \dot{a}) \\
\frac{\partial R}{\partial x_2} &= \dot{\tau}\left[\frac{1}{R}(\dot{x}_1 - \dot{a})\right] + \ddot{\tau}\left[\frac{1}{R}\left(\dot{x}_2 - \dot{b}\right)\right]
\end{aligned}$$

Use of (6) and (12) in (13) yields

$$\begin{aligned}
\eta(\mathbf{x}_0) g^{1/2}(\mathbf{x}_0) \phi(\mathbf{x}_0) &= \int_{\partial\Omega} \left\{ \left[g^{1/2}(\mathbf{x}) \Gamma(\mathbf{x}, \mathbf{x}_0) - P_g(\mathbf{x}) \Phi(\mathbf{x}, \mathbf{x}_0)\right] \phi(\mathbf{x}) \right. \\
&\quad \left. - \left[g^{-1/2}(\mathbf{x}) \Phi(\mathbf{x}, \mathbf{x}_0)\right] P(\mathbf{x}) \right\} ds(\mathbf{x})
\end{aligned} \tag{15}$$

This equation provides a boundary integral equation for determining ϕ and P at all points of Ω.

4. Numerical examples

Some particular boundary value problems will be solved numerically by employing the integral equation (15). Standard boundary element method is employed to obtain numerical results. The integrals in equation (15) are evaluated numerically using the Bode's quadrature (see Abramowitz and Stegun [15]). The main aim is to show the validity of the analysis used above for deriving the boundary integral equation (15) and the appropriateness of the BEM in solving the problems through the derived boundary integral equation (15).

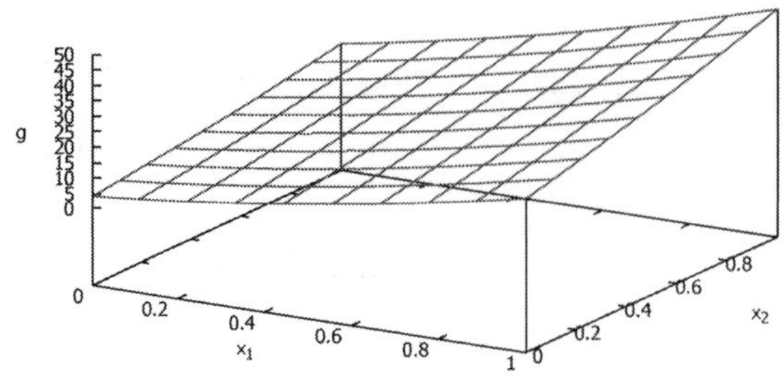

Figure 1. A quadratic inhomogeneity function $g(\mathbf{x}) = [2\,(1 + 1.5x_1 + x_2)]^2$

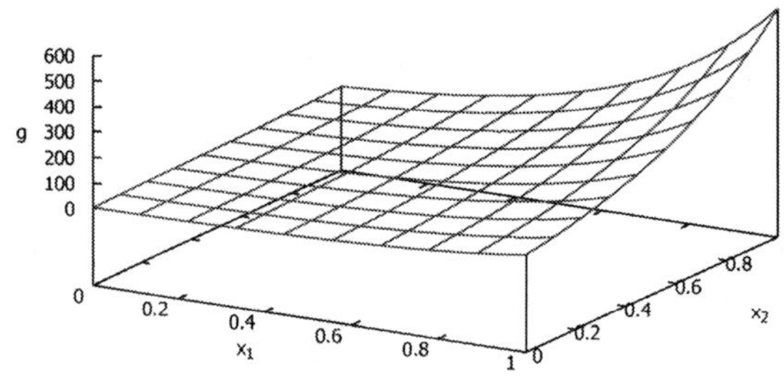

Figure 2. An exponential inhomogeneity function $g(\mathbf{x}) = [2\exp\,(1.5x_1 + x_2)]^2$

4.1. Examples with analytical solutions

In order to see the convergence and accuracy of the BEM we will consider some examples of problems with analytical solutions. Two possible multiparameter forms of the inhomogeneity function $g(\mathbf{x})$ satisfying (8) and (10) are quadratical and exponential functions respectively. For all problems considered in this section we take quadratical and exponential functions $g(\mathbf{x})$ respectively as

$$g(\mathbf{x}) = [A\,(\alpha_0 + \alpha_1 x_1 + \alpha_2 x_2)]^2 \quad (16)$$

$$g(\mathbf{x}) = [A\exp\,(\alpha_1 x_1 + \alpha_2 x_2)]^2 \quad \overline{\beta}^2 = \overline{\lambda}_{11}\alpha_1^2 + 2\overline{\lambda}_{12}\alpha_1\alpha_2 + \overline{\lambda}_{22}\alpha_2^2 \quad (17)$$

with $A = 2, \alpha_0 = 1, \alpha_1 = 1.5, \alpha_2 = 1$. Plots of $g(\mathbf{x})$ are shown in Figures 1–2. The geometry of the region Ω and the boundary conditions are as depicted in Figure 3. The values of the constant coefficients $\overline{\lambda}_{ij}$ for the governing equation (1) are

$$\overline{\lambda}_{11} = 1, \overline{\lambda}_{12} = 0.5, \overline{\lambda}_{22} = 2$$

For the case when $g(\mathbf{x})$ satisfies equation (8), $\psi(\mathbf{x})$ must satisfy (9). For the analytical solution ϕ we will take

$$\psi(\mathbf{x}) = B\,[\cos\,(\gamma_1 x_1 + \gamma_2 x_2) + \sin\,(\gamma_1 x_1 + \gamma_2 x_2)] \quad \overline{\beta}^2 = \overline{\lambda}_{11}\gamma_1^2 + 2\overline{\lambda}_{12}\gamma_1\gamma_2 + \overline{\lambda}_{22}\gamma_2^2$$

$$\psi(\mathbf{x}) = B\,(\gamma_0 + \gamma_1 x_1 + \gamma_2 x_2) \quad \overline{\beta}^2 = 0$$

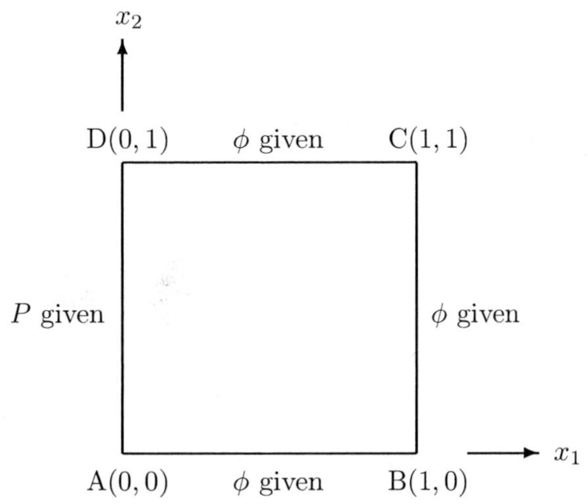

Figure 3. The geometry of all problems in Section 4.1

Whereas if $g(\mathbf{x})$ satisfies equation (10) then $\psi(\mathbf{x})$ satisfies (11). And we will take

$$\psi(\mathbf{x}) = B\left(\gamma_0 + \gamma_1 x_1 + \gamma_2 x_2\right)$$

Then the analytical solution ϕ can be obtained from equation (6). The parameters for ψ are taken to be

$$B = 2.5, \gamma_0 = 1, \gamma_1 = 1, \gamma_2 = 1.5$$

4.1.1. Quadratically graded media: $\mathbf{g(x)}$ *is of the form* (16)

Problem 4.1.1.1: $\overline{\boldsymbol{\beta}}^{\mathbf{2}} > \mathbf{0}$ *in equation* (9). The analytical solution is

$$\phi\left(\mathbf{x}\right) = \frac{B\left[\cos\left(\gamma_1 x_1 + \gamma_2 x_2\right) + \sin\left(\gamma_1 x_1 + \gamma_2 x_2\right)\right]}{A\left(\alpha_0 + \alpha_1 x_1 + \alpha_2 x_2\right)}$$

so that $\overline{\beta}^2 = \overline{\lambda}_{ij}\gamma_i\gamma_j = 7$. The results are shown in Table 1. The BEM solution converges to the analytical solution as the number of elements increases.

Problem 4.1.1.2: $\overline{\boldsymbol{\beta}}^{\mathbf{2}} = \mathbf{0}$ *in equation* (9). Now we choose analytical solution

$$\phi\left(\mathbf{x}\right) = \frac{B\left(\gamma_0 + \gamma_1 x_1 + \gamma_2 x_2\right)}{A\left(\alpha_0 + \alpha_1 x_1 + \alpha_2 x_2\right)}$$

The results are shown in Table 2. Again, the BEM solution converges to the analytical solution as the number of elements increases.

4.1.2. Exponentially graded media: $\mathbf{g(x)}$ *is of the form* (17)

Problem 4.1.2. We take the analytical solution

$$\phi\left(\mathbf{x}\right) = \frac{B\left(\gamma_0 + \gamma_1 x_1 + \gamma_2 x_2\right)}{A\exp\left(\alpha_1 x_1 + \alpha_2 x_2\right)}$$

Table 3 shows the results of the analytical and BEM solutions with 160, 320 and 640 elements of equal length. The BEM solution converges to the analytical solution as the number of elements increases.

Table 1. BEM and analytical solutions for Problem 4.1.1.1

(x_1, x_2)	ϕ	$\partial\phi/\partial x_1$	$\partial\phi/\partial x_2$	ϕ	$\partial\phi/\partial x_1$	$\partial\phi/\partial x_2$
	BEM 160 elements			BEM 320 elements		
(0.1,0.5)	1.0670	-1.0382	-0.7508	1.0681	-1.0399	-0.7511
(0.3,0.5)	0.8734	-0.9078	-0.8044	0.8742	-0.9090	-0.8043
(0.5,0.5)	0.7012	-0.8184	-0.8408	0.7018	-0.8193	-0.8405
(0.7,0.5)	0.5448	-0.7469	-0.8561	0.5453	-0.7477	-0.8557
(0.9,0.5)	0.4020	-0.6823	-0.8495	0.4023	-0.6831	-0.8494
(0.5,0.1)	0.9463	-0.6379	-0.3210	0.9465	-0.6379	-0.3194
(0.5,0.3)	0.8497	-0.7617	-0.6285	0.8502	-0.7627	-0.6276
(0.5,0.7)	0.5195	-0.8168	-0.9614	0.5201	-0.8176	-0.9616
(0.5,0.9)	0.3223	-0.7655	-0.9966	0.3229	-0.7663	-0.9966
	BEM 640 elements			Analytical		
(0.1,0.5)	1.0686	-1.0405	-0.7514	1.0691	-1.0411	-0.7517
(0.3,0.5)	0.8746	-0.9096	-0.8043	0.8750	-0.9102	-0.8043
(0.5,0.5)	0.7021	-0.8198	-0.8403	0.7024	-0.8203	-0.8402
(0.7,0.5)	0.5455	-0.7481	-0.8555	0.5457	-0.7486	-0.8553
(0.9,0.5)	0.4024	-0.6834	-0.8493	0.4025	-0.6838	-0.8491
(0.5,0.1)	0.9466	-0.6382	-0.3189	0.9468	-0.6387	-0.3183
(0.5,0.3)	0.8504	-0.7632	-0.6272	0.8507	-0.7637	-0.6269
(0.5,0.7)	0.5204	-0.8180	-0.9616	0.5207	-0.8183	-0.9618
(0.5,0.9)	0.3232	-0.7664	-0.9969	0.3234	-0.7665	-0.9972

Table 2. BEM and analytical solutions for Problem 4.1.1.2

(x_1, x_2)	ϕ	$\partial\phi/\partial x_1$	$\partial\phi/\partial x_2$	ϕ	$\partial\phi/\partial x_1$	$\partial\phi/\partial x_2$
	BEM 160 elements			BEM 320 elements		
(0.1,0.5)	1.4017	-0.5173	0.2893	1.4016	-0.5170	0.2881
(0.3,0.5)	1.3142	-0.3701	0.2896	1.3141	-0.3700	0.2886
(0.5,0.5)	1.2501	-0.2778	0.2794	1.2500	-0.2778	0.2786
(0.7,0.5)	1.2011	-0.2160	0.2655	1.2010	-0.2161	0.2649
(0.9,0.5)	1.1625	-0.1723	0.2506	1.1624	-0.1727	0.2503
(0.5,0.1)	1.1141	-0.2272	0.4136	1.1145	-0.2281	0.4122
(0.5,0.3)	1.1887	-0.2599	0.3366	1.1889	-0.2601	0.3356
(0.5,0.7)	1.3014	-0.2866	0.2356	1.3012	-0.2865	0.2349
(0.5,0.9)	1.3449	-0.2897	0.2016	1.3446	-0.2894	0.2008
	BEM 640 elements			Analytical		
(0.1,0.5)	1.4016	-0.5168	0.2875	1.4015	-0.5165	0.2870
(0.3,0.5)	1.3141	-0.3699	0.2881	1.3141	-0.3698	0.2876
(0.5,0.5)	1.2500	-0.2778	0.2782	1.2500	-0.2778	0.2778
(0.7,0.5)	1.2010	-0.2162	0.2646	1.2010	-0.2163	0.2643
(0.9,0.5)	1.1623	-0.1729	0.2502	1.1623	-0.1731	0.2501
(0.5,0.1)	1.1147	-0.2281	0.4116	1.1149	-0.2283	0.4109
(0.5,0.3)	1.1889	-0.2602	0.3351	1.1890	-0.2603	0.3346
(0.5,0.7)	1.3011	-0.2864	0.2346	1.3010	-0.2863	0.2343
(0.5,0.9)	1.3445	-0.2893	0.2005	1.3443	-0.2892	0.2002

Table 3. BEM and analytical solutions for Problem 4.1.2

(x_1, x_2)	ϕ	$\partial\phi/\partial x_1$	$\partial\phi/\partial x_2$	ϕ	$\partial\phi/\partial x_1$	$\partial\phi/\partial x_2$
	BEM 160 elements			BEM 320 elements		
(0.1,0.5)	1.2074	-1.1589	-0.2267	1.2073	-1.1586	-0.2275
(0.3,0.5)	0.9911	-1.0035	-0.2645	0.9910	-1.0032	-0.2652
(0.5,0.5)	0.8058	-0.8507	-0.2677	0.8058	-0.8506	-0.2681
(0.7,0.5)	0.6501	-0.7095	-0.2515	0.6500	-0.7097	-0.2517
(0.9,0.5)	0.5209	-0.5853	-0.2249	0.5209	-0.5844	-0.2260
(0.5,0.1)	0.8810	-0.7844	-0.0791	0.8812	-0.7872	-0.0789
(0.5,0.3)	0.8528	-0.8417	-0.1954	0.8529	-0.8418	-0.1961
(0.5,0.7)	0.7479	-0.8288	-0.3073	0.7478	-0.8285	-0.3076
(0.5,0.9)	0.6844	-0.7868	-0.3247	0.6843	-0.7860	-0.3239
	BEM 640 elements			Analytical		
(0.1,0.5)	1.2073	-1.1585	-0.2280	1.2072	-1.1583	-0.2284
(0.3,0.5)	0.9910	-1.0032	-0.2655	0.9910	-1.0031	-0.2659
(0.5,0.5)	0.8058	-0.8506	-0.2683	0.8058	-0.8506	-0.2686
(0.7,0.5)	0.6500	-0.7097	-0.2519	0.6500	-0.7097	-0.2520
(0.9,0.5)	0.5209	-0.5846	-0.2260	0.5208	-0.5847	-0.2260
(0.5,0.1)	0.8814	-0.7877	-0.0794	0.8815	-0.7880	-0.0801
(0.5,0.3)	0.8529	-0.8419	-0.1964	0.8530	-0.8420	-0.1968
(0.5,0.7)	0.7477	-0.8284	-0.3077	0.7477	-0.8283	-0.3079
(0.5,0.9)	0.6842	-0.7862	-0.3240	0.6842	-0.7862	-0.3241

4.2. Examples without analytical solutions

In this section we will consider some examples of problems without simple analytical solutions. We setup some problems for a homogeneous isotropic material by taking $g(\mathbf{x})$ constant, $\overline{\lambda}_{11} = \overline{\lambda}_{22} = 1, \overline{\lambda}_{12} = 0$ and with symmetrical boundary conditions. This function $g(\mathbf{x})$ satisfies equation (8) and so the corresponding $\psi(\mathbf{x})$ should satisfy (9). The aim is to see the consistency of the BEM of whether it produces symmetrical solutions.

4.2.1. Problem 4.2.1: $\overline{\boldsymbol{\beta}}^2 > \mathbf{0}$ *in equation* (9). We take $g(\mathbf{x}) = 4, \overline{\beta}^2 = 1$ and boundary conditions are as shown in Figure 4. Table 4 shows the results of the BEM solution using 80, 160, 320 and 640 elements of equal length. As expected, the results converge as the number of elements increases and also they are symmetrical about the axes $x_2 = 0.5$.

4.2.2. Problem 4.2.2: $\overline{\boldsymbol{\beta}}^2 = \mathbf{0}$ *in equation* (9). We consider a problem with $g(\mathbf{x}) = 9, \overline{\beta}^2 = 0$ and the boundary conditions are as shown in Figure 5. Table 5 shows the results of the BEM solution using 80, 160, 320 and 640 elements of equal length. The results converge as the number of elements increases and also they are symmetrical about the axes $x_1 = 0.5$.

5. Conclusion

The Helmholtz type governing equation (1) is sometimes used for modeling physical problems such as acoustic problems (when $\beta^2 > 0$), and antiplane strain in elastostatics and plane thermostatic problems (when $\beta^2 = 0$). The boundary integral equation (15) is derived from this governing equation (1) and then from (15) a BEM is constructed for calculation of numerical solutions to the problems for anisotropic functionally graded media including quadratically and exponentially graded media. The results show that the BEM solution gives a convergence, consistency, and accuracy. Therefore the results also prove that the analysis used for deriving

IOP Conf. Series: Earth and Environmental Science **279** (2019) 012008 doi:10.1088/1755-1315/279/1/012008

Table 4. BEM solution for Problem 4.2.1

(x_1, x_2)	ϕ	$\partial\phi/\partial x_1$	$\partial\phi/\partial x_2$	ϕ	$\partial\phi/\partial x_1$	$\partial\phi/\partial x_2$
	BEM 80 elements			BEM 160 elements		
(0.1,0.5)	0.3612	-0.2875	-0.0000	0.3619	-0.2876	-0.0000
(0.3,0.5)	0.2969	-0.3534	-0.0000	0.2975	-0.3537	-0.0000
(0.5,0.5)	0.2207	-0.4053	-0.0000	0.2213	-0.4058	-0.0000
(0.7,0.5)	0.1358	-0.4410	-0.0000	0.1363	-0.4416	-0.0000
(0.9,0.5)	0.0455	-0.4590	-0.0000	0.0459	-0.4598	-0.0000
(0.5,0.1)	0.2207	-0.4056	0.0000	0.2213	-0.4059	-0.0000
(0.5,0.3)	0.2207	-0.4054	0.0001	0.2213	-0.4058	0.0000
(0.5,0.7)	0.2207	-0.4054	-0.0001	0.2213	-0.4058	-0.0000
(0.5,0.9)	0.2207	-0.4056	-0.0000	0.2213	-0.4059	0.0000
	BEM 320 elements			BEM 640 elements		
(0.1,0.5)	0.3622	-0.2876	0.0000	0.3623	-0.2876	0.0000
(0.3,0.5)	0.2978	-0.3538	0.0000	0.2980	-0.3539	0.0000
(0.5,0.5)	0.2216	-0.4059	0.0000	0.2217	-0.4060	0.0000
(0.7,0.5)	0.1365	-0.4419	0.0000	0.1366	-0.4420	0.0000
(0.9,0.5)	0.0460	-0.4601	0.0000	0.0461	-0.4603	0.0000
(0.5,0.1)	0.2216	-0.4060	-0.0000	0.2217	-0.4060	-0.0000
(0.5,0.3)	0.2216	-0.4059	0.0000	0.2217	-0.4060	-0.0000
(0.5,0.7)	0.2216	-0.4059	-0.0000	0.2217	-0.4060	0.0000
(0.5,0.9)	0.2216	-0.4060	0.0000	0.2217	-0.4060	0.0000

Table 5. BEM solution for Problem 4.2.2

(x_1, x_2)	ϕ	$\partial\phi/\partial x_1$	$\partial\phi/\partial x_2$	ϕ	$\partial\phi/\partial x_1$	$\partial\phi/\partial x_2$
	BEM 80 elements			BEM 160 elements		
(0.1,0.5)	0.5221	0.2153	0.9230	0.5222	0.2155	0.9222
(0.3,0.5)	0.5576	0.1288	0.8031	0.5578	0.1288	0.8020
(0.5,0.5)	0.5708	0.0000	0.7607	0.5710	0.0000	0.7595
(0.7,0.5)	0.5576	0.1288	0.8031	0.5578	-0.1288	0.8020
(0.9,0.5)	0.5221	0.2154	0.9230	0.5222	-0.2155	0.9222
(0.5,0.1)	0.3472	0.0000	0.2866	0.3479	0.0000	0.2855
(0.5,0.3)	0.4353	0.0000	0.5773	0.4357	0.0000	0.5761
(0.5,0.7)	0.7339	0.0000	0.8588	0.7339	0.0000	0.8575
(0.5,0.9)	0.9106	0.0000	0.9005	0.9103	-0.0000	0.8990
	BEM 320 elements			BEM 640 elements		
(0.1,0.5)	0.5223	0.2156	0.9218	0.5224	0.2157	0.9217
(0.3,0.5)	0.5579	0.1289	0.8014	0.5579	0.1289	0.8011
(0.5,0.5)	0.5711	-0.0000	0.7589	0.5711	-0.0000	0.7586
(0.7,0.5)	0.5579	-0.1289	0.8014	0.5579	-0.1289	0.8011
(0.9,0.5)	0.5223	-0.2156	0.9218	0.5224	-0.2157	0.9217
(0.5,0.1)	0.3482	-0.0000	0.2851	0.3484	0.0000	0.2849
(0.5,0.3)	0.4360	-0.0000	0.5754	0.4361	0.0000	0.5751
(0.5,0.7)	0.7338	-0.0000	0.8568	0.7338	-0.0000	0.8565
(0.5,0.9)	0.9101	-0.0000	0.8983	0.9100	-0.0000	0.8979

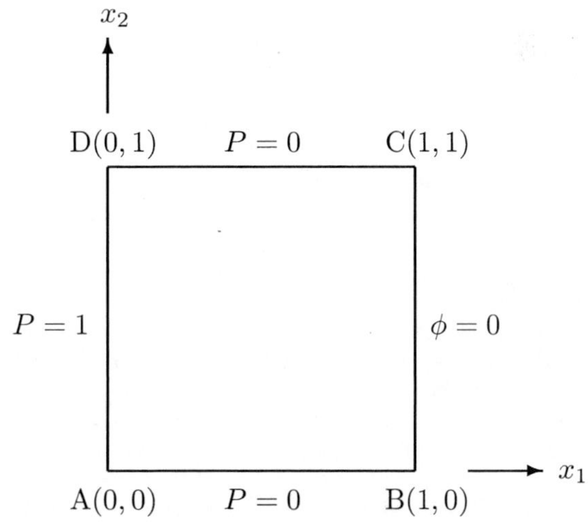

Figure 4. The geometry of Problem 4.2.1

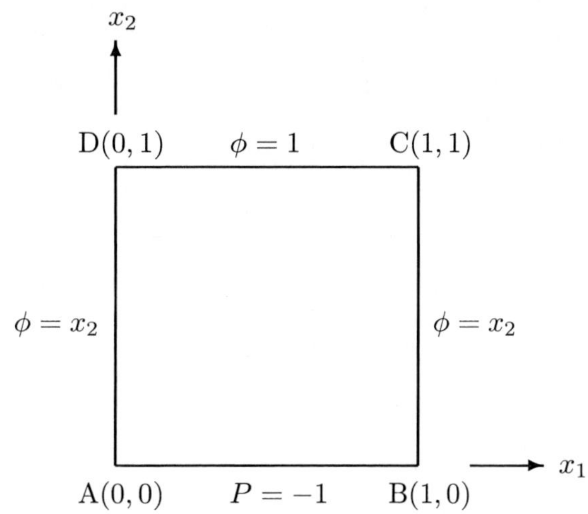

Figure 5. The geometry of Problem 4.2.2

the boundary integral equation (15) is valid. Together with its ease in implementation, it may be concluded that BEM is a useful numerical method for solving such kind of problems.

Acknowledgements
This work was supported by Hasanuddin University and The Ministry of Higher Education of Indonesia.

References

[1] Azis M I, Kasbawati, Haddade A and Thamrin S A 2018 On some examples of pollutant transport problems solved numerically using the boundary element method *Journal of Physics: Conference Series* **979** 1
[2] Azis M I, Asrul L, Khaeruddin and Paharuddin 2018 BEM solutions for unsteady transport problems in anisotropic media *JP Journal of Heat and Mass Transfer* **15** 4
[3] Haddade A, Salam N, Khaeruddin and Azis M I 2017 A Boundary Element Method for 2D Diffusion-Convection Problems in Anisotropic Media *Far East Journal of Mathematical Sciences* **102** 8
[4] Azis M I 2019 Numerical solutions for the Helmholtz boundary value problems of anisotropic media *Journal of Computational Physics* **381** 42

[5] Cheng A H-D 1984 Darcy's Flow with Variable Permeability: A Boundary Integral Solution *Water Resources Research* **20** 980
[6] Clements D L and Azis M I 2000 A Note on a Boundary Element Method for the Numerical Solution of Boundary Value Problems in Isotropic Inhomogeneous Elasticity *Journal of the Chinese Institute of Engineers* **23** 3 261
[7] Azis M I and Clements D L 2014 On some problems concerning deformations of functionally graded anisotropic elastic materials *Far East Journal of Mathematical Sciences*, **87** 2 173
[8] Azis M I and Clements D L 2014 A Boundary Element Method for Transient Heat Conduction Problem of Nonhomogeneous Anisotropic Materials *Far East Journal of Mathematical Sciences*, **89** 1 51
[9] Azis M I and Clements D L 2008 Nonlinear transient heat conduction problems for a class of inhomogeneous anisotropic materials by BEM *Engineering Analysis with Boundary Elements* **32** 1054
[10] Azis M I, Toaha S, Bahri M and Ilyas N 2018 A boundary element method with analytical integration for deformation of inhomogeneous elastic materials *Journal of Physics: Conference Series* **979** 1
[11] Salam N, Haddade A, Clements D L and Azis M I 2017 A boundary element method for a class of elliptic boundary value problems of functionally graded media *Engineering Analysis with Boundary Elements* **84** 186 doi: 10.1016/j.enganabound.2017.08.017
[12] Barucq H, Bendali A, Fares M, Mattesi V and Tordeux S 2017 A symmetric Trefftz-DG formulation based on a local boundary element method for the solution of the Helmholtz equation *Journal of Computational Physics* **330** 1069
[13] Loeffler C F, Mansur W J, Barcelos H-d-M and Bulcão A 2015 Solving Helmholtz problems with the boundary element method using direct radial basis function interpolation *Engineering Analysis with Boundary Elements* **61** 218
[14] Azis M I 2017 Fundamental solutions to two types of 2D boundary value problems of anisotropic materials *Far East Journal of Mathematical Sciences* **101** 11 2405
[15] Abramowitz M and Stegun I A 1972 *Handbook of mathematical functions: with formulas, graphs and mathematical tables* Dover Publications Washington

Using Geospatial Information Technology for Regional Assessment of Food Crop Land in South Sulawesi

Sumbangan Baja[1,3], Risma Neswati[1,3], Samsu Arif[2,3]

[1]Department of Soil Science, Hasanuddin University Makassar; (corresponding author, email: sbja02@yahoo.com.au)
[1]Department of Soil Science, Hasanuddin University Makassar; (email: neswati76@gmail.com)
[2]Department of Geophysics, Hasanuddin University Makassar; (email: samsu_arif@)unhas.ac.id)
[3]Regional Development and Spatial Information, Hasanuddin University, (email: puslitbangwitaris@unhas.ac.id)

Abstract. Recently, geospatial information on land suitability and land availability for specific purposes has become a prerequisite for strategic planning in regional development. Although geospatial data on land resources and their uses have become more available and detailed, in some circumstances, their geographical representations for spatial decision making are often considerably intricate. Accordingly, it requires advanced geospatial methods and approaches that could deal with those complex phenomena in an effective manner. The principal aim of this paper is to utilize Geographic Information Systems (GIS) and Remote Sensing (RS) to assess food crop land suitability at a regional scale. We fully utilize integrated techniques of GIS and RS in mapping land areas of three regencies, namely Takalar and Jeneponto, where such regencies have to date become strategic areas for food crop development, especially maize. Our preliminary finding as show that although Jeneponto district has long been recognized as a very dry area (where some parts of the region have less than 1,000 mm annual rainfall), this area has a very large potential suitability for maize cultivation, followed by Takalar. Such analysis has also been confirmed by the results of our geospatial data analysis following ground visits and local government statistical data. The results of this study not only present the suitability of future development of food crop especially maize, but also depict spatially the distribution of land class fitness for a given crop in such three regions; and thus provide the basis for decision making in future land use management.

Keywords: geospatial information, land suitability, food crop, South Sulawesi.

1. Introduction

The maize productivity was still classified as low, was only 4.20 ton/ha, this productivity is lower than Vietnam's which reaches 4.60 ton/ha and even Uni Europe has reached 6.80 ton/ha. As a result, Indonesia should import maize as much 800.000 ton on 2010, dan proximately 900.000 ton on 2013 for fulfill domestic needed that has average about 5 million per year [1]. The southeast of South Sulawesi region has big potential for food crop development, especially for maize, both for agroecology and culture of society in maize cultivation. Nevertheless, this potential has been not explored and has not used yet. Specifically, in South Sulawesi, the average of productivity was about 4.70 ton/ha, only a few on the above national average.

Available land was wide enough, and the farmers still have hereditary of maize cultivation. According to [2], maize was a flexible plant and has great tolerance with its environment, in require that growing period should be ice-free. Maize could growth on temperature 12-24°C, minimum temperature was 26-29 °C. rainfall for maize is about 500-5.000 mm. The optimal amount of water for maize was about 500-1.200 mm/growing period [3]. Maize could be possible growth on variety kind of soils. Maize was suitable with soil which has good drainage, good aeration, has loam and silt loam texture, and enough percentage of organic matter [3]. The land has bad drainage have to be avoided, groundwater level should be below 0,75 m from the surface. Maize could not be resistant if there is flooding on 5 weeks after planting [4]. In order dry region, poor capability of water retention and slight rainfall have to do cultivate in low density. The optimal acidity soil for maize growth was about 5.8-7.8.

Identification of potential, constraints, and land suitability for specific commodity development were done by survey and land evaluation. In a fact that various land use needed different requirements, thus it was used information about properties of land from various appropriate aspect with land use planning [5]. Land characteristics could be gained by land survey and other analysis. Different areas might have different soils type based on the affected condition. Different soils also have different properties. Difference of soil characteristics might cause different potential, constraints, and land suitability for certain commodity development.

Land suitability evaluation was to determine potential of land in order to certain purpose [6]. Basically, it was to evaluate land for certain land use, such as paddy field, maize, and others. Actual land suitability was assessment of land suitability before land improvement. Potential land suitability is assessment of land suitability after land improvement. For example there was land as actual was classified as marginally suitable (S3) due to nutrient deficiency could be change as moderetely suitable (S2) or even suitable (S1), after it has done by fertilization [7]. The aim of this study was to elevate effectivity of management system for Maize in Sulawesi region by spatial-based integrated model applying for biophysics, social-economy, and wisdom of cultivation at maize central area.

2. Methodology

The study was done in 3 (three) stages continually, consists of: (i) **stage I**: food crop's biophysics analyze (suitability and productivity of land); (ii) **stage II**: Social-economy analyze and management system of maize; and (iii) **stage III**: wisdom analyze and spatial integration between biophysics factor, social-economy, and wisdom factor to take decision for maize management. This study will be held over South Sulawesi to turn out indication of land suitability maps and to analyze 6 regencies by spatial integrated to get wisdom of maize management in South Sulawesi. Biophysics factor was also needed, consists of climate, topography, geology, and soil type.

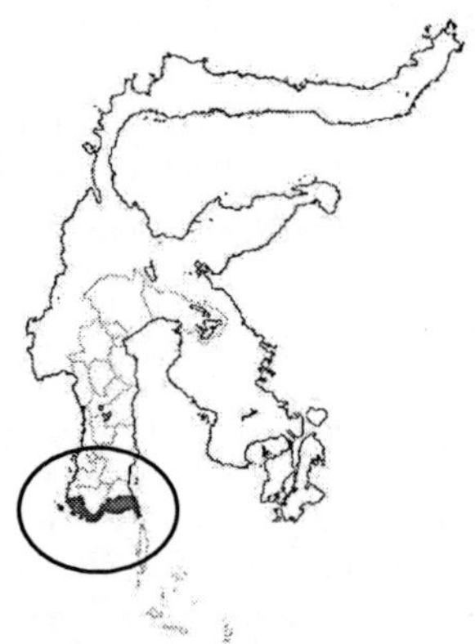

Figure 1. Study area

3. Results

3.1 Jeneponto regency

Jeneponto's rainfall was slightly if it was compared with other regency in South Sulawesi. Rainfall of Jeneponto have been noted on 2001-2010 were about 1.278 mm until 1.986 mm per year, mean of temperature is 24.01°C, mean of maximum and minimum temperature were 27.12°C and 22.71°C. The rainfall was often erratic and it was begun to pour regurarly at the end of October. Dry season starts in the middle of June until October. Rain season starts on November and it was getting down on the beginning of March until the end of May. Instead, temperature was more regularly and deviation between temperatures of different months was relative narrow both minimum or maximum temperature even for mean of temperature. The most maximum temperature was on October (33.5°C) and the most minimum was on February (26.13°C). Mean of relative humidity was 85.32% with length of radiation (n/N) was 0.63 and the highest was measured on September was 0.78.

Topography of Jeneponto could be classified as flat (0-3%) percentage as 5%, wavy (8-15%) as 31%, hilly (15-25%) as 32% and mountainous (>45%) as 31%. Tamalatea district was classified as flat (0-3%) percentage as 16.2%, undulating (3-8%) as much as 38.3%, rolling (8-15%) as 6.1%, hilly (8-15%) as much as 7.1%, and mountainous (>45%) with percentage as 32.3%.

As lithography, Jeneponto was formed from 4 rock types, those were: 1. Qac formation (Quarter alluvium coastal). This formation was arranged by alluvial deposit and coast (gravel, sand, clay, mud, corl and limestone); 2. Qlvb formation (Quarter breccia volcanic Lompobattang) which was consists of breccia, lava, and volcanic ash; 3. Tmc formation (Tersier miosen Camba) which was formed as marine sedimentary rock interbedded volcanic; 4. Temt formation (Tersier eosin miosen Tonasa) that was arranged from limestone [8]. The dominant formation that found in Jeneponto was Qlvb and following to Temt, Qac, and Tmc. Based on soil map of South Sulawesi scale 1:500.000 [9] that was overlaying with soil map based on soil taxonomy [10], there were 3 types of soils in Jeneponto regency, those were Alfisols (45%), Inceptisols (24,8%), dan Vertisols (29,8%).

Based on criteria of land suitability classification for maize that has been determined by [11] generally on study area in jeneponto classified as S2 as much as 7.067 Ha (8,9%) and S3 as much as 33.292 Ha (41,8%) with various limiting factors. The main limiting factor was climate specifically for rainfall on vegetative stage of maize, relative humidity on generative stage, length of radiation, soil fertility for base saturation and soil pH and other soil physics, those were course fragments and soil depth.

The results showed that variety of maize yield at area study was wide enough as 3.0– 5.2 ton/ha. Generally, average of hybrid maize yield that was gained is on dry area, in this case Jeneponto, which was represented by Tamalatea district, Kelara district, and West Bangkala district. The maize yield that are showed on attachment is result of maize yield average on study area for twice planting that measured using square area method 2.5 m x 2.5 m. If it adjusts with growing period in Jeneponto with assumption that using hybrid seed could be harvested 110 days after planting, so planting time that is proposed for Jeneponto regency if does in twice are first week of November until second week of February (MT1) and first week of March until second week of June (MT2). Determination of planting time in growing period was important in maize cultivation, in line with research of [12] and [13] who has results that appropriate planting time could be improve crop yield as long as growing period was long enough and relative humidity was in controlled. According to [14] appropriate planting time might avoid maize yield loss due to climate change. Furthermore, Nafziger [14] explained that in Illinois state which is the largest maize produced in USA, has been determined that planting time starts in the middle of April, if it does backwards will be cause yield loss, thus if its back out at the beginning of June will be cause yield loss up to 50% due to frost problem on generative stage. Thus, determination of planting time was urgent for maize cultivation which is adjusted with growing period. As for the proposed growing period has been showed on Table 1.

Table 1. The proposed planting time according to growing period on study area (Jeneponto)

Location		Planting time*
Jeneponto	Kelara	1.November-February (MT1) 2.February – May (MT2)
	Tamalatea	1.December – March (MT1)
	West Bangkala	1.November-February (MT1) 2.February-May (MT2)

3.2 Takalar regency

According to Oldeman climate classification, Takalar classified in climate types of D3, C3 and D4. On a year, mostly amount of wet season only 3 until 4 months, instead dry season was 3 until 5 months for D3, 4 until 6 months on wet season and 3 months for dry season for C3, and more than 5 months on dry season for D4. Even though the amount of dry season is more than wet season, rainfall of Takalar was still classified as high, is about 1.905 mm for Sanrobone station, 2.217 mm for Galesong station, and 1.979 mm for Biringkassi station.

The result showed that various of maize yield per ha on study area was wide enough, it was 4.0– 5.5 ton/ha. Average of hybrid maize yield on dry areas that were represented by 7 districts in Takalar, those were Polongbangkeng Utara, Polongbangkeng Selatan, Mangarabombang, Pattalassang, Galesong Utara, Galesong Selatan and Mappakasunggu. The maize yield that shown on Attachment 2 was results of mean maize yield on study area for twice planting which was measured using square area method 2.5 m x 2.5 m. If it adjusts with growing period in Takalar with assumption that using hybrid seed could be harvested 110 days after planting, so planting time that was proposed for Takalar regency if does in twice were first week of November until week 2 of February (MT1) and first week of February until fourth week of May (MT2). Determination of planting time in growing period was important in maize cultivation, in line with reseach of [12] and [13] who has results that appropriate planting time could be improve crop yield as long as growing period was long enough and relative humidity was in controlled. According to [14], appropriate planting time might avoid maize yield loss due to climate change. Furthermore, Nafziger [14]) explained that in Illinois state which is the largest maize produced in USA, has been determined that planting time starts in the middle of April, if it does backwards will be cause yield loss, thus if its back out at the beginning of June will be cause yield loss up to 50% due to frost problem on generative stage. Thus, determination of planting time was urgent for maize cultivation which was adjusted with growing period. As for the proposed growing period has been showed on Table 2.

Table 2. The proposed planting time in line with growing period on study area (Takalar)

Location	Planting time*
Takalar	1.November-February (MT1) 2.February-May (MT2)

4. Conclusion

The study results on the first year showed that:

1) Potential of biophysics consists of climate (rainfall, temperature, humidity, n/N), physical properties of soil (texture, drainage, soil depth, coarse fragments), and chemical properties soil (pH, amount of N, P, and K, Organic C, salinity, CEC) on South Sulawesi might support food crop development for maize.

2) Study areas was classified as S1 (very suitable); S2 (moderately suitable) and S3 (marginally suitable) with limiting factors, such as topography, soil depth, and soil fertility, climate (rainfall on growing period).
3) Maize development potency on South Sulawesi was high that was supported by heredity society policy and suitability of biophysics factor.
4) Measured of farmers production on study area showed that maize yield was about 3.0 until 5.5 ton/ha which was dominated by maize farmers who have been used hybrid and composite seeds.

References

[1] Badan Pusat Statistik (BPS) 2012. *Statistik Indonesia 2011* Badan Pusat Statistik: Jakarta.
[2] Sys C 1993 *Land Evaluation Part I, II, and III* University of Ghent: Ghent.
[3] Janick J 2010 *Plant Breeding Reviews* **20** 15–66.
[4] Djaenuddin D, Basuni, Hardjowigeno S, Subagyo H, Sukardi M, Ismangun, Marsudi D S, Suharta N, Hakim L, Widagdo, Dai J, Suwandi V, Bachri S, Jordens E R 1994 *Kesesuaian Lahan Untuk Tanaman Pertanian dan Tanaman Kehutanan*. Second Land Resource Evaluation and Planning Project. ADB Loan No. 1099 INO. Centre for Soil and Agroclimate Research, Bogor.
[5] Keleman A and Bellon M R 2009 *The Geographical Journal* **175** 52–70.
[6] Baja S 2009 Land use choice and land resource assessment in agriculture. *CAB Reviews: Perspectives in Agriculture, Veterinary Science, Nutrition and Natural Resources* **4** No. 015.
[7] Dent D and Young A 1981 *Soil Survey and Land Evaluation* George Allen & Unwin (*publishers*) Ltd. London p. 277.
[8] Soekamto R and Supriatna 1982 *Geological Map of Sulawesi* Geological Research and Development Centre.
[9] LPT 1967 *Peta Tanah Tindjau Skala 1:500.000* Lembaga Penelitian Tanah: Bogor.
[10] Soil Survey Staff 2010 *Keys to Soil Taxonomy Eleventh Edition*. United States Department of Agriculture Natural Resources Conversation Service.
[11] Sys C, Van Ranst E, Debaveye J, Beernaert F 1993 *Land Evaluation, Part III* Crop Requirements. Agricultural Publications-No7. State University of Ghent, Belgium General Administration for Development Cooperation Place du Champ de Mars 5 bte 57-1050 Brussels-Belgium.
[12] Kucharik C J 2008 *Agronomy Journal* **100** (2) 328-336.
[13] Osafo D M 1976 *Ghana Journal Agricultural Science* **9** 25-34.
[14] Nafziger E 2013 *Corn Plant Development* http:/www. extension.cropsci.illinois.edu/handbook/pdfs/chapters02.pdf (accessed on 20 Agustus 2016).

IOP Conf. Series: Earth and Environmental Science **279** (2019) 012010 doi:10.1088/1755-1315/279/1/012010

BEM solutions to BVPs governed by the anisotropic modified Helmholtz equation for quadratically graded media

M. I. Azis[1,*], **R. Syam**[2], **S. Hamzah**[3]

[1]Department of Mathematics, Hasanuddin University, Makassar, Indonesia
[2]Department of Mechanical Engineering, Hasanuddin University, Makassar, Indonesia
[3]Department of Civil Engineering, Hasanuddin University, Makassar, Indonesia

E-mail: mohivanazis@yahoo.co.id

Abstract. A Boundary Element Method (BEM) is used for obtaining solutions to anisotropic quadratically graded media (FGM) boundary value problems (BVPs) governed by the modified Helmholtz type equation. A technique of transforming the variable coefficient governing equation to a constant coefficient equation is utilized for deriving a boundary integral equation. Some particular problems are considered to illustrate the application of the BEM. The results show the convergence, consistency, and accuracy of the BEM solutions.

1. Introduction

The BEM has been successfully used for solving many types of problems of homogeneous media. Some works on homogeneous media problems have been recently done by Azis et. al [1, 2] and Haddade et. al [3] in which the authors considered pollutant transport problems governed by 2D diffusion-convection, and also Azis [4] in which numerical solutions for the Helmholtz boundary value problems were obtained.

However, this is generally not the case for FGM. There are two techniques usually used to deal with problems of FGM. The first one uses a technique of deriving a relevant Green function or fundamental solution to the FGM problem. Cheng [5] had applied this technique. The second technique is by transforming the variable coefficient governing equation to a constant coefficient equation. Some progress on using the second technique has been made. For examples, Clements and Azis [6] considered the case for isotropic FGM. For anisotropic FGM some works have been studied by Azis and Clements [7, 8, 9] and Azis et. al [10]. In addition to this, recently Salam et. al [11] have been working on a class of elliptic problems for anisotropic FGM.

This paper discusses derivation of a boundary integral equation for numerically solving 2D boundary value problems governed by the dimensionless modified Helmholtz type equation of the form

$$\frac{\partial}{\partial x_i}\left[\lambda_{ij}(x_1,x_2)\frac{\partial \phi(x_1,x_2)}{\partial x_j}\right] - \beta^2(x_1,x_2)\,\phi(x_1,x_2) = 0 \qquad (1)$$

where the coefficients λ_{ij} and β^2 depend on x_1 and x_2 and the repeated summation convention (summing from 1 to 2) is employed.

IOP Conf. Series: Earth and Environmental Science **279** (2019) 012010 doi:10.1088/1755-1315/279/1/012010

For (1) to be a second order elliptic partial differential equation, we assume that the matrix of coefficients $[\lambda_{ij}]$ is a real symmetric positive definite matrix. Explicitly equation (1) may be written as

$$\frac{\partial}{\partial x_1}\left(\lambda_{11}\frac{\partial \phi}{\partial x_1}\right)+2\frac{\partial}{\partial x_1}\left(\lambda_{12}\frac{\partial \phi}{\partial x_2}\right)+\frac{\partial}{\partial x_2}\left(\lambda_{22}\frac{\partial \phi}{\partial x_2}\right)-\beta^2\phi=0$$

Further, the coefficients λ_{ij} and β^2 are required to be twice differentiable functions of the two independent variables x_1 and x_2. The analysis is also relevant for isotropic media as a special case of anisotropic media which occurs when $\lambda_{11}=\lambda_{22}$ and $\lambda_{12}=0$.

A variety of problems of both isotropic and anisotropic inhomogeneous media are usually modeled with equation (1). Infiltration problems (when $\beta^2>0$), and antiplane strain in elastostatics and plane thermostatic problems (when $\beta^2=0$) are the areas for which the governing equation is of the type (1). Recently some works on the modified Helmholtz equation for isotropic and/or homogeneous media have been done. For example, Clements and Lobo in [12] and Solekhudin and Ang in [13] considered the modified Helmholtz equation as the governing equation for isotropic and homogeneous media which is a special case of the equation (1).

2. The boundary value problem

Referred to a Cartesian frame Ox_1x_2 a solution to (1) is sought which is valid in a region Ω in R^2 with boundary $\partial\Omega$ which consists of a finite number of piecewise smooth closed curves. On $\partial\Omega_1$ the dependent variable $\phi(\mathbf{x})$ $(\mathbf{x}=(x_1,x_2))$ is specified and on $\partial\Omega_2$

$$P(\mathbf{x})=\lambda_{ij}\left(\partial\phi/\partial x_j\right)n_i \qquad (2)$$

is specified where $\partial\Omega=\partial\Omega_1\cup\partial\Omega_2$ and $\mathbf{n}=(n_1,n_2)$ denotes the outward pointing normal to $\partial\Omega$.

3. The boundary integral equation

The boundary integral equation is derived by transforming the variable coefficient equation (1) to a constant coefficient equation. The coefficients λ_{ij} and β are required to take the form

$$\lambda_{ij}(\mathbf{x}) = \overline{\lambda}_{ij}g(\mathbf{x}) \qquad (3)$$

$$\beta^2(\mathbf{x}) = \overline{\beta}^2 g(\mathbf{x}) \qquad (4)$$

where the $\overline{\lambda}_{ij}$ and $\overline{\beta}$ are constants and g is an exponential function of $\mathbf{x}$. Use of (3) and (4) and in (1) yields

$$\overline{\lambda}_{ij}\frac{\partial}{\partial x_i}\left(g\frac{\partial\phi}{\partial x_j}\right)-\overline{\beta}^2 g\phi=0 \qquad (5)$$

Let

$$\phi(\mathbf{x})=g^{-1/2}(\mathbf{x})\,\psi(\mathbf{x}) \qquad (6)$$

so that (5) may be written in the form

$$\overline{\lambda}_{ij}\frac{\partial}{\partial x_i}\left[g\frac{\partial\left(g^{-1/2}\psi\right)}{\partial x_j}\right]-\overline{\beta}^2 g^{1/2}\psi=0$$

That is

$$\overline{\lambda}_{ij}\left[\left(\frac{1}{4}g^{-3/2}\frac{\partial g}{\partial x_i}\frac{\partial g}{\partial x_j}-\frac{1}{2}g^{-1/2}\frac{\partial^2 g}{\partial x_i\partial x_j}\right)\psi+g^{1/2}\frac{\partial^2\psi}{\partial x_i\partial x_j}\right]-\overline{\beta}^2 g^{1/2}\psi=0 \qquad (7)$$

IOP Conf. Series: Earth and Environmental Science **279** (2019) 012010 doi:10.1088/1755-1315/279/1/012010

Use of the identity

$$\frac{\partial^2 g^{1/2}}{\partial x_i \partial x_j} = -\frac{1}{4} g^{-3/2} \frac{\partial g}{\partial x_i} \frac{\partial g}{\partial x_j} + \frac{1}{2} g^{-1/2} \frac{\partial^2 g}{\partial x_i \partial x_j}$$

permits (7) to be written in the form

$$g^{1/2} \overline{\lambda}_{ij} \frac{\partial^2 \psi}{\partial x_i \partial x_j} - \psi \overline{\lambda}_{ij} \frac{\partial^2 g^{1/2}}{\partial x_i \partial x_j} - \overline{\beta}^2 g^{1/2} \psi = 0 \tag{8}$$

If we further restrict the function $g(\mathbf{x})$ to take the exponential form

$$g(\mathbf{x}) = \left[A\left(\alpha_0 + \alpha_1 x_1 + \alpha_2 x_2\right)\right]^2 \tag{9}$$

where α_m are constant, then

$$\overline{\lambda}_{ij} \frac{\partial^2 g^{1/2}}{\partial x_i \partial x_j} = 0 \tag{10}$$

Subtitution (10) into (8) implies a constant coefficients equation

$$\overline{\lambda}_{ij} \frac{\partial^2 \psi}{\partial x_i \partial x_j} - \overline{\beta}^2 \psi = 0 \tag{11}$$

Also, substitution of (3) and (6) into (2) gives

$$P = -P_g \psi + P_\psi g^{1/2} \tag{12}$$

where

$$P_g(\mathbf{x}) = \overline{\lambda}_{ij} \frac{\partial g^{1/2}}{\partial x_j} n_i \qquad P_\psi(\mathbf{x}) = \overline{\lambda}_{ij} \frac{\partial \psi}{\partial x_j} n_i$$

A boundary integral equation for the solution of (11) is given in the form

$$\eta(\mathbf{x}_0)\,\psi(\mathbf{x}_0) = \int_{\partial\Omega} \left[\Gamma(\mathbf{x}, \mathbf{x}_0)\,\psi(\mathbf{x}) - \Phi(\mathbf{x}, \mathbf{x}_0)\,P_\psi(\mathbf{x})\right] ds(\mathbf{x}) \tag{13}$$

where $\mathbf{x}_0 = (a, b)$, $\eta = 0$ if $(a, b) \notin \Omega \cup \partial\Omega$, $\eta = 1$ if $(a, b) \in \Omega$, $\eta = \frac{1}{2}$ if $(a, b) \in \partial\Omega$ and $\partial\Omega$ has a continuously turning tangent at (a, b).

The so called fundamental solution Φ in (13) is any solution of the equation

$$\overline{\lambda}_{ij} \frac{\partial^2 \Phi}{\partial x_i \partial x_j} - \overline{\beta}^2 \Phi = \delta(\mathbf{x} - \mathbf{x}_0)$$

and the Γ is given by

$$\Gamma(\mathbf{x}, \mathbf{x}_0) = \overline{\lambda}_{ij} \frac{\partial \Phi(\mathbf{x}, \mathbf{x}_0)}{\partial x_j} n_i$$

where δ is the Dirac delta function. Following Azis [14], for two-dimensional problems Φ and Γ are given by

$$\begin{aligned} \Phi(\mathbf{x}, \mathbf{x}_0) &= \begin{cases} \frac{K}{2\pi} \ln R & \text{if } \overline{\beta}^2 = 0 \\ \frac{-K}{2\pi} K_0(\omega R) & \text{if } \overline{\beta}^2 > 0 \end{cases} \\ \Gamma(\mathbf{x}, \mathbf{x}_0) &= \begin{cases} \frac{K}{2\pi} \frac{1}{R} \overline{\lambda}_{ij} \frac{\partial R}{\partial x_j} n_i & \text{if } \overline{\beta}^2 = 0 \\ \frac{K\omega}{2\pi} K_1(\omega R)\, \overline{\lambda}_{ij} \frac{\partial R}{\partial x_j} n_i & \text{if } \overline{\beta}^2 > 0 \end{cases} \end{aligned} \tag{14}$$

where

$$\begin{aligned}
K &= \ddot{\tau}/\zeta \\
\omega &= \sqrt{|\overline{\beta}^2/\zeta|} \\
\zeta &= \left[\overline{\lambda}_{11} + 2\overline{\lambda}_{12}\dot{\tau} + \overline{\lambda}_{22}\left(\dot{\tau}^2 + \ddot{\tau}^2\right)\right]/2 \\
R &= \sqrt{(\dot{x}_1 - \dot{a})^2 + (\dot{x}_2 - \dot{b})^2} \\
\dot{x}_1 &= x_1 + \dot{\tau}x_2 \\
\dot{a} &= a + \dot{\tau}b \\
\dot{x}_2 &= \ddot{\tau}x_2 \\
\dot{b} &= \ddot{\tau}b
\end{aligned}$$

where $\dot{\tau}$ and $\ddot{\tau}$ are respectively the real and the positive imaginary parts of the complex root τ of the quadratic

$$\overline{\lambda}_{11} + 2\overline{\lambda}_{12}\tau + \overline{\lambda}_{22}\tau^2 = 0$$

and K_0, K_1 denote the modified Bessel function of order zero and order one respectively. The derivatives $\partial R/\partial x_j$ needed for the calculation of the Γ in (14) are given by

$$\begin{aligned}
\frac{\partial R}{\partial x_1} &= \frac{1}{R}(\dot{x}_1 - \dot{a}) \\
\frac{\partial R}{\partial x_2} &= \dot{\tau}\left[\frac{1}{R}(\dot{x}_1 - \dot{a})\right] + \ddot{\tau}\left[\frac{1}{R}\left(\dot{x}_2 - \dot{b}\right)\right]
\end{aligned}$$

Use of (6) and (12) in (13) yields

$$\begin{aligned}
\eta(\mathbf{x}_0)\, g^{1/2}(\mathbf{x}_0)\,\phi(\mathbf{x}_0) &= \int_{\partial\Omega}\left\{\left[g^{1/2}(\mathbf{x})\,\Gamma(\mathbf{x},\mathbf{x}_0) - P_g(\mathbf{x})\,\Phi(\mathbf{x},\mathbf{x}_0)\right]\phi(\mathbf{x})\right. \\
&\quad \left. - \left[g^{-1/2}(\mathbf{x})\,\Phi(\mathbf{x},\mathbf{x}_0)\right]P(\mathbf{x})\right\} ds(\mathbf{x}) \qquad (15)
\end{aligned}$$

This equation provides a boundary integral equation for determining ϕ and P at all points of Ω.

4. Numerical examples

Some particular boundary value problems will be solved numerically by employing the integral equation (15). The main aim is to show the validity of the analysis used above for deriving the boundary integral equation (15) and the appropriateness of the BEM in solving the problems through the derived boundary integral equation (15). Standard boundary element method is employed to obtain numerical results. The integrals in equation (15) are evaluated numerically using the Bode's quadrature (see Abramowitz and Stegun [15]).

4.1. Problem 1

We will consider a problem with analytical solution. The aim is to verify the convergence and accuracy of the BEM. We choose an anisotropic medium with

$$\overline{\lambda}_{11} = 1, \overline{\lambda}_{12} = 0.5, \overline{\lambda}_{22} = 1$$

and we take an inhomogeneity function $g(\mathbf{x})$ satisfying (9) of the form

$$g(\mathbf{x}) = \left[2\left(1 + 0.5x_1 + 0.75x_2\right)\right]^2$$

IOP Conf. Series: Earth and Environmental Science **279** (2019) 012010 doi:10.1088/1755-1315/279/1/012010

Plot of $g(\mathbf{x})$ is shown in Figure 1. For the analytical solution ϕ we will take

$$\psi(\mathbf{x}) = 3.5 \exp(0.75x_1 + 0.25x_2)$$

so that from (11) $\overline{\beta}^2 = 0.8125$. Therefore from (6) the analytical solution

$$\phi(\mathbf{x}) = \frac{3.5 \exp(0.75x_1 + 0.25x_2)}{2(1 + 0.5x_1 + 0.75x_2)}$$

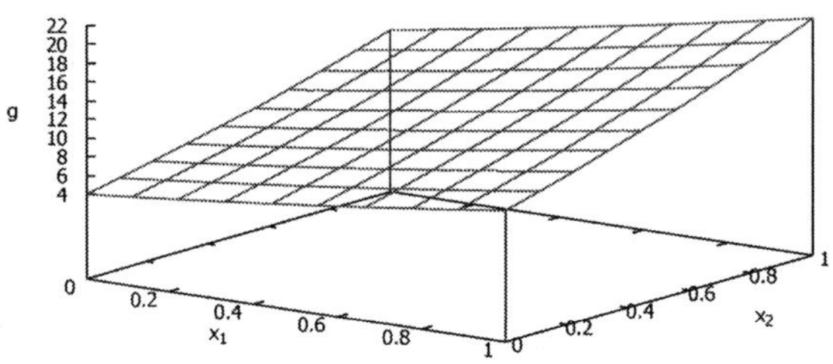

Figure 1. A quadratical inhomogeneity function $g(\mathbf{x}) = [2(1 + 0.5x_1 + 0.75x_2)]^2$

The geometry of the region Ω and the boundary conditions are as depicted in Figure 2.

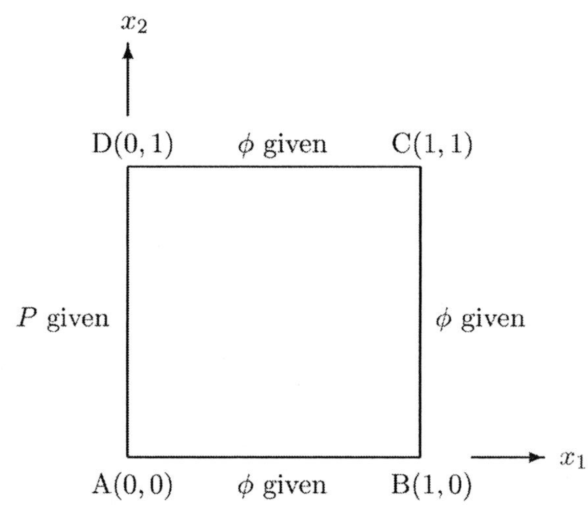

Figure 2. The geometry of Problem 1

The results are shown in Table 1. The BEM solution converges to the analytical solution as the number of elements increases. The accuracy of the BEM solutions lie on the fourth figure of the decimals.

Table 1. BEM and analytical solutions for Problem 1

(x_1, x_2)	ϕ	$\partial\phi/\partial x_1$	$\partial\phi/\partial x_2$	ϕ	$\partial\phi/\partial x_1$	$\partial\phi/\partial x_2$
	BEM 160 elements			BEM 320 elements		
(0.1,0.5)	1.5005	0.5981	-0.4138	1.5002	0.5984	-0.4141
(0.3,0.5)	1.6289	0.6873	-0.3929	1.6287	0.6874	-0.3933
(0.5,0.5)	1.7761	0.7858	-0.3747	1.7758	0.7856	-0.3751
(0.7,0.5)	1.9440	0.8954	-0.3583	1.9437	0.8948	-0.3587
(0.9,0.5)	2.1351	1.0179	-0.3433	2.1346	1.0169	-0.3433
(0.5,0.1)	1.9700	0.7343	-0.6184	1.9702	0.7344	-0.6207
(0.5,0.3)	1.8610	0.7653	-0.4789	1.8608	0.7651	-0.4799
(0.5,0.7)	1.7095	0.8007	-0.2946	1.7092	0.8005	-0.2947
(0.5,0.9)	1.6571	0.8123	-0.2311	1.6568	0.8122	-0.2313
	BEM 640 elements			Analytical		
(0.1,0.5)	1.5001	0.5985	-0.4143	1.5000	0.5987	-0.4145
(0.3,0.5)	1.6286	0.6874	-0.3935	1.6284	0.6874	-0.3938
(0.5,0.5)	1.7757	0.7855	-0.3754	1.7755	0.7853	-0.3756
(0.7,0.5)	1.9435	0.8945	-0.3589	1.9433	0.8942	-0.3591
(0.9,0.5)	2.1343	1.0164	-0.3434	2.1341	1.0159	-0.3435
(0.5,0.1)	1.9702	0.7343	-0.6217	1.9703	0.7342	-0.6227
(0.5,0.3)	1.8608	0.7649	-0.4804	1.8607	0.7648	-0.4809
(0.5,0.7)	1.7090	0.8004	-0.2948	1.7088	0.8003	-0.2948
(0.5,0.9)	1.6566	0.8122	-0.2313	1.6565	0.8121	-0.2313

4.2. Problem 2

We consider a problem for a homogeneous isotropic medium with $g(\mathbf{x}) = 25$, $\overline{\lambda}_{11} = \overline{\lambda}_{22} = 1, \overline{\lambda}_{12} = 0$ and $\overline{\beta}^2 = 1$. The geometry of the region Ω and the boundary conditions are as depicted in Figure 3. As shown in Figure 3 the boundary conditions are symmetrical about the axes $x_1 = 0.5$. Table 2 shows the results of the BEM solution using 80, 160, 320 and 640 elements of equal length. As expected, the results converge as the number of elements increases and also they are symmetrical about the axes $x_1 = 0.5$.

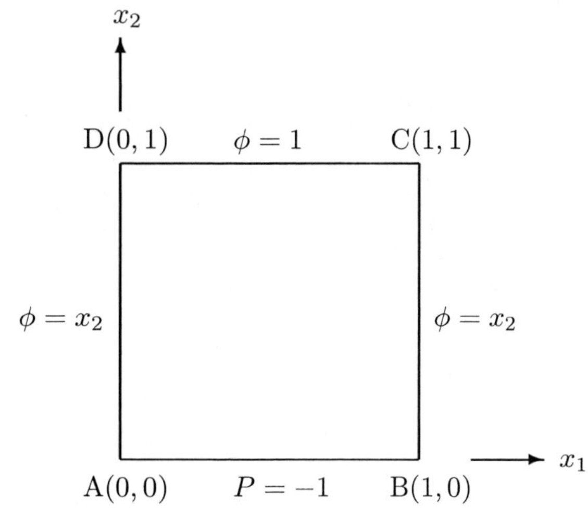

Figure 3. The geometry of Problem 2 and Problem 3

Table 2. BEM solution for Problem 2

(x_1, x_2)	ϕ	$\partial\phi/\partial x_1$	$\partial\phi/\partial x_2$	ϕ	$\partial\phi/\partial x_1$	$\partial\phi/\partial x_2$
	BEM 80 elements			BEM 160 elements		
(0.1,0.5)	0.5060	0.0748	0.9118	0.5060	0.0750	0.9110
(0.3,0.5)	0.5214	0.0647	0.7857	0.5214	0.0646	0.7845
(0.5,0.5)	0.5282	-0.0000	0.7435	0.5283	-0.0000	0.7423
(0.7,0.5)	0.5214	-0.0647	0.7857	0.5214	-0.0646	0.7845
(0.9,0.5)	0.5060	-0.0748	0.9118	0.5060	-0.0750	0.9110
(0.5,0.1)	0.3230	-0.0000	0.2217	0.3235	-0.0000	0.2207
(0.5,0.3)	0.3996	-0.0000	0.5291	0.3999	-0.0000	0.5279
(0.5,0.7)	0.6935	0.0000	0.9045	0.6933	-0.0000	0.9032
(0.5,0.9)	0.8901	0.0000	1.0658	0.8896	-0.0000	1.0643
	BEM 320 elements			BEM 640 elements		
(0.1,0.5)	0.5060	0.0751	0.9105	0.5061	0.0751	0.9103
(0.3,0.5)	0.5214	0.0646	0.7840	0.5215	0.0646	0.7837
(0.5,0.5)	0.5283	-0.0000	0.7417	0.5283	-0.0000	0.7414
(0.7,0.5)	0.5214	-0.0646	0.7840	0.5215	-0.0646	0.7837
(0.9,0.5)	0.5060	-0.0751	0.9105	0.5061	-0.0751	0.9103
(0.5,0.1)	0.3238	0.0000	0.2203	0.3239	0.0000	0.2201
(0.5,0.3)	0.4001	0.0000	0.5273	0.4001	0.0000	0.5270
(0.5,0.7)	0.6932	-0.0000	0.9026	0.6931	-0.0000	0.9022
(0.5,0.9)	0.8894	0.0000	1.0635	0.8893	0.0000	1.0631

4.3. Problem 3

Now we consider a problem for an exponentially graded isotropic material with $g(\mathbf{x}) = [2(1 + 0.5x_1 + 0.75x_2)]^2$, $\overline{\lambda}_{11} = \overline{\lambda}_{22} = 1, \overline{\lambda}_{12} = 0$ and $\overline{\beta}^2 = 1$. Again, the geometry of the region Ω and the boundary conditions are as depicted in Figure 3. Table 3 shows the results of the BEM solution using 80, 160, 320 and 640 elements of equal length. As expected, the results converge as the number of elements increases but they are not symmetrical anymore as the material is not homogeneous.

Table 3. BEM solution for Problem 3

(x_1, x_2)	ϕ	$\partial\phi/\partial x_1$	$\partial\phi/\partial x_2$	ϕ	$\partial\phi/\partial x_1$	$\partial\phi/\partial x_2$
	BEM 80 elements			BEM 160 elements		
(0.1,0.5)	0.5311	0.2748	0.9006	0.5313	0.2742	0.8993
(0.3,0.5)	0.5700	0.1161	0.7752	0.5700	0.1156	0.7740
(0.5,0.5)	0.5787	-0.0250	0.7438	0.5787	-0.0251	0.7426
(0.7,0.5)	0.5615	-0.1431	0.7941	0.5615	-0.1429	0.7929
(0.9,0.5)	0.5238	-0.2260	0.9191	0.5239	-0.2262	0.9171
(0.5,0.1)	0.3488	-0.0371	0.3342	0.3493	-0.0373	0.3335
(0.5,0.3)	0.4436	-0.0354	0.5930	0.4439	-0.0356	0.5919
(0.5,0.7)	0.7370	-0.0149	0.8326	0.7368	-0.0150	0.8312
(0.5,0.9)	0.9102	-0.0056	0.8973	0.9096	-0.0057	0.8957
	BEM 320 elements			BEM 640 elements		
(0.1,0.5)	0.5314	0.2739	0.8989	0.5314	0.2736	0.8987
(0.3,0.5)	0.5701	0.1154	0.7735	0.5701	0.1152	0.7732
(0.5,0.5)	0.5787	-0.0252	0.7420	0.5787	-0.0252	0.7417
(0.7,0.5)	0.5615	-0.1429	0.7923	0.5615	-0.1428	0.7920
(0.9,0.5)	0.5239	-0.2262	0.9166	0.5239	-0.2261	0.9164
(0.5,0.1)	0.3495	-0.0373	0.3331	0.3496	-0.0373	0.3329
(0.5,0.3)	0.4440	-0.0356	0.5913	0.4440	-0.0356	0.5910
(0.5,0.7)	0.7367	-0.0151	0.8305	0.7366	-0.0151	0.8301
(0.5,0.9)	0.9094	-0.0056	0.8948	0.9092	-0.0056	0.8944

5. Conclusion

The modified Helmholtz type governing equation (1) is sometimes used for modeling physical problems such as steady infiltration problems (when $\beta^2 > 0$), and antiplane strain in elastostatics and plane thermostatic problems (when $\beta^2 = 0$). The boundary integral equation (15) is derived from this governing equation (1) and then from (15) a BEM is constructed for calculation of numerical solutions to the problems for anisotropic quadratically graded media. The results show that the BEM solution gives a convergence, consistency, and accuracy. Therefore the results also prove that the analysis used for deriving the boundary integral equation (15) is valid. Together with its ease in implementation, it may be concluded that BEM is a useful numerical method for solving such kind of problems.

Acknowledgements

This work was supported by Hasanuddin University and The Ministry of Higher Education of Indonesia.

References

[1] Azis M I, Kasbawati, Haddade A and Thamrin S A 2018 On some examples of pollutant transport problems solved numerically using the boundary element method *Journal of Physics: Conference Series* **979** 1

[2] Azis M I, Asrul L, Khaeruddin and Paharuddin 2018 BEM solutions for unsteady transport problems in anisotropic media *JP Journal of Heat and Mass Transfer* **15** 4

[3] Haddade A, Salam N, Khaeruddin and Azis M I 2017 A Boundary Element Method for 2D Diffusion-Convection Problems in Anisotropic Media *Far East Journal of Mathematical Sciences* **102** 8

[4] Azis M I 2019 Numerical solutions for the Helmholtz boundary value problems of anisotropic media *Journal of Computational Physics* **381** 42

[5] Cheng A H-D 1984 Darcy's Flow with Variable Permeability: A Boundary Integral Solution *Water Resources Research* **20** 980

[6] Clements D L and Azis M I 2000 A Note on a Boundary Element Method for the Numerical Solution of Boundary Value Problems in Isotropic Inhomogeneous Elasticity *Journal of the Chinese Institute of Engineers* **23** 3 261
[7] Azis M I and Clements D L 2014 On some problems concerning deformations of functionally graded anisotropic elastic materials *Far East Journal of Mathematical Sciences*, **87** 2 173
[8] Azis M I and Clements D L 2014 A Boundary Element Method for Transient Heat Conduction Problem of Nonhomogeneous Anisotropic Materials *Far East Journal of Mathematical Sciences*, **89** 1 51
[9] Azis M I and Clements D L 2008 Nonlinear transient heat conduction problems for a class of inhomogeneous anisotropic materials by BEM *Engineering Analysis with Boundary Elements* **32** 1054
[10] Azis M I, Toaha S, Bahri M and Ilyas N 2018 A boundary element method with analytical integration for deformation of inhomogeneous elastic materials *Journal of Physics: Conference Series* **979** 1
[11] Salam N, Haddade A, Clements D L and Azis M I 2017 A boundary element method for a class of elliptic boundary value problems of functionally graded media *Engineering Analysis with Boundary Elements* **84** 186 doi: 10.1016/j.enganabound.2017.08.017
[12] Clements D L and Lobo M 2010 A BEM for time dependent infiltration from an irrigation channel *Engineering Analysis with Boundary Elements* **34** 1100
[13] Solekhudin I and Ang K-C 2012 A DRBEM with a predictor–corrector scheme for steady infiltration from periodic channels with root-water uptake *Engineering Analysis with Boundary Elements* **36** 1199
[14] Azis M I 2017 Fundamental solutions to two types of 2D boundary value problems of anisotropic materials *Far East Journal of Mathematical Sciences* **101** 11 2405
[15] Abramowitz M and Stegun I A 1972 *Handbook of mathematical functions: with formulas, graphs and mathematical tables* Dover Publications Washington

IOP Conf. Series: Earth and Environmental Science **279** (2019) 012011 doi:10.1088/1755-1315/279/1/012011

Using GIS for Integrated Assessment of Agriculture Land Suitability and Food Security in Small Islands

R Neswati, S Baja, A Ramlan and S Arif

Laboratory of Geospatial Information and Land Use Planning, Department of Soil Science, Faculty of Agriculture, Hasanuddin University,Jl. Perintis Kemerdekaan KM 10, Makassar, 90245, Indonesia. Phone +62-411 587076, Fax +62-411 587076

E-mail: neswati76@gmail.com; sbja02@yahoo.com.au (corresponding author); ramlanmartin@yahoo.co.id; samsu_arif@)unhas.ac.id

Abstract. Food security has recently become one of important issues in Indonesia's small islands, since increasing human populations and declining land quality have continued to occur in such areas. With limited new agricultural land in small islands to develop, meeting the food needs of future populations in such areas will require an increase of crop yields, through appropriate decisions on land use and land management. The principal aim of this paper is to construct a spatial decision making mechanism using Geographic Information Systems (GIS) to perform land suitability analysis and to assess food security in a very limited agricultural land of a small island. We were use spatially integrated method in GIS and Analytical Hierarchy Process (AHP) to generate priority of food crop development that may give insights into some important criteria of food security. In our analysis, we use five types of food crops (dryland rice, maize, sweet potato, taro plant, and cassava), with five generic criteria: land potential (LAND), economic aspect (ECON), culture aspect (CULT), infrastructure availability (INFR), and local government policy (POLY). Our preliminary findings show that maize was the most important food crop to be developed in this area, followed by cassava, dryland rice, sweet potato, and taro plant, although each criterion selected gives a different rank for the five crop types. The results of this analysis could provide the basis for local government to strengthen its decision making in the allocation of food crops from the perspectives of the small island's food security.

Key words: agricultural land suitability, food security, GIS, small islands

[1] Introduction

The main problem that is often faced even failure in agricultural land development is that the selected commodities are less or even not in accordance with the local soil and climate conditions. Thus, the genetic potential of cultivated plants cannot be optimized due to a mismatch between the requirements for plant growth and the characteristics of the soil and the local climate. Young (1988) argues that, based on many experiences, there is a risk level of 30-50 percent for a large-scale agricultural enterprise if it is not based on detailed information about the nature of the soil and climate. In the long-term perspective, these non-conformities not only have an impact on risks such as low productivity of plants, but also pose a threat to damage to land resources.

Indonesia is an archipelago with thousands of large islands and small islands. Each region, especially in areas with small islands, will vary in terms of the success of responding to the implementation of the self-sufficiency program mentioned above. Food needs and approaches to fulfillment will differ from one another, depending on geographical typology, land area and quality, climate, level of vulnerability to natural disasters, and community culture. Generally, the small islands in Indonesia face the problem of limited agricultural land and / or at the same time the limited infrastructure that connects the region with the enclaves of agricultural food production.

Evaluation of agroecological suitability (land suitability) will emphasize the analysis of the diversity of land characteristics, because it will determine the type of commodity that can be cultivated, which in its sequence determines the level of land productivity (Baja, 2001). Meanwhile, socio-economic studies can provide direction on the economic feasibility of developing certain commodities (Rossiter, 1995), including the ability of farmers to manage land within a particular region (Webb and Thiha, 2002). The development of agricultural commodities in areas that are in accordance with the requirements of plant propaganda (climate, soil, and topography) and socio-economic conditions, coupled with management based on land characteristics, will provide optimal and sustainable results (Doran et al, 2002). This requires continuous and integrated research.

Research objectives
The purpose of this study is to know and solve the problems of developing food commodities in small island areas, from the aspects of quality and availability of land (biophysics), public interest and preferences, and policies implemented.

[2] Methodology

This research was conducted at Kaledupa Island, Wakatobi Regency, Southeast Sulawesi Province Indonesia from May to August 2018 (Fig. 1). Analysis of soil samples in the laboratory of the Department of Soil Science, Faculty of Agriculture, Hasanuddin University, Makassar, Indonesia.

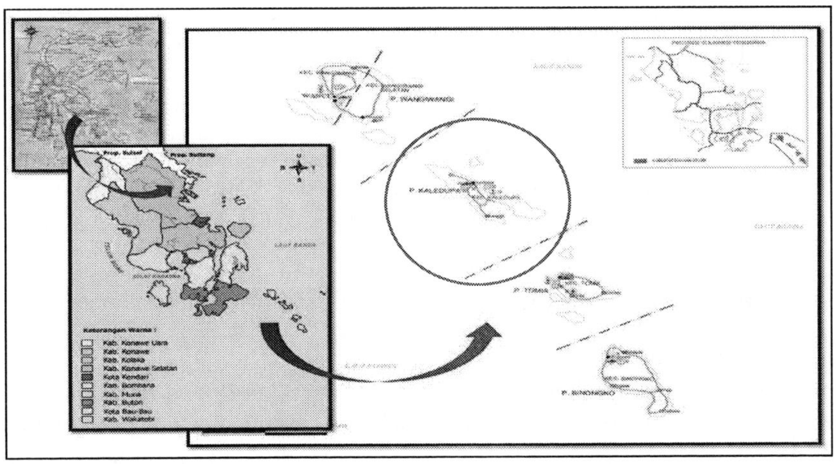

Figure 1. Map of Research Site

The base map was built from a 1: 50,000 topographic maps. The base map or work map was used to make the land unit and the land suitability map. The work or base map was digitized and stored in a spatial database [1]. Soil characteristics reached from land systems map at scale 1:250,000 [3], and based on land system map, 19 land units were built for field observation and soil sampling. Soil observation in the field was conducted by transect [2,4]. Soil classification was determined according to Soil Survey Staff [5]. Field observation and soil sampling were conducted on the 19 land units. The field survey to identification of land characteristics such as soil depth, soil drainage, slope, soil surface rock, flood hazard, and others. In this stage included analysis of climate data. Soil sample analysis was conducted primarily for land evaluation. Soil characteristics analysis consist of: a). soil texture, b). drainage, c). soil depth, d). rock surface, and e). inundation (flooding). Properties of soil chemical are: a). soil acidity (pH), b). base saturation, c). organic matter, d). cation exchange capacity (CEC), and e). salinity. For land suitability classification using GIS software ArcGIS 10.2 (licensed by ESRI Indonesia). Land suitability classification was analyzed based on FAO system with simple limitations method [6, 7] for 5 crops consist of dryland rice, maize, sweet potato, taro plant, and cassava.

[3] Results and Discussion

There are 2 quite broad sub-districts, namely Wangi-Wangi and South Wangi-Wangi sub-districts covering 54, 43% of the total area of Wakatobi Regency. Whereas Kaledupa District is only 5.53% and South Kaledupa is only 7.11% of Wakatobi Regency. However, Kaledupa and South Kaledupa Subdistricts are two sub-districts that have a extensive land area, so the development of agricultural commodities on Kaledupa Island was very potential to be carried out.

The result of land suitability analysis showed the level of suitability of the land unit for specific crops. The class of suitability consists of; S1 = highly suitable; S2 = moderately suitable; S3 = marginally suitable, N1 = currently not suitable and N2 = not suitable permanently, and the most potential crops to be developed were dryland rice (S2) 2182.7 ha, maize (S2) 2.067,4 ha, and taro plant (S2) 2.067,4 ha.

Table 1. Land suitability and development area of food crops in Kaledupa Island (ha)

Crops	S1	S2	S3	Land potency (ha)	Development potency (ha)
1. Dryland rice	-	2.182,7	2.250,1	4.432,8	4.432,8
2. Maize	-	2.067,4	2.365,4	4.432,8	4.432,8
3. Cassava	-	1.906,0	2.526,0	4.432,8	4.432,8
4. Sweet potato	-	1.906,0	2.018,0	3.924,0	3.924,0
5. Taro plant	-	2.067,4	2.365,4	4.432,8	4.432,8

Soil analysis results also showed that soil characteristics mainly soil chemistry and fertility classified as quite good. Soil acidity, cation exchange capacity, C-organic and base saturation were generally moderate while potassium and nitrogen were at medium to high level. The level of limiting factor of soil chemical was relatively mild, while the degree of limiting factors from soil physics was classified moderate to severe and it's needed more efforts to overcome the limitations such as soil depth, percentage of surface stoniness, coral, drainage of land unit. The mild constraints of soil fertility (pH, CEC, C-organic,) could be improved by fertilization and calcification [12, 13]. Dryland rice is the most prospective commodity, followed by maize and taro plant based on the land suitability class and land potency for every crop. The limiting factors that should be considered in land development were the soil depth and stoniness in the surface.

Based on land suitability analysis, dryland rice classified as moderately suitable (S2) and marginally suitable (S3) in area 4.432,8 ha, which consist of 2.182,7 ha in S2 and 2.250,1 ha in S3. Maize classified as

S2 located at 2.067,4 ha, S3 2.365,4 Ha and potentially to be developed become 4.432,8 ha. Cassava was suitable to be cultivated at 1.906 ha (S2) and 2.526 ha (S3). Sweet potato has potentially to be cultivated toward 4.432,8 ha which consist of 1.906 ha as S2 and 2.018 ha (S3). The suitable area for taro plant located class S2 was 2.067,4 ha, class S3 2.365,4 ha. The examples of land suitability distribution in Kaledupa showed in the Fig. 2 for dryland rice and sweet potato.

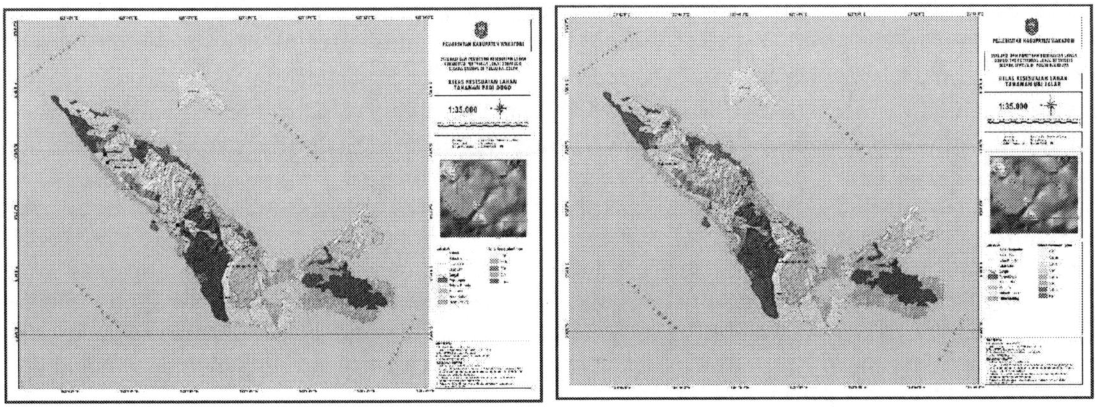

Figure 2. Map of land suitability of dryland rice (a) and (b) sweet potato in the research area

[4] Conclusion

The preliminary findings show that dryland rice was the most important food crop to be developed in this area, followed by maize, cassava, dryland rice, sweet potato, and taro plant, although each criterion selected gives a different rank for the five crop types. The results of this analysis could provide the basis for local government to strengthen its decision making in the allocation of food crops from the perspectives of the small island's food security. The main limiting factors to the development of food crops in this area are soil fertility, rock outcrops, and topography.

Acknowledgments

The authors deeply thankful to The Ministry of Research, Technology, and Higher Education, Republic of Indonesia, through an BMIS research scheme and the Department of Soil Science, Faculty of Agriculture, Hasanuddin University, for supporting data and facilities for this research.

References

[1] Baja S, Chapman D M, Dragovich D 2002 *Environmental Management* **29** 647-61.

[2] Davidson D A and Jones G E 1986 *App Geog* **6** (3) 255-265.

[3] Regional Physical Planning Project For Transmigration (RePPProT) 1988 *Land System Map* Indonesia: Department of Transmigration Republic of Indonesia.

[4] Steers C A and Hajek B F 1979 *Soil Sci. Soc. Am. J.* **431** 56-60 doi:10.2136/sssaj1979.03615995004300010030x.

[5] Soil Survey Staff 2014 *Keys to Soil Taxonomy* United States: United States Department of Agriculture, Natural Resources Conservation Service.

[6] Djaenudin D, Marwan H Suharta 2000 *Kriteria Kesesuaian Lahan untuk Komoditas Pertanian. Badan Penelitian dan Pengembangan Pertanian* Bogor: Pusat Penelitian dan Agroklimat.

[7] Soil Research Center and Agroclimate 2001 *Technical Guidelines for Preparation of Agricultural Commodity Map based on Agro Ecology Zone (ZAE) Scale 1: 50,000* Bogor: Center for Soil and Agro-climate Research, Balitbang Pertanian, Departemen Pertanian.

[8] Schmidt F H and Ferguson J H A 1951 *Rainfall types based on wet and dry period rations for*

Indonesia with Western New Guinea Jakarta: Ministry of transportation of meteorology and geophysics.
[9] Sikumbang N and Sanyoto P 1981 *Peta Geologi Lembar Buton dan Muna, Sulawesi Tenggara, Skala 1 : 250.000, S 81 – 4.*
[10] Indonesia Soil Research Center 1985 *Soil map of Muna Island and Buton Island, Southeast Sulawesi*.
[11] Mahanty B, Kim S, Kim C G 2013 *Microbial Eco* **65** (3) 679.
[12] Sikirić B, Mrvić V, Stevanović D, Maksimović S, Stajković O, Bogdanović D. 2009 *Agrochimica* **53** (4) 250-259 ref.37.

IOP Conf. Series: Earth and Environmental Science **279** (2019) 012012 doi:10.1088/1755-1315/279/1/012012

Estimation Mean by the Bayesian Approach on the Exponentially Weighted Moving Average Control Chart

M Siruddin, E T Herdiani and S A Thamrin

Statistics, Department of Mathematics, Hasanuddin University, Makassar, Indonesia

E-mail: meylinas27@gmail.com

Abstract. A control chart is one of the primary techniques of statistical quality control. Exponentially Weighted Moving Average (EWMA) control chart is one of the control charts used to detect a small shift in the mean process. In the quality control process, the control chart is often constructed by ignoring parameter uncertainty. This can affect the long-term performance of the control chart in a controlled or out-of-control state, so parameter estimation is necessary. The parameter estimation method used in this paper is the Bayesian approach. The Bayesian approach is used to develop an EWMA control chart in monitoring a process. The performance of the EWMA control chart with the Bayesian approach has been evaluated using Average Run Length (ARL). Based on the comparison of the ARL value from the EWMA control chart with the Bayesian approach and the Classical EWMA control chart, it shows that the EWMA control chart with the Bayesian approach is more sensitive in detecting the shift in the mean process.

1. Introduction

In statistics, there is a statistical quality control system that aims to provide a new tool that makes the inspection process more effective [1]. In statistical quality control, it is known that control chart is used to monitor process output and detect when changes in the inputs are required to bring the process back to an in-control state. The control chart was first developed by Dr. Walter A. Shewhart in 1930 was used to detect large shifts in a process. This control chart is also known as Shewhart's control chart. According to [2] a major disadvantage of a Shewhart control chart is that it is relatively insensitive to small process shifts. This is because the Shewhart control chart uses only information contained in the last sample observation and it ignores the information provided by previous samples. Based on these reasons, cumulative-sum (CUSUM) control chart and exponentially weighted moving average (EWMA) control chart are developed.

In 1959 Robert developed EWMA control chart which was then widely developed by Crowder (1989) and Lucas & Saccucci (1990). This control chart combines information from several samples, causing the EWMA control chart to be more sensitive in monitoring small process shifts when compared to Shewhart control chart. But in practice, the process parameters are unknown and the control chart is built by ignoring the parameter uncertainty using estimates as a substitute for parameters. This is because the problem of parameter estimation only receives relatively simple attention in terms of control chart so that it affects the long-term performance of the control chart in of control or out of control conditions.

Ignoring parameter uncertainty involved in the quality control process can lead to errors. Therefore, it is important to make precise estimates of the parameters related to make the statistical process performance of the control chart accurate. In parameter estimation, there are two methods of approach, namely the classical approach method and the Bayesian approach method. In the Bayesian approach, the parameters are considered as variables that describe the initial information about the parameters expressed in the prior distribution. This approach combines sample information with previous information expressed in a distribution to overcome the parameters uncertainty. This method has been widely applied in every scientific discipline including statistical quality control and is extensively used in the framework of a control chart.

Research on control chart in the Bayesian framework has been previously carried out. In 2013, [3] developed an EWMA control chart for the mean and variance of the normal distribution using the Bayesian approach. Also based on the Bayesian approach the EWMA control chart is designed for monitor variance and for combine monitoring of mean and variance of a process [4]. Next, [5] using the Bayesian approach based on the loss function in the EWMA control chart to detect small shifts in a process. In his research, the EWMA control chart was developed by using posterior predictive distribution to monitor the mean of a process considering different loss functions. Based on the review, this paper discusses the comparison of EWMA control chart with the Bayesian approach and Classical EWMA control chart for detecting process mean shift.

The paper is organized as follows. We explained methods in section 2 such as the Bayesian approach, Classical EWMA control chart, EWMA control chart with the Bayesian approach, and Average Run Length. Section 3, we present the results. Then, section 4 concludes this paper with a summary.

2. Methods

2.1. Bayesian Approach

In the parameter estimation theory, there are two methods of approach, namely the classical approach method and the Bayesian approach method. In the classical approach more dependent on inferential processes on sample data from a population [6]. This method of approach views the parameter μ as a parameter with a fixed quantity that is unknown in price [7]. While the Bayesian approach in addition to utilizing information from sample data, it also takes account of previous information about unknown parameter probability distributions known as prior distributions. [6]. The Bayesian approach views the μ parameter as a random variable that describes the previous information about the parameters stated in a prior distribution from which we can determine the posterior distribution to obtain the Bayesian estimator [7].

The posterior distribution $f(\mu|x)$ is a combination of the likelihood function $f(x|\mu)$ and the prior distribution $f(\mu)$, where the prior selected is the non-informative prior. One of the approaches used to define non-informative prior distributions is Jeffrey's method, so it is also called Jeffrey's prior. The Jeffrey's prior is defined as follows [8]:

$$f(\mu) = [J(\mu)]^{\frac{1}{2}} \tag{1}$$

where $J(\mu)$ is Fisher's information. The Jeffrey's prior distribution is defined as follows:

$$p(\mu) = \sqrt{n/\sigma^2} = constan \tag{2}$$

where n and σ^2 are constants, so Jeffrey's prior is constant.

Therefore, the posterior distribution for continuous random variables can be written as follows:

$$f(\mu|x) = \frac{f(x|\mu)f(\mu)}{\int f(x|\mu)f(\mu)\,d\mu} = [2\pi(\sigma^2/n)]^{-1/2} \exp\left[-\frac{1}{2\sigma^2/n}(\mu - \bar{x})^2\right] \tag{3}$$

An estimator minimizing the risk of Bayes can be obtained by minimizing the posterior risk defined as $R(\hat{\mu},\pi) = \int L(\mu,\hat{\mu}) \times p(\mu|x)\,d\mu$, where $L(\mu,\hat{\mu}) = (\mu - \hat{\mu})^2$ is squared error loss function (SELF) [9]. SELF is a loss function that is widely used because it is easy to calculate and is symmetric loss function [5]. Therefore, the Bayes estimator value based on the posterior distribution $\mu|x$ are $\hat{\mu}_{SELF} = \bar{x}$ and $\sigma^2_{SELF} = \sigma^2/n$ or can be written $\mu|X \sim N(\hat{\mu}_{SELF}, \sigma^2_{SELF})$.

In the Bayesian method, posterior predictive distribution is the main tool for handling prediction problems in statistics so that making it easier to make predictions [9]. Posterior predictive distribution is defined as follows:

$$f(y|x) = \int f(y|\mu) \times f(\mu|x)d\mu \tag{4}$$

Let $y_1, y_2, \dots, y_n$ are the next n observations with the posterior predictive distribution $Y|X \sim N(\hat{\mu}_{SELF}, \sigma^2_{Y|X})$ where $\sigma^2_{Y|X} = \sigma^2 + (\sigma^2/n)$. For $\bar{y}_i = (y_{i1} + y_{i2} + \dots + y_{ik})/k$ then $\bar{Y}|X \sim N(\hat{\mu}_{SELF}, \psi^2)$ where $\psi^2 = (\sigma^2/k) + (\sigma^2/n)$.

2.2. *Classical EWMA Control Chart*

EWMA control chart is one of an alternative to Shewhart's control chart to monitoring small shifts in a process. The Classical EWMA statistics is defined as [2]:

$$z_i = \lambda\bar{x}_i + (1-\lambda)z_{i-1} \tag{5}$$

where $0 < \lambda < 1$ which is the determinant of the value of the weight given in the current sample. The initial value z_0 is usually specified as the target value of a process so that $z_0 = \mu_0$. The center line and control limits for the Classical EWMA control chart are defined as follows

$$UCL = \mu_0 + 3\left[\sigma_{x_i}/\sqrt{n}\right]\sqrt{[\lambda/(2-\lambda)][1-(1-\lambda)^{2i}]} \tag{6}$$

$$CL = \mu_0 \tag{7}$$

$$LCL = \mu_0 - 3\left[\sigma_{x_i}/\sqrt{n}\right]\sqrt{[\lambda/(2-\lambda)][1-(1-\lambda)^{2i}]} \tag{8}$$

The form $\left(1-(1-\lambda)^{2i}\right)$ in equations (6) and (8) will converge to 1 when the value of i gets larger. This means that the control limits will approach steady-state values given by

$$UCL = \mu_0 + 3\left[\sigma_{x_i}/\sqrt{n}\right]\sqrt{\lambda/(2-\lambda)} \tag{9}$$

$$CL = \mu_0 \tag{10}$$

$$LCL = \mu_0 - 3\left[\sigma_{x_i}/\sqrt{n}\right]\sqrt{\lambda/(2-\lambda)} \tag{11}$$

2.3. *EWMA Control Chart with the Bayesian Approach*

EWMA control chart with the Bayesian approach is also called Bayesian EWMA control chart. The EWMA statistics based on Bayesian estimator $y|x$ is defined as [5]:

$$z_i = \lambda(y|x_i) + (1-\lambda)z_{i-1} \tag{12}$$

The center line and control limits for the Bayesian EWMA control chart on the steady-state is as follows

$$UCL = \hat{\mu}_{SELF} + 3\psi\sqrt{\lambda/(2-\lambda)} \tag{13}$$

$$CL = \hat{\mu}_{SELF} \tag{14}$$

$$LCL = \hat{\mu}_{SELF} - 3\psi\sqrt{\lambda/(2-\lambda)} \tag{15}$$

2.4. *Average Run Length*

Average Run Length (ARL) is the average number of points that must be plotted before a point indicates an out-of-control condition. The control charts with a smaller ARL value indicate that the smaller the number of samples taken to detect a shift in the process [2].

One of the methods that are widely used in determining the ARL value on the CUSUM and EWMA control chart is the Markov chain method developed by Brook and Evans (1972). The procedure in the Markov chain method divides the interval between the UCL and LCL values into $k = 2m + 1$ subinterval with width 2δ, where the width of each interval is obtained by the formula $w = (UCL - LCL)/k$. The midpoint of each interval is defined $S_i = LCL + (i - 0.5)w$ for $i = -m, -m+1, \dots, m-1, m$. The transition probability matrix represented in partitioned matrix form is given by [11]:

$$\boldsymbol{P} = \begin{pmatrix} \boldsymbol{R} & (\boldsymbol{I}-\boldsymbol{R})\boldsymbol{1} \\ \boldsymbol{0}^T & 1 \end{pmatrix} \tag{16}$$

where the sub-matrix $\boldsymbol{R} = [r_{ij}]$ of $k \times k$ size is a controlled transition probability expressed by $r_{ij} = \Phi\left[\left((S_j + \delta) - (1-\lambda)S_i - \lambda\mu\right)/\lambda\sigma\right] - \Phi\left[\left((S_j - \delta) - (1-\lambda)S_i - \lambda\mu\right)/\lambda\sigma\right]$ for $i, j = -m, -m+1, \dots, m-1, m$. Therefore, ARL on the EWMA control chart for the mean process is given by:

$$ARL = \boldsymbol{q}^T(\boldsymbol{I} - \boldsymbol{R})^{-1}\boldsymbol{1} \quad (17)$$

where $\boldsymbol{I}$ is the $k \times k$ identity matrix, and $\boldsymbol{1}$ is the $k \times 1$ column vector of ones. We substitute $\boldsymbol{q} = \boldsymbol{e}_i$ in equation (17), where $\boldsymbol{e}_i$ is the ith unit vector with all elements zero except the ith element which is 1. In equation (17), we use $i = (k+1)/2$ (12).

3. Results

Here, we demonstrate the performance and implementation of the EWMA control chart based on the Bayesian approach and its comparison with the Classical EWMA control chart. Dataset is taken from European J. Industrial Engineering in the paper [13]. The data contained 5 results of the weight measurement of yogurt cups in each subgroup, where the average of the weight measurement of yogurt cups in each subgroup can be seen in Figure 1.

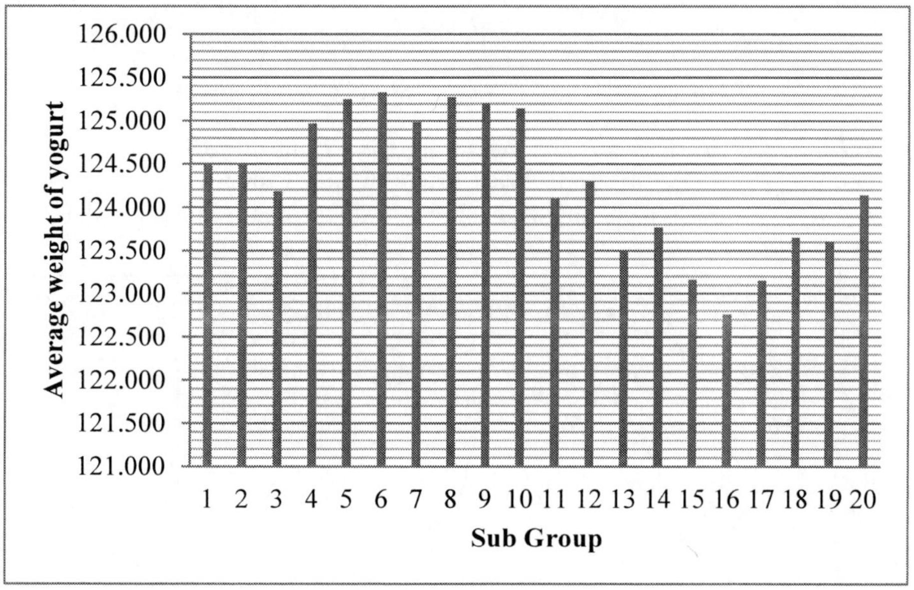

Figure 1. The average weight of yogurt cups in 20 sample subgroup

Classical EWMA control chart for data measuring the weight of yogurt cups shown in Figure 2 with the mean of data is $\mu_0 = \left(\sum_{i=1}^{20} \bar{x}_i\right)/20 = 124{,}2748$ and the estimated standard deviation is $\sigma = \bar{S}/c_4 = 0{,}7398$, where $c_4 = [2/(k-1)\,]^{1/2}[\Gamma(k/2)]\left[\Gamma\left((k-1)/2\right)\right]^{-1}$. Figure 2 shows that there are 10 from 20 averages data of weight measurement yogurt that are outside the control limits so that the product can be said uncontrolled.

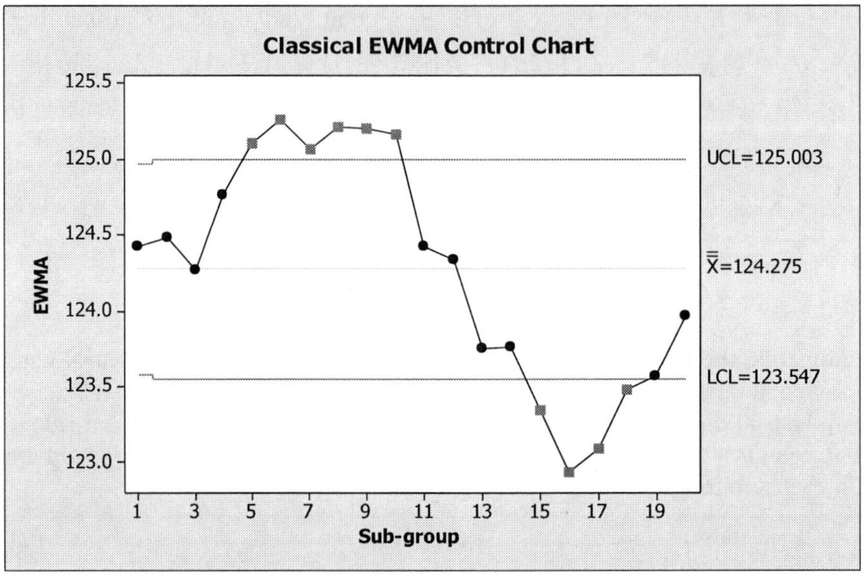

Figure 2. Classical EWMA Control Chart at $\lambda = 0.7$

Bayesian EWMA control chart for data measuring the weight of yogurt cups shown in Figure 3 with the values $\hat{\mu}_{SELF} = 124{,}2748$ and $\psi = 0{,}3187$. Figure 3 shows that there are 11 from 20 averages data of weight measurement yogurt that are outside the control limits so that the product can be said uncontrolled.

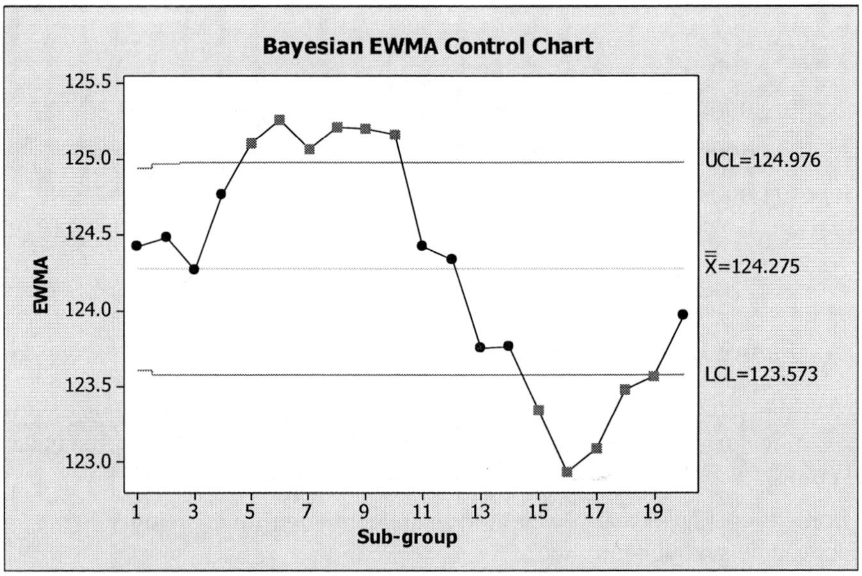

Figure 3. Bayesian EWMA Control Chart at $\lambda = 0.7$

Based on the value $0 < \lambda < 1$ and the number of data that is outside the control limit show that Bayesian EWMA control chart is faster to detect data out of control than the Classical EWMA control chart. However, to be more precise will be used the calculation of the ARL value to detect the accuracy of the control chart. The ARL value in the Classical EWMA and Bayesian EWMA control charts can be seen in Table 1, where the ARL Bayesian EWMA control chart is smaller than the ARL Classical EWMA control chart.

IOP Conf. Series: Earth and Environmental Science **279** (2019) 012012 doi:10.1088/1755-1315/279/1/012012

Table 1. ARL for some values λ in Classical EWMA and Bayesian EWMA control charts

λ	ARL Classical EWMA	ARL Bayesian EWMA	λ	ARL Classical EWMA	ARL Bayesian EWMA	λ	ARL Classical EWMA	ARL Bayesian EWMA
0.01	57.0883	52.1042	0.34	3.2882	3.1888	0.67	2.1414	2.1134
0.02	28.4407	26.8096	0.35	3.2246	3.1291	0.68	2.1226	2.0958
0.03	20.6010	19.5125	0.36	3.1643	3.0726	0.69	2.1043	2.0785
0.04	16.4750	15.6214	0.37	3.1071	3.0189	0.70	2.0863	2.0617
0.05	13.8452	13.1382	0.38	3.0526	2.9678	0.71	2.0689	2.0453
0.06	12.0082	11.4045	0.39	3.0008	2.9192	0.72	2.0518	2.0293
0.07	10.6476	10.1209	0.40	2.9514	2.8729	0.73	2.0351	2.0136
0.08	9.5967	9.1296	0.41	2.9042	2.8287	0.74	2.0188	1.9982
0.09	8.7587	8.3394	0.42	2.8591	2.7864	0.75	2.0028	1.9832
0.10	8.0738	7.6935	0.43	2.8160	2.7460	0.76	1.9872	1.9685
0.11	7.5027	7.1551	0.44	2.7747	2.7073	0.77	1.9719	1.9541
0.12	7.0188	6.6989	0.45	2.7351	2.6701	0.78	1.9569	1.9400
0.13	6.6031	6.3070	0.46	2.6971	2.6345	0.79	1.9422	1.9262
0.14	6.2418	5.9667	0.47	2.6605	2.6003	0.80	1.9278	1.9126
0.15	5.9249	5.6681	0.48	2.6254	2.5673	0.81	1.9137	1.8993
0.16	5.6444	5.4039	0.49	2.5916	2.5356	0.82	1.8998	1.8862
0.17	5.3943	5.1685	0.50	2.5590	2.5051	0.83	1.8862	1.8734
0.18	5.1699	4.9572	0.51	2.5276	2.4757	0.84	1.8728	1.8608
0.19	4.9673	4.7666	0.52	2.4973	2.4473	0.85	1.8597	1.8484
0.20	4.7835	4.5936	0.53	2.4680	2.4198	0.86	1.8468	1.8363
0.21	4.6159	4.4360	0.54	2.4398	2.3933	0.87	1.8341	1.8243
0.22	4.4625	4.2917	0.55	2.4124	2.3677	0.88	1.8216	1.8125
0.23	4.3214	4.1591	0.56	2.3859	2.3428	0.89	1.8093	1.8009
0.24	4.1913	4.0368	0.57	2.3603	2.3188	0.90	1.7972	1.7895
0.25	4.0709	3.9236	0.58	2.3354	2.2955	0.91	1.7853	1.7782
0.26	3.9591	3.8186	0.59	2.3113	2.2729	0.92	1.7736	1.7671
0.27	3.8551	3.7209	0.60	2.2879	2.2509	0.93	1.7621	1.7562
0.28	3.7579	3.6297	0.61	2.2652	2.2296	0.94	1.7507	1.7454
0.29	3.6671	3.5444	0.62	2.2431	2.2089	0.95	1.7394	1.7348
0.30	3.5819	3.4644	0.63	2.2217	2.1888	0.96	1.7284	1.7243
0.31	3.5018	3.3893	0.64	2.2008	2.1692	0.97	1.7174	1.7140
0.32	3.4265	3.3185	0.65	2.1805	2.1501	0.98	1.7067	1.7037
0.33	3.3554	3.2518	0.66	2.1607	2.1315	0.99	1.6960	1.6936

4. Conclusion

In this paper, we have presented the EWMA control chart with the Bayesian approach and compared the ARL values obtained with the ARL values Classical EWMA control chart. The results obtained show that the EWMA control chart with the Bayesian approach is more sensitive in detecting the shift in the process mean compared to the Classical EWMA control chart.

References

[1] Grant E L and Richard S L 1993 *Pengendalian Mutu Statistik* (Jakarta: Erlangga)
[2] Montgomery D C 2009 *Introduction to Statistical Quality Control Sixth Edition* (Amerika: John Wiley & Sons, Inc.)
[3] Menzefricke U 2013a Combined exponentially weighted moving average chart for the mean and variance based on the predictive distribution *Communications in Statistics-Theory and Methods* **42** (22) 4003-4016
[4] Menzefricke U 2013b Control charts for mean and variance based on changepoint methodology *Communications in Statistics-Theory and Methods* **42** (6) 988-1007
[5] Riaz S, Riaz M, Hussain Z and Abbas T 2017 Monitoring the performance of Bayesian EWMA control chart using loss functions *Computers and Industrial* Engineering **112** 426-436
[6] Walpole R E and Myers R H 1995 *Ilmu Peluang dan Statistika untuk Insinyur dan Ilmuan Edisi Keempat* translation by Sembiring R K (Bandung: Penerbit ITB)
[7] Bolstad W M 2007 *Introduction to Bayesian Statistics Second Edition* (Amerika: A John Wiley & Sons. Inc)
[8] Gelman A, Carlin J B, Stern H S, Dunson D B, Vehtari A and Rubin D B 2014 *Bayesian Data Analysis Third Edition* (New York: CRC Press, Taylor & Francis Group)
[9] Robert C P 2007 *The Bayesian Choice From Decision-Theoretic Foundations to Computational Implementation Second Edition* (Paris : Springer Texts in Statistics)
[10] Barbieri M M 2015 Posterior predictive distribution *Wiley StatsRef: Statistics Reference Online* https://doi.org/10.1002/9781118445112.stat07839 Accessed 17 Mei 2018
[11] Patel A K and Divecha J 2011 Modified exponentially weighted moving average (EWMA) control chart for an analytical process data *Journal of Chemical Engineering and Materials Science* **2** (1) 12-20
[12] Serel D A 2009 Economic design of EWMA control charts based on loss function *Mathematical and Computer Modelling* **49** 745-759
[13] Hu X, Castagliola P, Sun J, Khoo M B C 2016 Effect of measurement errors on the VSI $\bar{X}$ chart *European Journal Industrial Engineering* **10** (2) 224-242

IOP Conf. Series: Earth and Environmental Science **279** (2019) 012013 doi:10.1088/1755-1315/279/1/012013

Geographical Mapping of Dengue Fever Incidence 2012-2016 in Makassar, Indonesia

Sri Astuti Thamrin* and **Alimun**

Statistics Study Program, Hasanuddin University, Makassar, Indonesia

*tuti@unhas.ac.id

Abstract. To present the geographical map of Dengue fever (DF) in Makassar city, Indonesia, we used the registered DF data to compute the relative risk (RR) of DF in different areas of Makassar city, Indonesia, for a five-year period (from 2012 to 2016). This paper aim to identify the the spatial distribution of DF occurrence to provide the map of risk area in Makassar City through the geographical map. To estimate RR for DF in Makassar City, Indonesia, the Besag York Mollie (BYM) model was proposed. The Bayesian estimates of DF relative risk were utilized for geographical mapping of DF. The results of the present study reveals that DF is still a serious health problem in Makassar city, Indonesia. The distribution of high and low risk areas of DF occurrences for all sub-districts in Makassar City can be identified in the risk geographical map to allow the better allocation and risk assessment.

1. Introduction

One of the infectious diseases that can cause death is dengue fever (DF). This disease is generally transmitted by the Aedes Aegypti mosquito infected with the Dengue virus. The level (stage) of DF disease is 2, namely early stage and advanced stage [1]. What distinguishes these two stages is the absence of plasma leakage from cells in blood vessels [2]. In advanced stage DF disease plasma leakage occurs, which is commonly referred to as Dengue Hemorrhagic Fever (DHF) disease while patients with early stage DF disease have not experienced this.

According to [3], aedes mosquitoes can only breed in tropical regions with temperatures above 16℃ and at altitudes less than 1,000 meters above sea level. Topographically, the city of Makassar has temperatures ranging between 26.2℃ and 29.8℃ and located at between 1- 25 meters above sea level (city mks 1). Therefore, the city of Makassar is an area frequented by aedes mosquitoes to breed, then many cases of DF disease in the city occur. Tthe rainfall rate in Makassar city ranges from an average of 2729 mm with the number of rainy days ranging from an average of 144 days [4].

Relative risk analysis of the spread of DF disease in the city of Makassar needs to be done to see which is sub-districts in the Makassar city are at high risk of developing DF disease. This is very closely related to national development problems, especially in the Makassar city, South Sulawesi, because one of the basic capital in the implementation of development is a good health community.

Determination of the relative risk value for the spread of DF disease has been carried out by using the Standardized Morbidity Ratio (SMR) model [5]. In that model, the number of cases of DF is seen as data on counts, not as a random variable. Meanwhile, [6] used the Poisson Gamma and the Besag, York, and Mollie (BYM) models to estimate the relative risk of the three stages of DF. In this study, the relative risk value of the spread of DF disease will be calculated using a method based on the BYM model. Under this model, disease mapping assumes that the number of people with DF disease is a Poisson random variable [7]. With this BYM model, the risk level of DF disease every year for each sub-district in Makassar city can be determined. Therefore, this paper aims to identify the risk area of DF in Makassar city through the geographical map. In the first section, the paper presents BYM model. Then, the application and the result of the study is presented in the next section. The conclusion is presented at the last.

2. Material and Methods

2.1 Dengue Disease and Its Situation in Makassar

In this paper, the observed count DF data for 14 sub-districts in Makassar, Indonesia, from 2011 to 2016 was used. This data was provided by the Board of Health of Makassar. The map of Makassar city is displayed in Figure 1. Sub-district of Rappocini had the highest total of DF cases, 158 cases, followed by Manggala with 116 cases. Ujung tanah, Wajo and Ujung Pandang are the only three sub-districts had less than 20 cases. These can be seen in Figure 2.

2.2. Besag-York-Mollie (BYM) Model

The relative risks with Bayesian approach was introduced by [8]. Then, [9] was developed to a fully Bayesian and it is commonly known as the BYM model. This model can be used to map the disease and it used hierarchical Bayesian spatial model [10].

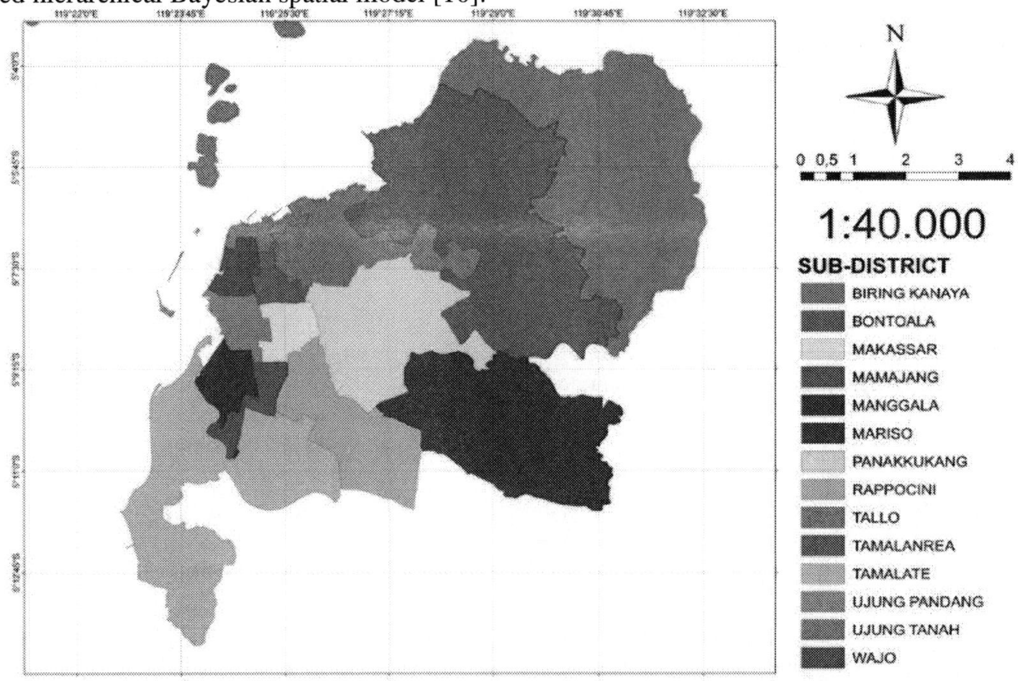

Figure 1. The map of Makassar city with all sub-districrs in South Sulawesi, Indonesia

BYM model was used to estimate the relative risk (RR). To do it, random effects and spatially structured heterogeneity is incorporated into the log-linear model. Random effects are the correlated heterogeneity that is unstructured. This effects need to be inserted in order to find the smoothing of RR at the level global and local.

If the observed number of DF cases in area *I*, called y_i and assumed to follow a Poisson distribution, then it can be written as follows:

$$y_i \sim \text{Poisson}\,(e_i\,\theta_i\,),$$

where $e_i\,\theta_i$ is the mean of Poisson distribution. The expected number of cases in the ith geographic unit is denoted by e_i. Meanwhile, the unknown relative risk in area *i* that is θ_i. This θ_i is need to be estimated. There are three componens of the variability of the log relative risk log θ_i. It can be written as log $\theta_i = \alpha + u_i + v_i$, where an overall level of the relative risk, called α. The correlated heterogeneity or spatial random effect is denoted by u_i and the uncorrelated heterogeneity was represented by random effect v_i.

Before applying the Bayesian modelling, the prior distribution for each components needs to be specified. The prior of uncorrelated heterogeneity, v_i does not depend on geographic location and it follows a normal distribution, with zero mean and a common variance τ_v^2 (precision parameter) as follows:

$$v_i \sim N(0, \tau_v^2)$$

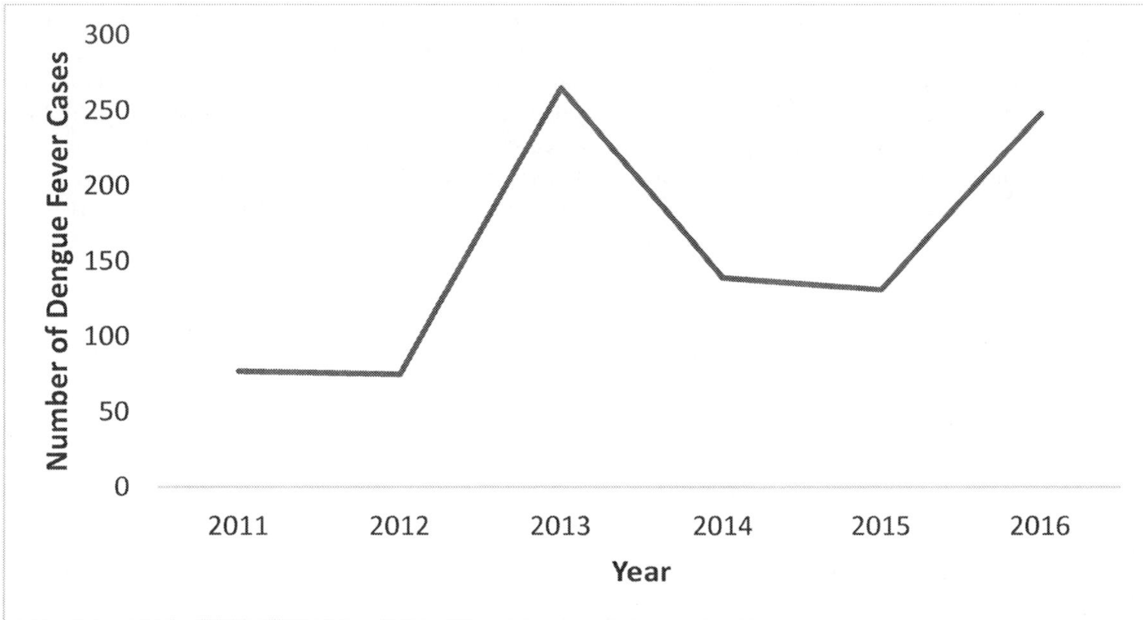

Figure 2. Total Number of DF Cases from 2011 to 2016 in All Sub-districrs in Makassar, Indonesia

The spatially structured heterogeneity is used for the clustering component. It means that when the risk in any area is estimated, the neighbouring areas is need to be considered. Lastly, the average value of its neighbouring areas influenced by a value of a parameter in one area is need to be fitted. This average value is called as the conditional autoregressive (CAR). In this spatial structure model, there is an additional variability, that is a conditional variance that depends on the number of neighbours. This is can be written as:

$$[u_i | u_j, i \neq j, \tau_u^2] \sim N(\bar{u}_i, \tau_i^2),$$

where $\bar{u}_i$ is the areas bordering area *i*, that is $\bar{u}_i = \frac{1}{\sum_j \omega_{ij}} \sum_j u_j \omega_{ij}$ and $\tau_i^2 = \frac{\tau_u^2}{\sum_j \omega_{ij}}$, where the relationship between the area *i* and *j* can be presented as ω_{ij}. As $\omega_{ij} = 1$ and *i*, *j* are adjacent, then the first order binary weighthing occurs. Meanwhile, $\omega_{ij} = 0$ then *i*, *j* are not adjacent. When the other u_j has a weighted average and $i \neq j$ is the prior mean of each u_i, then, the amount of variability of random effects v and u is controlled by the precision parameter τ_v^2 and τ_u^2.

The prior distributions for each precision parameters τ_v^2 and τ_u^2 need to be specified as well. If there are not prior estimation for precisions of the random effects, a large variance is considered. The prior distribution for τ_v^2 and τ_u^2 is the less informative and are followed by gamma distributions, respectively

[11]. In this case, the prior information is dominated by the likelihood data. As a result, the inference of relative risks is influenced by the prior minimumly.

To estimate the relative risk of the DF disease in Makassar city, the model of BYM was then be used. In this model, the Markov Chain Monte Carlo (MCMC) methods is implemented. The model for each data set including the weights was fitted using R version 3.3.3. In this paper, the relative risk of DF disease was categorised based on the number of cases, the five different levels of risks used to categorize the relative risk values were very low, low, medium, high and very high with the intervals of (<0.5, [0.5-1.0), [1.0-1.5), [1.5-2.0) and >2.0), respectively. The darkest shade in the map showing the area is in very high risk and the lightest shade region is considered as very low risk area [12].

3. Results

In this section, the result of relative risk estimation based on the application of Besag-York- Mollie (BYM) model which was applied on the observed number cases of DF disease transmission in Makassar, Indonesia is displays in the form of graphs, table and maps. Figure 3 depicts the time series plot of DF disease based on the number of cases for each sub-district in Makassar, Indonesia from 2011 to 2016. From the plot, Manggala had the highest number of cases. Among these years, most of the cases were above 1for all sub-district. However, the number of cases in Ujung tanah is zero for 2014 and increased again in 2015 with 2 cases. While for Wajo, Bontoala, and Ujung tanah, most had less than 10 cases throughout the years.

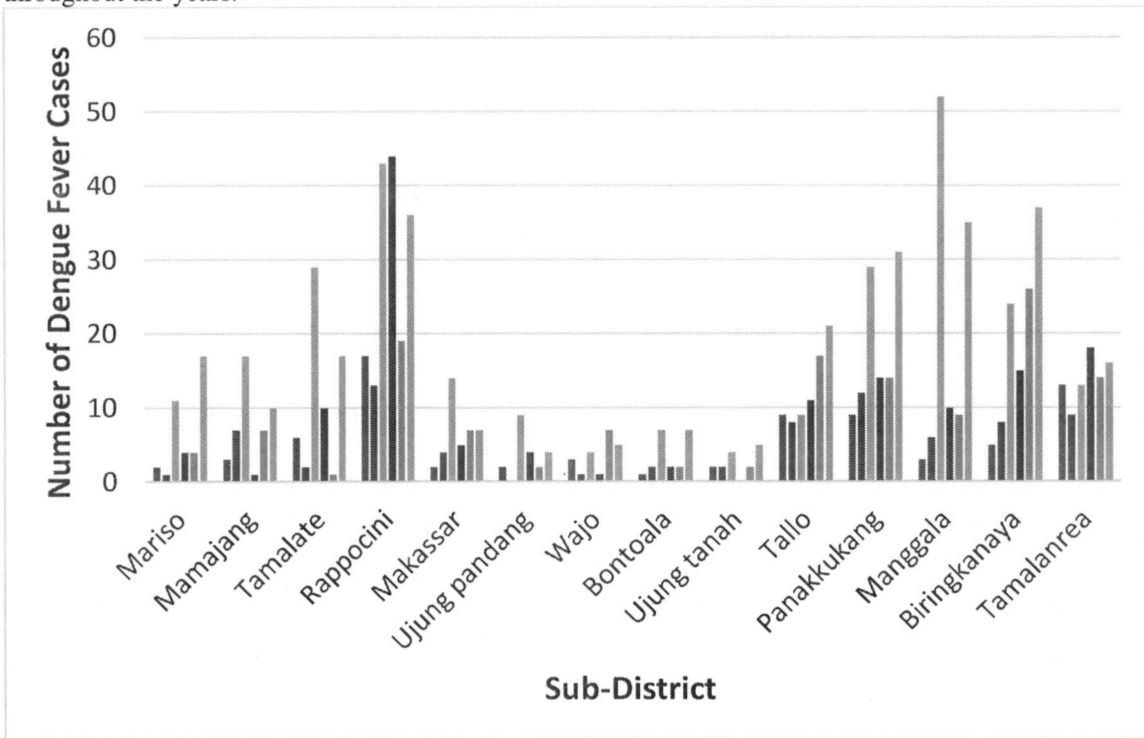

Figure 3. Time Series Plot based on Number of Cases for Each Sub-district in Makassar, Indonesia from 2011 to 2016.

Tabel 1. Relative risk of DF disease for each sub-district in Makassar city, 2011 to 2016.

Sub-district	Relative risk					
	2011	2012	2013	2014	2015	2016
Ujung Tanah	0.9895666	0.9964079	0.7715359	0.4733532	0.8303137	0.9374488
Ujung Pandang	1.0207386	0.9909359	1.1693975	1.1588194	0.9901662	0.9700467
Bontoala	0.9719477	0.9975187	0.8135825	0.6156551	0.8286892	0.9510518
Mariso	0.9762723	0.9780361	1.0278355	0.7864179	0.8603418	1.1001221
Makassar	0.964145	1.0038125	0.9741512	0.7147738	0.9581804	0.9041086
Mamajang	0.9949323	1.0272643	1.2995752	0.5103333	1.050105	0.9817329
Wajo	1.0366089	0.9961071	0.8711148	0.6730201	1.4514642	0.9780208
Manggala	0.9587311	1.0114916	1.9197258	0.8148877	0.8821492	1.1533659
Tamalate	0.9534338	0.9523744	0.9125008	0.6248958	0.4962983	0.8517659
Panakukkang	1.0216283	1.0363048	1.0403053	0.9672972	1.0350458	1.0642859
Tamalanrea	1.1286118	1.0358617	0.7506715	1.5090788	1.2269703	0.9784727
Tallo	1.0303018	1.01267	0.5493764	0.84027	1.2046229	0.97006
Rappocini	1.1286118	1.0342938	1.3905733	2.5542036	1.1310579	1.0796329
Biring Kanaya	0.9363576	1.0012052	1.6761441	0.8280359	1.13393381	1.0396603

Table 1 presented the estimated relative risk based on BYM model for DF disease in all sub-districts of Makassar city in year 2011 to 2016. According to Table 1, Rappocini had the highest relative risk (above 2) in 2014. Manggala almost had relative risks above 2 in 2013 and Biring kanaya had high relative risk in 2014. These indicated that people live in these states are more likely to be infected by Dengue fever disease compared to people in overall population. However, in 2014 and 2015, Ujung tanah and Tamalate had the very low relative risk of dengue fever. These mean that people in these states are less likely to contract with dengue fever disease compared to people in overall population.

Based on the results of relative risk estimation based on the application of Besag-York- Mollie (BYM) model in Table 1, the risk map is constructed. The DF risk maps cluster and identify the regions with different levels of risks displays in Figure 4. This figure depicts that Manggala sub-district were in very high risk for infective DF cases in 2013 and 2016. The figure also showed Tamalanrea as a high risk area in 2011, followed by Rappocini as a medium risk and Panakukang, Tallo, Ujung Tanah and Wajo as a low risk area. Other states were considered as very low risk areas in 2011. The sub-districts Panakukang, Rappocine and Wajo reached a very high risk for infective DF cases in 2012, 2014 and 2015, respectively. From the map, interestingly, Tamalate sub-district have a low risk for infective DF cases during 2011-2016. Because of the total number of population in every sub-district are different in the calculation of expected cases, then Figures 4 give different levels of risks result as well.

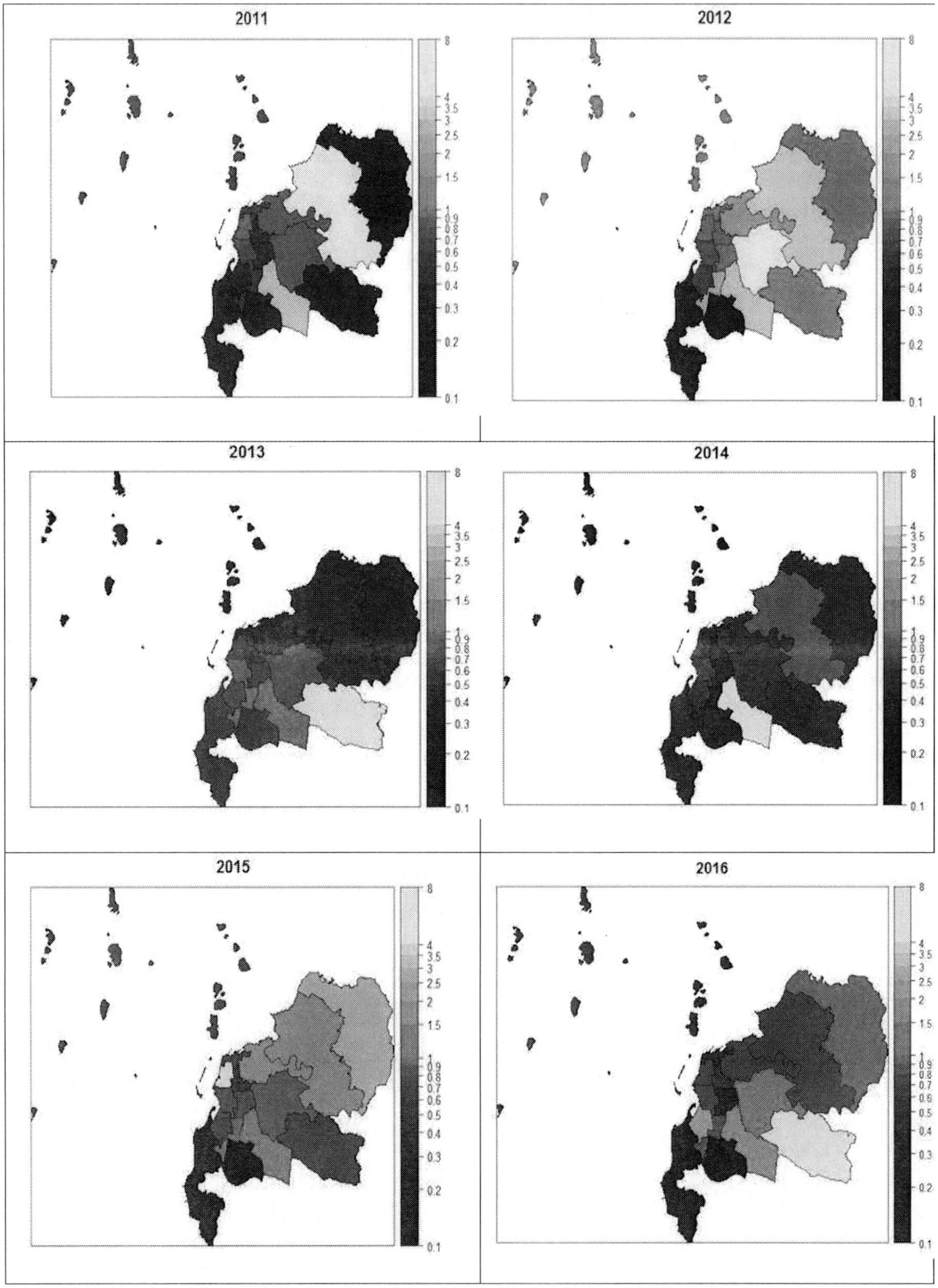

Figure 4. Risk Map for DF based on BYM Model for Year 2011 to 2016 in Makassar city, Indonesia.

4. Conclusion and Future Works

In this paper, BYM was used to estimate the relative risk of DF disease. In the risk geographical map, the distribution of high and low risk areas of DF occurrences for all sub-districts in Makassar City can be identified in order to allow the better allocation and risk assessment. The sub-districts which is vulnerable to the transmission of DF in Makassar, suggesting these sub-districts need to be pay more attention by both the city government and local communities. The BYM model can be easily applied in public health. Nevertheless, further improvement must be done to improve the current model and or compared to the others methods to increase the accuracy of the risk map. It could be made for more detailed comparisons by including more confounding factors.

Acknowledgement

This work was supported by the PDUPT (Penelitian Dasar Unggulan Perguruan Tinggi) Grant number 1623/UN4.21/PL.00.00/2018 funded by the Indonesian Government (Ristek-Dikti), Indonesia.

References

[1] World Health Organization 2012 *Handbook for Clinical Management of Dengue* Switzerland.
[2] Chen K, Pohan HT and Sinto R 2009 *MEDICINUS – Scientific Journal of Pharmaceutical Development and Medical Application* **22**(1) 3-7.
[3] Keman S 2007 *Jurnal Kesehatan Lingkungan* **3**(2) 195-204.
[4] Climate Data 2018 *Iklim Makassar* https://id.climate-data.org/asia/indonesia/south-sulawesi/makassar-3646/ [Accessed online 29 October 2018].
[5] Irawan R, Yong B and Kristiani F 2015 *Penentuan Risiko Relatif untuk Penyebaran Penyakit Demam Dengue di Kota Makassar pada Tahun 2013 dengan Menggunakan Model SMR* Prosiding Seminar Nasional Matematika UNPAR, 10.
[6] Kristiani F, Yong B and Irawan R 2016 *Jurnal Teknologi* **78**(11) 57-64.
[7] Wake eld J and Elliot P 1999 *Statistics in Medicine* **18** 2377-2399.
[8] Clayton D and Kaldor J 1987 *Biometrics* 43(3) 671-681.
[9] Besag J, York JC and Molli ́EA 1991 *Annals of the Institute Statistical Mathematics* **43** 1–59.
[10] Samat NA and Mey LW 2017 *Journal of physics: conference series* **890**(012167) 1-8.
[11] Bernardinelli L, Clayton D, Pascutto C, Montomoli C, Ghislandi M and Songini M 1995 *Statistics in Medicine* **14**(21) 2433-2443.
[12] Samat NA and Maarof SHMI 2013 *International Journal of Mathematical, Computational, Natural, and Physical Engineering* **7**(8) 785-789.

Optimal harvesting of prey-predator fishery modeling in a two patch environment and harvesting in unprotected area

S Toaha [1,a] and Kasbawati [1,b]

[1] Department of Mathematics, Hasanuddin University,
Jln. Perintis Kemerdekaan, KM 10, 90245, Makassar, Indonesia

[a] syamsuddint@gmail.com, [b] kasbawati@gmail.com

Abstract. This article deals with the dynamics of prey and predator populations in a two patch environment, a protected area and an unprotected area for fishing. The prey disperses between the two patches and migrates easily. There are two predators, one is in the protected area and another is in the unprotected area. The predators cannot migrate. Both prey and predator in unprotected area are harvested with constant efforts. The dynamical behavior of the populations is stated as a system of differential equations. The existence of a positive equilibrium point and its stability are investigated. We discuss the local stability of the positive equilibrium point. The stable equilibrium point is then associated with optimal harvesting problems. Based on the analysis, we found that there exist a stable positive equilibrium point when there is no harvesting. For model with constant efforts for both prey and predator, we found that over fishing will maximize the profit but the predator in the unprotected area will be extinct. With the help of Pontryagin's maximum principle in maximizing the present value of revenues, we found the extremal of the efforts that maximize the present value of revenues. This means that both prey and predator in the protected area as well as the prey and predator in the unprotected area are possibly coexist although the prey and the predator in the unprotected area are harvested with constant efforts. Some numerical simulations area given to confirm the result of analysis.

Keywords: Predator-prey, two patch environment, stability, protected area, present value of revenue

1. Introduction

The study of dynamical behavior of prey and predator populations does not only focus on how to control the population from the extinction. Fishery industries need some strategies to manage the population in order to get the optimal harvesting and the population will not lead to the extinction. In the exploitation activities, there are many things that need to be controlled so that the populations as valuable stocks can be well managed.

In the study of prey and predator models, some authors considered many factors in the models in the various perspectives. Some models with harvesting, migration, diffusion, stage structure, or harvesting and tax, have been studied in details by many authors. Some researchers considered the models with one prey and one predator, two preys and one predator, or prey with stage structure and one predator, with and without harvesting.

The prey-predator model with stage structure for predator in a two patch environment with harvesting in the unprotected area was discussed in [1] and determined a certain conditions to get an optimal harvesting. A model with harvesting in reserve area was considered and inserted tax as a control to prevent over explotation of the population [2] and harvesting to get the maximum value of present value of revenues [3,4].

Becuase of the complexity of dynamics of prey and predator fisheries, some researchers considered the selective harvesting in the models. Those who have considered selective harvesting of the prey, for examples in [2, 5, 6]. The studies of prey and predator models with selective harvesting of the predator, for examples in [7, 8, 9, 10]. The studies of prey and predator models with both prey and predator are harvested, for examples in [4, 11]. The dynamics of prey and predator fishery models in a two patch environment, protected area and unprotected area for any kind of fishing, with harvesting in the unprotected area and problems of maximizing the profit as well as present value of revenues have been discussed in detail by many authors, for examples in [1, 2, 3, 12, 13, 14, 15].

In this article, we consider the dynamical behavior of prey and predator populations in a two patch environment, namely a protected area and an unprotected area for fishing. The prey disperses and migrates in the two areas. There are two kinds of predators, one is in the protected area and the other is in the unprotected area. We suppose the predators can not migrate. The model includes four differential equations. Both prey and predator populations in the unprotected area are then harvested with constant efforts. We analyse the existence and the stability of the positive equilibrium point. The stable equilibrium point is then associated with the problem of maximizing profit as well as present value of revenues. The critical points of the efforts are determined in order to get the maximum profit as well as maximum present value of revenues. The Routh-Hurwitz criteria are referred to determine the stability of the equilibrium point. With the help of the Pontryagin's maximum principle, the optimal harvesting policy of present value of revenues can be done.

2. The dynamics of prey and predator in a two patch environment

We consider a prey-predator fishery management in a two patch environment, namely an unprotected area for fishing and another is a protected area for any kind of fishing. The prey population disperses in the two areas and can migrate easily from one area to another area. There is a predator preys the prey in the unprotected area and there exists also another predator preys the prey in the protected area. The two patches of environment are supposed to have the same characteristics. The growth rate of the prey is assumed to be logistic. The dynamical behavior of prey and predator populations is constituted as a system of four differential equations

$$\begin{aligned}
\frac{dx}{dt} &= rx\left(1-\frac{x}{K}\right)-\tau_1 x+\tau_2 y-\alpha_1 xz \\
\frac{dy}{dt} &= sy\left(1-\frac{y}{L}\right)+\tau_1 x-\tau_2 y-\alpha_2 yw \\
\frac{dz}{dt} &= \beta_1\alpha_1 xz-k_1 z \\
\frac{dw}{dt} &= \beta_2\alpha_2 yw-k_2 w.
\end{aligned} \tag{1}$$

The symbols $x=x(t)$ and $y=y(t)$ constitute the size of preys population in the unprotected area and in the protected area at time *t*, respectively. The symbols $z=z(t)$ and $w=w(t)$ constitute the size of predators population in the unprotected area and in the protected area at time *t*. Parameters r and s state the intrinsic growth rate for the preys in the unprotected area and in the protected area. Parameters K and L state the carrying capacity of the environment for the preys in the unprotected area and in the protected area. Parameters α_1 and α_1 measure the intensity interaction between the prey and predator in the unprotected area and in the protected area. The values of β_1 and β_2 which take the value from zero to one measure the effect of predation to the predator in the unprotected area and in the protected area. Parameter τ_1 defines the migration rate from the prey in the unprotected area to the prey in the protected area and parameter τ_2 defines otherwise. Parameters k_1 and k_2 denote the mortality rate for the predators in the unprotected area and in the protected area. All parameters of the model are supposed to be positive. Since the model constitutes the dynamics of the prey and predator populations then we just consider the value of populations as $x(t)\geq 0$, $y(t)\geq 0$, $z(t)\geq 0$, and $w(t)\geq 0$.

Under consideration that the prey and predator populations are valuable stocks, the prey and predator in the unprotected area are exploited with constant efforts. The dynamical behavior of the prey and predator populations is then extended and written in the form of

$$\begin{aligned}\frac{dx}{dt}&=rx\left(1-\frac{x}{K}\right)-\tau_1x+\tau_2y-\alpha_1xz-q_1E_1x\\ \frac{dy}{dt}&=sy\left(1-\frac{y}{L}\right)+\tau_1x-\tau_2y-\alpha_2yw\\ \frac{dz}{dt}&=\beta_1\alpha_1xz-k_1z-q_2E_2z\\ \frac{dw}{dt}&=\beta_2\alpha_2yw-k_2w.\end{aligned}\qquad(2)$$

In the model (2), parameters q_1 and q_2 state the cathability coefficient for the prey and predator populations respectively. Parameters E_1 and E_2 state the constant efforts of harvesting which satisfy $0\leq E_i\leq E_{i\max}$ for $i=1,2$ and a certain value of $E_{i\max}$. There are five equilibrium points suitable to be analysed, namely $(0,0,0,0)$, $(x_+, y_+, 0, 0)$, $(x_+, y_+, z_+, 0)$, (x_+, y_+, z_+, w_+), and $(x_+, y_+, 0, w_+)$. There is only one possibe positive equilibrium point for the model (2), as well as for model (1). Here, only the positive equilibrium point will be analysed. The possible positive equilibrium point for model (2) is written as $EQ=(x_1, y_1, z_1, w_1)$, where $x_1=\dfrac{k_1+q_2E_2}{\alpha_1\beta_1}$, $y_1=\dfrac{k_2}{\alpha_2\beta_2}$, $z_1=\dfrac{rx_1K-rx_1^2-\tau_1x_1K+\tau_2y_1K-q_1E_1x_1K}{\alpha_1x_1K}$, and $w_1=\dfrac{sy_1L-sy_1^2+\tau_1x_1L-\tau_2y_1L}{\alpha_2y_1L}$. The equilibrium point $EQ=(x_1, y_1, z_1, w_1)$ become a positive equilibrium point when $rx_1K-rx_1^2-\tau_1x_1K+\tau_2y_1K-q_1E_1x_1K>0$ and $sy_1L-sy_1^2+\tau_1x_1L-\tau_2y_1L>0$.

From the first variation of model (2), we get the Jacobian matrix as

$$A=\begin{pmatrix} r-\frac{2rx}{K}-\tau_1-\alpha_1z-q_1E_1 & \tau_2 & -\alpha_1x & 0\\ \tau_1 & s-\frac{2sy}{L}-\tau_2-\alpha_2w & 0 & -\alpha_2y\\ \alpha_1\beta_1z & 0 & \alpha_1\beta_1x-k_1-q_2E_2 & 0\\ 0 & \alpha_2\beta_2w & 0 & \alpha_2\beta_2y-k_2\end{pmatrix}.$$

After substituting the equilibrium point $EQ=(x_1, y_1, z_1, w_1)$ into the Jacobian matrix A we get

$$A_E=\begin{pmatrix} d_1 & \tau_2 & -d_2 & 0\\ \tau_1 & d_3 & 0 & -d_4\\ d_5 & 0 & d_6 & 0\\ 0 & d_7 & 0 & d_8\end{pmatrix},$$

where $d_1 = r - \frac{2rx_1}{K} - \tau_1 - \alpha_1 z_1 - q_1 E_1$, $d_2 = \alpha x_1$, $d_3 = s - \frac{2sy_1}{L} - \tau_2 - \alpha_2 w_1$, $d_4 = \alpha_2 y_1$, $d_5 = \alpha_1 \beta_1 z_1$, $d_6 = \alpha_1 \beta_1 x_1 - k_1 - q_2 E_2$, $d_7 = \alpha_2 \beta_2 w_2$, and $d_8 = \alpha_2 \beta_2 y_2 - k_2$.

The characteristic equation associated with the Jacobian matrix A_E is given by $f(\lambda) = \det(\lambda I - A_E)$, i.e. $f(\lambda) = \lambda^4 + b_3\lambda^3 + b_2\lambda^2 + b_1\lambda + b_0$, where
$b_3 = -(d_1 + d_3 + d_6 + d_8)$, $b_2 = (d_3 + d_6 + d_8)d_1 + d_2 d_5 + d_3 d_6 - \tau_1\tau_2 + d_4 d_7 + d_3 d_8 + d_6 d_8$,
$b_1 = (-d_3 d_6 - d_3 d_8 - d_4 d_7 - d_6 d_8)d_1 + (\tau_1\tau_2 - d_3 d_8 - d_4 d_7)d_6 + (\tau_1\tau_2 - d_2 d_5)d_8 - d_2 d_3 d_5$,
and $b_0 = (d_3 d_6 d_8 + d_4 d_6 d_7)d_1 + d_2 d_4 d_5 d_7 - \tau_1\tau_2 d_6 d_8 + d_2 d_3 d_5 d_8$.

The equilibrium point $EQ = (x_1, y_1, z_1, w_1)$ is locally asymptotically stable when the Routh-Hurwitz stability test [16], i.e. $b_0 > 0$, $b_1 > 0$, $b_2 > 0$, $b_3 > 0$, and $b_3 b_2 b_1 - b_1^2 - b_0 b_3^2 > 0$ are satisfied.

Example 1. Suppose the parameter values for model (1) are given as $r = 1.5$, $s = 1.5$, $\tau_1 = 0.25$, $\tau_2 = 0.50$, $\beta_1 = 0.0511$, $\beta_2 = 0.0512$, $K = 1000$, $\alpha_1 = 0.0521$, $\alpha_2 = 0.0522$, $L = 1000$, $k_1 = 0.51$, and $k_2 = 0.52$ in appropriate units. Then we have the positive equilibrium point $EQ = (191.56297, 194.56418, 28.22436, 18.28155)$. The characteristic equation associated with Jacobian matrix evaluated at the equilibrium point EQ is $f(\lambda) = \lambda^4 + 1.33316\lambda^3 + 1.54898\lambda^2 + 0.79806\lambda + 0.37215$. From the characteristic equation, it is easy to check that the Routh-Hurwitz criterion follows and the eigen values are $-0.50382 \pm 0.6163i$ and $-0.16277 \pm 0.74886i$. All of real parts of the eigen values are negative, then the equilibrium point EQ is locally asymptotically stable. This means the two preys and the two predators will be sustainable for a long period of time.

3. Optimal harvesting policies

The stable positive equilibrium point EQ of the model (2) is associated with the maximum profit problem. Under consideration that both prey and predator populations in the unprotected area for fishing are economically valuable, then the populations in that area are harvested with constant efforts. Exploitation activities require cost function and gives revenue consequences. Then we define the total cost function as $TC = cE$, where c denotes the unit cost of exploitation and E is the constant harvesting effort. While the total revenue function is denoted as $TR = pY(E)$, where p is the unit price of population stock (N). The yield of exploitation is then denoted by $Y(E, N) = qEN$, where q is the cathability coefficient of the population. Therefore we get the profit function as $\pi = TR - TC$. Since the positive equilibrium point $EQ = (x_1, y_1, z_1, w_1)$ depends on the efforts then the profit function depends also on the efforts and then it is written as $\pi(E) = TR(E) - TC(E)$.

In order to get the equilibrium point $EQ = (x_1, y_1, z_1, w_1)$ become a positive point, the values of harvesting efforts E_1 and E_2 have to follow the conditions (i) $rx_1 K - rx_1^2 - \tau_1 x_1 K + \tau_2 y_1 K - q_1 E_1 x_1 K > 0$ and (ii) $sy_1 L - sy_1^2 + \tau_1 x_1 L - \tau_2 y_1 L > 0$. Condition (i) is written as

$$E_1 \le f(E_2) = \frac{1}{-q_1 K \alpha_2 \beta_2 \beta_1 \alpha_1 (k_1 + q_2 E_2)} \Big\{ rk_1^2 \alpha_2 \beta_2 + rq_2^2 E_2^2 \alpha_2 \beta_2 + 2rk_1 \alpha_2 \beta_2 q_2 E_2 - rK\alpha_2\beta_2\beta_1\alpha_1 k_1 - rK\alpha_2\beta_2\beta_1\alpha_1 q_2 E_2 - K\alpha_1^2\beta_1^2 k_2 \tau_2 + \tau_1 K\alpha_2\beta_2\alpha_1\beta_1 k_1 + \tau_1 K\alpha_2\beta_2\alpha_1\beta_1 q_2 E_2 \Big\}.$$

Condition (ii) can be written as

$$E_2 > E_{2l} = \frac{-\beta_1\alpha_1 k_2 sL\beta_2\alpha_2 + \beta_1\alpha_1 k_2^2 s + \beta_1\alpha_1 k_2\tau_2 L\beta_2\alpha_2 - k_1\tau_1 L\beta_2^2\alpha_2^2}{q_2\tau_1 L\beta_2^2\alpha_2^2}.$$

The equilibrium point EQ becomes a positive point when $(E_1, E_2) \in D$, where $D = \{(E_1, E_2): 0 \le E_1 \le \min\{E_{1\max}, f(E_2)\}, \max\{0, E_{2l}\} \le E_2 \le E_{2\max}\}$.
The profit function from the exploitation of both prey and predator in the unprotected area at the equilibrium point $EQ = (x_1, y_1, z_1, w_1)$ is given by

$$\pi(E_1, E_2) = p_1 q_1 x_1 E_1 + p_2 q_2 z_1 E_2 - (c_1 E_1 + c_2 E_2). \tag{3}$$

The profit function (3) will be maximized in the feasible region of D for some values of $E_{1\max}$ and $E_{2\max}$. The critical values of the efforts E_1 and E_2 are determined by considering the first partial derivatives and the critical values of the efforts at the boundary of D.

The biological equilibrium is determined by evaluating the system of equations $\frac{dx}{dt} = 0$, $\frac{dy}{dt} = 0$, $\frac{dz}{dt} = 0$, and $\frac{dw}{dt} = 0$, simultaneously. The economic equilibrium is reached when the total revenue is the same with the total cost. The profit function for the exploited prey and predator populations is written as $\pi(E_1, E_2) = p_1 q_1 x E_1 + p_2 q_2 z E_2 - c_1 E_1 - c_2 E_2$. Our objective is to maximize the present value of net revenues for infinite horizon problem which is given by

$$J = \int_0^\infty e^{-\delta t}\{(p_1 q_1 x - c_1)E_1(t) + (p_2 q_2 z - c_2)E_2(t)\}\, dt. \tag{4}$$

The symbol δ states the discount rate of the net revenue. Our goal is to maximize the present value of J subject to the system of equations (2) with the help of the Pontryagin's maximum principle, as stated in [17]. The control variables $E_1(t)$ and $E_2(t)$ are subject to the conditions $0 \le E_i(t) \le E_{i\max}$ for $i = 1, 2$.

From the problem and improper integral equation (4) we write the Hamiltonian function as

$$\begin{aligned} H &= e^{-\delta t}\{(p_1 q_1 x - c_1)E_1 + (p_2 q_2 z - c_2)E_2\} + \lambda_1\left\{ rx - \frac{r}{K}x^2 - \tau_1 x + \tau_2 y - \alpha_1 xz - q_1 E_1 x \right\} \\ &+ \lambda_2\left\{ sy - \frac{s}{L}y^2 + \tau_1 x - \tau_2 y - \alpha_2 yw \right\} + \lambda_3\{\beta_1\alpha_1 xz - k_1 z - q_2 E_2 z\} + \lambda_4\{\beta_2\alpha_2 yw - k_2 w\}, \end{aligned} \tag{5}$$

where the variables $\lambda_1(t)$, $\lambda_2(t)$, $\lambda_3(t)$, and $\lambda_4(t)$ denote the adjoints of the problem. We set $\frac{\partial H}{\partial E_1} = 0$ and $\frac{\partial H}{\partial E_2} = 0$ as the necessary conditions for the control variables E_1 and E_2 to be optimal. From the Hamiltonian equation (5), we have $\frac{\partial H}{\partial E_1} = e^{-\delta t}(p_1 q_1 x - c_1) - \lambda_1 q_1 x = 0$ and $\frac{\partial H}{\partial E_2} = e^{-\delta t}(p_2 q_2 z - c_2) - \lambda_3 q_2 z = 0$. Then we get $\lambda_1 = \frac{e^{-\delta t}(p_1 q_1 x - c_1)}{q_1 x}$ and $\lambda_3 = \frac{e^{-\delta t}(p_2 q_2 z - c_2)}{q_2 z}$.

From the Pontryagin's maximum principle $\dot{\lambda}_1 = -\frac{\partial H}{\partial x}$, $\dot{\lambda}_2 = -\frac{\partial H}{\partial y}$, $\dot{\lambda}_3 = -\frac{\partial H}{\partial z}$, and $\dot{\lambda}_4 = -\frac{\partial H}{\partial w}$ and adjoint variables $\lambda_1(t) = \frac{e^{-\delta t}(p_1 q_1 x - c_1)}{q_1 x}$ and $\lambda_3(t) = \frac{e^{-\delta t}(p_2 q_2 z - c_2)}{q_2 z}$ then solving them to find the remains adjoint variables $\lambda_2(t)$ and $\lambda_4(t)$. The adjoint variables $\lambda_2(t)$ and $\lambda_4(t)$ are determined by solving the system of differential equations $\dot{\lambda}_2+\left(s-\frac{2sy}{L}-\tau_2-\alpha_2 w\right)\lambda_2+\beta_2\alpha_2 w\lambda_4=-\tau_2\lambda_1$ and $\dot{\lambda}_4+\left(\beta_2\alpha_2 y-k_2\right)\lambda_4-\alpha_2 y\lambda_2=0$ which satisfy the transversality conditions $\lambda_2(t)=0$ and $\lambda_4(t)=0$, as t tends to infinity.

Substitute $\lambda_1(t)$, $\lambda_2(t)$, $\lambda_3(t)$, and $\lambda_4(t)$ into the system of equations $\dot{\lambda}_1=-\frac{\partial H}{\partial x}$ and $\dot{\lambda}_3=-\frac{\partial H}{\partial z}$ and solving them to get E_1 and E_2. Now, the control variables E_1 and E_2 still depend on x, y, z, and w, i.e., $E_1=E_1(x,y,z,w)$ and $E_2=E_2(x,y,z,w)$. By substituting $x=x_1=\frac{k_1+q_2E_2}{\alpha_1\beta_1}$, $y=y_1=\frac{k_2}{\alpha_2\beta_2}$, $z=z_1=\frac{rx_1K-rx_1^2-\tau_1x_1K+\tau_2y_1K-q_1E_1x_1K}{\alpha_1x_1K}$, and $w=w_1=\frac{sy_1L-sy_1^2+\tau_1x_1L-\tau_2y_1L}{\alpha_2y_1L}$ we possibly get the suitable values of control variables E_1 and E_2. The values of E_1, E_2, x_1, and z_1 maximize the present value of revenues J.

Example 2. Suppose the parameter values for model (2) are given as $r=3.5$, $s=3.5$, $\tau_1=0.25$, $\tau_2=0.50$, $\beta_1=0.1511$, $\beta_2=0.1512$, $K=100000$, $\alpha_1=0.1221$, $\alpha_2=0.1222$, $L=1000$, $k_1=0.61$, $k_2=0.62$, $q_1=0.5$, and $q_2=0.5$ in appropriate units. Take $p_1=10$, $p_2=12$, $c_1=5$, and $c_2=6$. Then we get equilibrium point $EQ=(x_1,y_1,z_1,w_1)$, where $x_1=33.06357+27.10128E_2$, $y_1=33.55585$,

$$z_1=\frac{7.7205\cdot10^{16}+5.8678\cdot10^{16}E_2-1.7138\cdot10^{13}E_2^2-1.1021\cdot10^{16}E_1-9.0338\cdot10^{13}E_1E_2}{4.5132\cdot10^{16}+3.6993\cdot10^{16}E_2},$$

and $w_1=28.60195+1.652304E_2$.

Suppose that $E_{1\max}=1$ and $E_{2\max}=1$, then we get the feasible region of $D=\{(E_1,E_2):0\le E_1\le\min\{1,f(E_2)\},\max\{0,E_{2l}\}\le E_2\le1\}$. After doing a few calculations we get the value of E_{2l} as $E_{2l}=-17.31034$ and

$$f(E_2)=\frac{0.02600\left(-62181E_2^2+2.12899\cdot10^8E_2+2.80121\cdot10^8\right)}{8.52203\cdot10^5E_2+1.039688\cdot10^6}.$$

It is easy to check that $f(E_2)\ge6.77465$ for $E_2\in[0,1]$. Then the region of D becomes $D=\{(E_1,E_2):0\le E_1\le1,0\le E_2\le1\}$. The profit function is now written as

$$\pi(E_1,E_2)=\frac{5.000\cdot10^{-10}}{2.018\cdot10^6+1.654\cdot10^6E_2}\Big\{6.472\cdot10^{17}E_1+9.783\cdot10^{17}E_1E_2+3.671E_1E_2^2+6.706\cdot10^{17}E_2+5.082\cdot10^{17}E_2^2-1.542\cdot10^{14}E_2^3\Big\}.$$

The critical values of the efforts E_1 and E_2 are determined by considering the first partial derivatives and the values at the boundary of D and we get a pair of the critical value of efforts

$\left(E_1^*, E_2^*\right) = (1, 1)$ that maximizes the profit function with the value of $\pi(1,1) = 431.7077$. The value of efforts $E_1 = 1$ and $E_2 = 1$ are at the maximum level. The value pairs of the efforts lies at the boundary of D. The critical value $\left(E_1^*, E_2^*\right) = (1, 1)$ gives a positive equilibrium point $EQ = (60.165, 33.556, 23.647, 30.254)$. The eigen values associated with the positive equilibrium point are $-0.03487 \pm 1.75043i$ and $-0.26061 \pm 1.52602i$. This means that the equilibrium point EQ is locally asymptotically stable. In this situation, the populations will coexist for a long period of time although the prey and the predator populations in the unprotected area for fishing are harvested at the maximum level of efforts.

Example 3. Suppose the parameter values for model (2) are given as $r = 1.5$, $s = 1.5$, $\tau_1 = 0.25$, $\tau_2 = 0.50$, $\beta_1 = 0.0511$, $\beta_2 = 0.0512$, $K = 1000$, $\alpha_1 = 0.0521$, $\alpha_2 = 0.0522$, $L = 1000$, $k_1 = 0.51$, $k_2 = 0.52$, $q_1 = 0.8$, and $q_2 = 0.5$ in appropriate units. Take $p_1 = 10$, $p_2 = 12$, $c_1 = 5$, $c_2 = 6$, and $\delta = 0.005$ in appropriate units. Then we have equilibrium point $EQ = (x_1, y_1, z_1, w_1)$, where $x_1 = 191.56296 + 186.80683E_2$, $y_1 = 194.56418$, $z_1 = \dfrac{0.01919\left(1250x_1 + 1.5x_1^2 - 97282.08810 + 800x_1E_1\right)}{x_1}$, and $w_1 = 18.28156 + 4.62294E_2$.

With the help of the Pontryagin's maximum principle and by considering the transversality condition for the infinite horizon problem we get $E_1 = 1.07984$ and $E_2 = 1.18462$. Then the positive equilibrium point is $EQ = (414.0431, 194.5642, 0.0003, 23.7580)$ with the eigen values -1.00773, $-0.33606 \pm 0.65943i$, and -0.00002. Under these conditions, the positive equilibrium point EQ is locally asymptotically stable. The adjoint variables are written as $\lambda_1 = 9.98490e^{-0.005t}$, $\lambda_2 = -0.03846\, e^{-0.005t}$, $\lambda_3 = -41653.55154\, e^{-0.005t}$, and $\lambda_4 = 78.12952\, e^{-0.005t}$. Then we get the maximum present value of the revenues $J = \int_0^\infty 3564.317259\, e^{-0.005t}\, dt = 7.128634518 \cdot 10^5$.

4. Conclusions

We have studied the dynamical behavior of prey and predator in a two patch environment. The model for dynamics of prey and predator without harvesting in the protected area for fishing as well as in the unprotected area for fishing possibly has a positive equilibrium point. Under a certain conditions of parameter values, the equilibrium point is locally asymptotically stable which means that the prey and predator populations can live in coexistence for a long period of time.

For the model with harvesting of prey and predator in the unprotected area for fishing, the stable equilibrium point is related to the problems of maximizing profit and present value of revenues. When the prey and predator are harvested at the maximum level of the efforts, the exploitation activity maximizes the profit function, but over exploitation will lead to the extinction for harvested predator. For the problem of maximizing the present value of revenues, there exists the extremal for the efforts maximizing the present value of revenues. Beside that, the prey and predator populations will remain coexistence for a long period of time.

Acknowledgements
This research is supported by Ministry of Research and Higher Education of Republic Indonesia via LP2M Hasanuddin University grant competition of Post Graduate Program Research Scheme with contract number: 1660/UN4.21/PL.00.00/2018.

References

[1] Srinivas M N, Srinivas M A S, Das K and Gazi N H 2011 Prey-predator fishery model with stage structure in two patchy marine aquatic environment *Applied Mathematics* **2** 1405-1416

[2] Hou H F, Jiang H M and Meng X Y 2012 A dynamic model for fishery resource with reserve area and taxation *Journal of Applied Mathematics* http://dx.doi.org/10.1155/2012/794719

[3] Yunfei L, Yuang R and Pei Y 2013 A prey-predator model with harvesting for fishery resource with reserve area *Applied Mathematical Modelling* **37**(5) 3048-3062

[4] Toaha S, Kusuma J, Khaeruddin and Mawardi 2014 Stabilty analysis and optimal harvesting policy of prey-predator model with stage structure for predator *Applied Mathematical Sciences* **8(**159) 7923-7934

[5] Kar T K 2010 A dynamic reaction model of a prey-predator system with stage-structure for predator *Modern Applied Science* **4**(5) 183-195

[6] Gupta R P, Banerjee M and Chandra P 2012 Bifurcation analysis and control of Leslie–Gower predator–prey model with Michaelis–Menten type prey-harvesting *Journal Differential Equations and Dynamical Systems* **22**(3) 339-366

[7] Qu Y and Wei J 2010 Bifurcation analysis in a predator-prey system with stage stucture and harvesting *Journal of Franklin Institute* **347**(7) 1096-1113

[8] Liu C and Tang W 2011 The dynamics and control of a harvested differential-algebraic prey-predator model *International Journal of Information and System Sciences* **7**(1) 103-113

[9] Chakraborty K, Jana S and Kar T 2012 Global dynamics and bifurcation in a stage structured prey-predator fishery model with harvesting *Applied Mathematics and Computation* **218**(5) 9271-9290

[10] Laksmi G M V, Gunasekaran M and Vijaya S 2014 Bifurcation analysis of predator model with harvested predator *International Journal of Engineering Research and Development* **10**(6) 42-51

[11] Chakraborty K, Chakraborty M and Kar T K 2011 Optimal control of harvest and bifurcation of prey-predator model with stage structure *Applied Mathematics and Computation* **217** 8778-8792

[12] Ghosh M 2010 Modeling prey-predator type fishery with reserve area *Int. J. Biomath.* **03**(03) 351-365

[13] Chakraborty K, Das K and Kar T K 2013 An ecological perspective on marine reserves in prey-predator dynamics *Journal of Biological Physics* **39**(4) 749-776

[14] Toaha S and Rustam 2017 Optimal harvesting policy of predator-prey model with free fishing and reserve zones AIP Conference Proceedings **1825**, 020023 (2017), https://doi.org/10.1063/1.4978992

[15] Yang H and Jia J 2017 Harvesting of a predator-prey model with reserve area for prey and in the presence of toxicity *Journal of Applied Mathematics and Computing* **53**(1-2) 693-708

[16] Keshet L E 2015 *Mathematical models in biology* SIAM Philadelphia 231-234

[17] Grass D, Caulkins J P, Feichtinger G, Tragler G and Behrens D A 2003 *Optimal control of nonlinear processes* Springer-Verlag Berlin Heidelberg 122-138

Optimal harvesting and stability of predator prey model with Monod-Haldane predation response function and stage structure for predator

R A Pratama [1,a], S Toaha [1,b,*] and Kasbawati [1,c]

[1] Department of Mathematics, Hasanuddin University,
Jln. Perintis Kemerdekaan, KM 10, 90245, Makassar, Indonesia

[a] rianadepratama@gmail.com, [b] syamsuddint@gmail.com,
[c] kasbawati@gmail.com, * corresponding author

Abstract. This article deals with growth rate predator prey population model with stage structure for predator. The predator population is divided into two stages, mature and immature predator. Predation functional response in the model follows the Monod-Haldane type. Under consideration that the population is a valuable stock, the mature predator and prey population are then harvested with constant effort. This model is then analysed by found the equilibrium point and stabilities. There exists four equilibrium points and stability analysed is focused only for interior point. The stable interior point is related to the maximum profit problem. From the analyses we found that there exists a condition for the effort of harvesting so that the interior point is still stable and also obtained maximum profit.

Keywords: Predator prey model, Monod-Haldane function, stage structure, optimal harvesting

1. Introduction

In study of population dynamics, the growth rate of predator prey is one model that widely studied by some researchers. When there are two or more populations living in the same environment, there will be interaction between the populations and the interaction that occur most often is predation, that is, one population as a predator and the other as a prey. In the case of frequencies of interaction between prey and predator very often, the prey population will be extinct. One strategy to protect the prey from extinction due to predation is by controlling the growth rate of prey and its predator, [1].

The dynamics of predator prey models have been studied and developed by many researchers. There are researchers who consider more than two populations in the model, some researchers consider economic aspects and other aspects in the model. Prey as well as predator can be considered into two stages, immature and mature populations. Regarding to the stage structured for predator prey model. Research on predator prey models with stage structure was also studied in [2] with stage structure in prey population. Study the predator pretty model with stage structure in [3] was found that the coexistence of both populations are strongly influenced by the population on the stage structure. Study of population dynamics of predator prey with stage structure and some extensions can be found, for example in [4, 5, 6].

In ecological modeling, the population as a useful stock is then exploited. The predator prey model with harvesting also become a special topic. The predator prey model with harvesting has been studied in [7] and model with stage structure and harvesting has been studied in [8]. In this article, we develop a model proposed in [3] by incorporating selective harvesting factor to the prey and mature predator population. We analyze the existence of the interior equilibrium point and its stability. The stable equilibrium point is then linked to how maximizing the profit from exploitation of the populations.

2. Predator prey model with harvesting and stage structure

A predator prey model with stage structure for predator in [3] has been studied. The model follows predation functions with Monod-Haldane type. In this model, mature predator population is assumed has interaction with prey population. Whereas for immature predator has not interaction directly with the prey. The immature predator has not reproductive abilities and it depends on the interaction of mature predator and prey population. The model is

$$\begin{aligned} \frac{dx}{dt} &= \rho_1 x\left(1-\frac{x}{k_1}\right)-\frac{\beta\, x\, z}{\phi+\tau\, x^2} \\ \frac{dy}{dt} &= \frac{c\,\beta\, x\, z}{\phi+\tau\, x^2}-(\alpha+\delta_1)\, y \\ \frac{dz}{dt} &= \alpha\, y-\delta_2\, z. \end{aligned} \tag{1}$$

The symbols $\frac{dx}{dt}, \frac{dy}{dt}$, and $\frac{dz}{dt}$ define the growth rate of prey, immature predator, and mature predator respectively. Constant ρ_1 is the intrinsic rate of prey population and k_1 is carrying capacity for the prey when there is no predator population. A simplification functional response of Monod Haldane is $\frac{\beta\, x}{\phi+\tau\, x^2}$, [9]. The symbols β and c state the rate of predation and efficiency of predation for growth rate of immature predator. The symbol α is the rate from immature predator becomes mature predator. The symbols δ_1 and δ_2 denote the rate mortality for immature and mature predator respectively.

Under consideration that the populations in the model are beneficial, then harvesting factor is incorporated in the growth rate of the population. We improve model (1) by considering selective harvesting in prey and mature predator populations. The model becomes

$$\begin{aligned} \frac{dx}{dt} &= \rho_1 x\left(1-\frac{x}{k_1}\right)-\frac{\beta\, x\, z}{\phi+\tau\, x^2}-q_1\, E_1\, x \\ \frac{dy}{dt} &= \frac{c\,\beta\, x\, z}{\phi+\tau\, x^2}-(\alpha+\delta_1)\, y \\ \frac{dz}{dt} &= \alpha\, y-\delta_2\, z-q_2\, E_2\, z. \end{aligned} \tag{2}$$

The symbols q_1 and q_2 denote catchability coefficient for prey and mature predator population respectively. The symbols E_1 and E_2 denote constant efforts of harvesting. There are four suitable equilibrium points to be analysed, namely adalah $T_0=(0,0,0)$, $T_1=(K,0,0)$, $T_2=(x_*,\ y_*,\ z_*)$, and $T_3=(x^*,\ y^*,\ z^*)$,

where $K=\frac{k_1(\rho_1-q_1E_1)}{\rho_1}$,

$$x_*=\frac{(\alpha c\beta)+\sqrt{A}}{2\,(\alpha+\delta_1)(q_2\,E_2+\delta_2)\tau},$$

$$y_*=\frac{c\,(\rho_1-q_1E_1)}{(\alpha+\delta_1)}\left(\frac{(\alpha c\beta)+\sqrt{A}}{2\,(\alpha+\delta_1)(q_2\,E_2+\delta_2)\tau}\right)$$

$$-\frac{\rho_1\, c\, \alpha}{((\alpha+\delta_1)(q_2\, E_2+\delta_2))\, k_1}\left(\frac{(\alpha c\beta)+\sqrt{A}}{2\,(\alpha+\delta_1)(q_2\, E_2+\delta_2)\tau}\right)^2,$$

$$z_* = \frac{c\,\alpha(\,\rho_1 - q_1 E_1)}{((\alpha+\delta_1)(q_2\, E_2+\delta_2))}\left(\frac{(\alpha c\beta)+\sqrt{A}}{2\,(\alpha+\delta_1)(q_2\, E_2+\delta_2)\tau}\right) -\frac{\rho_1\, c\, \alpha}{((\alpha+\delta_1)(q_2\, E_2+\delta_2))\, k_1}\left(\frac{(\alpha c\beta)+\sqrt{A}}{2\,(\alpha+\delta_1)(q_2\, E_2+\delta_2)\tau}\right)^2.$$

$$x^* = \frac{(\alpha c\beta)-\sqrt{A}}{2\,(\alpha+\delta_1)(q_2\, E_2+\delta_2)\tau},$$

$$y^* = \frac{c\,(\,\rho_1 - q_1 E_1)}{(\alpha+\delta_1)}\left(\frac{(\alpha c\beta)-\sqrt{A}}{2\,(\alpha+\delta_1)(q_2\, E_2+\delta_2)\tau}\right) -\frac{\rho_1\, c\, \alpha}{((\alpha+\delta_1)(q_2\, E_2+\delta_2))\, k_1}\left(\frac{(\alpha c\beta)-\sqrt{A}}{2\,(\alpha+\delta_1)(q_2\, E_2+\delta_2)\tau}\right)^2,$$

$$z^* = \frac{c\,\alpha(\,\rho_1 - q_1 E_1)}{((\alpha+\delta_1)(q_2\, E_2+\delta_2))}\left(\frac{(\alpha c\beta)-\sqrt{A}}{2\,(\alpha+\delta_1)(q_2\, E_2+\delta_2)\tau}\right) -\frac{\rho_1\, c\, \alpha}{((\alpha+\delta_1)(q_2\, E_2+\delta_2))\, k_1}\left(\frac{(\alpha c\beta)-\sqrt{A}}{2\,(\alpha+\delta_1)(q_2\, E_2+\delta_2)\tau}\right)^2,$$

and

$$A = (\alpha c\beta)^2 - 4\,(\alpha+\delta_1)(q_2\, E_2+\delta_2)\tau\,\phi\,(\alpha+\delta_1)(q_2\, E_2+\delta_2).$$

There is only one possibe positive equilibrium point for the model (2), as well as for model (1). Here, only the positive equilibrium point will be analysed. The possible positive equilibrium point and also stable for model (2) is written as $T_3 = (x^*,\ y^*,\ z^*)$. The Jacobian matrix for model evaluated at equilibrium point T_3 is given by

$$J = \begin{bmatrix} J_{11} & 0 & J_{13} \\ J_{21} & J_{22} & J_{23} \\ 0 & J_{32} & J_{33} \end{bmatrix},$$

where $J_{11} = \rho_1 - 2\frac{\rho_1 x^*}{k_1} - \frac{\beta\, z^*}{\phi+\tau\, x^{*2}} + \frac{2\,\beta\, z^*\,\tau\, x^{*2}}{(\phi+\tau\, x^{*2})^2} - q_1 E_1$, $J_{13} = -\frac{\beta\, x^*}{\phi+\tau\, x^{*2}}$, $J_{21} = \frac{c\,\beta\, z^*}{\phi+\tau\, x^{*2}} - \frac{2\, c\,\beta\, z^*\,\tau\, x^{*2}}{(\phi+\tau\, x^{*2})^2}$, $J_{22} = -\alpha - \delta_1$, $J_{23} = \frac{c\,\beta\, x^*}{\phi+\tau\, x^{*2}}$, $J_{32} = \alpha$, $J_{33} = -q_2 E_2 - \delta_2$

The characteristic equation associated with the Jacobian matrix J is given by

$$f(\lambda) = \lambda^3 + A_1\lambda^2 + A_2\lambda + A_3,$$

where $A_1 = -(J_{11} + J_{22} + J_{33})$, $A_2 = J_{11}J_{22} + J_{11}J_{33} + J_{22}J_{33} - J_{23}J_{32}$, and $A_3 = J_{11}J_{23}J_{32} - J_{11}J_{22}J_{33} - J_{13}J_{32}J_{21}$. According to Hurwitz stability test [10], the equilibrium point T_3 is asymptotically stable when the conditions $A_1 > 0$, $A_2 > 0$, $A_3 > 0$, and $A_1 A_2 > A_3$ are satisfied.

3. Maximum profit in population harvesting

The stable equilibrium point of the model is then related to maximizing the profit for exploitation of the populations. We suppose that the prey and mature predator population are harvested with constant effort. From the exploitation we consider revenue and cost function. We define total revenue as $TR = p_1\, x^*\, E_1 + p_2\, z^* E_2$ and total cost $TC = c_1 E_1 + c_2 E_2$. The Constants, while the constants c_1 and c_2 denote unit cost of harvesting for prey and mature predator respectively.

The equilibrium point T_3 becomes a positive equilibrium point when $(E_1, E_2)\ D$, where $D = \{(E_1, E_2)|0 \leq E_1 \leq E_{1max}, 0 \leq E_2 \leq E_{2max}\}$ for some value of E_{1max} and E_{2max}. The profit function for harvesting of prey and mature predator at equilibrium point from the exploitation at the equilibrium point T_3 is given by

$$\pi(E_1, E_2) = p_1 x^* E_1 + p_2 z^* E_2 - c_1 E_1 - c_2 E_2 . \quad (3)$$

We need to maximize the the profit function (3) in the feasible region of D for some values of E_{1max} and E_{2max}. The critical values of the efforts E_1 and E_2 are determined by considering the first partial derivatives and the critical values of the efforts at domain D. The profit function the equilibrium point T_3 is

$$\pi(E_1, E_2) = p_1 x^*\ E_1 + p_2 z^*\ E_2 - c_1 E_1 - c_2 E_2$$

$$= \frac{1}{2}\frac{p_1 E_1\left(\alpha c\beta - \sqrt{w\ E_2^2 + v\ E_2 + u}\right)}{t\ E_2\ +\ s}$$

$$+ \frac{p_2 E_2}{4\left(e_1\ E_2^3 + e_2\ E_2^2 + e_3\ E_2 + e_4\right)}\Big(a_1\ E_1\ E_2 +\ a_3\ E_2 - a_4\ E_2 + a_2 -\ 2\ a_5\ E_1\ E_2\sqrt{w\ E_2^2 + v\ E_2 + u} -$$

$$a_6\ - 2\ a_7\sqrt{w\ E_2^2 + v\ E_2 + u} - 2\ a_8\ E_1\sqrt{w\ E_2^2 + v\ E_2 + u} + 2\ a_9\ E_2\sqrt{w\ E_2^2 + v\ E_2 + u} +$$

$$2\ a_{10}\sqrt{w\ E_2^2 + v\ E_2 + u} + a_{11}\ E_2^2 + a_{12}\ E_2 + a_{13}\Big) - c_1 E_1 - c_2 E_2\ ,$$

where

$e_1 = \alpha\ t^2\ k_1 +\ t^2\ \delta_1\ k_1,\ e_2 = 2\ \alpha\ s\ t\ k_1 + \alpha\ t^2\ \delta_2\ k_1 + 2\ s\ t\ \delta_1\ k_1 + t^2\ \delta_1\ \delta_2\ k_1,$ $e_3 = \alpha\ s^2\ k_1 + 2\ \alpha\ s\ t\ \delta_2\ k_1 +\ s^2\ \delta_1\ k_1 + 2\ s\ t\ \delta_1\ \delta_2\ k_1,\ e_4 = \alpha\ s^2\ \delta_2\ k_1 + s^2 \delta_1\ \delta_2\ k_1,\ a_1 = 2\ \alpha^2\ c^2\ \beta\ \ t\ k_1,\ a_2 = \alpha^3 c^3 \beta^2 \rho_1,\ a_3 = 2\ \alpha^2\ c^2\ \beta\ s\ k_1,\ a_4 = 2\ \alpha^2\ c^2\ \beta\ t\ k_1\ \rho_1,\ a_5 = \alpha\ c\ t\ k_1,\ a_6 = 2\ \alpha^2\ c^2\ \beta\ s\ k_1\ \rho_1,\ a_7 = \alpha^2\ c^2\ \beta\ \rho_1,\ a_8 = \alpha\ c\ s\ k_1,\ \ a_9 = \alpha\ c\ t\ k_1\ \rho_1,\ a_{10} = \alpha\ c\ s\ k_1,\ a_{11} = w\ \alpha\ c\ \rho_1,\ a_{12} = v\ \alpha\ c\ \rho_1,\ a_{13} = u\ \alpha\ c\ \rho_1,\ \ s = \alpha\ \tau\ \delta_2 +\ \tau\ \delta_1\ \delta_2,\ t = \alpha\ \tau + \tau\ \delta_1,\ u = \alpha^2\ c^2\ \beta^2 - 4\ \alpha^2\ \phi\ \tau\ {\delta_2}^2 - 8\ \alpha\ \phi\ \tau\ \delta_1\ {\delta_2}^2 - 4\ \phi\ \tau\ {\delta_1}^2\ {\delta_2}^2,\ v = -\ 8\ \alpha^2\ \phi\ \ \tau\ \delta_2 - 16\ \alpha\ \phi\ \tau\ \delta_1\ \delta_2 - 8\ \phi\ \tau\ {\delta_1}^2 \delta_2,$ and $w = -\ 4\ \alpha^2\ \phi\ \ \tau - 8\ \alpha\ \phi\ \ \tau\ \delta_1 - 4\ \phi\ \tau\ {\delta_1}^2.$

4. Numerical simulations

For simulation on model (1) without harvesting, we take the parameters $\rho_1 = 1.5$, $k_1 = 100,\ \phi = 8,\ \beta = 0.16,\ c = 0.4,\ \alpha = 0.02, \tau = 0.001, \delta_1 = 0.01, \delta_2 = 0.2,\ E_1 = 0$ dan $E_2 = 0$ in appropriate units. We have positive equilibrium point $T_3 = (48.5480,\ 499.5783,\ 49.9578)$. The characteristic equation associated with Jacobian matrix evaluated at the equilibrium point is $f(\lambda) = \lambda^3 + 0.60696\ \lambda^2 + 0.86699\ \lambda + 0.25231$. From the characteristic equation, it is easy to check that the Routh-Hurwitz criterion follows and the eigen values are $\lambda_1 = -0.41089, \lambda_2 = -0.39129$, and $\lambda_3 = -0.15692$. All of eigen values are real and negative, then the equilibrium point T_3 is locally asymptotically stable. This means the prey and mature predator will sustain for a long period of time.

For simulation on model (2), we use the values of parameter $\rho_1 = 1.5$, $k_1 = 100,\ \phi = 8,\ \beta = 0.16,\ \ c = 0.4,\ \ \alpha = 0.02,\ \ \tau = 0.001,\ \ \delta_1 = 0.01,\ \delta_2 = 0.2,\ q_1 = 1,\ \ q_2 = 1$ in appropriate units. We get an interior equilibrium point $T_3 = (x^*, y^*, z^*)$, where

$$x^* = \frac{1}{2}\frac{0.00128 - \sqrt{A_1}}{0.000029\ E_2 + 0.0000058},$$

$$y^* = \frac{6.896551725\ (-E_1 + 1.5)\big(0.00128 - \sqrt{A_1}\big)}{0.000029\ E_2 + 0.0000058}$$

$$-\frac{0.001034482758\,\left(0.00128-\sqrt{A_1}\right)^2}{(E_2+0.2)(0.000029\,E_2+0.0000058)},$$

$$z^*=\frac{0.1379310345\,(-E_1+1.5)\left(0.00128-\sqrt{A}\right)}{(E_2+0.2)\,(0.000029\,E_2+0.0000058)}$$
$$-\frac{0.001034482758\,\left(0.00128-\sqrt{A_1}\right)^2}{(E_2+0.2)(0.000029\,E_2+0.0000058)},$$

where $A_1=-0.000026912\,E_2^2-0.0000107648\,E_2+0.00000056192.$
From equation (3) we get

$$\pi(E_1,E_2)=\frac{50\,E_1\left(0.00128-\sqrt{A_2}\right)}{0.00003\,E_2\,+\,0.000006}$$
$$-\frac{25\,E_2}{0.000000003\,E_2^3+0.0000000016\,E_2^2+0.00000000032\,E_2+0.000000000022}$$
$$\big(0.00000614\,E_1\,E_2+0.0000012288\,E_1-0.00000023\,E_2+0.00000000706$$
$$-\,0.000048\,E_1\,E_2\sqrt{A_2}$$
$$-\,0.00002112\sqrt{A_2}$$
$$-\,0.0000096\,E_1\sqrt{A_2}$$
$$+\,0.000072\,E_2\sqrt{A_2}+0.0000003456\,E_2^2\big)$$
$$-50\,(E_1+E_2)\,,$$

where $A_2=-0.0000288\,E_2^2-0.00001152\,E_2+0.0000004864.$

We consider the range of efforts as $D=\{(E_1,E_2)|0\le E_1\le 0.038, 0\le E_2\le 0.038\}$. The critical point of the efforts E_1 and E_2 are determined by using the first partial derivatives and the boundary of *D*. In domain *D* we get four critical points for $\pi(E_1,E_2)$, namely $P_1=(E_1,E_2)=(0.006485, 0.00372)$ as stationary point (saddle point), $P_2=(E_1,E_2)=(0, 0.02957)$, $P_3=(E_1,E_2)=(0.03800, 0)$, $P_4=(E_1,E_2)=(0.038, 0.03522)$, $P_5=(E_1,E_2)=(0.03800, 0.03800)$. Profit function which is evaluated at the critical point are given as follow;

1) $(E_1,E_2)=(0.00649, 0.00372)$ with profit $\pi=31.06066$.
2) $(E_1,E_2)=(0, 0.02956)$ with profit $\pi=148.87029$.
3) $(E_1,E_2)=(0.03800, 0)$ with profit $\pi=182.58245$.
4) $(E_1,E_2)=(0.03800, 0.03522)$ with profit $\pi=408.16992$, and
5) $(E_1,E_2)=(0.03800, 0.03800)$ with profit $\pi=398.33448$.

From these critical points we get profit maximum with profit $\pi=408.16992$. The equilibrium point associated with the critical point is $T_3=(75.6780, 329.7842, 28.0403)$. The characteristic equation from Jacobian matrix evaluated at this equilibrium point is $f(\lambda)=\lambda^3+1.12768\lambda^2+0.22874\,\lambda+0.000382$ which gives eigenvalues as $\lambda_1=-0.86319$, $\lambda_2=-0.26279$, and $\lambda_3=-0.00168$. This means that the equilibrium point T_3 is locally asymptotically stable. In this situation, the prey and mature predator populations will sustain for a long period of time.

For initial population around the equilibrium point T_3, we take $z(0)=45, y(0)=90$, and $z(0)=20$ we get trajectories for prey, immature, and mature predator with and without harvesting are given the following figures.

IOP Conf. Series: Earth and Environmental Science **279** (2019) 012015 doi:10.1088/1755-1315/279/1/012015

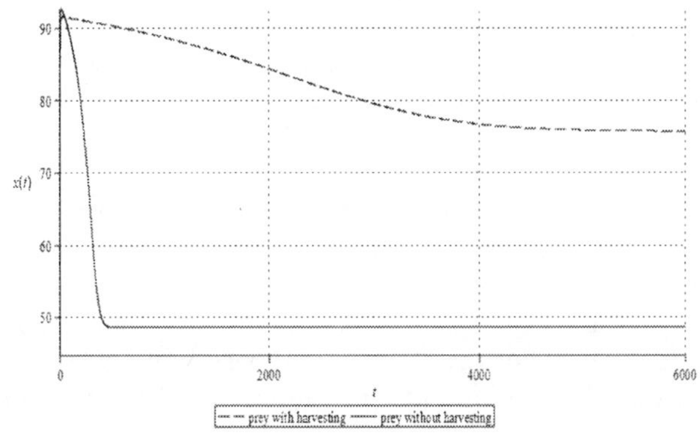

Figure 1. Trajectories for prey with and without harvesting

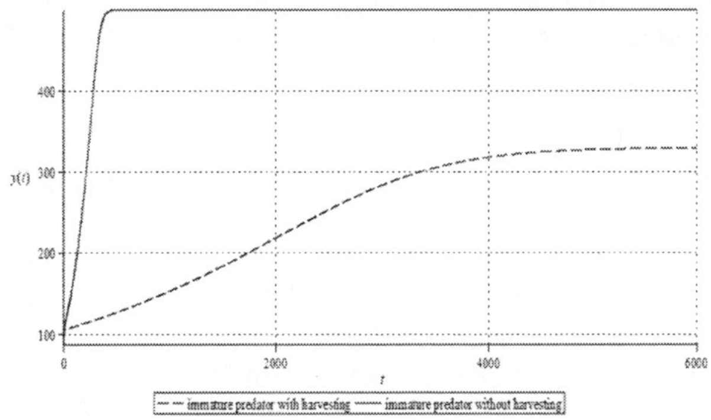

Figure 2. Trajectories for immature predator with and without harvesting

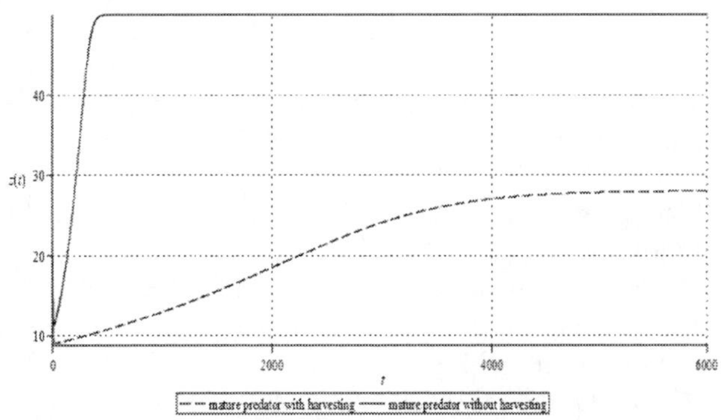

Figure 3. Trajectories for mature predator with and without harvesting

From Figure (1), (2), and (3) we know that all populations will sustain for a long time. However, there is a change in the value of the equilibrium point without harvesting $T_3 = (48.5480, 499.5783,\ 49.9578)$ and equilibrium point without harvesting $T_3 = (75.6780,\ 329.7842,\ 28.0403)$. The value of equilibrium point for prey and mature predator are decreasing but for immature predator is increasing in value.

5. Conclusions

One model of predator and prey populations with stage structure for predator population and Monod-Haldane functional response has been developed by adding harvesting factor to the prey and the mature predator populations. We found four equilibrium points, but only one is possible to become an interior equilibrium point. With suitable value of parameters, the interior equilibrium point without harvesting becomes locally asymptotically stable.

For the model with constant effort of harvesting, the stable equilibrium point is then linked to the problem of maximizing the profit function. We found a condition for the value of harvesting efforts that provides maximum profit and the interior equilibrium point remains stable. This means that the value of harvesting efforts keeps the prey and predator population sustainable and provides maximum profit for a long period of time.

Acknowledgements

This research is supported by Ministry of Research and Higher Education of Republic Indonesia via LP2M Hasanuddin University grant competition of Post Graduate Program Research Scheme with contract number: 1660/UN4.21/PL.00.00/2018.

References

[1] Li J 2014 Control schemes to reduce risk of extinction in the Lotka-Volterra predator prey model *Journal of Applied Mathematics and Physics* **2** 644-652
[2] Chakraborty K, Chakraborty M and Kar T K 2011 Optimal control of harvest and bifurcation of a prey-predator model with stage structure *Applied Mathematics and Computation* **217** 8788-8792
[3] Khajanchi S 2017 Modeling the dynamics of stage-structure predator-prey system with Monod-Haldane type response function *Applied Mathematics and Computation* **302** 122-143
[4] Wang W and Chen L 1997 A predator-prey system with stage-structure for predator *Computers and Mathematics with Applications* **33**(8) 83-91
[5] Zhang X, Chen L and Neumann A U 2000 The stage structured predator-prey model and optimal harvesting policy *Mathematical Biosciences* **168** 201-210
[6] Gourley S A and Kuang Y 2004 A stage structured predator-prey model and its dependence on maturation delay and death rate *Mathematical Biology* **49** 188-200
[7] Toaha S and Rustam 2017 Optimal harvesting policy of predator-prey model with free fishing and reserve zones AIP Conference Proceedings **1825**, 020023 (2017), https://doi.org/10.1063/1.4978992
[8] Srinivas M N, Srinivas M A S, Das K and Gazi N H 2011 Prey-predator fishery model with stage structure in two patchy marine aquatic environment *Applied Mathematics* **2** 1405-1416
[9] Xiao D and Ruan S 2001 Global dynamics of a ratio-depedent predator-prey system *J. Math. Biol.* **43** 268-290
[10] Keshet L E 2015 *Mathematical models in biology* SIAM Philadelphia 231-234

Present-day Tectonic Regime from Focal Mechanism Data in South Sulawesi

Muhammad Altin Massinai*, Muhammad Fawzy Ismullah Massinai, Mustakima, Erfan Syamsuddin

Geophysics Dept., Hasanuddin University

*altin@science.unhas.ac.id

Abstract. Focal mechanism used on tectonic regime study which is fault type determination in South Sulawesi. Fault parameter data consist of strike, dip and rake determined based on the first arrival polarity at earthquake waveform manually picked. The result were several beach-ball diagrams on 2010 – 2017 in South Sulawesi. Three reverse-oblique types, normal-oblique type and normal type caused by Palu-Koro Fault. In other pattern, there were Makassar Thrust: reverse type, two strike-slip types and four normal-oblique types. Walennae Fault signed by reverse type. Two reverse-oblique types and two normal-oblique types affected by Parang Loe Fault. All of them showed tectonic regime nowadays in South Sulawesi is complex.

Keyword. Focal mechanism; tectonic; South Sulawesi

1. Introduction

Sulawesi is located in the middle of the Indonesian archipelago which is an island that has impact of the tectonic activity. South Sulawesi is one of the arms of Sulawesi Island which is flanked by two active expansion zones, namely the Makassar Strait and Bone Bay, causing in a very high seismicity in this area [1].

In the area of South Sulawesi, besides some active regional faults, active subduction is also found. The existence of this active subduction can trigger the activity of the faults, so that the level of seismicity is high [2]. Active faults which are the source of earthquake zone include the Palu-Koro Fault in central Sulawesi, Walennae Fault in southern Sulawesi, Matano Fault and Lawanopo Fault in eastern Sulawesi, and the Gorontalo Fault in northern Sulawesi (Fig. 1). For overall, all of the faults which are the earthquake source zone are large-scale horizontal faults [3].

The South Arm Sulawesi tectonic model states that the Makassar Strait is interpreted as the foreland basin on both sides of the Sunda and Australian - New Guinea plates. Meanwhile the pre-Eocene oceanic crust (Lamasi Complex) to West Sulawesi occurred in the Late-Miocene oligocene while the Late Miocene West Sulawesi magmatic arc was thought to be the result of Continentals collisions [4]. This study using focal mechanism which known as one of common method used in seismicity. Energy released from the source of the earthquake and the determination of the type of fault that caused the earthquake based on the nodal plane from the results of the polarity of the hypocenter emitted waves.

Parameters in determining the orientation of the fault plane, [5]:

1. Strike (Φ) is the angle formed between fault direction in the north direction, measured clockwise ($0^{o} \leq \Phi \leq 360^{o}$).
2. Dip (δ) is the angle formed by the fault plane with the horizontal plane measured from the horizontal surface ($0^{o} \leq \delta \leq 90^{o}$).
3. Rake is the angle formed by the strike direction with the slip direction.
4. The relative shift (slip) (λ) is the relative shift in the fault measured from one block to another block in the fault area. Slip is a shift of points that previously coincided. Slip angle (λ), where the slip direction is based on the motion of the hanging-wall, ($-180^{o} \leq \lambda \leq 180^{o}$). If the fault is in the form of thrust fault $\lambda > 0^{o}$, whereas if the fault is normal fault then $\lambda < 0^{o}$.

How to identify the origin properties of this earthquake is called the earthquake focus mechanism. Through this technique every earthquake that occurs can be analyzed by a normal fault, reverse fault, or strike - slip fault. Each direction and slope can also be determined [6].

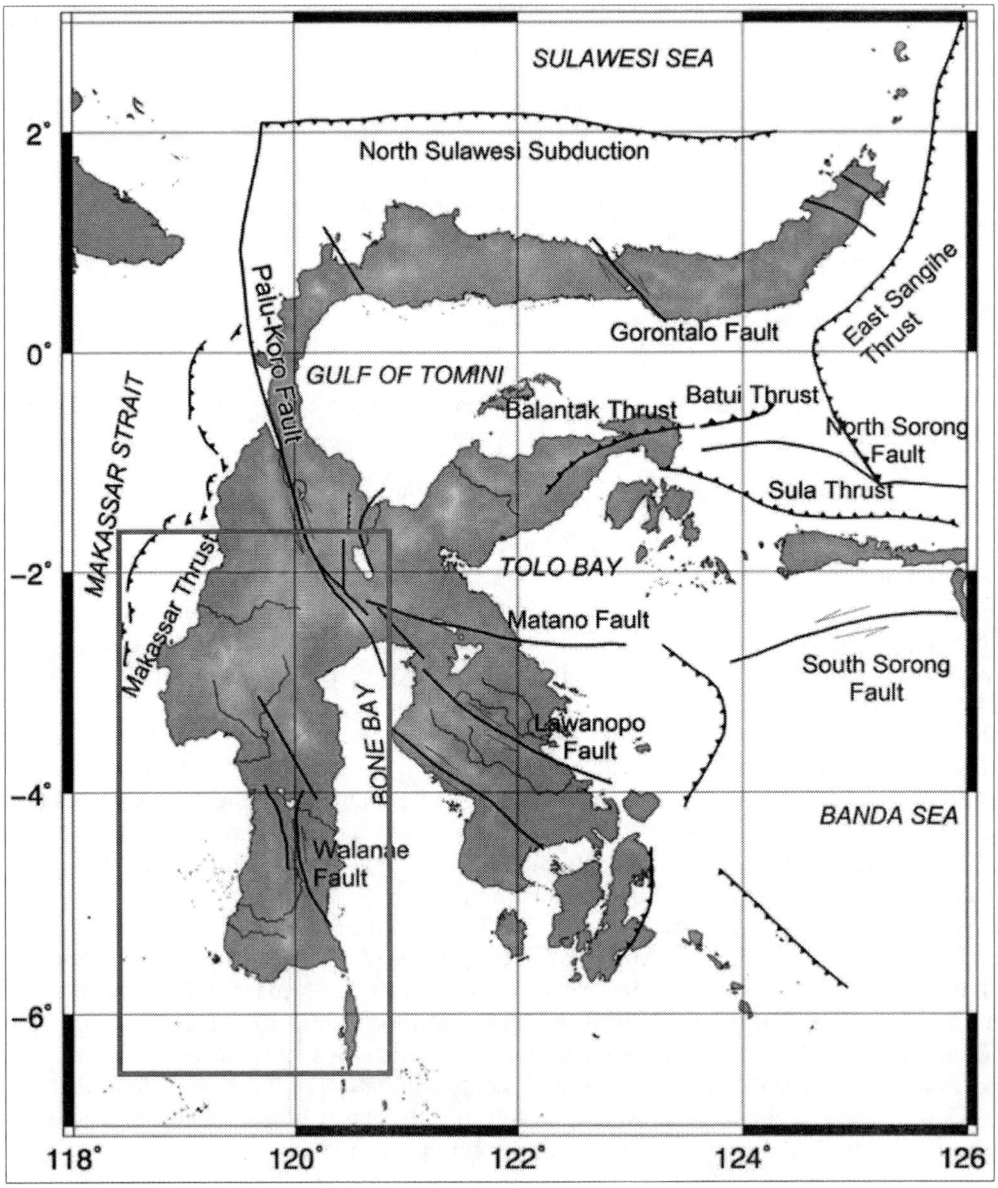

Figure 1. Several active fault in Sulawesi [7]. Red square is research area.

IOP Conf. Series: Earth and Environmental Science **279** (2019) 012016 doi:10.1088/1755-1315/279/1/012016

2. Methodology

Research location was conducted in Southern Sulawesi region which is located at the coordinates 2.38^0 - 5.69^0 S and 117.24^0 - 120.64^0 E.

Research data used were secondary data of earthquake events including origin time, hypocenter coordinates, depth, magnitude and seismogram of earthquake in the South Sulawesi region for the period 2010-2017 from BMKG catalogue.

Focal mechanism was determined using Azmtak [8] code. This code used amplitude polarity from first arrival P-wave. This information got by manual picking from seismogram (Fig. 2).

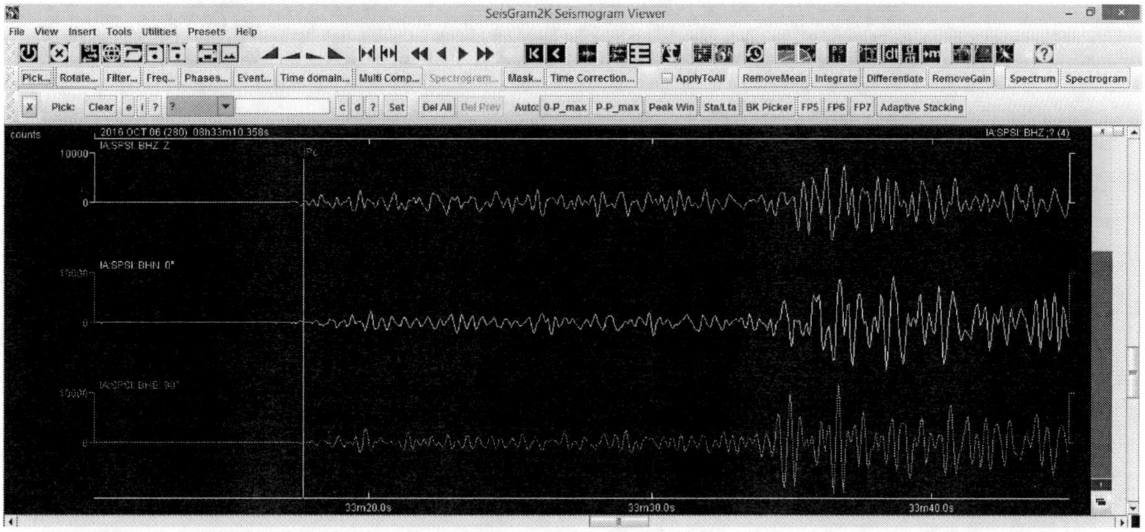

Figure 2. Example of seismogram manual picking which was recorded at SPSI station

Table 1. Example of manual picking result, as input to Azmtak code. + is amplitude have positive value and – is amplitude have negative value

Station	Amplitude
TTSI	+
SPSI	+
PMSI	+
BNSI	+
BKSI	+

Amplitude polarity value converted to numerical value as input to Azmtak code. + is 1 means compression and – is -1 means dilatation. Azmtak code will plot this information, with station coordinates, in spherical coordinate and give output such as strike, dip, rake and beachball diagram. This one result of this research (Fig. 3). In beachball diagram, red box is compression and empty box is dilatation. The validation of this diagram is given by consistent data. Consistent data show right amplitude polarity in right fault plane. If amplitude polarity is positive, so the station should be in the compression fault plane and vice versa. Compession fault plane is black area and white area is dilatation.

Normal, reverse/ thrust, strike-slip and oblique strike-slip fault have unique beachball diagram. Normal fault signed by dilatation fault plane in centroid and have only black-white-black pattern. Opposite with that pattern, white-black-white pattern sign compression fault plane in centroid and it is called reverse/ thrust fault. Strike-slip fault signed by no fault plane in centroid because is on line fault plane. It have four plane and have same wide on diagram. The last is oblique strike-slip fault, have same pattern with strike-slip fault and did not have same wide fault plane.

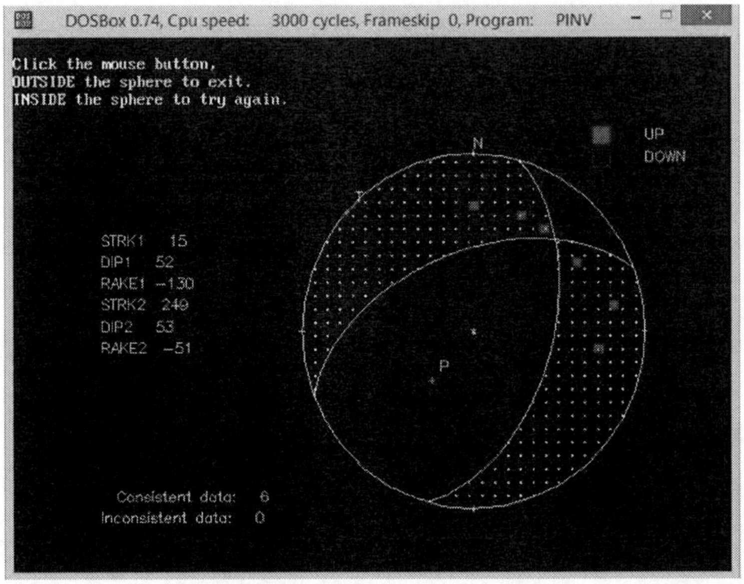

Figure 3. Azmtak code view beachball diagram. The diagram showed oblique strike-slip fault

3. Results and Discussion

3.1 Distribution of Earthquakes that occurred in 2010-2017 period in Southern Sulawesi Region

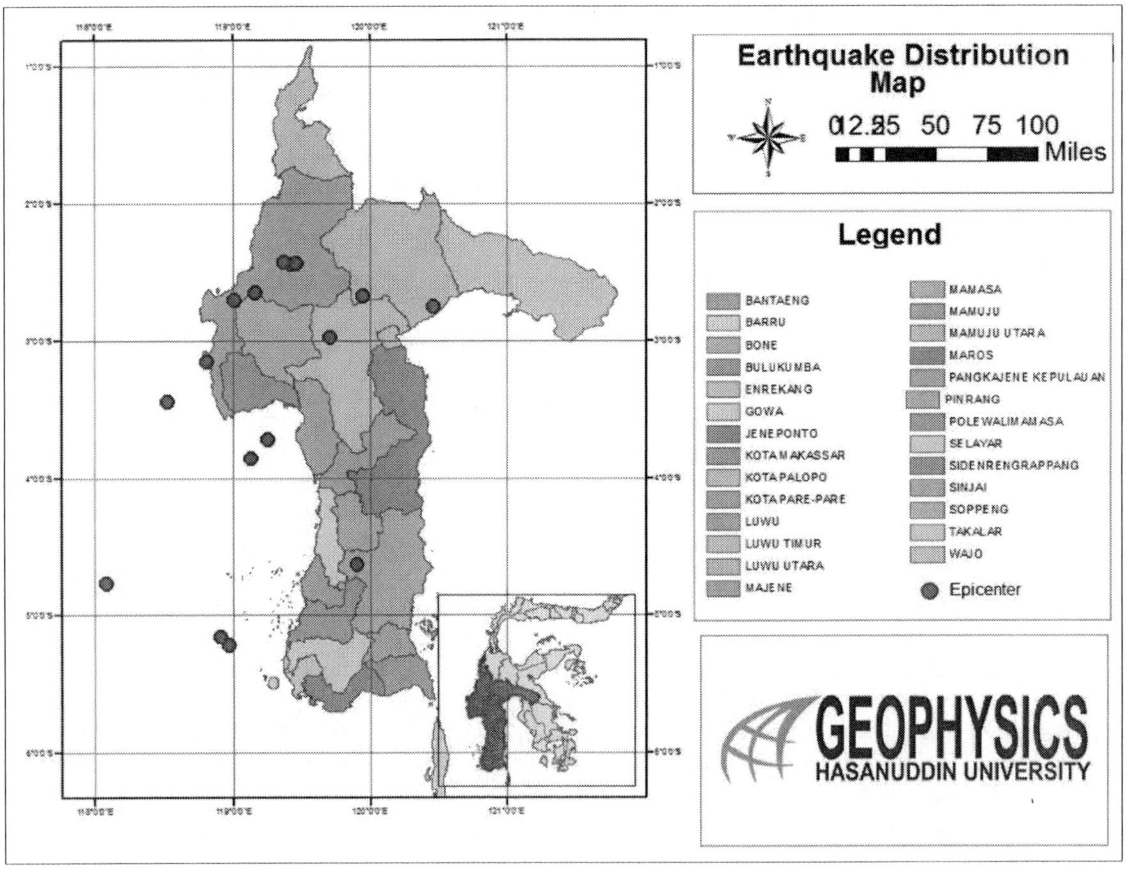

Figure 4. Map of Distribution of Earthquakes in Southern Sulawesi region 2010-2017 period

An earthquake distribution map in the South Sulawesi region obtained the distribution of earthquakes originating on land and ocean as in Fig. 4. There are 16 earthquake events which were shallow earthquakes with depth off to < 80 km in the Southern Sulawesi region. While based on the magnitude or strength of the earthquake is dominated by moderate earthquakes. Condition of seismicity it proves that this research area had characteristics of seismicity approached high frequency at shallow depths that are dominated by moderate earthquakes.

3.2 Distribution of Earthquake Focus Mechanisms in the Southern Sulawesi Region (Period 2010-2017)

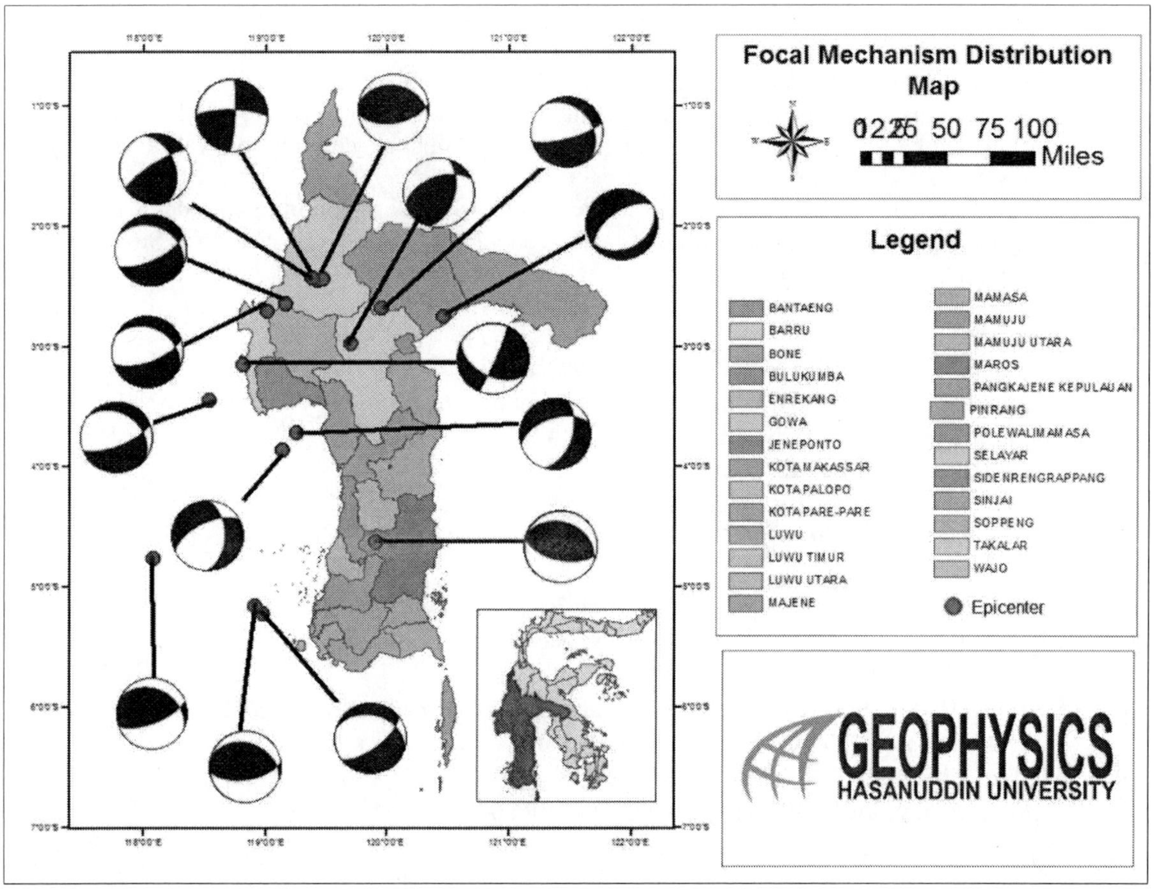

Figure 5. Distribution of Earthquake Focus Mechanisms in the Southern Sulawesi Region Period 2010-2017.

In Fig. 5 there are 16 earthquake focus mechanisms. Each focus mechanism of the earthquake shows that there are four forces that are perpendicular and as large as each earthquake. The black color of the earthquake focus mechanism shows that ground motion causes the earthquake to be polarized as repulsive or compressed, while the white color on the focus mechanism shows ground motion causing the earthquake to be polarized as an attractive or dilated motion.

Based on 16 earthquakes there were 3 strike - slip faults, 2 reverse faults, 1 normal fault, 7 oblique faults and 3 oblique faults. It can be seen that this earthquake is dominated by oblique faults. The analysis obtained from these 16 focals was that Southern Sulawesi was affected by oblique fault patterns, although there are other fault patterns obtained.

It can be seen in Figure 5 that the northside there are three earthquake points of the upper right are the reverse oblique fault, in addition normal oblique fault, and the normal fault affected by Palu - Koro

Fault which spread from Palu through northside of South Sulawesi to South Bone until Banda Sea. Whereas, the 7 upper left side earthquake points are reverse faults, 2 strike - slip faults, 4 normal oblique faults which are affected by Makassar Thrust (Makassar Strait) striped from coast of Mamuju beach, West Sulawesi diagonal cut across middle of South Sulawesi region then Southern South Sulawesi, namely Bulukumba City to the eastern area of Selayar Island and 1 point in central earthquake point, as a reverse fault affected by Walanae Fault. This agreed with previous research that East Walanae Fault is reverse fault [9]. Then 2 earthquake point in the central, namely each normal oblique fault and 3 earthquake point on lower, namely a reverse oblique fault, reverse oblique fault and a normal oblique fault which is affected by Parangloe fault.

4. Conclusion

Based on the results and discussion of this research it can be concluded that:

1. The earthquake that occurred in the Southern Sulawesi region in the period 2010-2017 was dominated by the type of moderate and shallow earthquakes with a high level of seismicity in the study area.
2. Fault types that occur in earthquakes in the South Sulawesi region in the period 2010-2017 are generally oblique fault pattern which is affected by Palu - Koro Fault, Makassar Strait, Walanae Fault and Parangloe Fault.

References

[1] Massinai M A and Nakir M N 2009 *Proc. PIT HAGI* p.12 – 15.
[2] Massinai M A, Sudrajat A, Lantu 2013 *Journal of Engineering and Technology* **3** 945 – 948.
[3] Massinai M F I, Lantu, Aswad S, Massinai M A 2015 *AIP Conference Proceedings* **1658** 030013-1 – 030013-10 http://dx.doi.org/10.1063/1.4915021
[4] Surono and Hartono U 2015 *Geologi Sulawesi* LIPI Press: Jakarta.
[5] Okal E A 2011 *Earthquake, Focal Mechanism* USA Springer Science+Business Media B. V. Evanston, II. 60208.
[6] Santoso D 2002 *Pengantar Teknik Geofisika* Bandung: Penerbit ITB.
[7] Cipta A, Robiana R, Griffin D, Horspool N, Hidayati S, Cummins P 2016 *Geol Soc London Spec Publ* **441** http://doi.org/10.1144/SP441.6
[8] Suetgusu D 1997 Source Mechanism Practice *Lecture notes International Institute of Seismology and Earthquake Engineering* Tsukuba Japan.
[9] Massinai M A and Massinai M F I 2018 *Journal of Physics: Conf. Series* **979** 012045 http://dx.doi.org/10.1088/1742-6596/979/1/012045

Mineralogy and Geochemistry Variation of Igneous Rocks from Ungaran and Muria Volcano and Its Processes related to Subduction Zone Magmatism of Sunda Arc

Jenian Marin, Tri Winarno, Dewi Mindasari
Departemen Teknik Geologi Fakultas Teknik Universitas Diponegoro

jenianmarin@gmail.com

Abstract. Sunda Arc represent Quaternary magmatism as one of the most significantly active volcanic arc in Indonesia resulted from subduction of oceanic plate beneath continental plate. Along this subduction system, there are systematic variations in subduction angle, depth, crust thickness, and composition contribute to geochemical characteristic of igneous rocks along and across the arc. This study compared mineralogy and geochemistry characteristics from two volcanoes which represent across-arc variation. Petrographic analysis have been conducted on rocks' thin section samples from Ungaran and Muria Volcano to characterize mineralogy of rocks. Selected samples were analyzed using X-Ray Fluorescence (XRF) for geochemistry of major and trace elements. Mineralogy of Ungaran rocks are dominated by plagioclase, clinopyroxene, and hornblende phenocrysts with composition range from andesite to basaltic andesite of high-K calc alkaline series. Samples from Muria Volcano are from alkaline series characterized by feldspathoid and alkali feldspar rich rocks with calcic plagioclase and pyroxene phenocrysts. This mineral assemblages, supported by major elements data suggest fractionation process in magma. Geochemical trend is systematically consistent with other island arc setting, where there is significant increase of total alkali from Ungaran to Muria. Trace element data shows decreasing in LILE/HFSE ratio further from arc. This suggests that Ungaran has more crustal and subducted slab influence, while Muria has higher degree of partial melting from deeper Benioff Zone.

Keywords: across-arc geochemistry, Sunda Arc, Muria, Ungaran

1. Introduction

Volcanic formation in subduction zones is created from complex magmatic processes, ranging from partial melting of the mantle and its associated crusts, magma rising to the earth's crust, and contamination with country rocks. This process of magmatic differentiation will produce variations in magma series and its igneous rocks. These variations can be observed based on mineralogical and geochemical composition of rocks. There have been widely reported that there is systematic geochemical variations in subduction setting such as the Izu-Oshima Islands Arc, Japan [1], the Southern Volcanic Zone, Chile [2], and the Andean Central Volcanic Zone [3].

Quaternary Java volcanic arc is formed due to the subduction of the Indo Australia Oceanic Plate to the Eurasian Continental Plate. There are variations in the composition of magma formed along the Sunda Arc due to the influence of the subducting plate and contamination of the crustal wall when the magma moves to the surface [4]. A study of lava in the volcanic arc of West Java showed similar variations in rock composition, but different from isotope parameters [5]. Rocks from the back-side arc are enriched by some trace elements compared to rocks from the front-side arc.

Ungaran Volcano and Muria Volcano are two Quaternary volcanoes that located at the back side of the volcanic arc in Central Java. The relationship between the two in the same volcanic arc can be seen from petrogenetic indicators, especially rock geochemistry. To be able to better explain the comparison of its characteristics; further mineralogy and geochemistry studies are needed.

2. Geological Setting

Sunda Arc is one of the most prominent Quaternary volcanic arc in Indonesia, consist of 80% of the total volcanoes in Indonesia. This arc extends from the northern tip of Sumatra to the southeast, continues eastward to Java, Bali, Lombok, Sumbawa and Timor. In north and west Sumatra, this arc is characterized by the emergence of small islands in the Indian Ocean; such as Nias Islands and Mentawai Islands. Although the tectonic setting is nearly the same, geochemical differences between the rocks from Sumatra and Java showing difference in the depth of Benioff Zone. Geophysical and geochemical data indicate the depletion of the crust to the east, where the thickness of the crust on Java is around 20-25 km which thinned to Flores around 15 km [6]. The Benioff Zone tilts north to a depth of 100 km as far as a horizontal distance of 200 km from the Java Trench. Under Java and the Java Sea, the slope increases by about 65° and reaches depths of up to 650 km (figure 1).

Some parts of the Quaternary volcanic arc in Indonesia have two or more volcanoes across the arc's continuity. This phenomenon is called a double volcanic arc, which is divided into front-side arc and the back-side arc. In Java, the volcanic chain on the back-side arc appears on the west side with an orientation of 105° - 110° E at a depth of surface slab subducting of about 200 km [7]. The chemical characteristics of Sunda Arc rocks in Java and Bali, characterize the magma series from toleitic to high-K calc alkaline above the Benioff zone with a depth of 120-250 km. The result is an increase in incompatible trace elements as the Benioff zone deepens. The most prominent parameter is the content of K which is also positively correlated with the depth of the Benioff zone. This change in chemical variation occurs due to the modification of the mantle source associated with the thermal structure of the subducted plate.

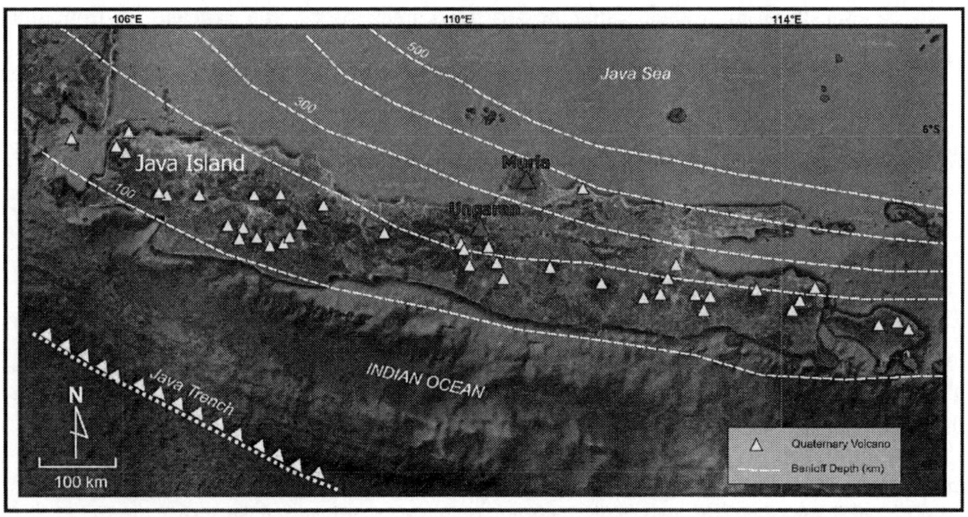

Figure 1. Tectonic setting of Ungaran and Muria Volcano in Sunda Arc. Benioff zone depth/ subducted slab contours are from [6]

The subduction zone on Java Island has undergone periodic changes since Cenozoic Age. Rock age data show that volcanic activity shifted north from Tertiary to Quaternary [8]. Most of the active volcanoes on the Java Island are stratovolcano, which lies extending in the middle of the island. The dominant rock composition is high-alumina basalt and pyroxene andesite. This active volcanic belt rises on older volcanic and volcaniclastic rocks from Oligocene and Neogene sediment formations [9].

IOP Conf. Series: Earth and Environmental Science **279** (2019) 012017 doi:10.1088/1755-1315/279/1/012017

3. Methodology

Field activities are needed for sampling fresh rock products of volcanic activity of Mount Ungaran and Mount Muria. The samples will be analyzed petrographically and geochemically as the primary data of the study. Petrographic sampling is carried out on nonfragmental rocks, both derived from lava flows and intrusions to represent the characteristics of rocks in each volcano. For petrographic analysis, fresh rock samples were prepared with a thin section of 30 μm which was attached to a glass plate and covered with thin glass. Preparations are observed under a polarizing microscope to identify mineralogy of rock and its texture. Detailed descriptions are made so that the rock petrogenesis can be determined. The classification used in petrographic observation is the classification of the International Union of Geological Science (IUGS) based on the composition of QAPF (Quartz - Alkali Feldspar - Plagioclase - Feldspathoid). Geochemistry of major and trace elements were analyzed using x-ray fluorescence (XRF) method at BPPTKG Yogyakarta. Before analyzed, LOI (loss on ignition) in all samples were determined using gravimetry method. All elements data were plotted to various diagrams to classify rocks, determine magma affinity, and construct petrogenetic models for the research area.

4. Results

4.1. Mineralogy of Rocks from Ungaran and Muria Volcano

Mineralogy of rocks from both volcanoes are significantly different, which is further explained by geochemistry. Samples were obtained from lavas, intrusions, and lithic fragments of andesitic breccia. Rocks from Ungaran are typical basaltic andesite - andesite with aphanitic - porphyritic texture. The groundmass consist of plagioclase and pyroxene microcrystals and volcanic glass. The most abundant phenocryst is plagioclase (25 – 45%) which composition tends to be more sodic in younger rocks. Other significant phenocrysts are clinopyroxene, hornblende, orthopyroxene, opaque minerals (Fe-Ti oxides). Zoned plagioclase is common in all samples with concentric, patchy, and spongy texture (figure 2). Some igneous textures featured subophitic, intergranular, glomeroporphyritic, reaction rim. Pyroxenes often occur as glomerocryst and have twinning habit. Olivine is rarely showed in the samples, while hornblende is more prominent in younger rocks and show distinctive habit of being altered in the rim (figure 3).

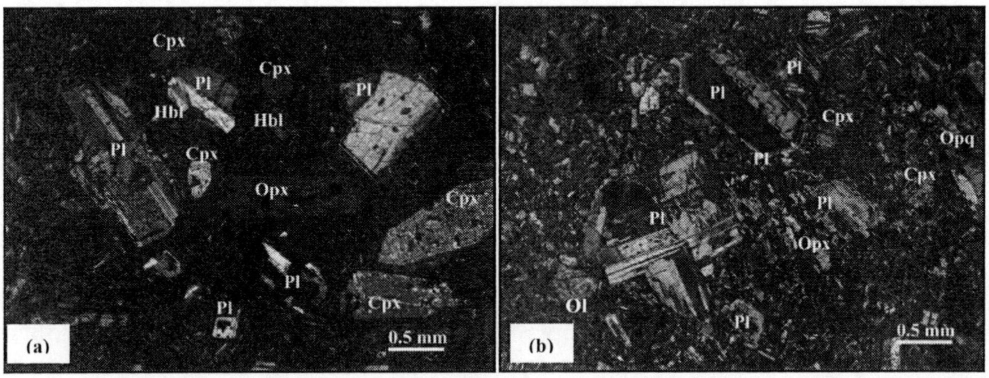

Figure 2. Mineralogy of rocks from Ungaran volcano, the abundant phenocryst of plagioclase, mostly accompanied with clinopyroxene, orthopyroxene (parallel extinction angle), and hornblende (a). Twinning and zoning in plagioclase as microtextural characteristic in basaltic andesite with minor olivine (b)

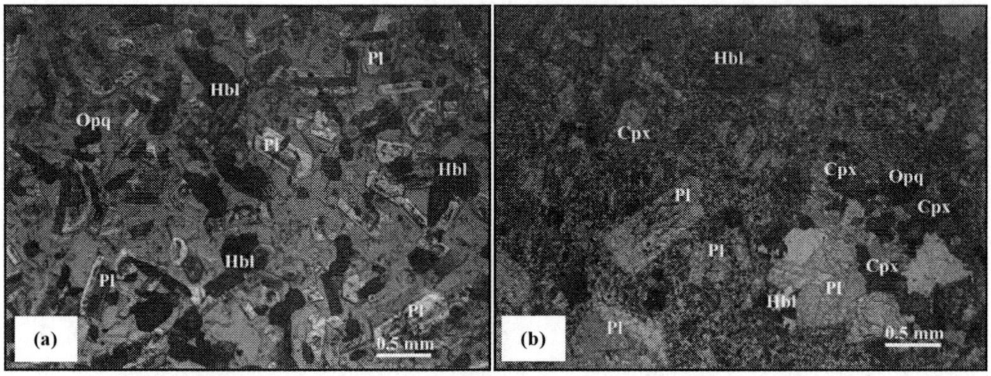

Figure 3. (a) Hornblende with altered rim is dominant in younger rocks with glassy groundmass (b) glomeroporphyritic texture with crystal aggregate of pyroxene, hornblende, and opaque minerals

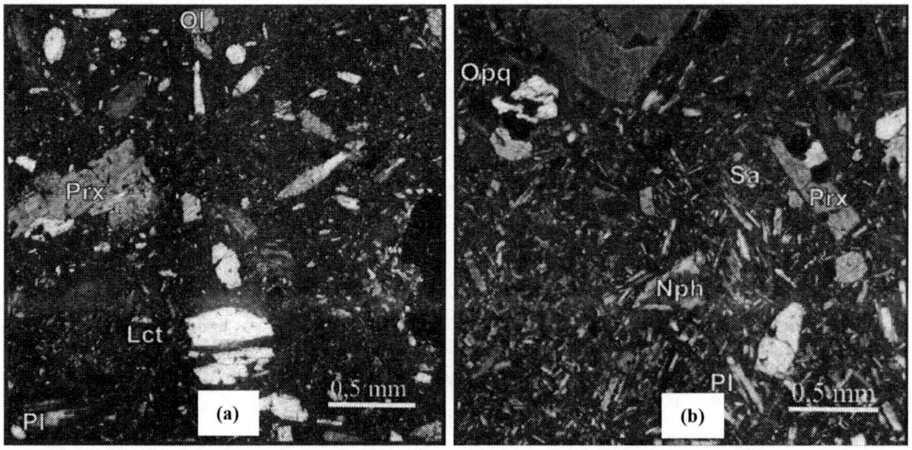

Figure 4. Petrographic images of rocks from Muria shows porphyritic texture, with plagioclase, pyroxene, feldspathoid, K-feldspar, olivine phenocryst (a). Abundant plagioclase as phenocryst and groundmass (b)

Rocks from Muria show different mineralogy than Ungaran based on lava and intrusive samples, which is more of alkali-basalt. Samples have hipocrystalline texture and porphyroaphanitic inequigranular. The major phenocryst is plagioclase (30-40%) with composition ranged calcic anorthite to more sodic andesine. Plagioclase also shows twinning and zoning. Nepheline and leucite from feldspathoid group appear at a significant amount (7-14%). Based on the type of twinning, it can distinguish plagioclase from K-feldspar (sanidine and microcline), which abundance is less than 5%. Ferromagnesian minerals dominated by clinopyroxene and orthopyroxene (8-10%), while olivine is just a minor constituent (less than 5%).Pyroxene crystal shows wide range of crystal size and habit, from coarse phenocryst to fine groundmass, and from prismatic-equant to lathlike crystals (figure 4).

4.2. Geochemistry of Rocks from Ungaran and Muria Volcano

Nine volcanic rock samples from Ungaran and two from Muria were analyzed both major and selected trace elements, combined with earlier data for Young Ungaran [10] and Muria [11].

4.2.1. Major elements. Some major elements are plotted on Harker diagram using SiO_2 as primary x-axis. The samples from two volcanoes belong to different magma series from calc-alkaline affinity. Based on alkalinity diagram [12], Ungaran rocks fall in the high-K alkaline, while Muria is more alkaline to shoshonitic series (figure 5). Plotting on TAS (Total Alkali Silica) diagram, rocks from Ungaran are classified predominantly as basaltic andesite and andesite, also trachyandesite. Younger samples tend to be more andesitic than the older and samples from parasitic lavas. Muria which is already has high K composition classified as basaltic trachyandesite and trachyandesite (figure 5). Total alkali for Ungaran is ranged 4,4 – 6,2 wt% and Muria is higher at more than 6 wt%. Other oxide that shows positive correlation is Al_2O_3, although in Muria it does not show clearly because of limited samples. This trend can be seen at Ungaran clearly on Harker diagram (figure 6). Negative correlations are happened with MgO, CaO, TiO_2 (slightly), and Fe_2O_3.

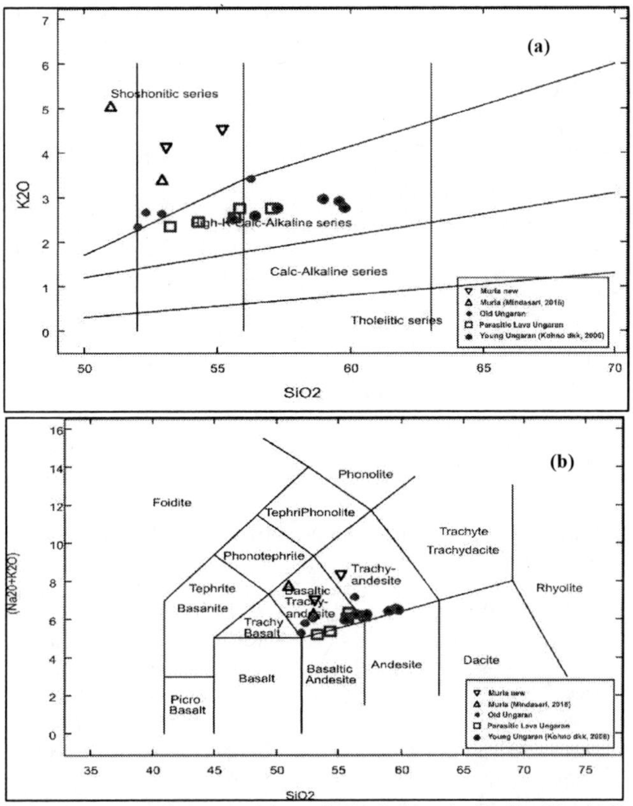

Figure 5. Alkalinity magma diagram (a) and TAS (total alkali silica) diagram (b) of Muria and Ungaran

4.2.2. Trace elements. Selected trace elements have been analyzed, not all are important indicator in petrogenetic aspect, only several incompatible element will be elaborated (table 1). Groups of LILE (large ion litophile element) includes barium, strontium, rubidium are good petrogenetic indicator. Rare earth elements are better indicator because of their stable phase. Niobium, zircon, mercury, and thorium are being used in this research. Some elements must be excellent indicator, if compared to other elements. Ratio of LILE/HFSE and HFSE/LREE will show more about magmatic processes produced the rocks. Based on calculation, Sr/Rb, Ba/Zr, Ba/Th, Pb/Ce ratio is higher in Ungaran than Muria and Nb/Zr is higher in Muria. For general, LILE is higher in Ungaran and HFSE is higher in Muria. Based on this characteristics, subduction and magmatic process can be interpreted.

IOP Conf. Series: Earth and Environmental Science **279** (2019) 012017 doi:10.1088/1755-1315/279/1/012017

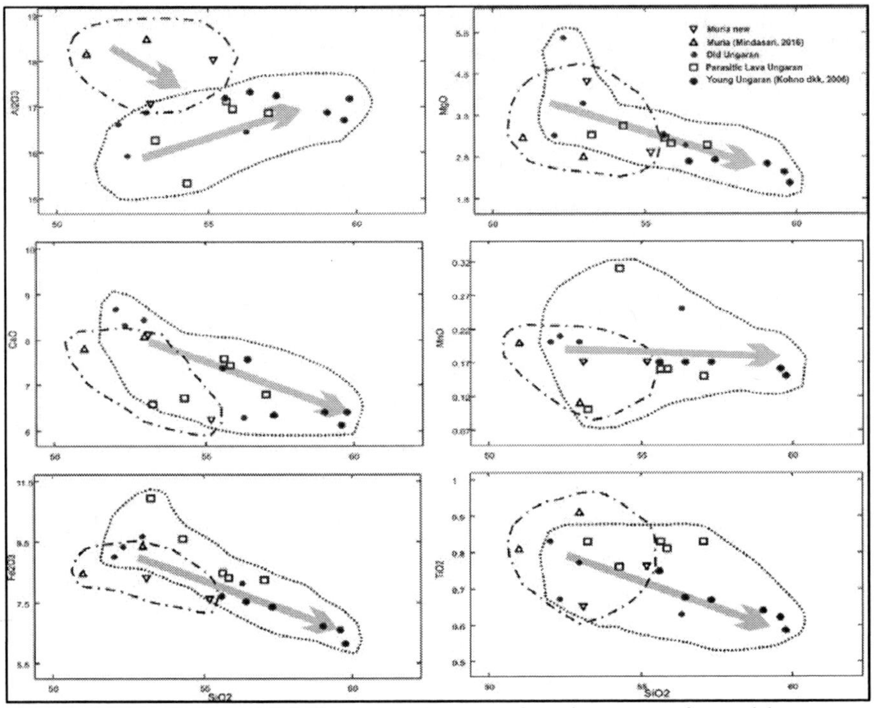

Figure 6. Harker variation diagram for SiO2 vs major oxides

Table 1. Trace element ratios of Ungaran and Muria samples. MU-01, MU-02 are from [11], UG05-UG12 are from [10]

Samples	SiO_2	Sr/Rb	Zr/Y	Ti/Y	Nb/Zr	Ba/Zr	Rb/Sr	Ba/Th	Pb/Ce
MU-01	53.08	1.87	11.00	194.85	0.12	3.05	0.53	44.67	0.05
MU-02	55.2	3.38	13.67	303.73	0.04	5.60	0.30	49.87	0.07
MU-03	50.99	4.02	9.55	220.73	0.02	0.87	0.25	8.27	-
MU-04	52.96	1.44	8.33	462.08	0.05	2.30	0.70	8.21	-
UN-01	54.30	8.10	5.47	196.13	0.07	5.12	0.12	56.25	0.25
UN-02	55.84	6.02	5.89	209.13	0.06	3.98	0.17	62.03	0.37
UN-03	57.04	5.34	6.17	210.22	0.07	3.75	0.19	55.57	0.42
UN-04	55.64	6.39	5.85	217.48	0.06	3.95	0.16	58.18	0.32
UN-05	53.25	6.05	5.28	210.94	0.09	4.47	0.17	42.27	0.37
UN-06	52.04	6.34	5.20	207.68	0.09	4.16	0.16	43.51	0.22
UN-07	56.32	9.21	5.18	160.66	0.04	8.59	0.11	116.89	0.35
UN-08	52.34	10.25	4.87	184.44	0.05	7.61	0.10	78.76	0.38
UN-09	52.97	10.66	4.81	192.09	0.05	8.81	0.09	90.16	0.43
UG - 05	57.31	4.67	6.30	174.65	0.07	3.31	0.21	34.29	-
UG - 06	59.59	4.69	6.52	161.61	0.07	3.42	0.21	32.06	-
UG - 08	55.61	6.69	6.08	187.33	0.06	3.64	0.15	37.93	-
UG - 09	59.79	6.10	6.78	153.78	0.06	3.46	0.16	36.00	-
UG - 11	59.02	1.65	6.63	159.88	0.06	3.45	0.61	36.60	-
UG - 12	56.45	6.08	6.41	185.32	0.06	3.37	0.16	39.58	-

5. Discussion

Geochemistry of rocks is a result of several complex processes, from tectonic setting, magma generation, magma processes and its interaction before igneous rocks crystallization. Magma generates under a specific tectonic setting, which interaction of every crustal component will be recorded in geochemical characteristics. Based on major and trace elements, Ungaran and Muria are product of calc-alkali basalt, which is produced in subduction zone setting (figure 7). As parental magma derives from mantle, in this setting, mantle wedges interact with two different crusts (oceanic-oceanic or oceanic-continental) and then produce more complex and evolved magma. In this setting, the occurrence of volcano depends on degree of partial melting underneath which controlled by Benioff Zone depth, heat flow, and crustal composition. Ungaran located nearer to the trench than Muria, hence the Benioff Zone depth is shallower at approximately 225 km. Muria located in the back of Sunda Arc approximately 325 km depth of Benioff Zone. While Muria is being one from just a few back arc volcano, the rock geochemistry is significantly different than other abundant front arc volcano. Being the nearest across arc volcano from Muria, Ungaran rock geochemistry slightly overlaps with Muria.

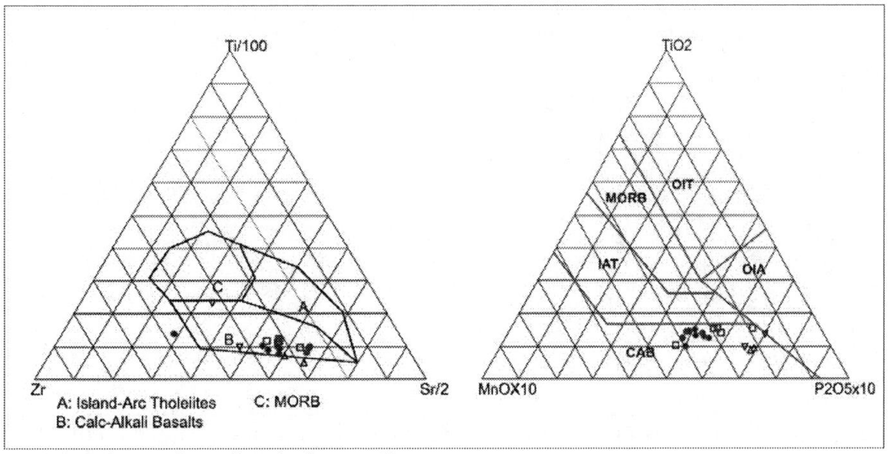

Figure 7. Tectonic setting discriminant diagram Ti-Zr-Sr and TiO_2-MnO-P_2O_5 [12] Ungaran and Muria plotted at CAB (calc-alkali basalt)

It has been determined that Ungaran rocks are product of high-K alkaline magma, while Muria rocks are product of alkaline magma. The mineralogy resulted from those product is a different mineral assemblage with slight overlap. Ungaran rocks' mineral assemblages mostly are silica saturated, with dominance of plagioclase, pyroxene, hornblende, and opaque minerals. Sometimes, an undersaturated melt still contain olivine in it. There are no quartz and little felsic minerals in all samples.

As Muria is product from deeper mantle, it should be expected that an undersaturated or near saturated mineral assemblages will appear in rocks. Mostly undersaturated melts resulted in such mineral assemblage: calcic to sodic plagioclase, pyroxene, feldspathoid, a little amount of K-feldspar and olivine, and opaque minerals. Leucite and nepheline are the most common feldspathoid here, which is much depleted in silica that has been used in K-feldspar.

It is also possible to configure magmatic processes (crystal fractionation, magma mixing, and assimilation) based on interpretation of geochemistry and mineralogy. Harker diagram using MgO as x-axis can be used to identify crystal fractionation in magmas correlating with geochemistry pattern. Normal differentiation will deplete MgO as it fractionate into ferromagnesian minerals. SiO_2 and Al_2O_3 are negatively correlated with MgO, while Fe_2O_3 and CaO are positively correlated. Samples contain olivine are from Muria, Old Ungaran, and its parasitic lava (figure 8). Those samples represent a lower degree of fractionation, with less silica content. Olivine is accompanied by abundant clinopyroxene and more calcic plagioclase. This process occurs when melts is high in MgO and lower silica content. Petrographic evidence showed mineral aggregates of pyroxenes, plagioclase, and opaque mineral which settled during magmatic processes. Thus, make fractionation process more effective as the ascending magma will be more evolved. Products from younger activity of Ungaran tend to have more hornblende/ amphibole group and also still abundant plagioclase. Therefore, the petrogenetic model for fractionation can be interpret. Diagram MgO vs Fe_2O_3 and MgO vs CaO display amphibole and pyroxene fractionation as MgO content decreases. Sometimes, fractionation in magma did not happened normally because of changes in physical condition (pressure and temperature fluctuation) and chemical condition,

IOP Conf. Series: Earth and Environmental Science **279** (2019) 012017 doi:10.1088/1755-1315/279/1/012017

in this case is change in magma composition because of mixing or assimilation. Zoned plagioclase is evident in suggesting this processes, although the exact changes cannot be determined without chemical analysis of single mineral.

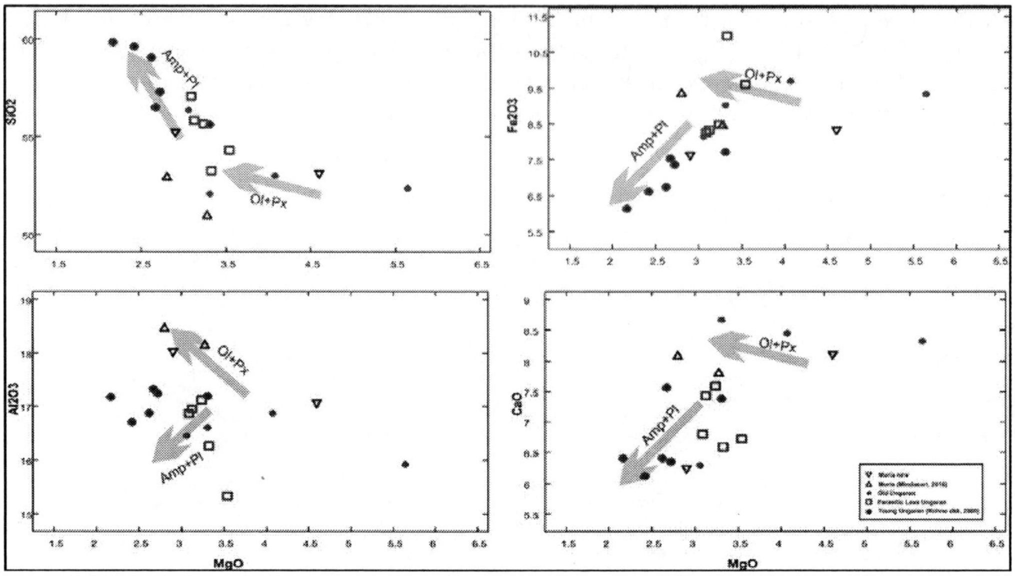

Figure 8. Harker diagram using MgO as x-axis, plotted with other major elements to identify correlation between mineralogy and rock chemistry.

Trace elements data provided more excellent evidence about magmatic processes. For this research, a comparison is needed to determine which influence is more prominent in one over another. LILE to HFSE ratios are decreases from Ungaran to Muria (Ba/Zr, Pb/Ce, Ba/Th, Nb/Zr). Those ratios indicate subducted slab derived fluid. The melting of subducted crust and sea sediment add LILE constituent to magma, and the amount will be decreased further away from trench. Diagram Ba/Zr vs Nb/Zr discriminate a mantle zone [4], where increasing in Ba content is enriched from subducted slab (figure 9). Crust contamination other than slab is the overlying crust and host rock surrounds magma chamber. Lower Sr/Rb in Muria suggests lower crustal contamination during magmatic process. Ratios of LILE/HFSE can be further confirmed by HFSE/LREE ratios [3]. For example, Ungaran Pb/Ce ratio is also higher than Muria, also indicates higher slab fluid.

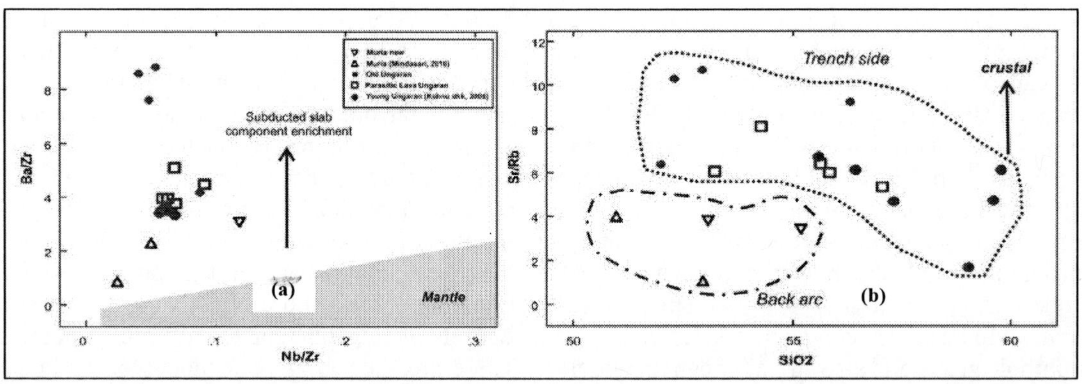

Figure 9. (a) Diagram Ba/Zr vs Nb/Zr ratio and (b) diagram SiO_2 vs Sr/Rb and the petrogenetic interpretation

Modification of magma composition or also mentioned as magma differentiation involves several processes. Even in the same tectonic setting and in the same arc, the degree of partial melting, subducted slab, and crustal contamination will contribute at different level. Therefore, systematic geochemical trend in a volcanic arc is a manifestation of those components. These trend can be observed in Muria and Ungaran too as part of Sunda Arc, along with other adjacent volcanoes. More detailed rare earth elements and isotopic data from whole rock or mineral geochemistry will be needed in the future to unravel more about this subduction magmatism in Java Island.

6. Conclusions

Geochemical and mineralogical evidence of igneous rocks from Ungaran and Muria volcano represent two different characteristics of subduction magmatism. Ungaran, which is located nearer from volcanic front, displays typical basaltic andesite – andesite with mostly silica saturated mineral assemblages. Plagioclase as main phenocryst, followed by pyroxene and hornblende. Less evolved magma also near saturated, with olivine occurrence. Meanwhile, Muria rocks composed by under saturated magma with plagioclase, pyroxene, feldspathoid minerals, and lesser K-feldspar and olivine. Magma serie change from high-K calc alkaline to alkaline magma. Geochemical trend is systematically consistent with other island arc setting, where there is significant increase of total alkali from Ungaran to Muria. Trace element data shows decreasing in LILE/HFSE ratio further from arc. This suggests that Ungaran has more crustal and subducted slab influence, while Muria has higher degree of partial melting from deeper Benioff Zone.

References

[1] Kimura J, Kent A, Rowe M, Katakuse M, Nakano F 2010 Origin of cross-chain geochemical variation in Quaternary lavas from the northern Izu arc: Using a quantitative mass balance approach to identify mantle sources and mantle wedge processes *Geochemistry Geophysics Geosystems Vol. 11, Number 10*

[2] Jacques G, Hoernle K, Gill J, Hauff F, Wehrmann H 2013 Across-arc geochemical constraint in the Southern Volcanic Zone, Chile: Constraints on mantle wedge and slab input compositions *Geochimica et Cosmochimica Acta 123*

[3] Michelfelder G, Feeley T, Wilder A, Klemetti E 2013 Modification of the continental crust by subduction zone magmatism and vice versa: across strike geochemical variations of silicic lavas from individual eruptive centers in the Andean Volcanic Zone *Geosciences 2013, 3*

[4] Handini E, Hasenaka T, Wibowo H, Shibata T, Harijoko A, Mori Y 2015 Petrogenesis batuan volkanik Kuarter Pulau Jawa Tengah: peran kontaminasi lempeng menunjam vs kerak dangkal *Proceeding Seminar Nasional Kebumian ke-8*

[5] Sendjaja Y and Kimura J Geochemical variation in Tertiary-Quaternary lavas of the West Java arc, Indonesia: Steady state subduction over the past 10 million years *Journal of Mineralogical and Petrological Sciences 105*

[6] Setijadji L 2010 Segmented volcanic arcs and its association with geothermal fields in Java Island, Indonesia *Proceedings World Geothermal Congress 2010*

[7] Tatsumi Y and Eggins S 1995 *Subduction Zone Magmatism* (USA: Blackwell Science) p 211

[8] Hamilton W 1979 Tectonic of Indonesian Region U.S. *Geological Survey Professional Paper 1078*

[9] Van Bemmelen R W 1979 *The Geology of Indonesia Volume 1A* (Netherlands: The Hague) p 732

[10] Kohno Y, Setijadji L, Utami P, Harijoko A, Pecskay Z, Imai A, Watanabe K 2005 Geochronology and petrogenetic aspects of Merapi-Merbabu-Telomoyo-Ungaran volcanoes, Central Jawa, Indonesia *Proceedings Joint Convention HAGI-IAGI-PERHAPI*

[11] Mindasari D 2016 Petrogenesis batuan beku di Gunungapi Lasem, Patiayam, dan Muria berdasarkan data mineralogi serta geokimia *Skripsi* (Semarang: Universitas Diponegoro)

[12] Rollinson H 1993 *Using geochemical data: Evaluation, Presentation, Interpretation* (London: Longman Group) p 351

Surface geological investigation as the initial stage of hydrocarbon exploration in Kei Besar Island, Southern Maluku

A Kurniasih, N Qadaryati, and R Setyawan

Department of Geological Engineering, Universitas Diponegoro, Semarang, Indonesia

aniskurniasihgeo@gmail.com, amy.qadaryati@gmail.com

Abstract. Kei Islands are known to have hydrocarbon potentials in shallow rock layers, trapped by folds and faults. This potential is characterized by the amount of gas seepage found in Kei Kecil, Kei Besar and Tayandu Island. The study was conducted in central and southern part of Kei Besar Island through outcrop observations and stratigraphic measurements aim to recognize the physical characteristics of rock. Kei Besar Island located in the easternmost part of Australian continental passive margin separated by the extension of Aru Through. Digital Elevation Model of Kei Besar Island shows that the northern part of research area has a steeper morphology than the southern. The stratigraphic characteristic in the central part of research area dominated by clastic limestone showing a shallowing-up stacking pattern. While in the southern part, the lithology dominated by coralline limestone contains many fragments of coral and large foraminifera. Stratigraphic analysis shows that lithology in the study area was formed in carbonate submarine fan which formed Elat Formation. This formation underlained by shallow carbonate rock which contain large foraminifera from Tamangil Formation. Tamangil Formation interfingered with lower part of Weduar Formation. Based on the analysis of surface data above, concluded that subsurface exploration is necessary to reconstruct the petroleum system in research area.

Keywords: Kei Besar Island, surface geological investigation, stratigraphy, limestone

1. Introduction

Kei Island is located in the east part of Banda Arc. The eastern-most of Kei Islands, Kei Besar, is actually the passive margin of Australia continental crust that separated due to Aru Trough extension. Nerong Strait in between Kei Besar and Kei Kecil was the result of thrust fault at fore-Banda Arc.

A fair hydrocarbon potential is identified from Kei Islands. A regional geology survey conducted by [1] mentioned that the potential hydrocarbon in Kei Islands is gas, formed in a shallow formation and trapped under folds and thrust faults. Hence, an encouraging hydrocarbon prospect awaits underneath. The complex regional structural geology of folds and thrust faults, that later act as structural trap in this petroleum system (Figure 1), was the result of several tectonic events. These facts also become the background of this research to conduct a more thorough observation and interpretation regarding the petroleum system in Kei Islands. Generally, the aim of this research is to obtain the formations characteristic as well as the stratigraphic relation, thus provides the initial stage reference for another next advanced stage exploration research.

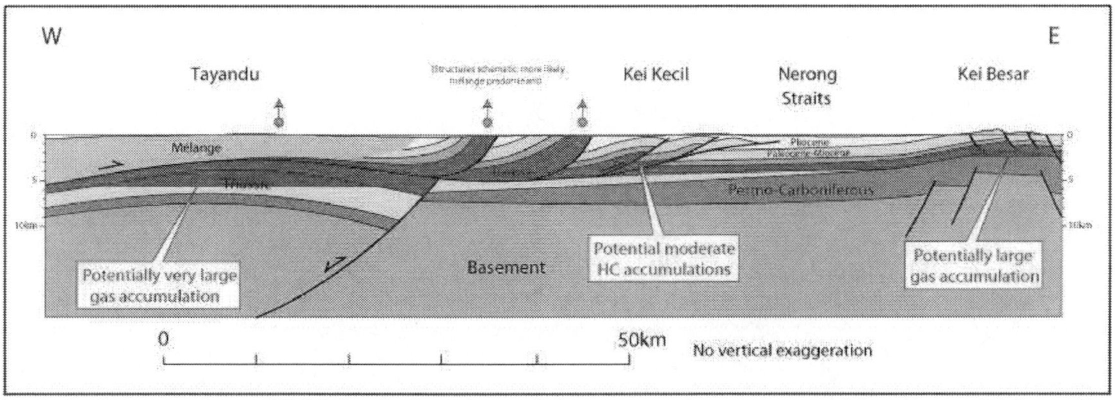

Figure 1. Cross section of Kei Islands that pictured the structural trap at several locations which confines the hydrocarbon [1].

In terms of stratigraphy, Kei Islands has rock formations that is dominated by carbonates, observed from plenty outcrops on the islands. The carbonates overlays on Metamorphic Rock Complex, consists of gneiss-feldspar-quartz-biotite-muscovite that is predicted as Pre-Cambrian products. The strata of the carbonates at Kei Islands itself begins from Elat Formation that is interfingering with Yamtimur Formation, formed at Early-Middle Eocene. Elat Formation consists of calcarenites that reaches 700 meters thick, exposed on the middle and south part of Kei Besar, while Yamtimur Formation is characterized by napal interbedded by calcarenites that thickens upward. Overlaying above those two formations are Tamangil Formation and Weduar Formation, which has interfingering relation. Tamangil Formation consists of limestone with dominant large foraminifera, 50 meters thick and exposed in Tamangil area, Kei Besar, ranging only from Middle to Late Oligocene. Weduar Formation continues to be formed until Middle Miocene, consists of reef limestone, chalk, calcarenite and napal. Weryahan Formation and Matot Tanjung Complex overlay Weduar Formation, composed of beds of clastic limestone, both are formed during Pliocene. The youngest carbonates are Ohoinol Formation and Kei Kecil (Pliocene) that showed the characteristic of shallow marine depositional environment.
The most complex geological structures are formed at Kei Besar [2], including anticlines, sinclines, faults and joints. Basically, the beds inclination does not exceed 35°. Plenty of mud volcanoes are found on islands at the middle part of Kei Islands. This area at least has experienced three orogeny events. The first orogeny took place at Late Eocene that folds Elat Formation and Yamtimur Formation. The second orogeny, during Late Miocene, folds the previous formations even tighter as well as Weduar Formation. Plio-Pleistocene is when the last orogeny took place that caused the folding of Weryahan Formation.

2. Research location

In general, the research took place Kei Besar, Southeast Maluku Regency. The observation sites were in Elat, Ohoilim, Werka, and Mata Holat (Figure 2) that represent the whole research area.

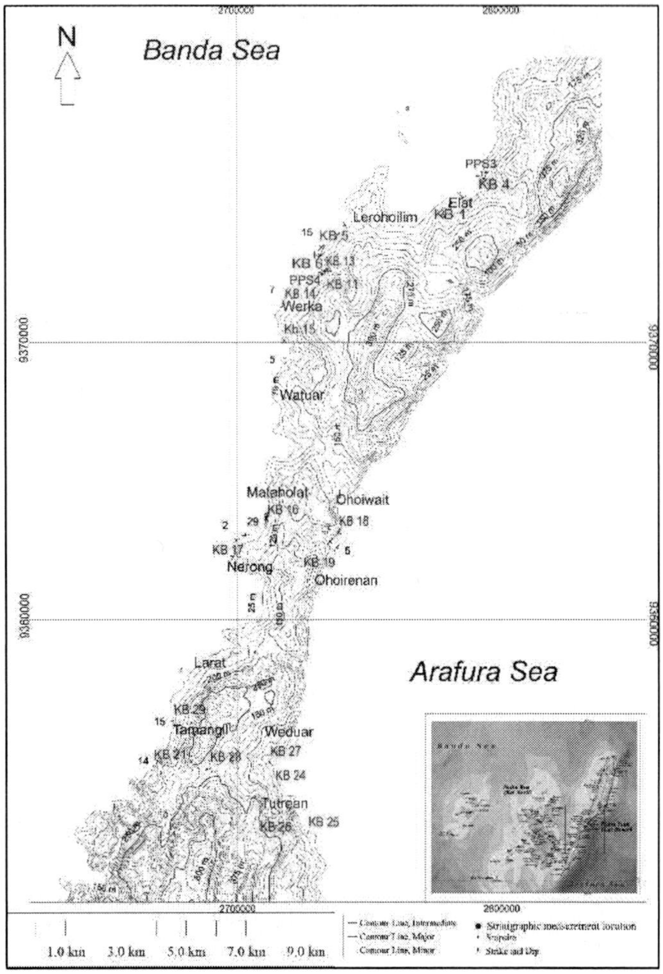

Figure 2. Research area coverage and sampling points.

3. Methods

3.1. Pace and compass

One of the methods to support measuring stratigraphic section is pace and compassing. Every length unit, distance and thickness, is measured by rope and recorded on the traverse section. Observing the lithology is also carried out in this method, including rock type, thickness, sedimentary structure, lithology contact and fossil assemblages.

3.2. Measured Stratigraphy Column

The stratigraphy column is built based on the strike and dip of lithology unit. This method will record the rock layers systematically, also provide the information of rock type, rock layer thickness and stacking patterns. Measuring stratigraphy was conducted in Elat and Ohoilim (Figure 2).

3.3. Facies analysis and Depositional environment

The analysis was initiated by determining the facies, utilizing the sedimentology aspect that is provided within the measured stratigraphy column. These facies association are compared to facies model [3] that can support the process to interpret the depositional environment of the formation in research area.

4. Results and Discussion

Banda Arc is sited at the border of three crusts, Indo-Australia, Eurasia and Pacific. Kei Islands is at the most outer margin part of Banda Arc. The outer part of Banda Arc consists of four segments, Timor, Tanimbar, Aru and Seram segments [4], and are thrusting westward.

Basically, the Timor, Tanimbar and Seram most likely have the same geological characteristic and structure, which represent the result of subduction process. On the other hand, Aru is the thrust-front part, thus shows different geological condition among the others. Considering that Kei Island is on the west part of Aru, hence Kei Island is part of the forearc as well. The extensional pattern of Aru Island shows that the thrust front section is situated to the west of Aru Island which possibly separates the Kei Besar from the other islands to the west (Kei Kecil and Tayandu) [5].

4.1. Geomorphology

Kei Besar Island is relatively elongated north-south, where the northern part is steeper than the southern. In the middle part, around Elat and Weduar, the island is narrowing eastward. Figure 3 show the west-east morphological sections from several area and represent the northern, middle and southern part of Kei Besar Island. The north and south area, the hills are steep they reach 500 m above sea level. Whilst, slope in the west is steeper than in the east. But, in the middle part of the island it shows the opposite, where the east slope is steeper than the west.

In terms of morphogenesis, Kei Besar Island is a karst hill. The manifestations are caves and underground water channel, speleothem, and springs.

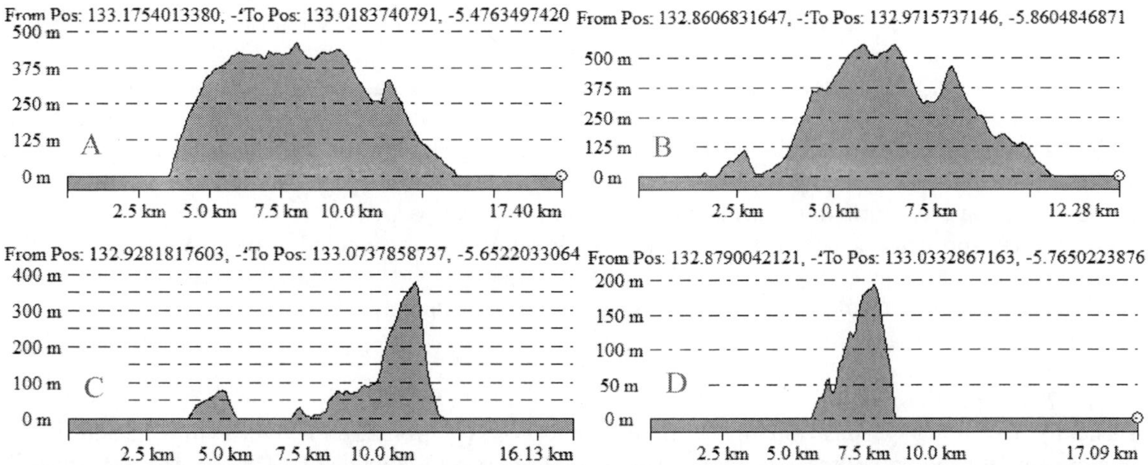

Figure 3. West-east morphological sections from several location. A. northern area and B. southern area of Kei Besar Island, while C. Elat and D. Weduar are in the middle part of Kei Besar Island.

4.2. Stratigraphy

The field observation was carried out from several area that are composed of Elat Formation, Tamangil Formation and Weduar Formation. Whilst, the measured stratigraphy section was conducted only from Elat Formation due to the vast distribution of this formation among the research area.

4.2.1. Elat Formation.

In general, Elat Formation is composed of intercalating calcarenite and calcilutite. Both facies are described below:

a. Calcarenite

The hand specimen of this facies is brownish-white and consist of fine to medium sand grain, it has parallel and wavy laminations. Nodules of massive clay grains are found on some fine sand intervals; the orientation is parallel to beds. The microfossil analysis mentioned that there are only small amounts of foraminifera, most of them are benthic foraminifera of Bolivina and Uvigerina. Rarely found planktonic foraminifera are from Globigerina and Orbulina genus.

b. Calcilutite

On a fresh outcrop, the facies shows greyish-white color, but on a weathered outcrop, it is brownish. This facies is composed of carbonate grains from silt to clay and laminated. Basically, the calcilutite is interbedded within the calcarenite. But, in some location, it was found more dominant and massive. There were no foraminifera found from the microfossil observation.

The measured stratigraphy section was obtained from two locations, PPS3 and PPS4, respectively from Wakatran Village, Elat and Udar Village, Ohoilim. Generally, the beds are dipping relatively to the south with less than 30°. Considering the strike and dip, the beds are most likely are younger southward, thus, stratigrahically PPS3 is older (under) than PPS4.

Since the fossil cannot be used as the depositional environment indicator, so it was determined from the stacking patterns which later are compared to the carbonate slope model [3]. On clastic carbonates depositional systems, sedimentary structures are not clearly defined because it was dismissed during diagenesis.

A typical of subaqueous slope is shown by an intercalating or interbedded of coarse grain and fine grain facies. Turbidite sequences are hardly observed, but the lower part of Bouma Sequence (Ta – Tb) was identified from debris sediments.

PPS3 shows that Elat Formation is composed of intercalating calcarenite and calcilutite. The calcarenite is 5-20 cm thick and consist of fine sand grain, it has a non-erosional contact with the underlaying rock. Beds, thin beds, lamination and wavy lamination are identified from this interval, where calcilutite is thinning upward. The stacking pattern is coarsening upward. Thus, Elat Formation in PPS3 is interpreted as a product of outer fan – carbonate submarine fan.

On the other hand, PPS4 shows calcilutite that interbed within the calcarenite. The calcarenite is 20-50 cm thick, it even reached more than 100 cm in several layers, where it has non-erosional contact. The facies is composed of fine grains and grading to coarser grain upward, and has bedding sediment structure. The calcilutite, as interbeds, at the lower part are very thin layers (less than 1 cm), thickenned at the middle part and turned into thin layers again at the upper part.

Basically, the formation is coarsening upward. The lower part is interpreted as the product of transition to middle fan-submarine carbonate fan, whilst the upper part is part of middle fan-submarine carbonate fan. Hence, stratigraphically, Elat Formation is shallowing upward deposited during Middle-Upper Eocene [2].

4.2.2. Tamangil Formation

Basically, the interfingering Tamangil Formation and Weduar Fromation is overlaying Elat Formation unconformably and both were deposited during Middle-Upper Oligocene [2]. The outcrops are found in Tamangil Nuhuten Village, one of the observation sites has a 5 m outcrop. The fresh outcrops show whitish-brown color, while the weathered samples are brown. The rock orientation was hardly identified. Large foraminifera are abundant, mostly 0.5-3 cm and elongated (Figure 4). It is also mentioned in a previous study [6] that Formasi Tamangil has high number of large foraminifera. Based on the petrography, these large foraminifera are Lepidocyclina. Calcareous algae are also present.

Figure 4. Outcrop sample of Tamangil Formation, Lepidocyclina is abundance.

4.2.3. Weduar Formation

The lower part of Weduar Formation is interfingering with Tamangil Formation. Regional geology map shows that Weduar Formation has direct contact to Elat Formation. Unfortunately, there were no exposed and distinct contact between those formations from the field, as well as the contact between Elat Formation and Tamangil.

Weduar Formation on the field is brownish-white of a compact reefal limestone (Figure 5 and 6). This outcrop shows the presence of coral fragments, algae and mollusc fragments. Chalky limestone can be found in Weduar and Tutrean area, it is white and consists of very fine grains. At the coastal part of Tamangil and Larat, there is outcrop of calcarenite. The outcrop shows beds with 30 cm average bed thickness, brownish-white, with bedding and cross-bedding. Another outcrop of calcarenite at at the border between Mata Holat and Nerong area. The calcarenite consist of fine-sand grain, shows wavy and swaley cross stratification (Figure 7) as the indicator of shallow marine environment that is influenced by fair weather-storm wave base. The bed is gently sloping for less than 20°. These described facies characteristics from the field are typical of lower part of Weduar Formation [6] and interpreted as the product of shallow marine depositional environment that was deposited during Middle Oligocene – Middle Miocene [2].

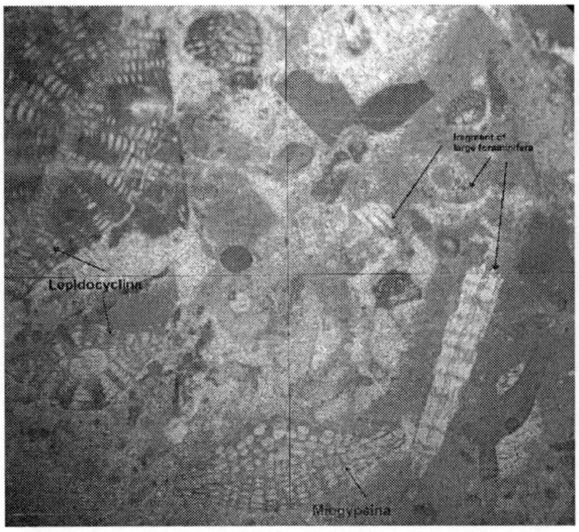

Figure 5. Thin section of Weduar Formation coralline limestone taken from Tamangil area, shows the abundance of large foraminifera.

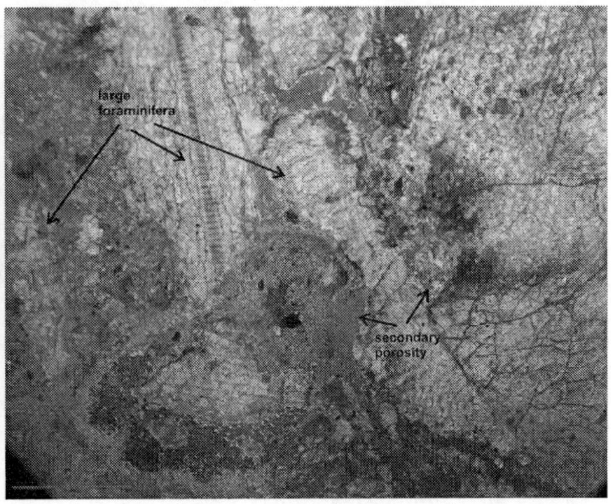

Figure 6. The intense secondary porosity shown in Weduar Formation rock sample taken from Weduar area.

IOP Conf. Series: Earth and Environmental Science **279** (2019) 012018 doi:10.1088/1755-1315/279/1/012018

4.3. Structural Geology

Several structural geology features, which are faults and folds, are identified during the field observation from Elat Formation. Normal fault, that also developed dragfold, was distinctly observed in Elat area, the fault plane orientation is at N 175°E/75° (Figure 7). Nonetheless, the shear joint is absent in the field, thus, fault determination is hardly done. Fold was identified in Mata Holat, it is known to be the product of the first period Eocene orogeny that folded Yamtimur Formation and Elat Formation [2]. Both anticline and syncline are clearly outcropped along the coast of Tanjung Mata Holat.

Study [6] has mentioned that the features of Post-Pliocene Orogeny in Kei Besar Island mostly are normal faults. The fault has cut through the Eocene to Pliocene rock intervals. This orogeny period is claimed to be the only massive deformation, at least since Eocene.

Figure 7. Swaley cross stratification within Weduar Formation shown on outcrop in Nerong area. Landscape outcrop view (left) and up-closed view (right). The feature is the indicator of a shallow marine product, in between fair weather – storm wave base zone.

Figure 8. Normal fault that also developed dragfold in Elat area.

Figure 9. Anticline and syncline feature that is found along the Tanjung Mata Holat Beach.

4.4. Petroleum System of Kei Besar Island

The pre-Tertiary and Tertiary sediments in several sediment basins in eastern Indonesia are proven hydrocarbon yielding rocks [7]. The reservoir rocks are mostly derived from the Mesozoic and Tertiary sediments that are composed of sandstones and carbonates. These reservoirs are mostly trapped by thrust faults and normal faults. Moreover, some reservoirs are also stratigraphically trapped as the buildup carbonates are sealed by pelagic shales.

Regarding the possibility of hydrocarbon prospect in Kei Islands, study [1] mentioned that there are gas seepages at some location in Kei Islands, the Tayandu and Yamtimur (Kei Besar Island) area. Outcrops in Kei Besar shows sequence of carbonate rocks that are shallowing upward and has the age of Paleogene through Quaternary. Based on the subsurface data interpretation [1], underlaying these carbonates are the Mesozoic and Paleozoic rocks that has typical characteristic of the Tanimbar stratigraphic sequences. These pre-Tertiary rocks are believed to have important roles in Kei Islands petroleum system, which are the source rock and happen to be the reservoir rock as well. The Miocene anticline and syncline of Kei Besar Island suits best as the structural trap in the petroleum system.

The Paleogene carbonates is also assumed as a potential in term of petroleum system. Deep marine limestones (Elat Formation) is a potential reservoir rock, considering the good secondary porosity (shown from thin section, figure 5&6), which is also the characteristic within reservoirs in sediment basins in eastern Indonesia. Further laboratorium analysis is required to confirm this assumption.

5. Conclusion

The carbonates in Kei Island, in terms of stratigraphy, basically is shallowing upward. Elat Middle-Upper Miocene Formation consist of intercalating calcarenite and calcilutite that is deposited on Outer-Middle Carbonate Sub-marine Fan. Tamangil Formation, that has characteristic of large foraminifera, is overlaying Elat Formation and also interfingering with the lower part of Weduar Formation. Both Tamangil Formation and Weduar Formation are typical product of shallow marine depositional environment. The upper part of Weduar Formation consists of reefal limestone and chalky limestone. Potential reservoirs are targeted within the thick Elat Formation which also has structural trap mechanism. It is highly recommended to conduct advance lab analysis and geophysical survey in Kei Island in order to observe the subsurface geological condition.

References

[1] Charlton TR 2008 The Petroleum potential of Kei Island *http://timcharlton.co.uk/banda-arc-prospectivity/kei-island* access 5th February 2014

[2] Achdan A and Turkandi T 1994 *Peta Geologi Kepulauan Kai dan Tayandu, Maluku* Pusat Penelitian dan Pengembangan Geologi Bandung

[3] Walker RG and James NP 1992 *Facies Model Response to Sea Level Change* Geological Association of Canada **352 – 355**

[4] Katili JA 1971 *A review of geotectonic theories and tectonic maps of Indonesia Earth-Sci. Rev.* 7 **143-163**

[5] Bowin C, Purdy GM, Johnston C, Shor GG, Lawver L, Hartono HMS and Jezek P 1980 Arc-continent collision in the Banda Sea region *Bull. Am. Assoc. Petrol. Geol.* **868-915** 64

[6] Charlton TR, Kaye SJ, Samodra H, and Sarjono 1991 Geology of the Kai Islands: implications for the evolution of the Aru Trough and Weber Basin, Banda Arc, Indonesia Mar. Petro. Geol. **62 – 69**

[7] Surjono SS and Wijayanti HDK 2011 Tectono-stratigraphic framework of Eartern Indonesia and its implication to petroleum systems ASEAN Eng. Jour. Part. C **111-124** 1

Acknowledgement

This research was financially supported by The Faculty of Engineering, Universitas Diponegoro, Indonesia through Strategic Research Grant 2018. The authors thanks to Kopassus TNI AD for the accommodation and transportation during the research. The thanks also go to parties who have been willing to provide corrections, criticisms, suggestions and input, so that the research and writing of this paper can be resolved.

IOP Conf. Series: Earth and Environmental Science **279** (2019) 012019 doi:10.1088/1755-1315/279/1/012019

Petrogenesis of Kalisonggo tertiary lava, Girimulyo Kulonprogo, Yogyakarta based on mineralogy and geochemical analysis

Y Aribowo[1], R K Ali[1] and A Najib[1]

[1] Geology Departement, Faculty of Engineering, Diponegoro University, Semarang, Central Java, Indonesia

Corresponding e-mail: rinal_khaidar@yahoo.com

Abstract. The andesite lava that exposed in Kalisonggo, Girimulyo, Kulonprogo, Yogyakarta is the product of Kulonprogo Volcano activity. Kulonprogo Volcano is an ancient volcano that has three eruption centers. This study was conducted to determine the petrogenesis of Kalisonggo lava including sources, tectonic setting and other processes that controlled the formation of the lava. The methods used in this research are petrographic analysis (6 rock samples) and geochemical analysis (major oxides and trace elements) by XRF. Petrographic analysis shows that in general, all samples are composed by plagioclase, clinopyroxene, orthopyroxene, opaque mineral, and glass. Hornblende occured in three samples from the south area only and no hornblende found in the sample from north area. The rocks have porphyritic, ophitic, glomeroporphyritic texture, and some plagioclase zoning. The XRF geochemical analysis result show that the rocks composition are andesitic-basaltic andesite. From SiO_2 vs K_2O + Na_2O diagrams and Co vs Th diagrams, the rocks is interpreted as product of Gadjah volcano eruption. This product characterized by calc-alkaline magma series. The magma affinity indicates that the magma generated in subduction zone or island arc.

[Keyword: Kalisonggo, Petrography, Geochemistry, Andesite lavas, Island arc]

1. Introduction

Most of Kulonprogo region was formed geologically by ancient magmatic-volcano activity, consist of three andesite-dacite intrusion body. (Rahardjo, 1977). The research was conducted in Kalisonggo area, which located at eastern part of Kulonprogo Mountains, where the lava outcrops as part of Old Andesite Formation are expossed. (Fig. 1). The aims of the study are to determine the lithology variation of the lavas, to determine physical, mineralogy, and chemical characteristics of the lavas; and to understand magmatic affinity, differentiation processes and tectonic setting that controlled the magmatism related to the lava formation.

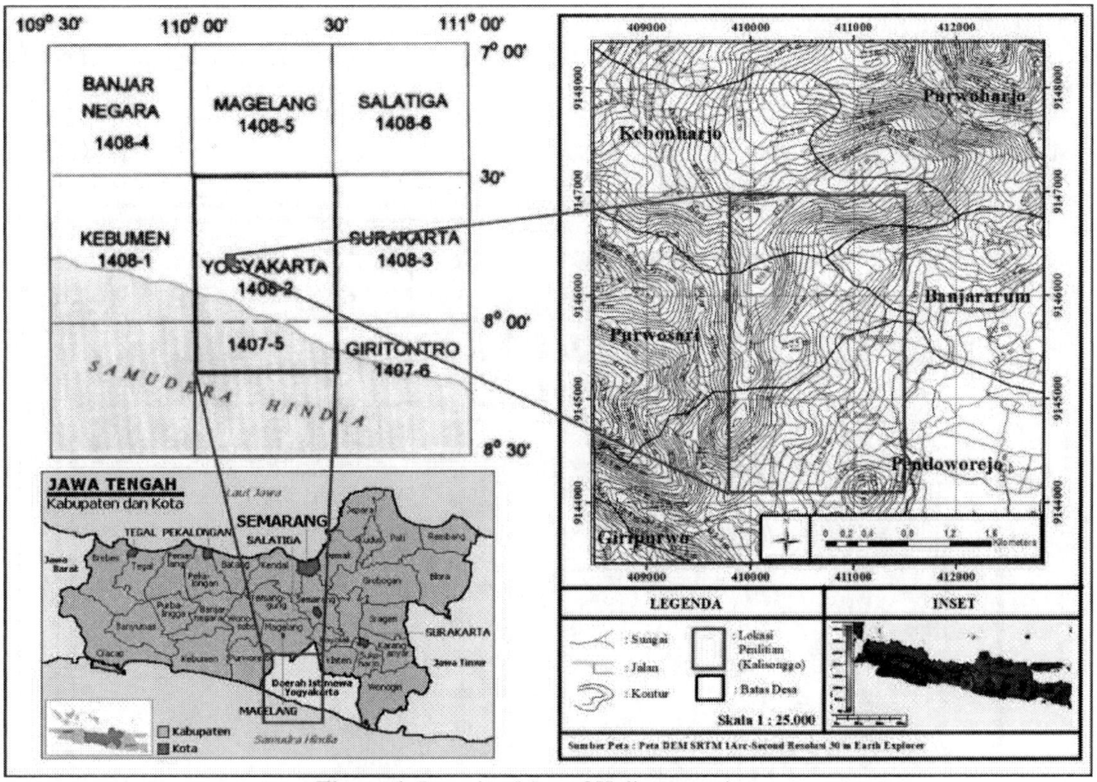

Figure 1. Location Map of Kalisonggo Area

2. Regional Geology

The research area located in Kalisonggo, Girimulyo Subdistric, Kulonprogo Regency, Yogyakarta. Physiograpically, Kulonprogo is part of Southern Central Java Zone, known as Jonggrangan Plateau (Van Bemmellen, 1949). This zone resulted by uplift that formed a large dome. The elongated dome extend about 30 km on the North-South direction, around 20 km wide. The dome well known as Oblong Dome.

Stratigraphy of Kulonprogo area consist of four formation, from the oldest: Nanggulan Formation, Old Andesite Formation (Kebo Butak Fm.), Jonggrangan Formation, Sentolo Formation and Quartenary deposit.

Nanggulan Formation consist of sandstone, lignite, sandy marl and claystone with limonite concretion, limestone and tuff, foraminifera and molusc rich. The thickness of this formation about 300 m. Based on planktonic foraminifers, the age of this formation is Middle Eocene-Upper Oligocene Hartono (1969).

Old Andesite Formation (OAF) consist of andesite breccia, lapili tuff, tuff, lapili breccia, agglomerate, lava and volcanoclastic sandstone. OAF unconformly depositd above Nanggulan Formation, with the thickness about 600 m, and the age of this formation is Oligocene – Miocene.

Jonggrangan Formation consist of tuff, marl, breccia, claystone with lignite, bioherm limestone and layered limestone in the upper part. The thickness of this formation about 2540 m, unconform above OAF and interfingering with Sentolo Formation. Based on biostratigraphy, Jonggrangan Formation age is Miocene. Sentolo Formation consist of tuffaceous marl in the lower part and marly sandstone and limestone in the upper part.It has thickness around 950 m, with geological age Lower Miocene - Pleistocene.

Quartenary deposit are formed along fluvial and coastal plain, consist of Alluvial and Colluvial deposit (gravel, sand, silt, clay and poorly sorted debris deposit), Sand dune deposit and Merapi Volcanic Deposit.

IOP Conf. Series: Earth and Environmental Science **279** (2019) 012019 doi:10.1088/1755-1315/279/1/012019

3. Methods and Sample Location

The methods used in this research are field mapping, rock sampling, petrographic analysis (6 samples), and geochemical analysis (X-Ray Fluoroscence). Petrographic analysis used to determine rock texture and mineral composition, to interpret the magmatic differentiation process during the crystalization. Geochemical analysis

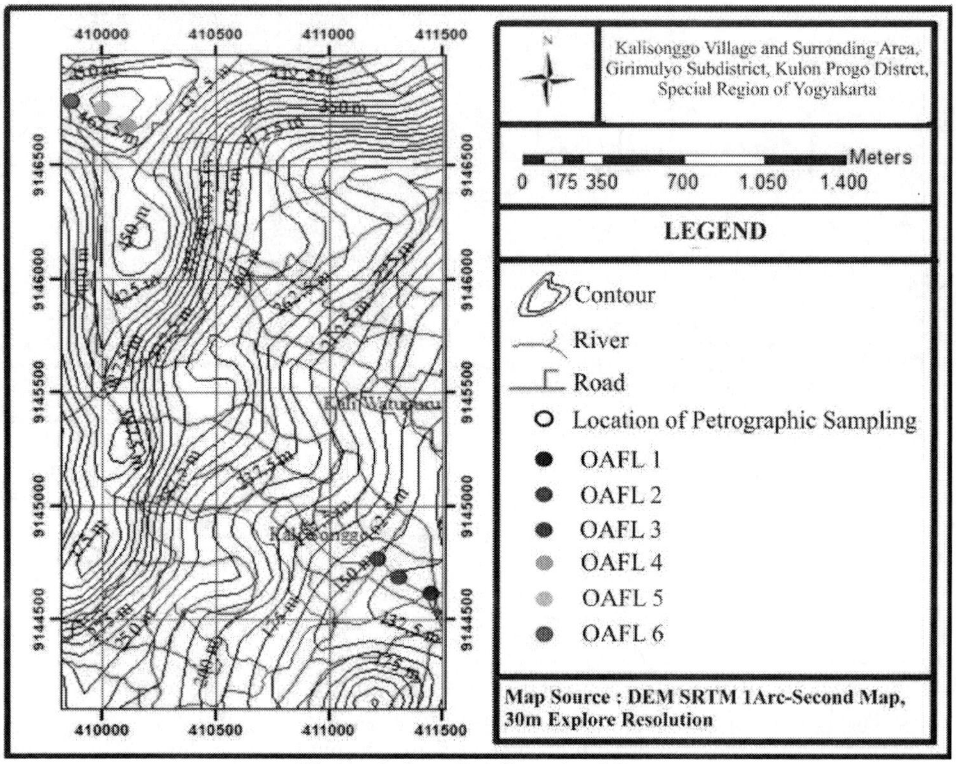

Figure 2. Sampling Location Map

4. Results and Discussion

4.1. Physical-optical properties of lava

Lava outcrops are expossed at two areas: southeast and northwest part of the research area. Physical properties of the lava outcrops are dark grey-black colour, fresh, sheeting joints (fig. 3- left), vesicular and autobrecciation (fig. 3- right).

Figure 3. sheeting joints (left), autobreccia (right). Kalisonggo lava

There are six petrographic samples, three samples from the southeast part (OAFL 1, OAFL 2 and OAFL 3) and three samples from northwest part (OAFL 4, OAFL5 dan OAFL 6). Based on texture and mineralogy, there are similarities and differences of the two sample groups. Major minerals of all samples are plagioclase (andesine and oligoclase) and clinopyroxene, but the pyroxene from the southeast samples are more abundant, and take part as groundmass. Northwest sample's pyroxenes occur as phenocrist and are associated with hornblendes, which are not found in southwest samples. (Table 1).

Table 1. Petrographic analysis resume

Code	*Coordinat* (UTM)	Mineral composition (%)					Specific textures				Rocks name
		Plg	Cpx	Opx	Hbl	Opq	*Z*	*I*	*P*	*G*	
OAFL-1	411451/9144617	60	35	-	-	5	V	V	-	-	Andesite
OAFL-2	411313/9144692	63	30	-	-	7	V	V	V	-	
OAFL-3	411218/9144765	65	28	-	-	7	V	V	V	-	
OAFL-4	410011/9146750	50	15	10	15	10	V	-	V	V	
OAFL-5	410113/9146667	45	25	10	12	8	V	-	V	V	
OAFL-6	409877/9146778	50	30	5	5	10	V	-	V	V	

Exp: Plg = plagioclase; Cpx = clinopiroxene; Opx = Orthopiroxene; Hbl = Hornblende; Opq = MineralOpak; *Z = Zoning*; *I = Intergranular*; *P = Poikilitic*; *G = Glomeroporphyritic.*

Specific textures that identified are zoning, intergranular, poikilitic dan glomeroporphyritic. Zoning are radial albite twinning of plagioclase, which developed by Ca-Na gradual composition change during crystal growth. The type of zoning being observed are oscillatory zoning (Fig. 4.A). Intergranular characterized by piroxene minerals surounded by elongated prismatic plagioclase (Fig 4.B). Intergranular texture is a result of plagioclase and pyroxene crystal growth at the same stage and relatively at the same temperature.

Glomeroporphyritic texture appear as intergrowth of plagioclase and piroxene as phenocrist surounded by finer groundmass. (Fig. 4.C). This type of texture formed by *synneusis;* cyrstal accumulation by surface tension. Poikilitic texture is spotted pyroxene enclosed by large plagioclase (Fig 4.D), it represent the different crystal growth pattern between plagioclase and pyroxene, where pyroxene crystal stop grow as plagioclase keep grow continually.

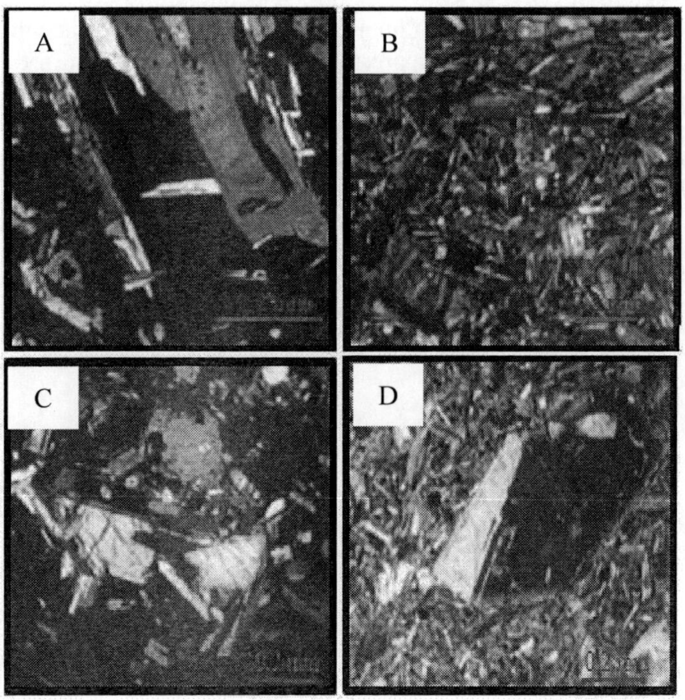

Figure 4. specific texture of rock samples: A) *oscillatory zoning,* B) intergranular, C) *glomeroporphyritic,* D) poikilitic

IOP Conf. Series: Earth and Environmental Science **279** (2019) 012019 doi:10.1088/1755-1315/279/1/012019

4.2. Geochemical analysis

6 rock samples had analyzed, the same samples analyzed by petrographic method: OAFL 1, OAFL 2, OAFL 3, OAFL 4, OAFL 5, dan OAFL 6. The result of XRF analysis are major oxides (Table 2) and trace elements concentration (Table 3).

Table 2.Major elements composition (weigt %).

Kode Sampel	Na_2O	MgO	Al2O3	SiO_2	P_2O_5	K_2O	CaO	TiO_2	Fe_2O_3
OAFL 1	3,74	5,828	18,33	55,43	0,527	1,1	6,83	1,15	9,595
OAFL 2	4,02	6,246	19,01	55,17	0,497	1,03	6,4	0,96	9,273
OAFL 3	3,4	8,247	16,82	55,07	0,457	1,04	6,55	0,95	10,11
OAFL 4	4,72	2,835	19,89	59,99	0,472	1,45	6,15	0,62	6,457
OAFL 5	4,75	2,687	20,1	60,48	0,432	1,47	6,11	0,59	6,015
OAFL 6	5,2	2,322	20,69	59,51	0,311	1,43	5,85	0,59	6,583

Table 3.Trace elements concentration (ppm)

Kode Sampel	Th	Nb	Ta	Sr	Zr	Hf	Y	Co
OAFL1	4,7	3,2856	0,8189	272,366	197,36	5,4	36,5	16,28
OAFL2	1,7	2,6564	0,8189	246,068	167,09	6,2	32,7	13,606
OAFL3	3,3	3,8448	0,8189	276,763	228,53	7,8	35,7	14,471
OAFL4	4,7	4,9633	0,8189	289,532	651,24	14,5	59,4	16,044
OAFL5	5,3	3,0059	0,8189	305,175	505,77	15,7	53,1	19,111
OAFL6	5,1	4,1943	0,8189	275,326	1366,6	9	107,3	21,313

Based on SiO2 concentration, sample from southeast part (OAFL 1,2,3) have lower concentration (near basaltic) than samples from northwest part (OAFL 4,5,6) around 55% compare to around 60%. Moreover, the MgO and Fe2O3 from southeast samples are higher than those of northwest samples: around 6-8% compare to around 2-3% MgO and 9-10% compare to 6% Fe2O3 (shaded data), those confirm that southeast samples (OAFL 1,2,3) more basaltic than northwest samples. The difference of SiO2, MgO, and Fe2O3 composition of the samples is evidence that the lava expossed at southeast genetically different from those expossed at the northwest side, or they were generated from different magma source both in term of time and/or place.

Volcanic rocks type determination using several discriminant/classification diagrams (fig. 5): A. SiO_2 vs $Na_2O + K_2O$ (Le Bas dkk., 1986), B. SiO_2 vs $Na_2O + K_2O$ (After Cox Bell Pank, 1979), C. SiO_2 vs K_2O (After Gill, 1981), D. SiO_2 vs K_2O (Ewart, 1982) serta E. Co vs Th (Hastie dkk., 2007). Based on data plot on those 5 diagrams, all samples consistently show similar pattern, where samples OAFL 1,2,3 classified as basaltic andesite, and OAFL 4,5,6 classified as andesite.

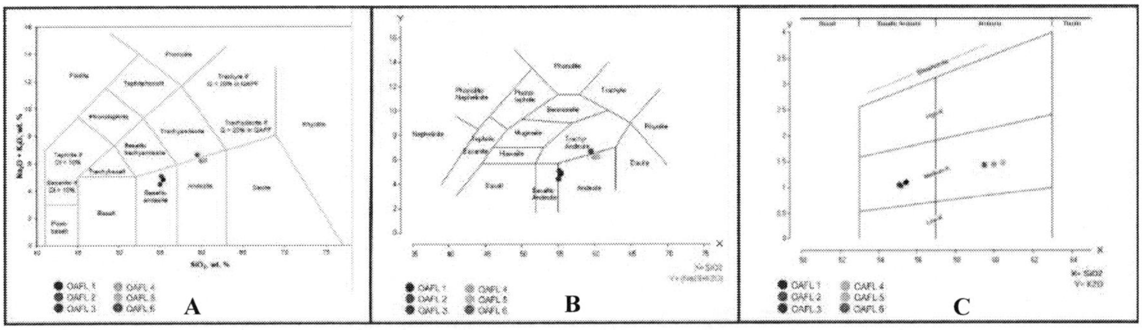

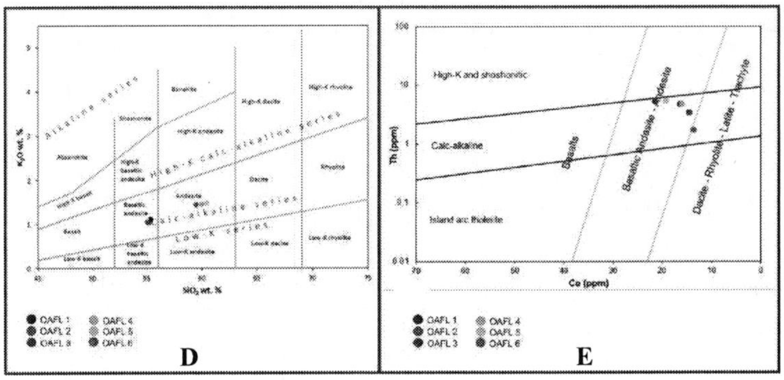

Figure 5. Volcanic Rocks Classification discriminant diagrams: A. SiO_2 vs $Na_2O + K_2O$ (Le Bas dkk., 1986), B. SiO_2 vs $Na_2O + K_2O$ (After Cox Bell Pank, 1979), C. SiO_2 vs K_2O (After Gill, 1981), D. SiO_2 vs K_2O (Ewart, 1982) serta E. Co vs Th (Hastie dkk., 2007)

Based on SiO2-($Na_2O + K_2O$) ratio (fig. 6) all samples classified in sub alkaline magmatic affinity or calc alkaline. Calc alkaline series magma could only generated in convergent plate margin setting (Wilson, 1989), thus confirmed that the tectonic setting of Kalisonggo lava magma generation is subduction zone.

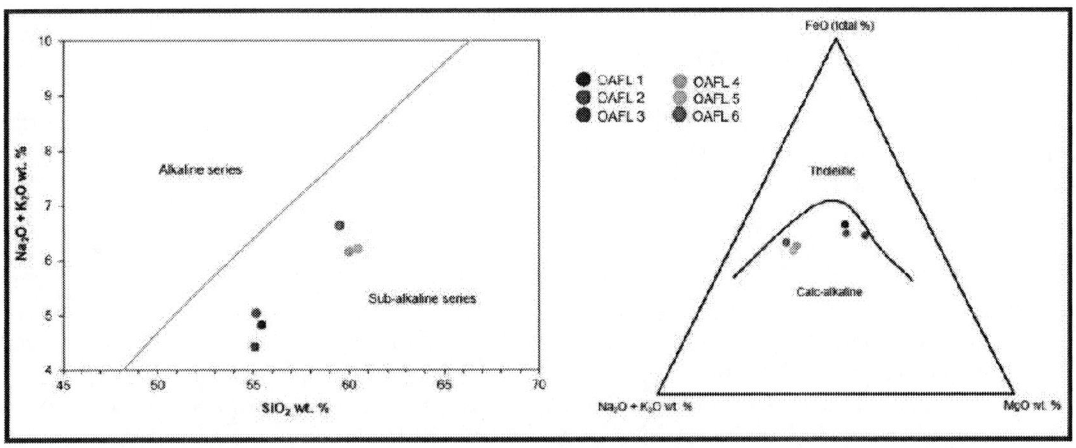

Figure 6. Binary discriminant chart SiO_2 vs $Na_2O + K_2O$ (left) and FeO_{Tot} vs $Na_2O + K_2O$ vs MgO (right) used to determine magmatic affinity (Irvine dan Baragar, 1971)

Tectonic setting determination using TiO_2-MnO-P_2O_5 diagram, all samples plotted into same area island arc calc alkaline basalt. Determination using ternary $FeO_{(Total)}$, MgO and Al_2O_3 all samples classified into island arc and active continental margin. Otherwise, data plot using trace elements ternary Hf/3 vs Th vs Ta; Hf/3, Th, Nb/16; and Zr/117, Th, Nb/16 confirm that the lava derived from magma source which generated in island arc setting (Fig. 7.A) with calc alkaline basalt type magma series (CAB) (Fig. 7.B).

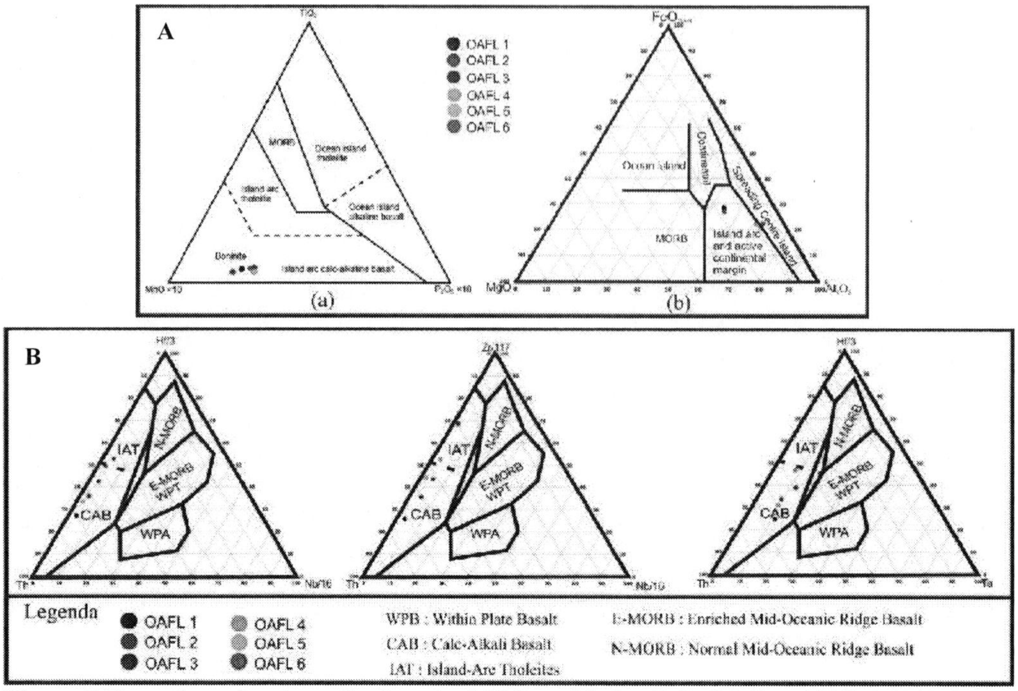

Figure 7. Tectonic setting determination: A. a) TiO_2 vs $MnO + P_2O_5$ (Mullen, 1983) A.b) FeO_{Tot} vs $Na_2O + K_2O$ vs MgO (Pearce dkk, 1971) and B. Hf/3 vs Th vs Ta; Hf/3, Th, Nb/16, dan Zr/117, Th, Nb/16 (Wood, 1980)

5. Conclusion

Kalisonggo Lava are classified into andesite (Streckeisen, 1967). Specific texture that identified are *intergranular, zoning, glomeroporphyritic* and *poikilitic*. Based on geochemical character, there are two type of lava/magma source, basaltic andesite of southeast outcrop and andesite of northwest outcrop. Magmatic affinity classified into calc-alkali, and the magma formed in Island Arc calc-alkaline basalt tectonic setting.

References

[1] Cox, K. G., Bell, J.D., Pankhurst, R. J. 1979. *The Interpretation of Igneous Rocks*. George, Allen and Unwin, London.

[2] Harjanto, A. 2011. *Vulkanostratigrafi di Daerah Kulonprogo dan Sekitarnya, Daerah Istimewa Yogyakarta*. Jurnal Ilmiah MTG, Vol.4.

[3] Hastie, A. R., Kerr, A.C., Pearce, J.A., Mitchell, S.F. 2007. *Classification of Altered Volcanic Island Arc Rocks Using Immobile Trace Elements: Development of the Th-Co discrimination diagram*. Cardiff University: UK

[4] Hulburt, Jr., Cornelius S., Klein C. 1977. *Manual of Mineralogy*. New York: Wiley

[5] Irvine, T. N., Baragar, W. R. A. (1971). A Guide to Chemical Clasiffication of the Common Volcanic Rock. Can. J. Earth Sci., 8, 523-548.

[6] Le Bas, M. J., LeMaitre, R.W., Streickeisen, A., Zanettin, B. 1986. *A Chemical Classification of Volcanic Rocks Based on the Total Alkali Silica Diagram*. Journal of Petrology, 27, 745-750p

[7] Mackenzie, W.S., Donaldson C.H. and Guilford C. 1982. *Atlas of Igneous Rocks and Their Texture.* Longman Group. United Kingdom.

[8] Mullen, E,D. 1983. *MnO/TiO2/P2O5 : a minor element discriminant for basaltic rocks of oceanic environment and its implications for petrogenesis*. Earth Planet Sci.

[9] Raharjo, W,. Sukandarrumidi & Rosidi. 1977. *Peta Geologi Lembar Yogyakarta, Jawa*, Bandung: Direktorat Geologi.

[10] Setijadji, L. D., Kajino, S., Imai, A., Watanabe, K. 2006. *Cenozoic Island Arc Magmatism in Java Island (Sunda Arc, Indonesia): Clues on Relathionships between Geodynamicsof Volcanic Centers and Ore Mineralization, Journal of Resource Geology*, vol. 56 no. 3, 267-292. New York.

[11] Setijadji, L. D., Watanabe, K. 2009. *Updated Age Data of Volcanic Center in the Southern Mountains of Central-East Java Island, Indonesia*, International Conference Earth Science and Technology, Yogyakarta..

[12] Streckeisen, A. L. 1976. *Classification of the Common Igneous Rocks by Means of Their Chemical Composition: a Provisional Attempt*. Neues Jarhbuch fur Mineralogie, Monatshefte.

[13] Subiyanto, 1989. *Calc-Alkaline Volcanic Rocks and Related Soils from West Progo, Yogyakarta (Java, Indonesia).* Thesis. Rijksuniversiteit Gent International Training Centre for Post Graduate Soil Scientists, Gent.

[14] Tatsumi, Y. and Eggins, S. 1995. *Subduction Zone Magmatim*. Blackwell Science, Boston, MA.

[15] Van Zuidam, R.A., Cancelado, F.I. 1979. *Terrain Analysis and Classification Using Aerial Photographs. A geomorphological approach*. ITC Textbook of Photo-interpretatiton. Enschede.

[16] Wilson, M. 1989. *Igneous Petrogenensis*, Springer, Dordrecht

[17] Winter, J. D. 2001. *An Introduction to Igneous and Metamorphic Petrology*, Prentice-Hall Inc., New Jersey

IOP Conf. Series: Earth and Environmental Science **279** (2019) 012020 doi:10.1088/1755-1315/279/1/012020

Behaviour of Mercury Around the Landfill of Tamangapa Antang, Makassar City, Indonesia

A. Artiningsih[1], H. Zubair[2], A.M. Imran[3], S. Widodo[4]
[1] Geological Engineering Department, Faculty of Engineering, Hasanuddin University 90245, Indonesia
[2] Agricultural Sciences, Faculty of Agricultural Science, Hasanuddin University 90245, Indonesia
[3] Geological Engineering Department, Faculty of Engineering, Hasanuddin University 90245, Indonesia
[4] Mining Engineering Department, Faculty of Engineering, Hasanuddin University 90245, Indonesia

andi.artiningsih@umi.ac.id

Abstract. Study of mercury behaviour has been conducted to know the spread of mercury (Hg) vertically and laterally around the Tamangapa Antang Landfill, Makassar City, Indonesia. Thirty-six samples (top soil) were collected in the area of Antang Landfill. The samples were devided into four lines, namely lines A, B, C, and D. Every line consist of three holes (line A: hole A1, A2, A3); (line B: hole B1, B2, B3); (line C: hole C1, C2, C3) and (line D: hole D1, D2, D3). Every hole was devided into three layers (Upper, Middle and Lower layer) with a thickness of layer 10 cm respectively. Furhermore all samples (top soil) were analyzed by means of cold vapour Atomic Absorption Spectrophotometer method (mercury determinition). The result of the study shows that the mercury in the study area has spread vertically and laterally around the Antang Landfill. The concentration of mercury in vertikal and lateral profile vary from 0.02 to 0.07 ppm. Eight top soil samples (A1-upper, A1-middle, A2-upper, A3-upper, A3-middle, D3- upper, D3-middle and D3-lower) were identified above the threshold level (according to State Minister for the Environment of Indonesia, number 51 2004 with threshold limit 0.05 ppm). Otherwise twenty-eight samples were found at the normal limit (less than 0.05 ppm). pH value in the studied area varies from 4.82 to 6 and it is shows an acidic condition. Soil pH significantly affects solubility and mobility of metals, since most metals dissolve in acid soils. So that proper remediation or precautions are needed to prevent risks to humans and the environment.

1. Introduction

Landfill of Tamangapa Antang uses Open Dumping method in its operation, where garbage or trash is heaped up and exposed to open air [1]. Garbage accumulation method in landfill can degrade environmental quality, such as air pollution as the result of solid waste burning, groundwater contamination as the result of leaching, and increased number of disease vectors that is harmful to the health of garbage collectors and communities around the landfill.

Soil is part of the cycle of heavy metal waste disposal, if the soil exceeds the ability to digest waste, this will result in soil pollution. The types of waste that have the potential to damage the environment are wastes that are included in Hazardous Toxic Substances (B3) which contain heavy metals [2], like the mercury (Hg). Heavy metal is one of the natural components of the earth that cannot be degraded or destroyed. In small concentrations, heavy metals can enter the body through food, drinks and air. According to [3], within the body of living things, heavy metals are trace minerals that are very small in number.

Landfill of Tamangapa Antang, is the only landfill located in Makassar that began operations in 1993. The Tamangapa Antang landfill which was originally designed for 10 years needs, but the fact is that until now the landfill is still being used, which means it is over from 20 years. By looking at this reality, it is known that it is probable that in the Tamangapa Antang landfill area, environmental pollution has occurred which can have an effect on environmental sanitation in this area. Along with its operation for decades, the Tamangapa Antang Landfill has been deemed inadequate and its capacity has been insufficient to accommodate the volume of waste in the next few years, not to mention the location that is very close to settlements which is very disturbing to the surrounding community. With a local population of around 1.5 million, Makassar City produces around 70 thousand tons per day or 1.4 million kilos per day of urban waste every day. In general, most of the waste produced is organic waste (wet waste), which covers 60-70% of the total volume of organic waste 86.4% and organic waste 13.6%. Most of the waste comes from population activities such as markets, trade centers, restaurants, and hotels [4],

Therefore, it can be assumed that the area of Landfill of Tamangapa Antang, in Makassar has been contaminated, namely soil, groundwater pollution which can be harmful to environmental sanitation and air pollution. This has a major impact on the quality of shallow ground water. Pollution of shallow ground water caused by the permeation of liquid waste from the landfill can also pollute the surrounding residential wells.

This study aims to analyze the mercury behavior around the Landfill of Tamangapa Antang in Makassar, Indonesia. The study focused on how the impacts that can be caused by the spread of mercury (Hg) around the landfill of Tamangapa Antang in Makassar, Indonesia. The samples were devided into four lines, namely lines A, B, C, and D. Every line consis of three holes for laterally (line A: hole A1, A2, A3); (line B: hole B1, B2, B3); (line C: hole C1, C2, C3) and (line D: hole D1, D2, D3). vertically for line A (0 to10 cm, 10 to 20 cm and 20 to 30 cm), for (line B: 0 to 10 cm, 10 to 20 cm and 20 to 30 cm), for (line C: 0 to 10 cm, 10 to 20 cm and 20 to 30 cm), and for (line D: 0 to 10 cm, 10 to 20 cm and 20 to 30 cm).

2. Research method

2.1. Research Area

Administratively, the study area situated in Bangkala, Manggala District, Makassar City, Indonesia. The map of the sampling points are shown in Figure 1.

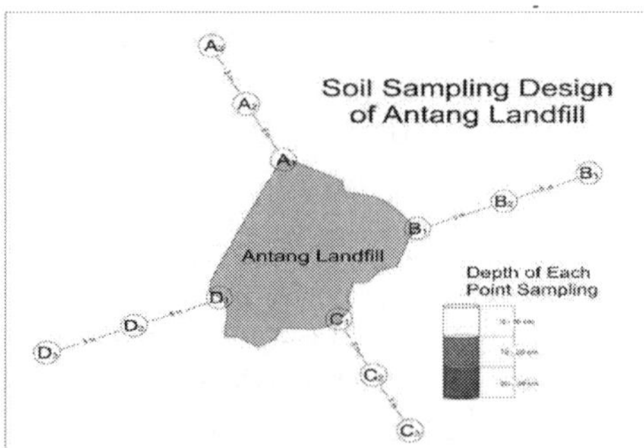

Figure 1. Location of Sampling Points, for line A, line B, line C and line D, lateral for lines A (hole A1, hole A2 and hole A3), for (line B: hole B1, hole B2 and hole B3), for (line C: hole C1, hole C2 and hole C3), and for (line D: hole D1, hole D2 and hole D3), and vertically for (line A: 0 to 10 cm, 10 to 20 cm and 20 to 30 cm), for (line B: 0 to 10 cm, 10 to 20 cm and 20 to 30 cm), for (line C: 0 to 10 cm, 10 to 20 cm and 20 to 30 cm), and for (line D: 0 to 10 cm, 10 to 20 cm and 20 to 30 cm).

2.2. Soil Sampling

The sample used in this study was topsoil collected from the Landfill of Tamangapa Antang in Makassar City, Indonesia. Determination of soil sampling points by dividing four lines. Laterally for Line A (hole A1, hole A2 and hole A3), for Line B (hole B1, hole B2 and hole B3), for Line C (hole C1, hole C2 and hole C3), and for Line D (hole D1, hole D2 and hole D3). Vertically for Line A with depth (0 to 10 cm, 10 to 20 cm and 20 to 30 cm), for Line B with depth (0 to 10 cm, 10 to 20 cm and 20 to 30 cm), for Line C with depth (0 to 10 cm, 10 to 20 cm and 20 to 30 cm), and for Line D with depth (0 to 10 cm, 10 to 20 cm and 20 to 30 cm).

2.3. Soil Sample Analysis

Soil samples was roasted in the oven for 48 hours with a temperature of 106°C ,[5], samples were then crushed with ceramic pestle and then sieved by passing a 2 mm filter [5]. Soil samples are then stored in desiccator and then the soil sample is taken to the laboratory for analysis [6].

2.4. Sample preparation and heavy metal analysis

A total of 5 grams of soil sample is put into a 300 ml polypropylene tube, then 200 ml of distilled water is added. It is then acidified with 10 ml HF solution, added with 5 ml $HCLO_4$, 2.5 ml HCl and 2.5 ml HNO_3, then homogenized for 16 hours, with a speed of 200-220 RPM, filtered using whatman filter paper (No. 42), stored in a bottle of acid, and ready to be analyzed by atomic absorption spectrophotometer [7].

3. Results and discussion

3.1. Results

The analysis Result of mercury heavy metal concentration (Hg) using the Atomic Absorption Spectrophotometer method for Line A, Line B, Line C and Line D, Laterally for Line A (hole A1, hole A2 and hole A3), for Line B (hole B1, hole B2 and hole B3), for Line C (hole C1, hole C2 and hole C3), and for Line D (hole D1, hole D2 and hole D3). Vertically for Line A with depth (0 to 10 cm, 10 to 20 cm and 20 to 30 cm), for Line B with depth (0 to 10 cm, 10 to 20 cm and 20 to 30 cm), for Line C with depth (0 to 10 cm, 10 to 20 cm and 20 to 30 cm), and for Line D with depth (0 to 10 cm, 10 to 20 cm and 20 to 30 cm). A graph of the results of analysis of mercury (Hg) vertically in holes A1, B1, C1 and D1 with the depths of 0 to 10 cm, 10 to 20 cm and 20 to 30 cm, around the area of Tamangapa Antang Landfill are shown in Figure 2.

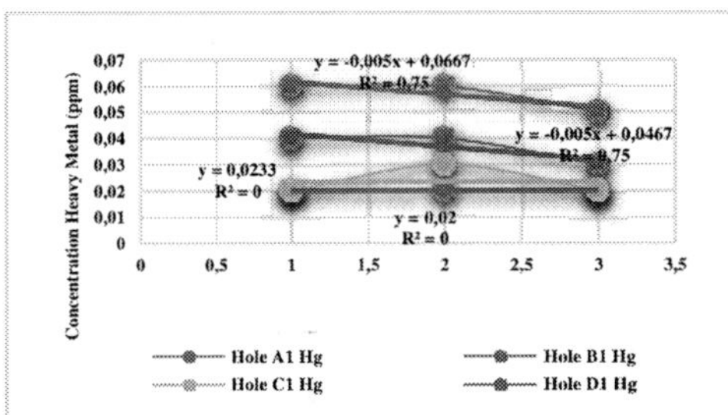

Figure 2. The graph of the results of analysis of mercury (Hg) vertically in holes A1, B1, C1 and D1, for at the depth of 0 to 10 cm (upper), 10 to 20 cm (middle) and 20 to 30 cm (lower) around Tamangapa Antang Landfill, Makassar City, Indonesia.

Figure 2 show that A1 hole at the depth of (0 to 10 cm) and (10 to 20 cm), the content of mercury is constant which value about 0.06 ppm, which has exceeded the threshold. While at a depth of 20 to 30 cm, the content of mercury decreases to 0.05 ppm, which is still in the normal range. B1 hole at a depth of 0-10 cm, 10 to 20 cm and 20 to 30 cm has constant content of around 0.02 ppm, which is still below the normal limit. The content in C1 hole at a depth of 0 to 10 cm ranged from0.02 ppm, at a depth of 10 to 20 cm the content of mercury increased to 0.03 ppm, while at a depth of 20 to 30 cm the content decreased to 0.02 ppm which is still below the normal limit. The content of mercury in D1 hole at the depth of 0 to 10 cm and 10 to 20 cm is constant that is around 0.04 ppm. While at a depth of 20 to 30 cm the content decreased to 0.03 ppm, which is still below the normal limit. A graph of the results of analysis of mercury (Hg) vertically in holes A2, B2, C2 and D2 with the depths of 0 to 10 cm, 10 to 20 cm and 20 to 30 cm, around of area of Tamangapa Antang Landfill are shown in Figure 3.

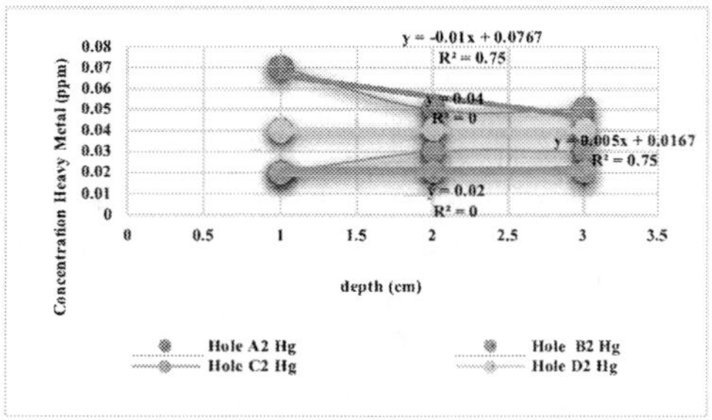

Figure 3. The graph of the results of analysis of mercury (Hg) vertically in holes A2, B2, C2 and D2, for the depth of 0 to 10 cm (upper), 10 to 20 cm (middle) and 20 to 30 cm (lower) around Tamangapa Antang Landfill, Makassar City, Indonesia.

A2 hole at a depth of 0 to 10 cm has a content of about 0.07 ppm, which is the concentration still at the normal limit (the value is illustrated in Figure 3). While at the position of 10 to 20 cm and 20 to 30 cm, the content decreased to 0.05 ppm which is still below normal limits. In B2 hole at the position of 0 to 10 cm, the content is about 0.02 ppm, while at the position of 10 to 20 cm and 20 to 30 cm the content is increasing to 0.03 ppm. The content of merkury in this hole is still below the normal limit (see Table 1 and Figure 3). C2 hole at the position of 0 to 10 cm, 10 to 20 cm and 20 to 30 cm have a constant content with a value of 0.02 ppm, which is still below the normal limit. Similar to C2 hole, D2 hole at the position of 0 to 10 cm, 10 to 20 cm and 20 to 30 cm have a constant content, and the value are 0.04 ppm respectively, which is still below the normal limit.

A graph of the results of mercury (Hg) heavy metal vertical analysis on holes A3, B3, C3 and D3 with the position of 0 to 10 cm, 10 to 20 cm and 20 to 30 cm, around Tamangapa Antang Landfill Site are shown in Figure 4.

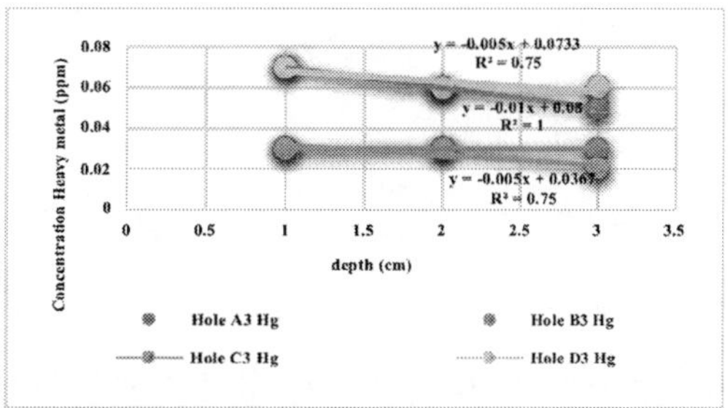

Figure 4. The graph of the results of analysis of mercury (Hg) vertically in holes A3, B3, C3 and D3, for at the depth of 0 to 10 cm (upper), 10 to 20 cm (middle) and 20 to 30 cm (lower) around Tamangapa Antang Landfill, Makassar City, Indonesia.

From Figure 4 it can be seen that the A3 hole at 0 to 10 cm depth have concentrations 0.07 ppm, at the position of 10 to 20 cm the content decreases to 0.06 ppm, which has exceeded the normal threshold. While at the position of 20 to 30 cm the content decreases again to 0.05 ppm, which is still within the normal range. B3 hole at the position of 0 to 10 cm, 10 to 20 cm and 20 to 30 cm has a constant content of about 0.03 ppm, which is still below the normal limit. C3 hole at the position of 0 to 10 cm and 10 to 20 cm have content ranging from 0.03 ppm and at the position of 20 to 30 cm the content decreases to 0.02 ppm, which is still below the normal limit. D3 hole at the position of 0 to 10 cm have content ranging from 0.07 ppm, while at the position of 10 to 20 cm and 20 to 30 cm the content decreases to 0.06 ppm, which has exceeded the normal threshold.

Graph of the results of laterally analysis of mercury (Hg) heavy metal in holes A1, A2 dan A3 (line A), with the position of 0 to 10 cm, 10 to 20 cm and 20 to 30 cm, around Tamangapa Antang Landfill is shown in Figure 5.

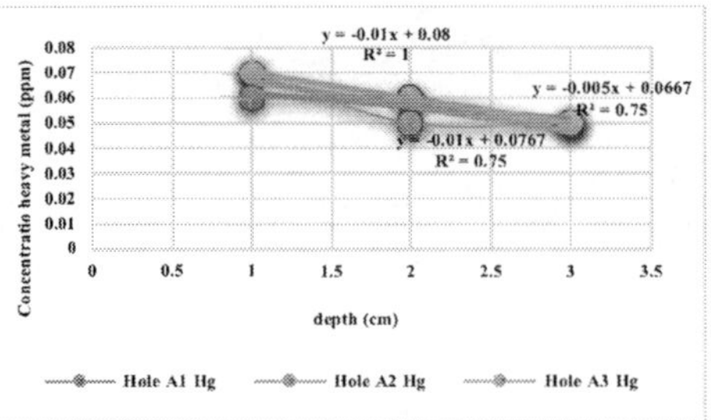

Figure 5. The graph of results of mercury (Hg) heavy metal laterally analysis in holes A1, A2, and A3, for the depth of 0 to 10 cm (upper), 10 to 20 cm (middle) and 20 to 30 cm (lower) around Tamangapa Antang Landfill, Makassar City, Indonesia.

From Figure 5 it can be seen that the A1 hole at the position of 0 to 10 cm and 10 to 20 cm have content ranging from 0.06 ppm, which has exceeded the average threshold. While at the position of 20 - 30 cm the content decreases to 0.05 ppm, which is still within the normal threshold. The A2 hole at the position of 0 to 10 cm has the content ranging from 0.07 ppm, which has exceeded the average threshold. While at the position of 10 to 20 cm and 20 to 30 cm the content decreased to 0.05 ppm, which is still at the normal limit. A3 Hole at the position of 0 to 10 cm and 10 to 20 cm has content ranging from 0.07 ppm and 0.06 ppm, which has exceeded the average threshold. While at position of 20 to 30 cm the content is about 0.05 ppm, which is still at the normal limit.

Graph of the results of lateral analysis of mercury (Hg) heavy metal in holes B1, B2 and B3 (line B), with the position of 0 to 10 cm, 10 to 20 cm and 20 to 30 cm, around Tamangapa Antang Landfill is shown in Figure 6.

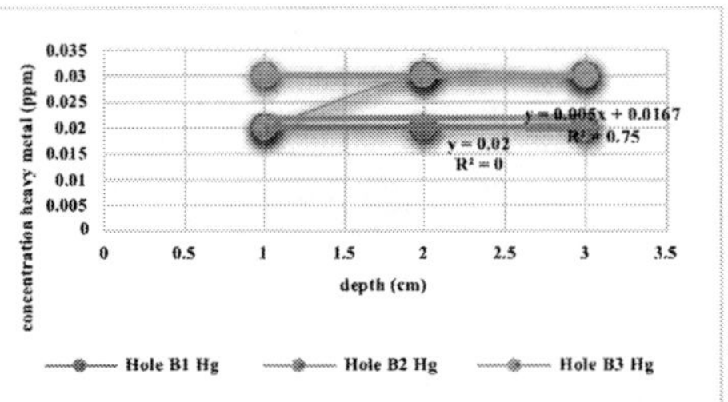

Figure 6. The graph of the results of analysis of mercury (Hg) laterally in holes B1, B2 dan B3 for at the position of 0 to 10 cm (upper), 10 to 20 cm (middle) and 20 to 30 cm (lower) around Tamangapa Antang Landfill, Makassar City, Indonesia.

From Figure 6 it can be seen that B1 hole at the position of 0 to 10 cm have content ranging from 0.02 ppm, at the position of 10 to 20 cm the content increases to 0.03 ppm, while at the position of 20 to 30 cm the content decreases to 0.02 ppm, which is still below the limit. B2 hole at the position of 0 to 10 cm has content ranging from 0.02 ppm. While at the position of 10 to 20 cm and 20 to 30 cm the content increased to 0.03 ppm, which is still below normal limit. Hole B3 at the position of 0 to 10 cm, 10 to 20 cm and 20 to 30 cm has a constant content of about 0.03 ppm, which is still below the normal limit. A graph of the result of lateral analysis of mercury (Hg) heavy metal in holes C1, C2 and C3 (line C), with the position of 0 to 10 cm, 10 to 20 cm and 20 to 30 cm, around Tamangapa Antang Landfill is shwon in Figure 7.

IOP Conf. Series: Earth and Environmental Science **279** (2019) 012020 doi:10.1088/1755-1315/279/1/012020

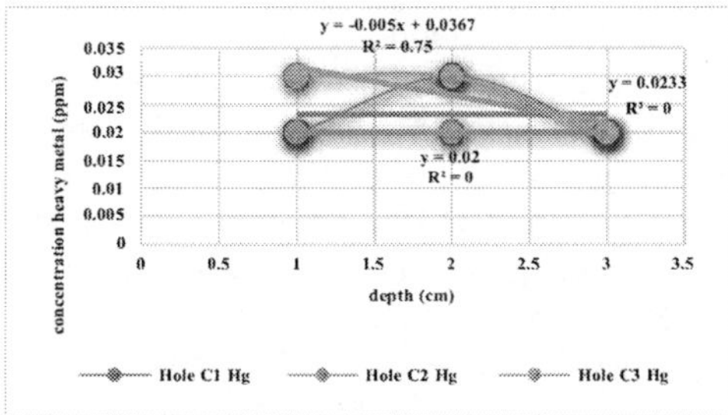

Figure 7. The graph of the results of analysis of mercury (Hg) laterally in holes C1, C2 dan C3 for at the position of 0 to 10 cm (upper), 10 to 20 cm (middle) and 20 to 30 cm (lower) around Tamangapa Antang Landfill, Makassar City, Indonesia.

From Figure 7 The data shows that C1 hole at the position of 0 to 10 cm have content ranging from 0.02 ppm, at the position of 10 to 20 cm the content increases to 0.03 ppm while at the position of 20 to 30 cm the content decreases to 0.02 ppm, which is still within normal limit. C2 hole at the position of 0 to 10 cm, 10 to 20 cm and 20-30 cm have a constant content of about 0.02 ppm, which is still below the normal limit. C3 hole at the position of 0 to 10 cm and 10 to 20 cm have a constant content of about 0.03 ppm, while at the position of 20 to 30 cm the content decreases to 0.02 ppm, which is still below normal limit.

A graph of the results of lateral analysis of mercury (Hg) heavy metal in holes D1, D2 and D3 (line D), with the position of 0 to 10 cm, 10 to 20 cm and 20 to 30 cm, around Tamangapa Antang Landfill is shown in Figure 8.

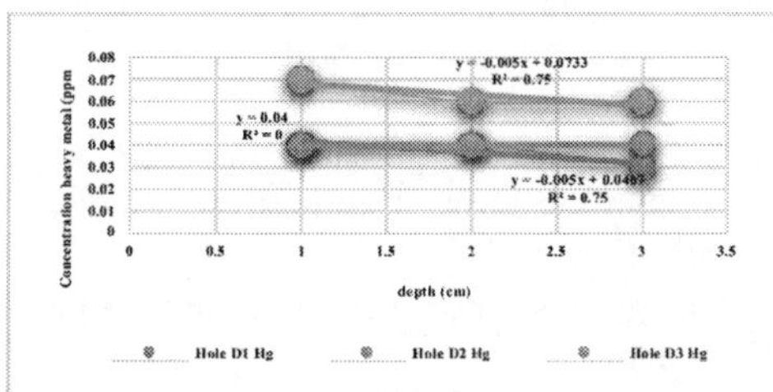

Figure 8. The graph of the results of analysis of mercury (Hg) laterally in holes D1, D2, and D3, for at the depth of 0 to 10 cm (upper), 10 to 20 cm (middle) and 20 to 30 cm (lower) around Tamangapa Antang landfill, Makassar City, Indonesia.

Figure 8 shows that D1 hole at the positiion of 0 to 10 cm and 10 to 20 cm have content ranging from 0.04 ppm, while at the position of 20 to 30 cm the content decreases to 0.03 ppm, which is still below the normal limit. D2 hole at the position of 0 to 10 cm, 10 to 20 cm and 20 to 30 cm has a constant content of about 0.04 ppm, which is still below the normal limit. D3 hole at the position of 0 to 10 cm have content ranging from 0.07 ppm, while at the position of 10 to 20 cm and 20 to 30 cm the content decreases to 0.06 ppm, which has exceeded the threshold.

3.2. Discussion

The distribution and abundance of heavy metals in the study site contained a lot of metallic mercury (Hg) which far exceeded the threshold set by [8]. The results of the analysis indicated that the total content of mercury (Hg) in Tamangapa Antang Landfill, Makassar City, has spread both vertically and laterally with a very varied amount at each distance and depth on the soil surface (topsoil), with the level of enrichment exceeding the threshold of the expected limit of the soil environment.

Enrichment of mercury (Hg) heavy metal element content on the surface of the topsoil around the Antang Landfill is still high. This is caused by anthropogenic influences, leachate, soil characteristics, the amount of organic waste and inorganic wastes that are discharged to the Tamangapa Antang Landfill in Makassar [9], [10]. exacerbated by very acidic pH, which ranges from 4.82 to 6, in acidic pH heavy metal in the soil are relatively high.

Low soil acidity can increase the accumulation of organic acids in the soil. The low retention or binding capacity of the soil against mercury (Hg) on line A, line B, line C and Line D is expected due to the low clay content. Heavy metal is carried by leachate and will settle in the soil, so that heavy metals such as mercury (Hg) will continue to accumulate in the soil [11]. The heavy metal will affect the environmental quality of the soil and groundwater around the landfill. This has a great impact on the lives of the communities around Tamangapa Antang landfill, because most of the people around the study area are still using dug wells as a source of drinking water [12]. Acidic pH is very dangerous for agricultural purposes, because plants take and accumulate heavy metals from contaminated soil [13]. Metal pollution in the soil can make organic complex reactions that dissolve at normal pH and then infiltrate with runoff into deeper soil layers.

The high mobility of mercury (Hg) on line A and line D is thought to be due to low soil pH (acid). This results in an increase in soil reduction due to the addition of organic matter, as well as an increase in the content of organic acids produced by organic matter. Mercury (Hg) in this study is mostly contained in the soil surface (topsoil). The mobility of metal mercury (Hg) in the soil depends on humus acid and the availability of organic matter because all heavy metals such as mercury (Hg) are very strong at absorbing organic matter on the soil. Mercury (Hg) can bind organic components, such as on line A and line D which contains a lot of mercury (Hg) above the average threshold is most likely because line A and line D are still actively receiving waste. When decomposing waste/garbage is still in process, a lot of waste/garbage is disposed of in the landfill without being sorted. Some an-organic waste is still in the process of decomposition on the soil surface compared to line B and Line C which is no longer actively receiving waste.

The Tamangapa Antang Landfill Site has ultisol soil (reddish soil that contains lots of clay layers), thus showing plasticity, encouraging flooding and surface water pollution [14]. The average soil pH in the analyzed Tamangapa Antang Landfill ranged from 4.82 to 6, acidic with an average value of 4.84, the average temperature ranged between 25°C to 31°C. pH affects metal mobility in the soil [5]. Low pH or acidic pH will help the availability of mobility and redistribution of heavy metal mercury (Hg) in various fractions due to an increase in ion solubility in acidic soil environments [5]. Relatively acidic soil has increased solubility and micronutrient mobility so that the concentration of heavy metals in the soil increases [14], [5].

Mercury (Hg) is a metal that is very toxic to organisms. In certain uses or activities mercury (Hg) will be distributed to the environment in the form of agriculture, medicine, paint, paper, mining and industrial waste. All forms of mercury (Hg), damage to hair and teeth, loss of memory and disruption of the nervous system. This is very dangerous for the health of the population [7], especially the people who live around the landfill environment.

4. Conclusion

The results showed that the soil around the Antang Landfill Site, Makassar City, Indonesia was contaminated by mercury heavy metals (Hg), with concentrations exceeding the threshold of each depth both vertically and laterally. Line A for A1-hole at 0 to 10 cm and 10 to 20 cm depth shows a concentration with a value of 0.06 ppm, has exceeded the threshold. At the position of 20 to 30 cm indicate a concentration with a value of 0.05 ppm. A2 hole at the position of 0 to10 cm the content of Hg show 0.07 ppm, has already exceeded the threshold. While at the position of 10 to 20 cm and 20 to 30 cm the conten of mercury shows a value of 0.05 ppm. A3 hole at the position of 0 to 10 cm and 10 to 20 cm, the mercury content is 0.05 ppm. B1 hole at the position of 0 to 10 cm 10 to 20 cm and 20 to 30 cm, the content shows a value of 0.02 ppm respectively. B2 hole at the position of 0 to 10 cm the content show 0.02 ppm, while at the position of 10 to 20 cm and 20 to 30 cm the content of mercury show a value of 0.03 ppm. B3 hole at the position of 0 to 10 cm, 10 to 20 cm and 20 to 30 cm the content is 0.03 ppm. C1 hole at the position of 0 to 10 cm and 10 to 20 cm the value show a content of 0.02 ppm. While at the position of 10 to 20 cm it ranges from 0.03 ppm. C2 hole at the position of 0 to 10 cm, 10 to 20 cm and 20 to 30 cm the value is 0.03 ppm. C3 hole at the position of 0 to 10 cm and 10 to 20 cm the content of mercury is 0.03 ppm. While at the position of 20 to 30 cm the content is 0.02 ppm. Line D for the D1 hole at the position of 0 to 10 cm and 10 to 20 cm shows a content value of 0.04 ppm. While at the position of 20 to 30 cm shows a mercury content value of ppm. D2 hole at the position of 0 to 10 cm, 10 to 20 cm and 20 to 30 cm the content value is ppm respectively. D3 hole at the position of 0 to 10 cm the content of mercury is 0.07 ppm, while at the position of 10 to 20 cm and 20 to 30 cm the content is 0.06. In general the content of metal mercury is still below of treshold (A1 hole at the position of 20 to 30 cm; A2 hole at the position of 10 to 20 cm and 20 to 30 cm; A3 hole at the position of 0 to 10 cm and 10 to 20 cm; B3 hole at the position of 0 to 10 cm, 10 to 20 cm and 20 to 30 cm; C3 hole at the position of 0 to 10 cm and 10 to 20 cm; C3 hole at the position of 20 to 30 cm; D1 hole at the position of 0 to 10 cm and 10 to 20 cm. D1 hole at the position of 20 to 30 cm. The pH of the soil is acidic which ranges from 4.82 to 6.0 and is one of the main factors that can affect the mobility of metal mercury (Hg), in the soil environment. Metal mercury (Hg) from landfills, can be damage to the environmental and health of human being due to its toxicity in the organism. Soil is very important for humans, because it can produce nutrients for plantations. Soil has the potential, to withstand additional heavy metal loads further under acidic pH conditions.

References

[1] Arbain N K, Mardana I, Sujana B 2011 *Journal of Environmental Science* 3(2) 55-60; ISSN: 1907-5626; pp: 61-66;.

[2] Dunya A, Al-Jibury, Essa S K 2016 *Journal of Environment and Earth Science* 6(10).

[3] Darmono 1995 *Logam Dalam Sistem Biologi Mahluk Hidup* Universitas Indonesia Jakarta.

[4] Parth V, Murthy N N, Saxena P R 2011 *E3 Journal of Environmental Research And Management* 2(2) 027-034.

[5] Sayyed, M.R.G., and Sayadi, M.H., 2011. Variations in the heavy metal accumulations within the surface soils from the Chitgar industrial area of Tehran, **1(1):36 46;** IAEES.

[6] Odu C T, Esurosu O, Nwaboshi I C, Ogunwale J A 1985 *Environmental study (Soil and Vegetation) of NigeriaAgip Oil Company Operation Area.* A report submitted to Nigeria Agip Oil Company Limited, Lagos, Nigeria. pp. 102-107.

[7] Su C, Jiang L, Zhang W 2014 *Environmental Skeptics and Critics* 3(2) 24-38.

[8] Direktorat Jenderal Pengawasan Obat dan Makanan 1989 *Lampiran Surat Keputusan Direktur Jenderal Pengawasan Obat dan Makanan* No. 03725/B/SK/VII/89

[9] Bityukova L, Shogenova A, Birke M 2000 *Environmental Geochemistry and Health* 22 173-193.

[10] Dunya A, Al-Jibury, Essa S K 2016 *Journal of Environment and Earth Science* 6(10).

[11] Sulaiman, Maigari M B A U, Danladi S 2016 *International Journal of Science, Environment* 5(5) 3059-3068.

[12] Arifin M F, Imran A M , Ramli M, Mukhsan Putra Hatta 2014 *Prosiding Seminar Nasional Geofisika.*

[13] Wei S, Zhou Q, Xin W., 2005. Identification of weed plants excluding the uptake of heavy metals. Environ. Intl. **31:829-834.**

[14] Peraturan daerah kota Makassar Nomor 6 Tahun 2006. Tentang Rencana tata Ruang Wilayah Kota Makassar 2005-2015.

IOP Conf. Series: Earth and Environmental Science **279** (2019) 012021 doi:10.1088/1755-1315/279/1/012021

Groundwater Conservation with Hole Infiltration of Biopore Cube

Muhammad Hamzah Syahruddin*, Amiruddin, Halmar Halide, Sakka, Makhrani

Geophysics, Hasanuddin University, Makassar, Indonesia

*hamzah@fmipa.unhas.ac.id

Abstract. Analysis of groundwater conservation with cube biopore infiltration holes using the value of soil permeability, the amount of rainfall, and the depth of the groundwater level. The groundwater conservation research method is carried out by making the cavity biopore infiltration holes to absorb the volume of rainwater and domestic waste water into the soil. Conservation is suitable for homes that have enough yard. The main reason is to use cube infiltration holes to accommodate large and small organic waste. Organic waste is trees, grass and domestic waste. For a small house yard, conservation of ground water enough with a cylindrical hole diameter of 10 cm. Benefits of biopore infiltration holes are reducing surface runoff, producing manure, fertilizing soil, reducing waste piles, and conserving ground water. The research objective was to analyze the cube biopore infiltration holes based on soil permeability, rainfall intensity, depth of groundwater level, and volume of organic waste. The results of the analysis can be obtained from the cube biopore infiltration holes which can absorb all the rainwater which can accommodate all organic waste. The results of permeability testing show that the average soil permeability is 0.00112 cm/s and the yield of maximum discharge runoff water in January is 97.57 cm3/second. The ground water level is two meters. While the volume of organic waste is one cubic per month. The result of the analysis showed that the area of cake absorption was 8.73 m2. When converted to a cube absorption field it takes approximately 8 meters wide by 0.5 meters and a depth of 0.5 meters. When divided into four cavity recharge holes LRB1 obtained area of 4 m2, LRB2 area of 2.7 m2, LRB3 area 2 m2, and LRB4 as reserve.

Keywords: Groundwater, Conservation, Cube LRB, Permeability, Rainfall, Organic Waste.

1. Introduction

The main problem of some parts of Indonesia is the water crisis during the dry season and flood in the rainy season. Does this crisis happen naturally or is it caused by human activity itself? Of course there are droughts and floods due to extreme weather. But there is also much environmental damage due to human activities resulting in drought and flooding. Some examples of human activities that can cause damage to the environment on land and in the ocean. Among them are excessive groundwater exploitation. Unplanned housing construction. Logging and burning forests. Sand mining in the sea.

Excessive use of chemicals and pesticides. Exploitation of marine resources. Use of motorized vehicles and others.

Makassar City residents still use a lot of ground water for their daily needs. As an illustration, especially in Mangala District, there are 62 drill wells. Of 62 drilled well data is taken at a distance of 50 meters [7]. If taken on a smaller space, more drilled wells will be found. In addition to drilling wells, the community also has a lot of digging wells. This shows that groundwater use in Makassar is still very massive.

Along with human growth and activity, the absorption of water into the soil becomes more difficult. Because human activity can cause a decrease in soil permeability [5]. Human activities by building infrastructure housing, roads, buildings and other facilities without taking into account the green open space resulted in increasingly narrow water absorption. The surface of the soil is getting denser so the pores of the soil get smaller. This causes the running water (run off) the greater. Another cause is the disposal of garbage in the flow of water bodies, so the water cannot flow smoothly that eventually lead to flooding.

To overcome the disruption of water absorption into the soil need technology. The technology needed is a simple technology that is easy for everyone to do. There are examples of simple technologies developed to cope with drought during the dry season and cope with flooding during the rainy season. Naturally keeping the forest is the solution, but since the forests are already bare deforested, it needs to be encouraged to replant trees for reforestation. In urban areas it is necessary to develop green open land and artificial lakes as recharge areas. In the blood of settlements developed technology of making hole biopori infiltration [3]. While deep groundwater exploitation through wellbore, restored by injection well [6].

In this research, cavity-shaped biopore infiltration holes are developed to absorb all the water in the soil surface. The water on the soil surface that is infiltration is rain water. The rain water covers the water that falls in the yard, and the water that falls from the roof of the house. In addition, water that needs to be infiltration is household waste water. As research material is taken case study of a house of resident in Antang unhas housing. The land area is 325 m2 and there is a house on it type 54.

2. Biopore Infiltration Holes (BIH)

Biopore is a continuous burrow macro pore that serves to accelerate the infiltration of water into the soil. The macro-shaped pores of this continuous shaft are formed by the activity of flora and fauna. The pores formed due to flora activity are decayed roots. While the pores formed by the soil fauna activity are worms, termites, and ants [2,3].

Biopore infiltration holes are holes that can be cylindrical or cube-shaped. The shape of the biopore infiltration hole depends on the available land. If the land is large enough, we are free to design a biopori infiltration hole according to the wishes of the owner. But if the land is narrow then simply made a cylindrical hole diameter of 10 cm. The depth of the biopore infiltration hole is not more than 100 cm from the soil surface. It should be noted that the depth of LRB should be adjusted to the depth of the groundwater level. The depth of the LRB is above groundwater depth.

The size of the biopore infiltration hole (BIH) depends on the amount of water discharge infiltration into the soil and the soil permeability value. If the water discharge is infused enough then the size of the BIH infiltration field is greater. Conversely, if the water is infiltrated slightly the area of BIH infiltration becomes smaller. What is more crucial is the soil permeability. If the soil permeability is large then the size of the LRB recharge field is smaller.

According to [4], for cylindrical holes, the size or number of biopore infiltration holes needed in certain areas of land is,

$$N = IxA / P \quad (1)$$

Information:
N = Total LRB
I = Rain intensity (mm / h)
A = Area of the impermeable area (m2)
P = Water absorption per hole (liter / hour).

IOP Conf. Series: Earth and Environmental Science **279** (2019) 012021 doi:10.1088/1755-1315/279/1/012021

Water infiltration in the soil occurs vertically and horizontally. Vertical water infiltration rate is smaller than the horizontal water infiltration rate. For fine-grained soils the rate of water impregnation is horizontally 10 times larger than the rate of vertical infiltration. From this is the significance of making the rainfall hole (LR) into the soil. Storage of water in the soil and the rate of infiltration of water into the soil will increase further after the formation of biopori. Biopori is formed due to the activity of small animals in the soil and the roots of trees that have been weathered.

The size of the absorption plane of a cubic-shaped biopore infiltration with a certain depth (h) can be calculated from the water discharge to be infiltration. The water discharge that is infiltration is the amount of rainfall multiplied by the land surface area where the rainwater falls (L). In order to absorb all water on the surface, the extent of the LRB infiltration of the cube must be greater than or equal to the extent of water absorption in the soil permeability. In other words the rate of water absorption into the soil is equal to or greater than the surface water rate. Thus surface water from rainwater and household wastewater can be absorbed all into the ground. The calculation area of the cube LRB infiltration area is based on the highest rainfall in January so that there is no running water left throughout the year. The calculation of the LRB recharge area of the cube uses equation 2 as follows.

$$A = (IxL / K) \qquad (2)$$

Information:

A = cubic LRB infiltration area (m^2)

I = rainfall intensity (m / day)

L = land surface area (m^2)

K = soil permeability (m/day)

3. Research Methods

This research is done by applying the biopore infiltration hole to settle rain water that falls in the yard and rain water that fall from the roof of the house. In addition, this biopori infiltration hole is also used to absorb household wastewater. Biopori infiltration holes are designed to absorb all of the water so there is no need to throw water into the ditch

The location of this research was conducted in Unhas Antang in one of the houses. The location of this study is at coordinates (-5.1691446S, 119.4801432E). The study area is tilted westward with a slope of 10°. There are two dug wells near the research location in the west. The depth of the groundwater is measured from the two wells. The measurement of the depth of the well is three meters. The land area is 325 m^2. There is a house on it type 54 with size (6 x 9) meters. The area of the yard is 271 m^2. The sketch and design of a cubical biopori infiltration hole can be seen in Figure 1.

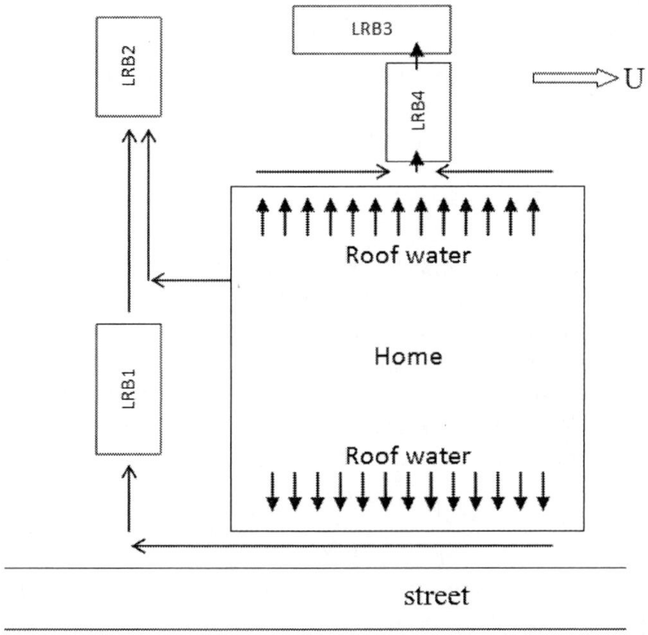

Figure 1. Sketch of Research Area LRB cube

In Figure 1 we can see that four LRBs are LRB1, LRB2, LRB3, LRB4. The size of each LRB cube is calculated based on the highest rainfall. Why is that, because the volume of water falling on the ground reaches the maximum in the highest rainfall. If maximum rainfall has been recovered all the soil can automatically absorb all water for smaller rainfall. Therefore the desired LRB design is LRB that can absorb the maximum volume of rainwater that falls on the ground surface.

The data required in the LRB cube analysis is the permeability of the soil research area, and the rainfall data of the research area. In addition, data on groundwater depth and organic waste are required. The depth of the groundwater level becomes a benchmark of LRB depth. Because the depth of LRB should be above groundwater depth. To enable LRB in organic trash required. Organic waste used is waste from household waste and from plants and trees contained in the research area.

4. Results and Discussion

To know the depth of ground water level in the research area using the existing dug wells around it. There are two dug wells near the research area in the west. The depth of the groundwater level is measured using a meter from the ground surface to the groundwater in the dug well. From the measurement results, it is known that the depth of groundwater in the study area is three meters. This data is taken from two wells close to the research site in July 2018. From the groundwater depth data, it can be made LRB depth more than one meter.

The long cube shaped LRB design aims to accommodate all forms of organic waste that exist. There is a large amount of organic waste collected in LRB and small ones. Large rubbish like tree trunks, tree stems, tree branches. Small-sized rubbish is leaves, grasses, household organic waste.

The research area covering an area of 325 m^2 contains many trees. Among them are four mango trees, two rambutan trees, and one durian tree. There are also plants such as moringa, srikaya, papaya, lombok, pariah, cereals, taro, galangal, turmeric etc. There are also many types of grasses. Garbage from these herbs and organic waste from household waste is used to fill LRB cubes. Volume of organic waste from household waste, trees, from plants not less than one cubic per month. This organic waste becomes food for small animals and some experience decay.

This organic waste quickly decomposes in LRB cubes because it is always moist so there are many biantang-animals such as termites, ants, earthworms etc. These animals develop quickly and make

IOP Conf. Series: Earth and Environmental Science **279** (2019) 012021 doi:10.1088/1755-1315/279/1/012021

more holes in the ground. The holes formed as soil pores can store and absorb more water in the soil. The remains of organic waste will break down into compost that fertilizes the soil.

LRB maintenance needs to be done regularly. The trick is to lift organic waste that has broken down into compost to the surface of the soil. Then spread to surrounding plants as fertilizer. Maintenance is carried out every six months. So the function of the Cube LRB is to fertilize the soil and overcome runaway water.

4.1. Soil permeability

To know the size of soil permeability in the research area, soil permeability test is done directly in the field (in-situ). The test was conducted at the beginning of July 2018. In July still often rain in the study area so that the soil conditions are still wet when testing. Soil permeability testing method is to make a cylindrical absorption hole with a diameter of 10 cm with a depth of 20 to 120 cm. The volume of water inserted into the cylindrical hole varies from 2 to 3.5 liters. Each hole is filled with a certain volume of water to be infiltration in the soil. Then calculated the time of impregnation with stopwatch. Permeability is calculated by equation 3.

$$K = V / A.t \tag{3}$$

Information:
K = permeability of soil
V = volume of infiltration water (cm^3)
A = Area of Cylindrical LSR recharge area (cm^2)
t = infiltration time (s)

The results of direct permeability testing in the field can be seen in Table 1.

Table 1. Results Permeability Test of Research Area

Parameter	LSR1	LSR2	LSR3	LSR4	LSR5	Average
Depth (cm)	20	25	30	65	120	52
Volume(cm^3)	1000	1500	2500	1000	1000	1400
Time (s)	1080	1320	2820	4980	7200	3480
Debit(cm^3/s)	0.9259	1.1363	0.8865	0.2008	0.1388	0.4023
Permeability(cm/s)	0.0019	0.0023	0.00082	0.00042	0.00014	0.00112

LSR: Infiltration cylindrical hole

The test results show that the mean permeability in the research area is 0.00112 cm / s. The soil's ability to absorb rainwater in the research area on average is 0.4023 cm3 / sec. If there are one hundred cylindrical absorption holes then it can be infused with rainwater of 0.4023 liters / sec.

The absorption of water into the soil in the study area varies with depth. The results of soil permeability testing show that the deeper the water hole seeps into the soil. The result of soil permeability test to depth can be seen in **Figure 2**.

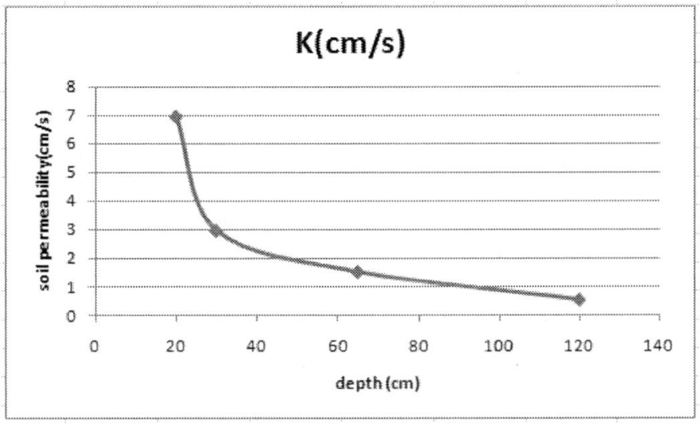

Figure 2. Changes in soil permeability to depth

4.2 Rainfall

Rainfall data used in this study is the rainfall data of Makassar City obtained from BMKG region VI Sulawesi. Rainfall data used is the last 12 years rainfall data is rainfall data year 2000 to 2011. Average annual rainfall in the city of Makassar in the last twelve years is 3060.04 mm / year, while the average monthly rainfall is 255,003 mm / month. Average rainfall per day in the last 12 years is 0.69864 mm / day while the average rainfall per second is 0.000000698 mm / s.

The highest rainfall occurred in January averaging 711 mm and the lowest in August 17.41 mm in the period of 12 year. High monthly rainfall in Makassar City on average within 12 years from 2000 to 2011 can be seen in **Figure 3**.

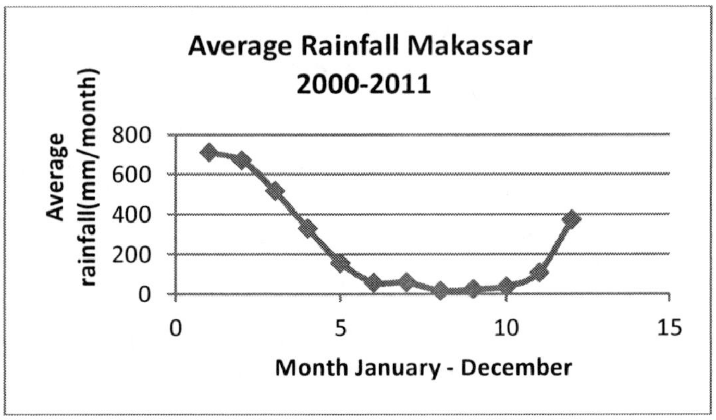

Figure 3. Average monthly rainfall of Makassar City from 2000-2011

In the analysis of LRB manufacture is used maximum rainfall that is in January. Because LRB optimization is based on maximum rainfall. If the maximum rainfall in January 711 mm then the average rainfall per day in January is 23 mm/day. This means that the rainwater infiltration in the soil per day in the study area of 325 m^2 is 7,475 m^3 or 86 cm^3/sec so that no surface water runoff occurs. **Table 1** shows the average soil permeability of 0.00112 cm/sec and the rain water infiltration in January 86 cm^3/second then the minimum area of rainwater infiltration is 77060 cm^2 or 7,706 m^2.

Another water source infiltration into the soil is household waste water. The results of a survey conducted by the Directorate of Drinking Water Development, Directorate General of Human Settlements in 2006 showed that every Indonesian consumes an average water of 144 liters per day. Use of clean water in Makassar City an average of 190 liters / person / day [1]. If the water usage per person per day is 200 liters / day then it takes 1000 liters of water or one cubic / day. If converted to

cm^3/sec then obtained 11.57 cm^3/sec. For household wastewater, the required area of impregnation is 10330 cm^2 or 1.033 m^2.

The maximum flow of running water is the rain water discharge plus the discharge of household waste water that is 97.57 cm^3/second. From the data of soil permeability and the running water debit can be obtained the extent of LRB recharge area of cube using equation (2). The result of this calculation shows that the total area of infiltration of LRB cube running water is 8.73 m^2. When converted to LRB cube then it takes approximately 8 meters wide by 0.5 meters and a depth of 0.5 meters. So based on the data permeability and rainfall data can be determined the size of the four areas of LRB recharge cube as in Figure 1 above. The size of the four LRB recharge areas is LRB1 4 m^2, LRB2 2.7 m^2, LRB3 2 m^2, LRB4 as reserve.

5. Conclusions

The results of soil permeability testing of the study area showed that the average soil permeability was 0.00112 cm/s. The volume of water infiltration from rainfall and domestic waste is 97.57, cm^3/sec. To absorb the water is required area of water absorption in LRB cube is 8.73 m^2. When converted to LRB cube size it takes approximately 8 meters wide by 0.5 meters and a depth of 0.5 meters.

Acknowledgments

Our high appreciation and appreciation are given to LPPM Unhas who has provided funds for community service in Pa'batangan Village and Lengkese Village, Takalar District.

References

[1] Directorate General of Human Settlements Ministry of Public Works 2002 *Technical Guidelines for Planning, Clean Water Supply System Engineering Design* Jakarta.
[2] Brata K R and Anne N 2008 *Biopore Infiltration Holes* Penebar Swadaya: Depok.
[3] Brata K R 2006 *Biopore Infiltration Holes* IPB Bogor.
[4] Rahmasari A F, Surupin, Sudarno 2015*journal Wahana Teknik Sipil* **20** (1).
[5] Rosyida E and Wirosoedarmo R 2013 *Journal Agritech UGM* **33**.
[6] Sunarto B 2007 *Journal JSDA* **3** (4).
[7] Syahruddin M H, Amiruddin, Syamsuddin, Aswad S 2016 *Modeling of Deep and Quality Groundwater in Makassar City* LP2M Hasanuddin University.

Analysis of Seismicity Level in Central Sulawesi Period 1997 - 2017

Alif'ka Cendani Putri*, Bambang Harimei, Muhammad Fawzy Ismullah Massinai

Solid Geophysics Laboratory, Geophysics Study Program, Faculty of Mathematics and Natural Science, Hasanuddin University, Makassar, Indonesia

*alifkaputri11.acp@gmail.com

Abstract. Indonesia is located at an active tectonic zone, due to the encounter of three main slabs in the world: Eurasia, Pacific, and Indo-Australian slabs. This lead to high earthquake level in Indonesia, one of them is in Central Sulawesi. The research was conducted in order to identify the a value, b value, and seismicity using the Maximum Likelihood method. Earthquake data from the USGS catalog and BMKG were used from 1997 – 2017 with magnitude (M) $\geq$ 3 and depth (h) $\leq$ 1000 km, located at -3^0 S – 0^0 S and 118^0 E – 124^0 E. The result showed that highest seismicity is located at the Eastern part of Central Sulawesi. The Mc value of 5,2, a value ranges 5 – 11,5 indicates high seismic activity, b value ranges 0,6 – 1,8 indicates high level of rock fragility and rock resistance levels to low stress. The result in this study effectively to observe the seismicity which can be used as guided for earthquake mitigation.

1. Introduction

Indonesia is an earthquake-prone area which passed by the meeting pathway of three tectonic plates, the Indo-Australian Plate, the Eurasian Plate, and the Pacific Plate. The Indo-Australian Plate moves relatively North-ward and breaks into the Eurasian Plate, while the Pacific Plate moves relative to the West [1]. Plate meeting paths are in the sea, so that the large earthquake with shallow depth will also potentially cause a tsunami [2].

The inter-plate encounters that took place in Indonesia led to the complexity of Indonesian tectonics and the high level of seismicity. One of the areas with high seismicity is Central Sulawesi, whereas located close to the source of earthquake both on land and sea. The earthquake source of the sea comes from subduction in the north side of Sulawesi Island, while the source of earthquake on land is from several active faults in Central Sulawesi mainland, which is Palu Koro Fault [3].

Central Sulawesi is composed by the Pompangeo Complex, the limestones and the ofiolit. The Pompangeo complex is consist of schist, graphite, slate, genes, serpentinite, quartz and limestone bamboo. Pompangeo Complex rocks are included in rocks that have high seismicity. Based on the map in Figure 1, Central Sulawesi region composed of several faults [4]:

- Palu Koro Fault
  This fault is found in South Sulawesi's North West-North-East Sulawesi.
- Poso Fault
  The Poso Fault establishes a curved valley alignment with the metamorphic rock complex at the boundary and the South Sulawesi Mandala.
- Wekuli Fault
  This fault is the boundary between the Complex Ofiolit of East Sulawesi and the metamorphic rock Complex of Central Sulawesi.
- Matano Fault

IOP Conf. Series: Earth and Environmental Science **279** (2019) 012022 doi:10.1088/1755-1315/279/1/012022

Matano Fault in land is characterized by the straightness of the valley that stretches from the Southeast Sulawesi Sea Beach, bypassing the Poso Rise in Central Sulawesi and finally joining the Palu Koro Fault.

- Banggai Sula Block

  Banggai Sula microcontinent is a fragment of the Australian continent thought to be separated from its parent continent in the Mesozoic era. Then, through the Sinistral mechanism of Sorong Fault, this microcontinent is transported westward. Until finally collide with East Sulawesi Arm in Central Miocene [5].

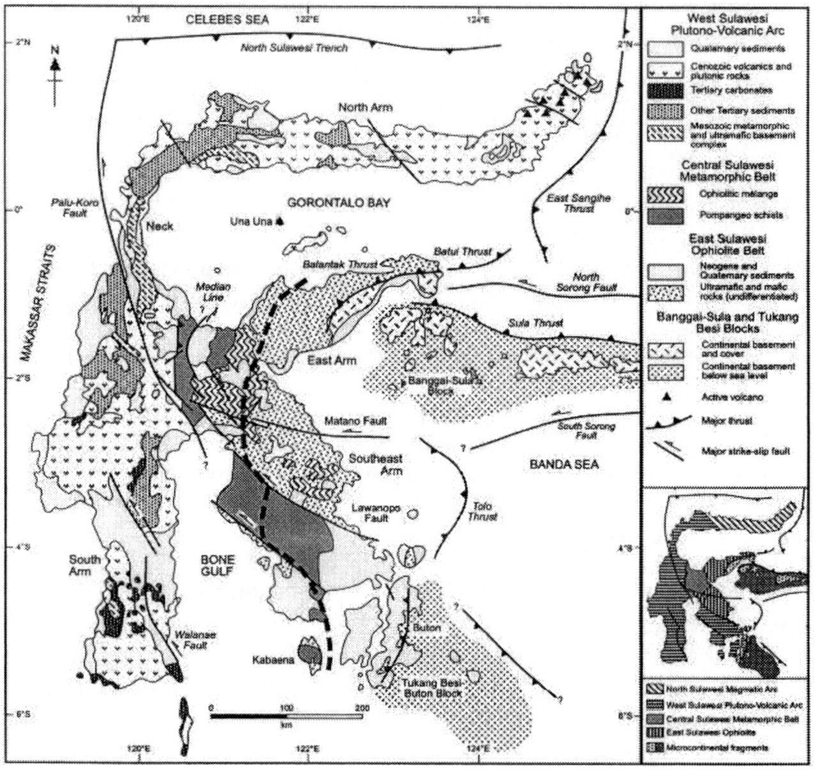

Figure 1. Geological Map of Sulawesi Island [6]

2. Methods and Data

The relation between frequency of earthquake occurrences with magnitude can be used to determine the number of earthquakes occurring in a region. This frequency and magnitude relationship produces **b** value (rock stress level) and **a** value (seismic activity). According to Gutenberg and Richter, The number of earthquakes (N) with magnitude (M) can be expressed by the following equation:

$$\log n(M) = a - bM \quad (1)$$

$$\log N(M) = \alpha - bM \quad (2)$$

With $\alpha = a - \log(b \ln 10)$

n (M) : the number of earthquakes that calculated from the occurence.

N(M) : the cumulative number of earthquakes is calculated on a certain scale.

a : A constant that depending on the period, wide of an area and the activity at the focus area.

b : Seismotectonic parameters, an area of an earthquake that depends on the nature of local rocks.

M : Magnitude

Seismic statistics are one of the research subjects that have been conducted for the past twenty years. In 1944, Gutenberg-Richter stated that the distribution of earthquakes in an area throughout the observed range of events would follow a formula known as the Gutenberg-Richter relation equation 2 [7].

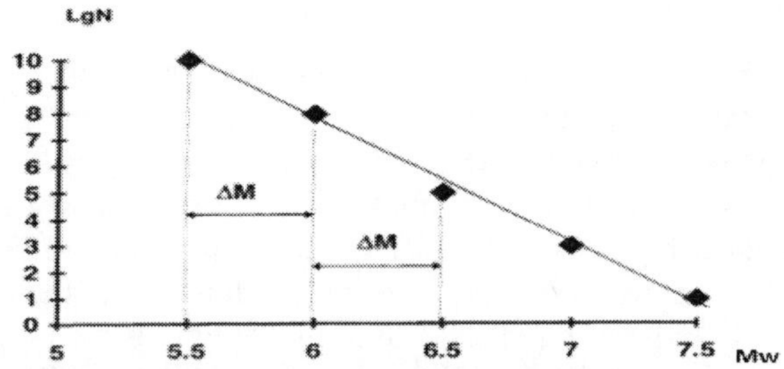

Figure 2. Relationship between Gutenberg-Richter which shows the relationship of logarithms of the number of earthquakes and magnitudes [7]

If the probability distribution function which depends on parameter b is f (M, b), then the likelihood function is defined as [8]:

$$L\,(M,b) = f(M_1,b).f(M_2,b).f(M_3,b)\ldots.(M_N,b) \quad (3)$$

$M_i$and b are the magnitude and gradient parameters of equation 2.

The estimated maximum likelihood value of b can be determined by the formula:

$$\frac{\partial ln\, L}{\partial b'} = 0 \quad (4)$$

The M_i probability distribution function in the magnitude-frequency distribution is:

$$f\,(M_i,b') = b'\, e^{-b'\,(M_i-M_0)}, M_i > M_0 \quad (5)$$

with b '= b. ln 10. Thus the sample likelihood function is:

$$L = (b')^N.exp\left[-b'\left(\textstyle\sum_i^N M_i - NM_0\right)\right] \quad (6)$$

Based on equation 6, estimation of maximum likelihood value of b is as follows [9]:

$$b = \frac{\log e}{\bar{M}-M_0} \quad (7)$$

The standard deviation of the value of b can be determined using the following formula [10]:

$$\delta b = 2.30b^2\sqrt{\sum_{i=1}^{n}\frac{(M_i-\bar{M})^2}{N\,(N-1)}} \quad (8)$$

Furthermore, the value of a can be estimated by the formula:

$$a = \log N(M) + b\,(\ln 10) + bM \quad (9)$$

ZMAP is an interactive software tool to investigate and visualize seismicity as a function of space and time made in 1994 by Stefan Wiemer. The main goal was to improve the resolution of potential seismicity anomalies and to study the overall characteristics of earthquake catalogs [11].

This research located at the Central Sulawesi region with coordinates -3^0 S - 0^0 S and 118^0 E - 124^0 E. Secondary data of earthquake occurrence in Central Sulawesi region at the period of 1997 - 2017 obtained from USGS and BMKG catalog.

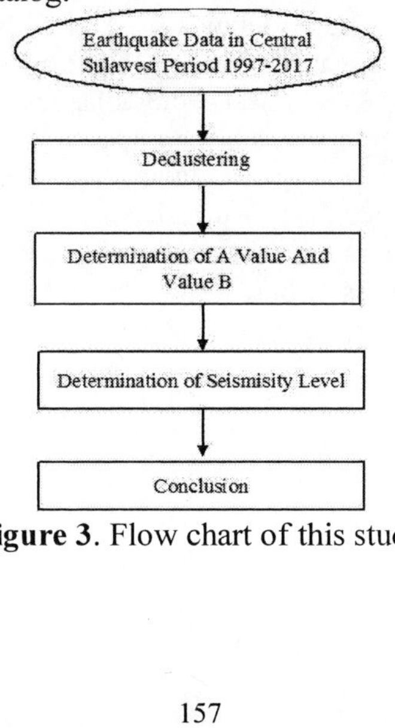

Figure 3. Flow chart of this study

3. Results and Discussion

Based on Figure 5 the Central Sulawesi region has a very high seismicity rate that is spread over the land and sea. The source of earthquake in Central Sulawesi region is located in the sea due to Sangihe Subduction and a group of active faults on land.

The dominant incidence of earthquake occurrence is with depth below 237 km in eastern area of research. The more dominant earthquakes in the eastern part of the study (122.70 E - 1240 E and 00 S – 1.50 S) follow structures such as Sangihe and fault of Sorong. It is also in the center of the study (119.70 E - 120.70 E and 0.20 S – 1.50 S) of earthquakes caused by the Palu Koro fault, Poso Fault, and Wekuli Fault. In the southeast of the study area (120.70 E - 122.60 E and 20 S - 30 S) the earthquake distribution was caused by the Matano Fault. An earthquake with a depth of 237 <Z <790 is scattered around Central Sulawesi.

The most important parameter in determining the value of a and b is the magnitude of completeness (Mc). Mc values indicate that the existing seismograph network is capable of recording earthquakes at such magnitudes well. The value of Magnitude of completeness (Mc) in the study area as Figure 4 is 5.2. This means that the USGS and BMKG catalogs are capable of recording earthquakes with the smallest magnitude of 5.2. Based on the distribution of frequency magnitude, the value of b is 1.7. This high b value means that the Central Sulawesi region has a low level of stress, meaning that it has the chance of frequent earthquakes. From the calculation also obtained a value of 11.5 which means that the region of Central Sulawesi has a high level of seismic activity.

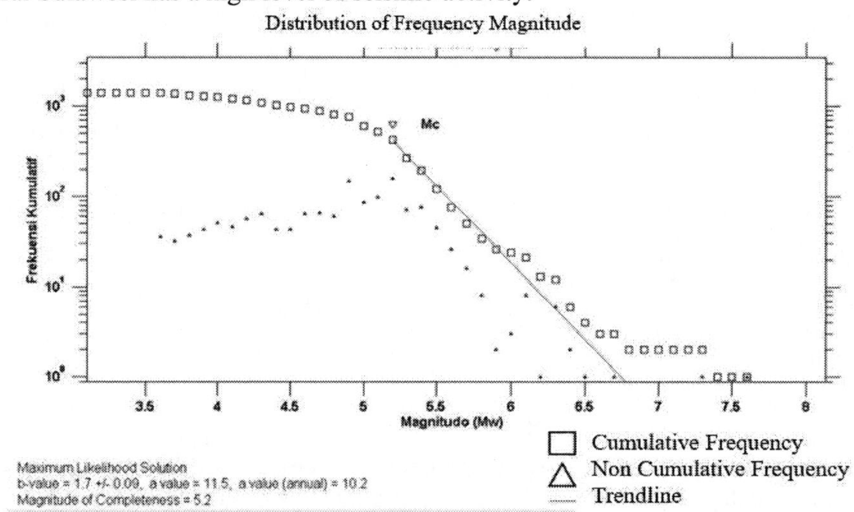

Figure 4. Distribution Frequency of Magnitude and Mc Value

Based on Figure 6, the southern Banggai Islands marked with dark blue and blue indicate that this area has low seismic activity. The areas of Palu, Majene, Morowali, East Luwu, West Banggai, Tojounauna, West Luwu Utara, East Poso, East Donggala, East North Mamuju are characterized by light blue and light green indicate that this area has high seismic activity. Yellow, orange and red in the South of Banggai, Banggai Islands, Eastern of Luwu Utara, Southwest Poso, South and Northwest Donggala, West North Mamuju indicate that the area has a high a value that indicated a highest level of seismicity. It has been proven that the number of earthquake events in this region over the past 20 years has been the magnitude of completeness of 5.2.

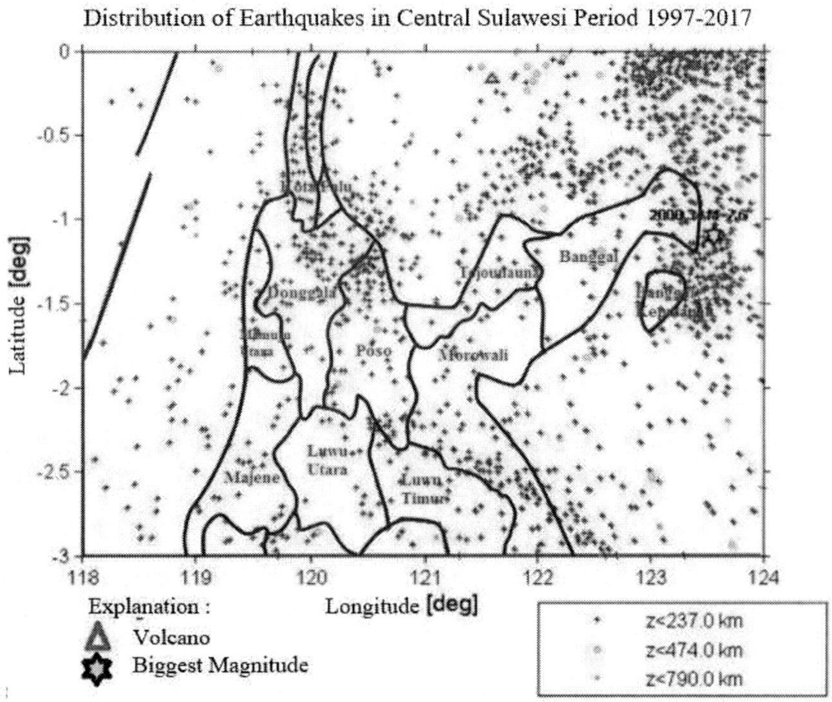

Figure 5. Distribution of Earthquake

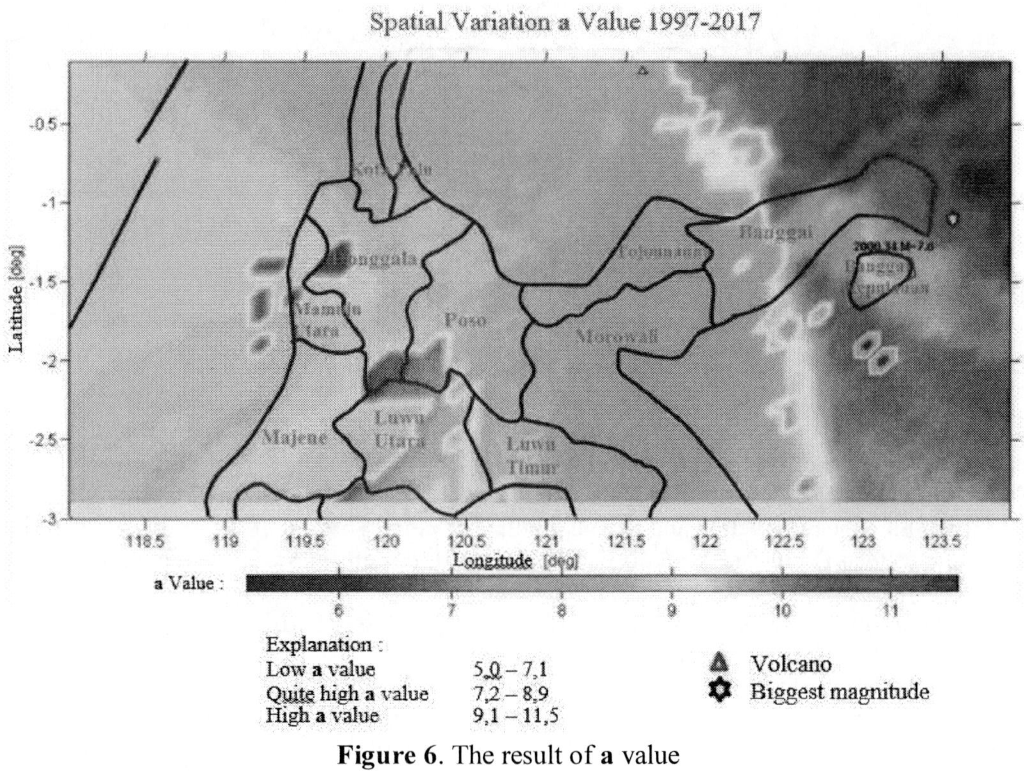

Figure 6. The result of **a** value

Based on Figure 7 yellow, orange and red in South Banggai, Banggai Islands, Eastern of North Luwu, Southwest Poso, South and Northwest Donggala, West North Mamuju indicate that the area has a high **b** value around 1.31 – 1.8 is caused by the present of volcanic rocks, plutonic rocks, rocks crust influenced by Palu-Koro Faults, Sorong Fault, and Sangihe Subduction.

The area of Palu, Majene, Morowali, East Luwu, West Banggai, Tojounauna, Western of North Luwu, East Poso, East Donggala, East North Mamuju characterized by light blue and light green indicates that this area has a fairly high b value of about 1.2 - 1.3 this is because in this area composed of sedimentary rocks, carbonate rocks, metamorphic, ultramafic, volcanic, schist, ofiolit passing by Palu-Koro active fault, Positive Fault Poso, Fault Matano shows high level of rock fragility and power level rock resistance to low stress. The southern Banggai islands marked with dark blue and blue indicate that this area has a low **b** value of about 0.6 to 1.1 is the result because the area is not passed by the faults.

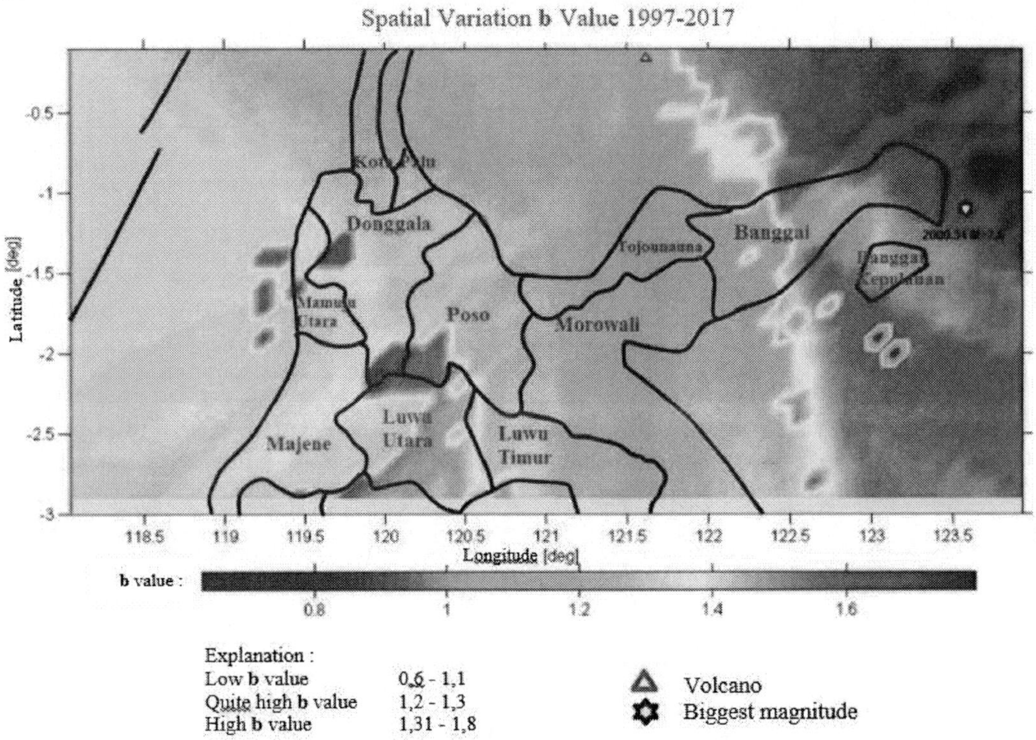

Figure 7. The result of **b** value

4. Conclusions

Based on the results and discussion, we had been conclude that:

1. The highest level of seismicity is in the eastern part of Central Sulawesi region, this is caused by the activity of Sangihe Subduction and Sorong Fault. The central region of the study was due to the Palu Koro Fault, Poso Fault, and Waikuli Fault. The south-eastern part of the study was caused by the Matano Fault.
2. Spatial variation of **a** value in the Central Sulawesi region ranges from 5 to 11.5, the region with the highest **a** value of South Banggai, Banggai Islands, Eastern of North Luwu, Southwest Poso, South and Northwest Donggala, Western of North Mamuju. The spatial variation of **b** value in the Central Sulawesi region ranges from 0.6 to 1.8, the region with the highest **b** value being South Banggai, Banggai Islands, Eastern of North Luwu, Southwest Poso, South and Northwest Donggala, Western of North Mamuju.

In the next step, the study will continue for determining hypocenter to get more accuracy in this result. The result in this study effectively to observe the seismicity which can be used as guided for earthquake mitigation.

Acknowledgements
We gratefully acknowledge the USGS and BMKG for data. We thank to Geophysics Study Program, Faculty of Mathematics and Natural Science Hasanuddin University (UNHAS) for supporting this research.

References
[1] Massinai M F I, Lantu, Aswad S and Massinai M A 2015 *AIP Conf. Proc.* **1658** 030012-1 – 030013-10 http://dx.doi.org/10.1063/1.4915021
[2] Climatology Meteorology and Geophysics Agency Region III Denpasar 2017 *Earthquake. Retrieved* at http://balai3.denpasar.bmkg.go.id/about-the earthquake.
[3] Supartoyo & Surono 2008 *Earthquake Damage Catalog Indonesia Year 1629 - 2006 (Fourth Edition)* Center for Volcanology and Geological Hazard Mitigation, Geological Agency, Ministry of Energy and Mineral Resources.
[4] Surono & Hartono Udi 2013 *Geology Sulawesi* LIPI Press. Jakarta.
[5] Natawidjaja D H, Sunardi, Kadarusman A, Ashari P 1996 *Tectonic Condition of the Collision Area Between Banggai Sula Microcontinent and East Sulawesi Arms* Research and Development of Mineral, Energy, Water and Land Resources. LIPI Geotechnology Research Center: Bandung.
[6] Hall R & Wilson M E J 2000 *Journal of Asian Earth Sciences* **18** 781 – 808.
[7] Rohadi, S. 2009. *Studi Seismotektonik Sebagai Indikator Potensi Gempabumi di Wilayah Indonesia*. Jurnal Meteorologi dan Geofisika, Vol. 10, No. 2, 111-120.
[8] Wandono, Sri Widiyantoro, Gunawan Ibrahim, Edy Soewono. 2004. *Analisis Hubungan Frekuensi-Magnitudo Gempabumi di Bali dan Sekitarnya*. Jurnal Matematika dan Sains Vol. 9 No. 3, 273-277.
[9] Utsu, T. 1965. *a Method for Determining The Value of b in a Formula Log N =a-bM Showing the Magnitude-Frequency Relation for Earthquakes*. Geophys. Bull. Hokkaido Univ., 13, 99-103.
[10] Shi, Y., and B.A. Bolt. 1982. *The Standard Error of The Magnitude-Frequency b-value, Bull.* Seismol. Soc. Am., 72, 1677-1687.
[11] Wiemer, Stefan. 2001. *A Software Package to Analyze Seismicity: ZMAP*. Seismol. Res. Lett., 72(2): 373-382.

IOP Conf. Series: Earth and Environmental Science **279** (2019) 012023 doi:10.1088/1755-1315/279/1/012023

The effectiveness of vacuum preloading on eliminate secondary settlement; case study in summarecon city bandung area's development project

A L Wahyu[1] **and I B Mochtar**

[1] Corresponding author

Department of Civil Engineering, Faculty of Civil, Environmental, and Geo Engineering, Institut Teknologi Sepuluh Nopember, Indonesia

azmilisaniwahyu@gmail.com

Abstract. Summarecon City Bandung Area's Development Project is used to use embankment preloading with Prefabricated Vertical Drain (PVD) to improve the soil, but due to weather problems, used another alternative by replacing embankment preloading to vacuum preloading. Vacuum preloading with PVD is only intended for accelerate primary settlement, while secondary settlement still exists. Secondary settlement in this project needs to be paid attention because according to the reality in the field, secondary settlement happened quite large because the soil mix with organic soil. Secondary settlement could cause differential settlement which can damage the infrastructure and make the infrastructure drop until lower than flood water level. So far, there has not been a study about how to eliminate secondary settlement in order to complete that settlement before the infrastructure is built, so that differential settlement does not happen. Therefore, a study about eliminate secondary settlement in order to be completed before the infrastructure is built is needed with remove it along with primary settlement. In this case, secondary settlement is going to be eliminated by adding additional load in the form of embankment preloading. Overall, this study can be a suggestion and consideration for the decision-making of Summarecon on selecting the soil improvement; embankment preloading or vacuum preloading.

1. Introduction

About 20 million hectares or more than 10% of the land area in Indonesia are soft soil which consist of soft clay and peat soil. For the development of infrastructure in Indonesia, the possibility of infrastructure must be built in areas where the soil condition is soft clay. The main problems in soft clay are the soil has low bearing capacity, so that the load that able to borne by the soil is limited and the soil settlement is relatively large and also occurs for a long time. Therefore, it is necessary to improve the soil in order to prevent the failure of the infrastructure before its service age.
The soil improvement which suitable for soft clay is preloading methods. Preloading methods include embankment preloading and vacuum preloading. Embankment preloading and vacuum preloading method's principle are the same, namely removing all or most of the soil settlement and increasing the bearing capacity of the soil by giving an initial load that greater or equal to the planned infrastructure load. In order to accelerate the time of soil settlement, both methods are usually combined with Prefabricated Vertical Drain (PVD).

Vacuum preloading was first introduced in Sweden by Kjellman (1942). Since then, this method has often been used as soil improvement to accelerate soil settlement in areas with soft soil in some countries. In Indonesia, vacuum preloading has been applied to Summarecon City Bandung Area's Development Project. The project is located in Gedebage, Bandung and built on soil that dominant of soft clay mix with organic soil. In the beginning, Summarecon City Bandung Area's Development Project used embankment preloading with PVD for the soil improvement. However, embankment preloading in this project has been obstructed due to weather problems which effected on project schedule. For this reason, another alternative is used to replace embankment preloading to vacuum preloading. As has been known that vacuum preloading does not require a lot of embankment materials, so that it is expected to still be able to run when facing weather problems again. Generally, vacuum preloading with PVD is only intended to accelerate primary settlement, while secondary settlement still exists. Both types of settlement take place at different times, starting with primary settlement and continues with secondary settlement. Under normal conditions, primary settlement occurred up to hundreds of years. With PVD, primary settlement's time becomes shorter which takes place monthly, so that secondary settlement occurs earlier. Most of the contractors pay little attention to secondary settlement because they consider this settlement is very small compared to primary settlement, even though the cases of bumpy roads and cracked buildings often happened due to differential settlement caused by secondary settlement which its magnitude is not the same. Secondary settlement is also feared causing the infrastructure drop until lower than flood water level if it is not eliminated. Secondary settlement in Summarecon Bandung City Area's Development Project needs to be considered because according to the reality on the field, secondary settlement occurred quite large because the soil mix with organic soil. Therefore, a study about how to eliminate secondary settlement is needed and will be discussed in this paper. Later on, this study can be a suggestion and consideration for the decision-making of Summarecon on selecting the soil improvement; embankment preloading or vacuum preloading.

2. Design of the project

Vacuum preloading is used in Dyanti cluster area which can be seen in figure 1. Later on, this cluster is going to be built for house. The cluster that has an area of 0.4 hectares has already been planned with vacuum pressure of 74 kPa which equivalent to 4.4 m of embankment, backfill of 2.8 m for levelling, triangular pattern of PVD with distance of 1.2 m and 20 m length, Prefabricated Horizontal Drain (PHD) as drainage, Non-Woven Geotextile for the protection layer, and also HDPE Geomembrane for impermeability layer. In order to observe the stability, the soil settlement, and pore water pressure as results of vacuum preloading, used geotechnical instrument devices. Geotechnical instrument devices consist of Vacuum Gauge to measure vacuum degree, Settlement Plates and Extensometers are used to monitor the soil settlement, Vibrating Wire Piezometer to monitor pore water pressure, Inclinometer to monitor lateral displacement of soil, and Observation Wall to measure ground water level. This project has already run since 2017. For more detail, vacuum preloading method in this project is presented in figure 2.

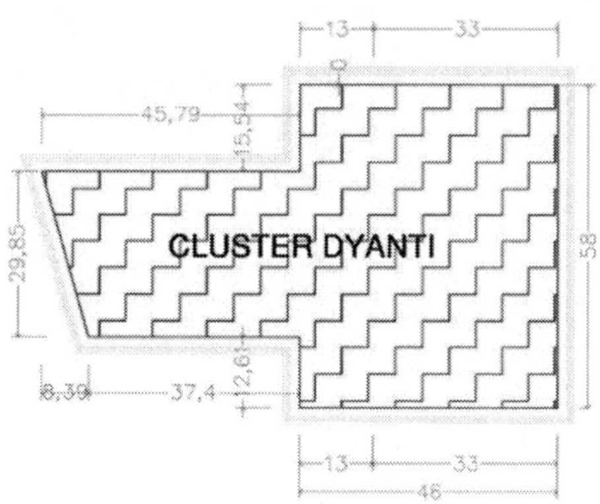

Figure 1. Dyanti cluster area.

IOP Conf. Series: Earth and Environmental Science **279** (2019) 012023 doi:10.1088/1755-1315/279/1/012023

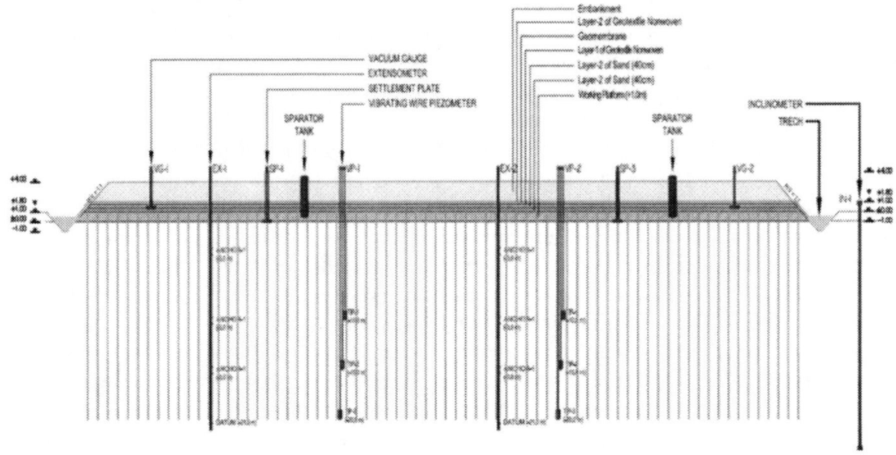

Figure 2. Vacuum preloading method in summarecon bandung city area's development project.

3. Experimental program

Study in this paper uses combination of vacuum preloading and embankment preloading that combination of both methods are expected to eliminate secondary settlement at the same time with primary settlement effectively. Vacuum preloading method is usually combined with embankment preloading method as backfill for levelling. The height of embankment preloading that must be added to eliminate secondary settlement using combination of vacuum preloading and embankment preloading will be compared to the height of embankment preloading when using only embankment preloading, where the parameters that needed to calculate primary settlement and secondary settlement is obtained from soil consolidation test.

This study was conducted based on field investigation tests include Cone Penetration Test (CPT) and Standard Penetration Test (SPT) to provide information on properties of soil where located around the vacuum area. The location of CPT named S-A and S-B and also SPT named BH-A can be seen in figure 3. This study was also conducted based on experimental study in a laboratory using undisturbed soil consisting of Volumetric and Gravimetric Test, Laboratory Vane Shear Test, and also Oedometer Test.

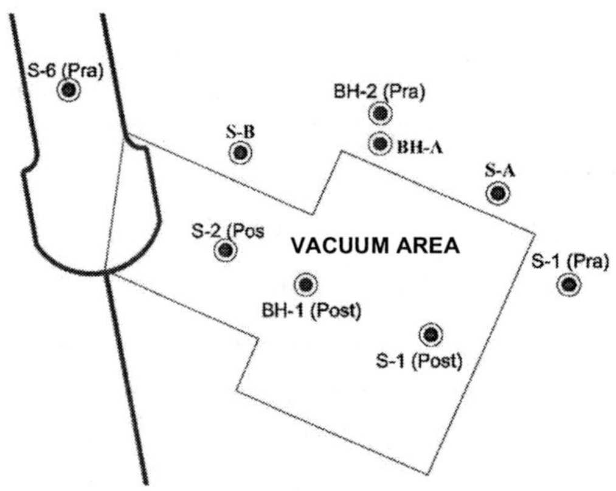

Figure 3. The location of soil investigation tests.

IOP Conf. Series: Earth and Environmental Science **279** (2019) 012023 doi:10.1088/1755-1315/279/1/012023

4. Results and discussion

4.1. Properties of soil based on field investigation tests

Both of CPT and SPT are done to 30 m depth. The result of CPT is the soil in Summarecon Bandung City Area's Development Project is dominant of clayey silt with thickness up to depth of 20 m. Above 22 m depth, the dominant soil is sand. As well as the result from SPT, the dominant soil is clayey silt with thickness up to depth of 26 m and sand with thickness from 26 m to 30 m. It can be concluded that the thickness of compressible layer is 20 m depth.

4.2. Properties of soil based on laboratory tests

The samples of soil for laboratory tests are obtained from undisturbed soil that has been taken from borehole in BH-A. Undisturbed soil has been gotten from 5 different depths. First, Laboratory Vane Shear Test is performed to determine the value of undrained shear strength (C_u). Then, Volumetric and Gravimetric Test to determine physical properties and also Oedometer Test to determine parameters of primary settlement and secondary settlement are performed. Oedometer Test was conducted in stages with load: 0.25; 0.5; 1; 2; 4 kg/cm^2 with loading time of 24 hours/loading. Properties of soil based on laboratory tests are listed on table 1. According to table 1, the soil in this project has larger value of water content (w_c) and void ratio (e_0) than common soft clay. It indicates that the soil mix with organic soil as said previously. Organic soil has high water content and void ratio. It is assumed that organic content in the soil only effect properties of soil, but the behaviour is still like mineral soils (organic silt and clay).

Table 1. Properties of soil based on laboratory tests.

	Depth (m)	Soil Consistencies	C_u (kPa)	w_c (%)	γ_{sat} (g/cm^3)	e_0	C_c	C_s	C_v (cm^2/s)	C_α
UDS 1	5-5.5 m	Very Soft	8.5	136.968	1.223	3.966	1.514	0.175	1.45x10^{-3}	0.087
UDS 2	9-9.55 m	Very Soft	10	210.239	1.259	5.313	2.282	0.170	5.05x10^{-4}	0.032
UDS 3	15-15.55 m	Very Soft	8	111.467	1.376	2.939	1.674	0.147	4.17x10^{-4}	0.031
UDS 4	19-19.55 m	Soft	22	184.164	1.215	4.992	3.016	0.348	4.80x10^{-4}	0.029
UDS 5	25-25.55 m	Soft	17	213.975	1.199	5.709	3.199	0.287	5.05x10^{-4}	0.048

4.3. Eliminate secondary settlement using embankment preloading

In this study, secondary settlement is eliminated at the same time with primary settlement by adding an additional load (Δq) that charged to primary settlement and causes more settlement with the magnitude which expected to equals secondary settlement. Then, Δq can be removed at the end of primary settlement. So that, after the soil improvement was completed and infrastructure has been built, there is no more settlement that caused by primary settlement and secondary settlement. To calculate primary settlement, determine some variations of specified constant of load embankment (q). q is load embankment after the soil was compressed. An embankment after undergoing the compression on the ground will undergo a weight change, because during the settlement some parts of embankment are under groundwater. After that, find out the magnitude of primary settlement (S_c), the initial height of embankment (H_i), and the final height of embankment (H_f) that caused by q.

Afterwards, calculate secondary settlement (S_s) with the reviewed time of occurrence is 20 years. Both of primary settlement and secondary settlement are calculated for compressible layer with thickness of 20 m. The calculation results of primary settlement, the initial height of embankment, the final height of embankment, and secondary settlement are listed on table 2. Then, from table 2, create a graph of the relationship between settlement (S) with load of embankment that shown in figure 4, a graph of the relationship between load of embankment with the initial height of embankment that shown in figure 5, and also a graph of the relationship between the initial height of embankment and the final height of embankment that shown in figure 6 below.

Table 2. The results of S_c, H_i, H_f, and S_s.

q (t/m^2)	S_c (m)	H_i (m)	H_f (m)	S_s (m)	S_t (m)
3	0.616	2.127	1.511	0.344	0.960
5	1.423	3.778	2.355	0.361	1.784
7	2.069	5.335	3.266	0.376	2.445
9	2.611	6.380	4.219	0.389	3.000
11	3.077	8.281	5.203	0.400	3.478

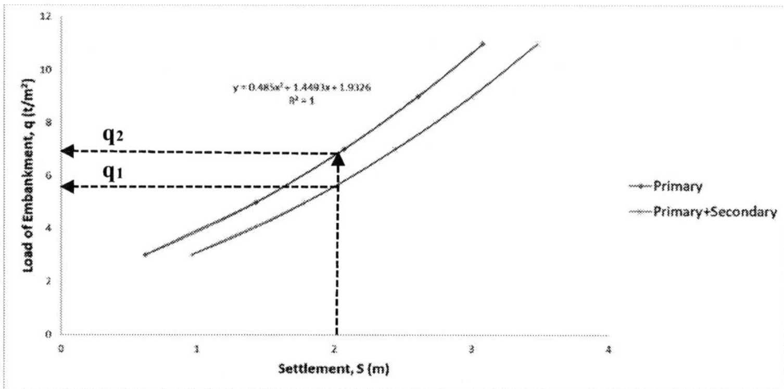

Figure 4. The relationship between S and q.

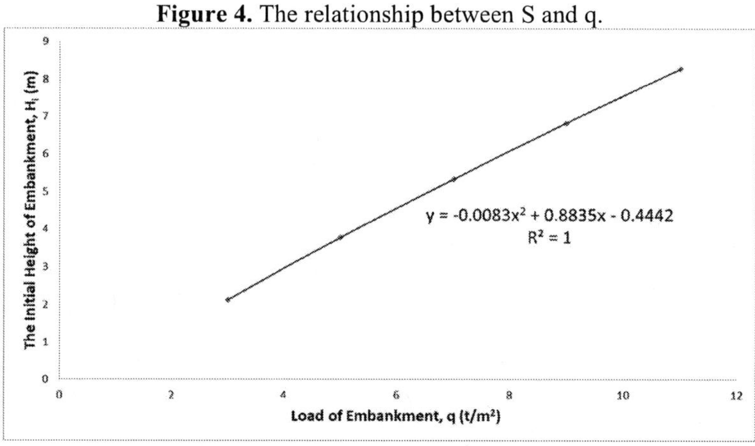

Figure 5. The relationship between q and H_i.

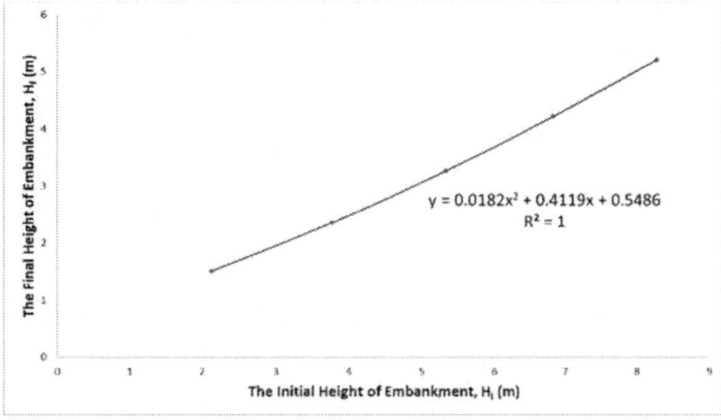

Figure 6. The relationship between H_i and H_f.

The following step is computing q_2 using the equation that shown in figure 4, so that the initial height of embankment and the final height of embankment have changed because there is Δq (q_2-q) refers to additional load that equals secondary settlement. Δq is going to be removed along with preload at the end of primary settlement. The new initial height of embankment that includes secondary settlement ($H_{i\,(p+s)}$) can be calculated by using equation that shown in figure 5, while the new final height of embankment that also includes secondary settlement ($H_{f\,(p+s)}$) can be calculated by using equation that shown in figure 6. The last is calculating the final height of embankment after primary settlement and secondary settlement were completed ($H_{f\,(field)}$). The complete calculation can be seen in table 3. From the result of calculation in table 3, a graph of the relationship between the final height of embankment after primary settlement and secondary settlement were completed and the new initial height of embankment is obtained as presented in figure 7.

Table 3. The results of q_2, Δq, $H_{i\,(p+s)}$, $H_{f\,(p+s)}$, and $H_{f\,(field)}$.

q_1 (t/m^2)	S_t (m)	q_2 (t/m^2)	Δq (t/m^2)	$H_{i\,(p+s)}$ (m)	$H_{f\,(p+s)}$ (m)	$H_{f\,(field)}$ (m)
3	0.960	3.77	0.77	2.769	1.829	0.787
5	1.784	6.06	1.06	4.606	2.832	1.619
7	2.445	8.38	1.38	6.374	3.913	2.516
9	3.000	10.64	1.64	8.019	5.022	3.467
11	3.478	12.84	1.84	9.531	6.128	4.458

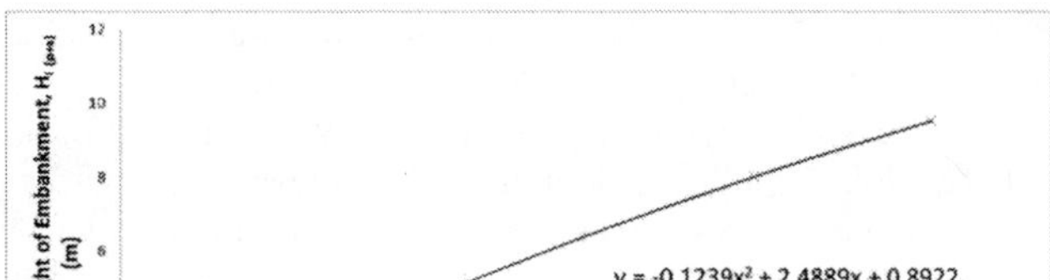

Figure 7. The relationship between $H_{f\,(field)}$ and $H_{i\,(p+s)}$.

A graph in figure 7 is later used as a guide for determining the initial height of embankment in embankment preloading method to eliminate secondary settlement at the same time with primary settlement. By inputting value of $H_{f\,(field)}$ on the equation that shown in figure 7, $H_{i\,(p+s)}$ will be obtained. $H_{f\,(field)}$ is also the height of embankment that designed on the planned elevation in the field. In Summarecon Bandung City Area's Development Project, the value of $H_{f\,(field)}$ is 3 m. Hence, in order to eliminate secondary settlement along with primary settlement, the height of $H_{i\,(p+s)}$ must be 7.2 m.

4.4. Eliminate secondary settlement using combination of vacuum preloading and embankment preloading

Unlike embankment preloading, vacuum preloading does not need lots of embankment. Vacuum preloading utilizes atmospheric pressure as surcharge load to accelerate the soil settlement. Vacuum preloading method must apply the pressure at least reach 80% of atmospheric pressure for effectiveness, but atmospheric pressure depends on elevation of location. In order to measure elevation of location from sea level, used a device named Altimeter, so that atmospheric pressure is known. Bandung is located at +660 m and the maximum atmospheric pressure in the area is 92.5 kPa, so that this project is planned with vacuum pressure of 74 kPa which the value is 80% of 92.5 kPa.

As previously mentioned that this project is going to be built for house. A house is equal to 10 kPa which meant the vacuum pressure of 74 kPa is overload. The excess of 64 kPa as surcharge load is expected to add more settlement that includes secondary settlement. As a result of 74 kPa, the soil undergoes settlement in total amount of 3.4 m. The settlement that will happen is very large, so that the ground seems to be at below of 0.0 m. In order to backfill for levelling according to planned elevation which is 3 m height, additional embankment as height as 4.6 m is needed. It means that 4.6 m is higher than the vacuum pressure which equal to 4.4 m and feared that the soil will undergo further settlement.

In order to prevent the case above, the solution is doing some stages of loading and still compressing the soil up to 3.4 m. For more detail, the stages of loading are explained below as follows:

- First stage is placing the embankment as height as critical height of embankment. The critical height of embankment is 2.1 m. Then, seal the embankment with Geomembrane and apply vacuum pressure of 74 kPa. In this project, the vacuum degree takes a month or 4 weeks to reach the vacuum pressure of 74 kPa. Therefore, the loading must be waited for 4 weeks. Based on the calculation, the settlement that will happen in week 4 is 1.5 m and the rest of embankment is 0.65 m. Afterwards, turn off the vacuum pump and open the sealing. In the field, in order to know the settlement that has happened, used Settlement Plate.
- Second stage is placing the rest of embankment from first stage, then add more embankment as height as 2.1 m. The height of embankment is independent variable. The important thing is compressing the soil up to 3.4 m with anything height of load. Then, seal the embankment with Geomembrane and apply vacuum pressure of 74 kPa, and also wait for 4 weeks as done on the first stage. Based on the calculation, the settlement that will happen is 1.6 m and the rest of embankment is 1.2 m. Then, when the settlement has already reached 1.6 m or fit to the calculation, turn the vacuum pump off and open the sealing.
- In the way above, the soil undergo settlement in total amount of 3.1 m. Therefore, after the second stage, backfill for levelling is done. In order to levelling according to planned elevation which is 3 m height, additional embankment as height as 1.8 m is needed. Hence, in order to eliminate secondary settlement along with primary settlement, the total height of embankment preloading that must be added using combination of vacuum preloading and embankment preloading is 6 m.

5. Conclusion

Based on the explanation above, it can be concluded that the height of embankment preloading using embankment preloading to eliminate secondary settlement at the same time with primary settlement is 7.2 m, while the total height of embankment preloading using combination of vacuum preloading and embankment preloading is 6 m. Vacuum preloading which should use less embankment material, in fact in this project, this method needs embankment preloading which is almost the same with embankment preloading. The height of embankment of both methods is only 1.2 m in difference. Actually, vacuum preloading is more effective on eliminate secondary settlement along with primary settlement than embankment preloading because vacuum preloading utilizes atmospheric pressure as surcharge load that expected to add more settlement that includes secondary settlement with require less embankment material. If viewed from the side of soil stability, eliminating secondary settlement using vacuum preloading method does not cause failure because this method applies atmospheric pressure that causes isotropic settlement which eliminates the risk of failure, while eliminating secondary settlement using embankment preloading method by placing 7.2 m of embankment material can cause failure because based on discussion previously, the critical height of embankment is only 2.1 m. Also if place some stages of 7.2 m of embankment material, it needs longer time. Hence, vacuum preloading should be an innovation on handling the problem of limited embankment material in infrastructure development in Indonesia and becomes an environmental friendly soil improvement because this method also applies no chemical admixtures.

In this project, vacuum preloading method is the only way in order to eliminate secondary settlement although this method is more expensive. However, it does not rule out the possibility, a further study about the effectiveness vacuum preloading on eliminate secondary settlement is done. A further study must be done in different area which dominant of pure soft clay. Therefore, the study is expected eliminate secondary settlement more effectively because pure soft clay is assumed undergo smaller settlement than organic soft clay and need less embankment material.

6. References

[1] Alihudien A and Mochtar I B 2009 *Usulan Perumusan Pemampatan Konsolidasi Sekunder untuk Tanah Lempung Prosiding Pertemuan Ilmiah Tahunan XIII 2009 Himpunan Ahli Teknik Tanah Indonesia* (Denpasar) ISBN: 978-979-96668-7-1

[2] Chai J C, Cater J P and Hayashi S 2005 *Ground Deformation Induced by Vacuum Consolidation Journal of Geotechnical and Geoenvironmental Engineering* ASCE 131(12): 1552-1561

[3] Chai J C, Hayashi S and Carter J P 2005 *Characteristics of Vacuum Consolidation Proceedings of the 16th International Conference on Soil Mechanics and Geotechnical Engineering* (Osaka)

[4] Gouw T L and Yu L 2012 *Soil Improvement by Vacuum Preloading for a Power Plant Project in Vietnam* HATTI-PIT-XVI 2012 (Hotel Borobudur Jakarta)

[5] Kjellman W 1952 *Consolidation of Clayey Soils by Atmospheric Pressure Proceedings of a Conference on Soil Stabilisation* (MIT Boston) pp 258–263

[6] Masse F et al 2001 *Vacuum Consolidation: A Review of 12 Years of Successful Development* http://www.menard-soltraitement.com

[7] Mochtar I B 2000 *Teknologi Perbaikan Tanah dan Alternatif Perencanaan Pada Tanah Bermasalah (Problematic Soils)* (Surabaya) Jurusan Teknik Sipil FTSP ITS

[8] Mochtar N E 2012 *Modul Ajar Metode Perbaikan Tanah (RC09-1402)* (Surabaya) Jurusan Teknik Sipil FTSP ITS

IOP Conf. Series: Earth and Environmental Science **279** (2019) 012024 doi:10.1088/1755-1315/279/1/012024

Interpretation of paleodepositional environment using biomarkers and carbon isotope ($\delta^{13}C$) in Talang Akar Formation, South Sumatra Basin, Indonesia.

Jamaluddin[1], Cheng Fuqi[1], Maria[2], Erfan Syamsuddin[2], Emi Prasetyawati Umar[3]
[1]Geological Engineering Department, School of Geosciences,
China University of Petroleum (East China), Qingdao, China.
[2] Geophysics Department, Science and Mathematic Faculty, Hasanuddin University, Makassar, Indonesia.
[3] Mining Engineering Department, Faculty of Industrial Technology,
Universitas Muslim Indonesia, Makassar, Indonesia.

Email: jamaljamaluddin1994@gmail.com; chengfuqi@upc.edu.cn

Abstract. Talang Akar Formation is a proven hydrocarbon source rock in South Sumatra basin. The formation contains dominant shale at the top, with some sandstone interbeds. This study uses 3 crude oil sample and 10 well-cutting obtained from two well (SMT-1 & SMT-2) in Prabumulih oil field. We used biomarkers and carbon isotopes ($\delta^{13}C$) data to determine the paleodepositional environment of Talang Akar Formation. The biomarkers data were obtained from gas chromatography (GC) and gas chromatography-mass spectrometry (GC-MS) analysis of normal alkanes, isoprenoids, triterpene, and steranes. Carbon isotopes ($\delta^{13}C$) include saturated and aromatic fractions. The pristine (Pr) to phytane (Ph) ratio is a good indicator of the depositional environment. Higher values of Pr/Ph ratio i.e. $\geq$3.0 indicate oxidizing conditions i.e. terrestrial while lower values in the range of 1.0-3.0 suggests siliciclastic-dominated marine conditions. However, very low values i.e. $\leq$ 1.0 indicate reducing conditions or fresh and brackish water conditions. The results in this research show that Pr/Ph ratios range from 7.90-16.66, characteristic of high wax crude oils, primarily originated in fluviatile and deltaic environment containing a significant amount of terrestrial organic matter. Similarly, the resultant ratios of Pr/n-C_{17} and Ph/n-C_{18} in SMT-1 and SMT-2 wells range from 0.91-10.72 and 0.11-1.29 respectively, which reflect that most of kerogen was derived from humic source and tend towards an oxidative environment of deposition. Cross-plot of carbon-13 isotopes ($\delta^{13}C$) shows saturated versus aromatic fraction. The resultant plot indicate a deltaic to marginal marine environment for SMT-2 well and a more marine environment for SMT-1 well. The oil/source rock correlation analysis using biomarker data shows that the oils in Prabumulih field is correlated with the oils in source rock of Talang Akar Formation. This study concludes that the source rock contains abundant humic organic matter that was deposited in a transitional (Fluvio-deltaic) to marginal marine environment under oxic conditions.

1. Introduction

Opportunities to increase petroleum exploration and exploitation, especially in the South Sumatra Basin, must be supported by integrated study such as geological, geophysical and geochemical studies to get better results. Geochemical research has been carried out to increase petroleum exploration and exploitation. Starting from research on the chemical composition, oil–oil and oil–source rock correlations. Organic geochemistry is generally used to evaluate basins, plays and prospects. As a prerequisite for understanding organic geochemistry, the formation of oil and gas beneath the earth's surface must be well understood [1].

Biomarkers are a highly developed part of organic geochemistry. Biomarkers are used to identify compounds contained in sediments that have a clear structural relationship with compounds of biological results. One of the uses of biomarkers in petroleum exploration is to identify the depositional environment and organic material. This study aims to characterize the samples and oil-source correlations by using biomarker were obtained from gas chromatography (GC), gas chromatography-mass spectrometry (GC-MS) and Carbon isotopes ($\delta^{13}C$) analyzes.

The South Sumatra Basin is a large basin divided into sub-basins as Jambi sub-basin (North Palembang), Central Palembang sub-basin, South Palembang sub-basin (Complex Palembang) [2][3]. The South Sumatra Basin includes proven active exploration lands as one of the large oil and gas producing regions except Kasai Formation, all formations including basement rocks have been proven potentially produce hydrocarbons. The South Sumatra Basin is an asymmetric basin bounded to the west and south by faults and uplifted exposures of pre-Tertiary rocks along the Barisan Mountains, to the north east by the sedimentary or depositional boundaries of the Sunda Shelf. The south-east boundary is represented by the Lampung High [4][5].

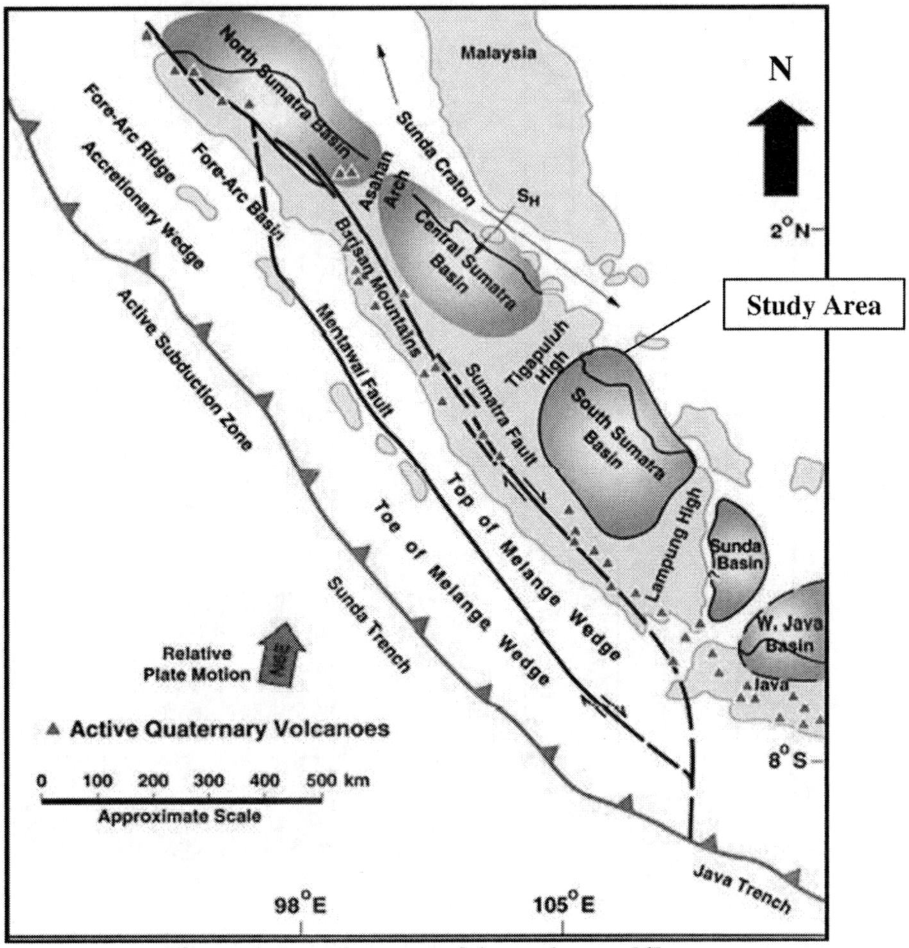

Figure 1. Location map of the study area [6].

2. Stratigraphy of South Sumatra Basin

Stratigraphy in South Sumatra Basin was control by process of megacycle system of trangresion and regression sedimentation in regionally, just like sedimentation in west Indonesia (Figure 2).

Talang Akar Formation (TRM) is dominated by shale, siltstone, sandstone, lime-stone streak, and coal. Coals are found along the upper part to the bottom. The Upper part is composed by shale, sandstone and limestone streak. Sandstones are medium porosity, white to colored, fine to medium hard, fine to medium grain, sub rounded to sub angular, medium sorted, calcareous cemented, containing glauconite minerals. Shale are brown to brown grey, medium hard, blocky, slight calcareous cemented and trace carbonate speek. Siltstone are light grey to grey, medium hard, sandy, consolidated, slight calcareous, trace carbon speek. The bottom part is dominated by shale, sandstone, minor siltstone, and coal layer. Shale are brown to dark brown, medium hard, blocky to platy, non-calcareous, silty, carbonaceous. Sandstone are medium porosity, colored to off white, fine to medium grain, trace coarse grain, sub rounded, medium sorted, carbonate speek [7].

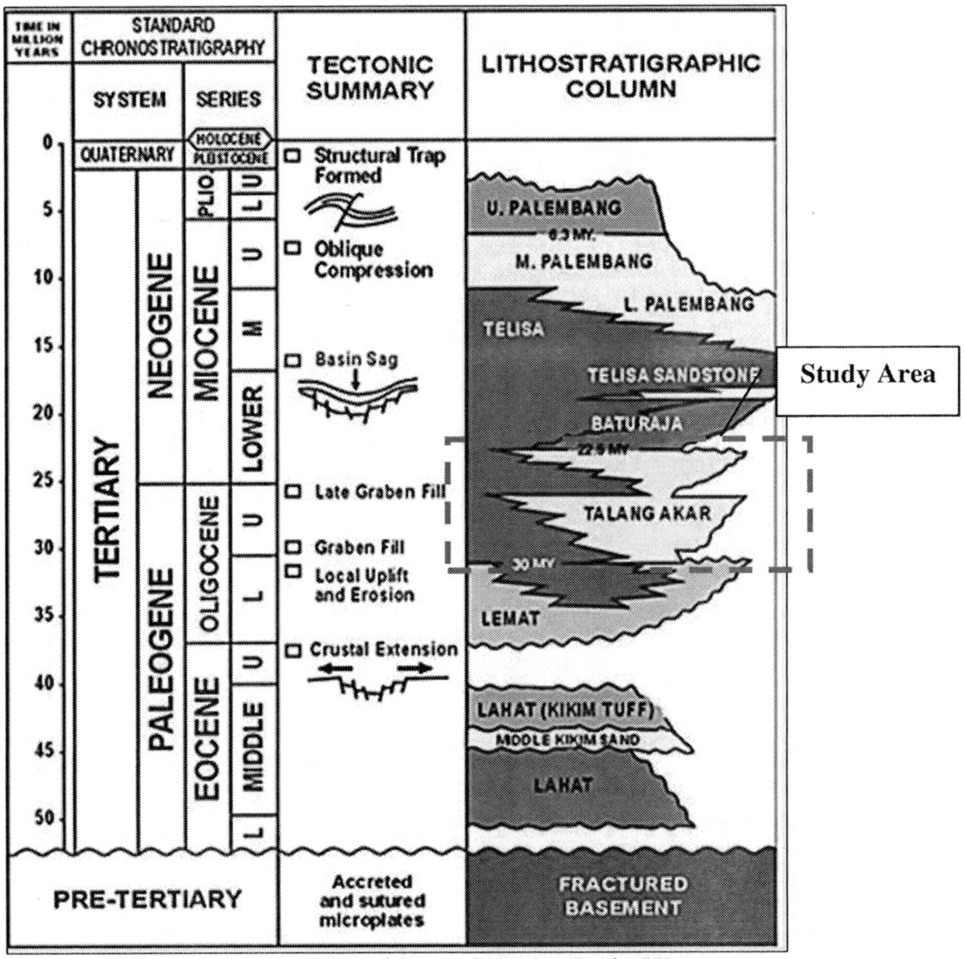

Figure 2. Stratigraphy of South Sumatra Basin [5].

The principal source rocks in the South Sumatra Basin are fluvio-deltaic marine, locally lacustrine and coaly facies of the Late Eocene to Middle Oligocene Lemat, and Late Oligocene to Early Miocene Talang Akar Formations. These rocks were deposited in the grabens and half grabens forms during the Late Cretaceous to Early Oligocene, and contribute to terrestrial derived oils. In addition, shales of the Telisa Formation provide a marine/paralic hydrocarbon source. Talang Akar sandstones constituted the main reservoir target and the potential traps are anticlinal structure that developed in response to Plio-Pleistocene compression. The Baturaja limestones are the other prime reservoir target; the traps are structural or combined structural and stratigraphic trap potential has also been indentified in the Baturaja Formation, with the potential trap dependent on facies change from porous reefal to tight platform carbonates [8].

3. Research Methods

This study uses 3 crude oil sample and 10 well-cutting from two well (SMT-1 & SMT-2) in Prabumulih oil field. Based on biomarkers and carbon isotopes ($\delta^{13}C$) data to determine the paleodepositional environment of Talang Akar Formation. The biomarkers data were obtained from gas chromatography (GC) and gas chromatography-mass spectrometry (GC-MS) analysis. Carbon isotopes ($\delta^{13}C$) include saturated and aromatic fractions.

4. Results and Discussion

From previous work about oil–oil and oil–source rock correlations of the South Sumatra Basin reservoirs compared four groups of oils that have different characters, namely oil derived from the sea or lacustrine, oil originating from a land deposition environment, petroleum derived from the depositional environment of lacustrine and biodegradable petroleum [9].

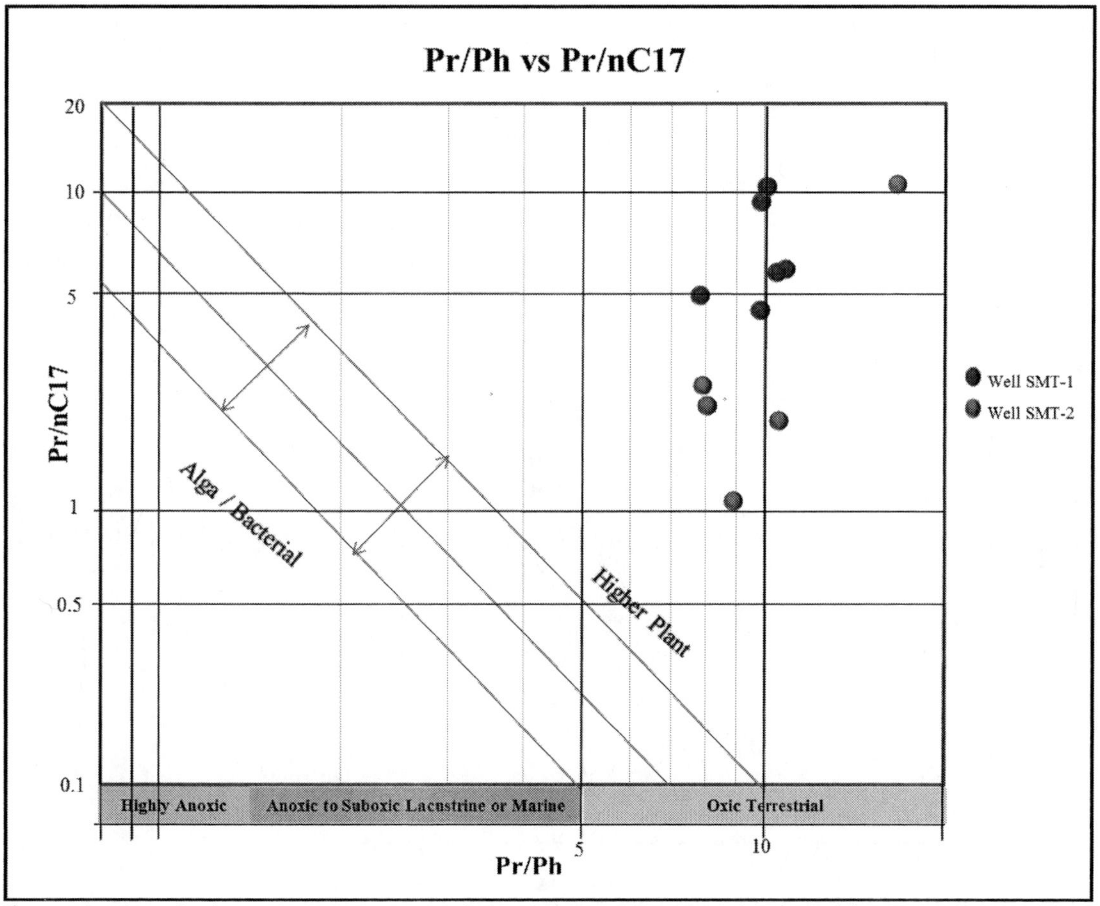

Figure 3. Plot of pristane/n-C17 versus pristane/phytane showing the depositional environment and the conditions of the studied rock samples.

Figure 3 shows a cross plot Pr/n-C_{17} and Pr/Ph, the results in this research show that Pr/Ph ratios range from 7.90-16.66, characteristic of high wax crude oils, primarily originated in fluviatile and deltaic environment containing a significant amount of terrestrial organic matter. The source rock samples in Prabumulih oil field are predominated by higher plant in oxic condition. Gas chromatography and gas chromatography-mass spectrometry analyses indicate that the oils in the South Sumatra Basin were derived from terrestrial higher plant material. These oils are characterized by high ratios of pristine to phytane, The comparison between Ph and Pr very high is a specific characteristic of coal. Ratio Pr/Ph is very high reflecting a condition the environment of deposition was very oxide. Three samples of coal shale and one other has a value of Pr/Ph is lower when compared with coal from the depths of 2885 m, although it is still quite high (> 3) to reflect an environment of deposition which is rich in oxygen. Oxidative environmental conditions that are reflected with ratio Pr/Ph > 3 the sediments analyzed deposited on areas that are associated with the delta and the coast. Thus it can be inferred that hydrocarbons produced by coal, petroleum or shale analyzed sediment is from a delta.

Lemat Formation has been considered as syn-rift sediments until now, consist of terrestrial sediment (fluvio-lacustrine), creating source rock with fluvio lacustrine characterization. While based on existing publications, showing that oil having fluvio-deltaic characterization and generated by source rock from Talang Akar Formation with fluvio-deltaic characterization. A number of potential reservoir rocks in the South Sumatra Basin occur within the regressive and transgressive sequences. The Muara Enim and Air Benakat Formations from the regressive sequences have good potential as reservoirs. The transgressive sequences are represented by the Talang Akar and Baturaja Formations.

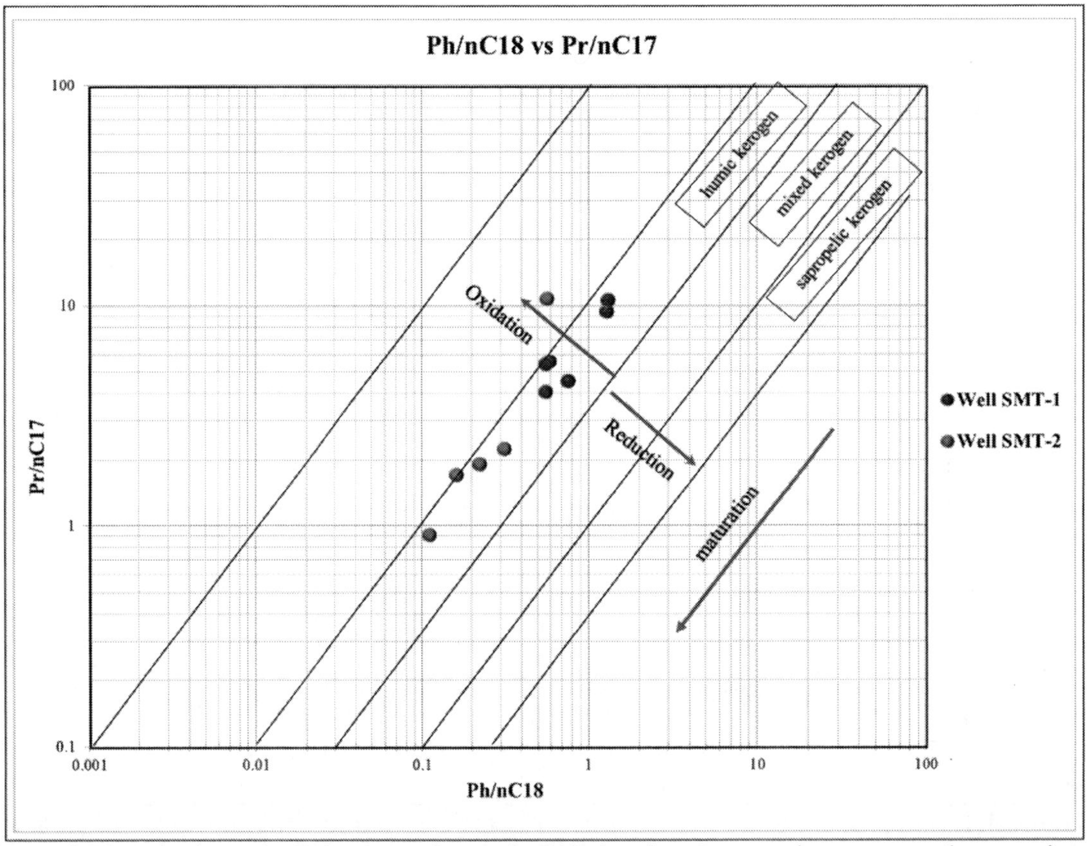

Figure 4. Plot of pristane/n-C17 versus phytane/n-C18 showing the organic sources and maturation of the studied rock samples.

The plot between Pr / n-C17 and Ph / n-C18 (Figure 4) obtained from GC data is done to analyze the oxidation conditions and type of organic material rock samples according to the classification, Pr/n-C17 and Ph/n-C18 ratios are used in addition to other parameters to determine source rock facies, maturity and the level of biodegradation of hydrocarbons. Low ratios indicate a more mature sample, because the isoprenoids will break down more readily than n-alkanes during maturation. The ratios can be used together with other parameters to rank related, non-biodegraded oils and bitumens based on thermal maturity. The ratio between the acyclic isoprenoids and n-alkanes can be used to estimate the origin of the organic matter and its maturity (Pr/n-C17 and Ph/n-C18). Lower plants, algae and bacteria contain smaller quantities of n-alkanes relative to acyclic isoprenoids, compared to the higher terrestrial plants but care should be taken because organic input and biodegradation may affect the ratio [10]. The resultant ratios of Pr/n-C_{17} and Ph/n-C_{18} in SMT-1 and SMT-2 wells range from 0.91-10.72 and 0.11-1.29 respectively, which reflect that most of kerogen was derived from humic source and tend towards an oxidative environment of deposition (Figure 4).

Cross-plot of carbon-13 isotopes ($\delta^{13}C$) shows saturated versus aromatic fraction indicate a deltaic to marginal marine environment for SMT-2 well and a more marine environment for SMT-1 well. Deposition of the Talang Akar Formation began in the Late Oligocene in the form of alluvial fan and braided stream environments filling topographic lows and depressions. Therefore, the Talang Akar Formation locally occurs overlying the pre-Tertiary rocks. This sedimentation continued in Early Miocene in a fluviatile, deltaic and marginal-shallow marine environment. During the time, the connection to open marine conditions became more significant and the sea gradually encroached into the basin.

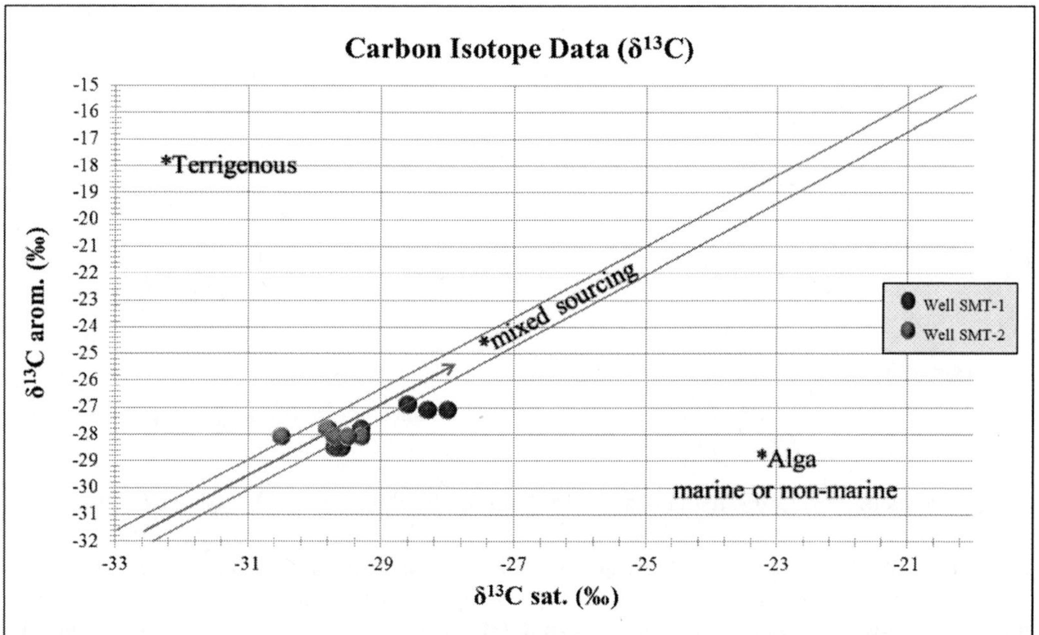

Figure 5 . Plot of carbon-13 isotopes ($\delta^{13}C$) showing the depositional environment of the rock samples.

All samples contain C27-C29 steranes. In some cases the relative abudance of C27-C29 steranes can be used as indicators of the nature of the photosynthetic biota, both terrestrial and aquatic, while triterpanes are usually indicators of depositional and diagenetic conditions [11]. Land plant inputs are usually inferred from a dominance of the C29 steranes. The C29 steranes of this type of crude were not derived from C29 sterols of land-plant origin (frequently proposed as the source of C29 steranes). The presence of these compounds shows a high contribution from terrestrial plants, especially from dipterocarpacceae species which produce resin. Biomarker compound with composition C29> C28 < C27, giving an indication of the dominant contribution of organic matter from terrestrial plants. The presence of diasterane compounds that are quite dominant indicates that the source rock of oil is deposited in an environment rich in clay minerals. The composition of styrene shows C29 > C28 < C27 which gives an indication of the origin of mixed organic matter between algae and terrestrial plants.

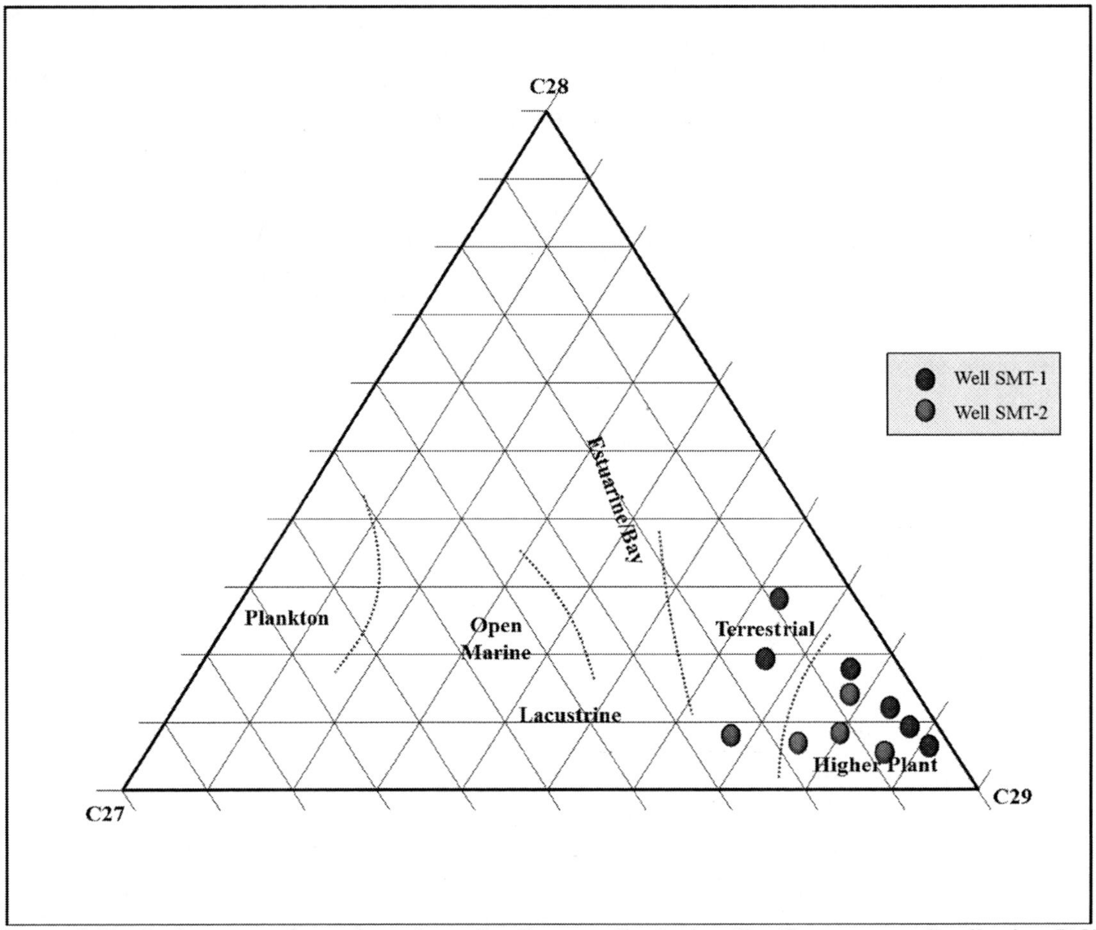

Figure 6. Facies interpretation using triangular diagram displaying C_{27}-C_{29} steranes distribution [12].

The depositional environment of the study area is a transitional environment for oxic conditions with material organic higher plants and marine algae. Based on oil–oil and oil–source rock correlations, Talang Akar Formation those were depositional environment (transitional) and precursor (terrestrial) [12].

5. Conclusions

Pristane to phytane ratios of the samples are relatively high ranging from 7.90-16.66. Higher values of Pr/Ph ratio i.e. ≥3.0 indicate oxidizing conditions i.e. terrestrial while lower values in the range of 1.0-3.0 suggests siliciclastic-dominated marine conditions. However, very low values i.e. ≤ 1.0 indicate reducing conditions or fresh and brackish water conditions

IOP Conf. Series: Earth and Environmental Science **279** (2019) 012024 doi:10.1088/1755-1315/279/1/012024

6. References

[1] Waples, D.W 1985 Geochemistry in Petroleum Exploration *International Human Resources Development Corporation*, Boston.

[2] Williams, H. H., Fowler, M., and Eubank, R. T. 1995 Characteristics of selected Palaeogene and Cretaceous lacustrine source basins of Southeast Asia, in Lambiase, J. J., ed., Hydrocarbon Habitat in Rift Basins: *Geological Society Special Publication*, **80**, p. 241-282.

[3] Suseno, P. H., Zakaria, Mujahindin, Nizar, and Subroto, E. A. 1992 Contribution of Lahat Formation as hydrocarbon source rock in South Palembang area, South Sumatera, Indonesia*: Proceedings Indonesian Petroleum Association 21st Annual Convention*, p. 325-337.

[4] Pulunggono, A. 1985 The changing pattern of ideas on Sundaland within the last hundred years, its implications to oil exploration: *Proceedings Indonesian Petroleum Association 14th Annual Convention*, p. 347- 378.

[5] de Coster, G. L. 1974 The geology of the Central and South Sumatra Basins: *Proceedings Indonesian Petroleum Association 3rd Annual Convention*, p. 77-110.

[6] Heidrick, T. L., and Aulia, K. 1993 A Structural and Tectonic Model of The Coastal Plains Block, Central Sumatra Basin, Indonesia. *Proceedings Indonesian Petroleum Association 22nd Annual Convention*, IPA.

[7] Spruyt, J.M., 1956 Subdivision and nomenclature of the Tertiary sediments of the Palembang Jambi Basin, EP-27168 and EP-27168B (Unpublished).

[8] Hutchison, C. S. 1996 South-East Asian Oil, Gas, Coal and Mineral Deposits: Clarendon Press Oxford.

[9] Kasim,S. A. and Armstrong, J., 2015, Oil- oil correlation of the South Sumatra Basin reservoirs. Journal of Petroleum and Gas Engineering. Vol. 6(5), p. 54-61

[10] Didyk, B.M., Simoneit, B.R.T., Brassell, S.C. and Eglinton, G 1978 Organic geochemical indicators of palaeoenvironmental conditions of sedimentation. *Nature*, **272**. 216-222.

[11] Huang, W.Y and Meinschein, W.G 1979 Sterols as ecological indicators. *Geochim. Cosmochim. Acta.*, **43**. p. 739-745.

[12] Setyawan, R, Nugroho, H, Krisna, W, dan Teguh S, Beiruny S 2011, Korelasi Geokimia Antara Minyak bumi dengan Batuan Induk Melalui Karakter Biomarker di Lapangan Namib, Mohave, Gobi dan Kalahari, Cekungan Sumatra Selatan S. N. K. F. S . *Universitas Diponegoro*, Semarang.

IOP Conf. Series: Earth and Environmental Science **279** (2019) 012025 doi:10.1088/1755-1315/279/1/012025

Poisson's Ratio Distribution in A Geothermal Field

Muhammad Fawzy Ismullah Massinai[1], Basdar Purwansah BS[2], Muhammad Yustin Kamah[3], Muhammad Altin Massinai[1]*

[1]Geophysics Dept., Hasanuddin University
[2]Professional Geoscience at PT. Pamapersada Nusantara
[3]Advisor at PT. Pertamina Geothermal Energy

*altin@science.unhas.ac.id

Abstract. Nowadays microearthquake or microseismic monitoring is powerful tool to investigate and evaluate the geothermal field. Each event of microseismic can be associated with permeable zone. One parameter can be used as a fracture indicator is Poisson's Ratio. The simple method to get information about Poisson's Ratio is Wadati Diagram. We applied this method in "A" Geothermal Field with 2 month recording microseismic data. The range of Poisson's Ratio distribution is 0.11197 – 0.468463. The lower Poisson's Ratio associated with un-fractured rocks and the higher indicate fractured rocks.

1. Introduction

Nowadays microearthquake or microseismic monitoring is powerful tool to investigate and evaluate the geothermal field. Many research used microseismic for that goal, such as: at Coso geothermal area, California [1]; at Lahendong Geothermal Field, North Sulawesi [2]; at Rotokawa Geothermal Field, New Zealand [3, 4, 5], at Okuaizu Geothermal Field, Japan [6]. Microseismic monitoring can give information about fracture characterization on geothermal field. Each event of microseismic can be associated with permeable zone. Understand fracture network characterization on geothermal field can give improvement on geothermal system. Good fracture networks signed by fracture connectivity until fluids migrate from injection well to reservoir or from reservoir to production well. Common microseismic mechanisms cited in literature are fluid pressure or temperature perturbations, reservoir volume changes, or chemical changes associated with extraction and/ or injection [5]; illustrated by Fig.1.

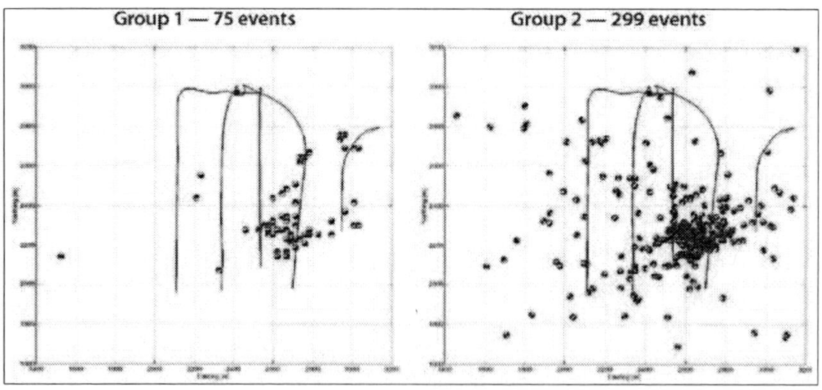

Figure 1. Group 2 is microseismic group after injection in Group 1. Dot is microseismic and line is wells [7].

As a result, this geophysical method is very helpful to develop a geothermal field. Poisson's ratio is one of the important parameters that can be used as a fracture indicator through simple calculation of Wadati Diagram. Whereas, well data like cores, minerals and loss circulation can provide reservoir permeability are very expensive and limited [2].

Wadati Diagram indicate different fracture intensity between shallow and deep reservoir. This paper explains Poisson's ratio derived from 2 month recording microseismic data at the "A" geothermal field.

2. The "A" Geothermal Field

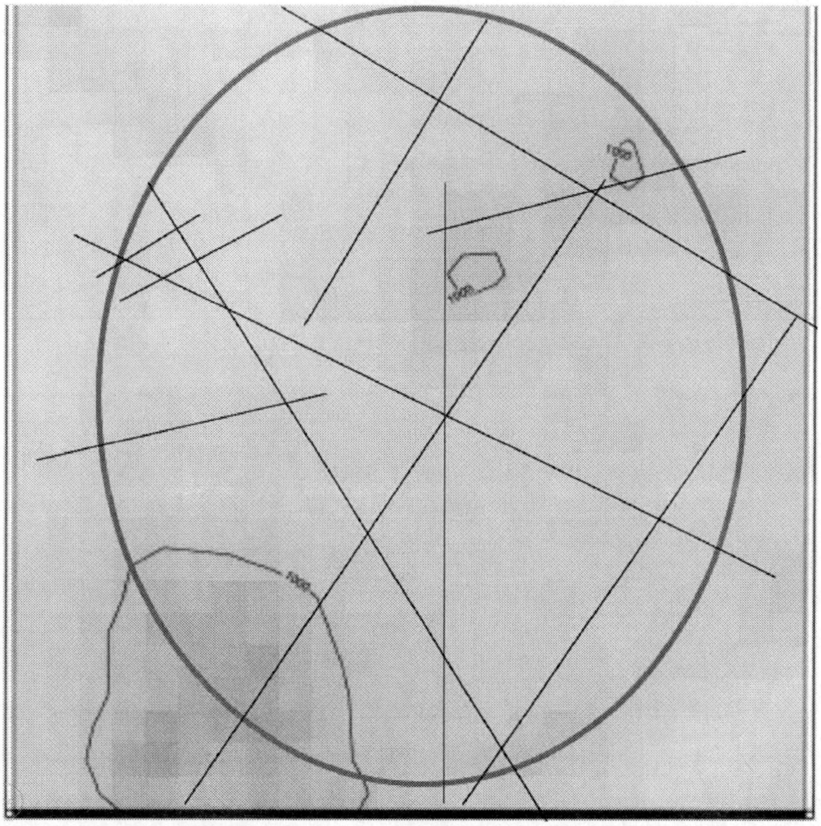

Figure 2. Map of "A" geothermal field. Red circle is location and black line are fault. Coordinate is hidden.

The "A" geothermal field is land at $\pm$ 750 m msl and as part of volcano-tectonic depression. Volcanoes around this geothermal with southwest to northeast. Geology main feature is caldera with 4 km diameter. This caldera limited by normal fault on its southwest [8]

Fault on research area dominated by normal fault such as R Fault, Li Fault, TP Fault, SK Fault, PU Fault, TU Fault and Le Fault (Fig. 2). Geothermal manifestation located around R Fault. This fault is normal fault and many feature indicated high alteration in this location. Other geothermal manifestation located around TP Fault. All fault may affected by regional tectonic which formed depression on this area [8].

Main factor of geothermal reservoir is permeability. Reservoir system of this geothermal field divided into two system; northern and southern. In southern have 300 – 350^{0}C temperature with 80% dry steam. In others, have lower around 250 – 280^{0}C temperature with 30% dry steam. They separated by N – S structure and assumed as pressure barrier. Well data indicated shallow reservoir. This reservoir found around 400 – 700 m depth. At W-1 and W-2, (Fig. 3) reservoir found around 650 m depth and W-4 and W-5 showed lateral current from shallow aquifer [8].

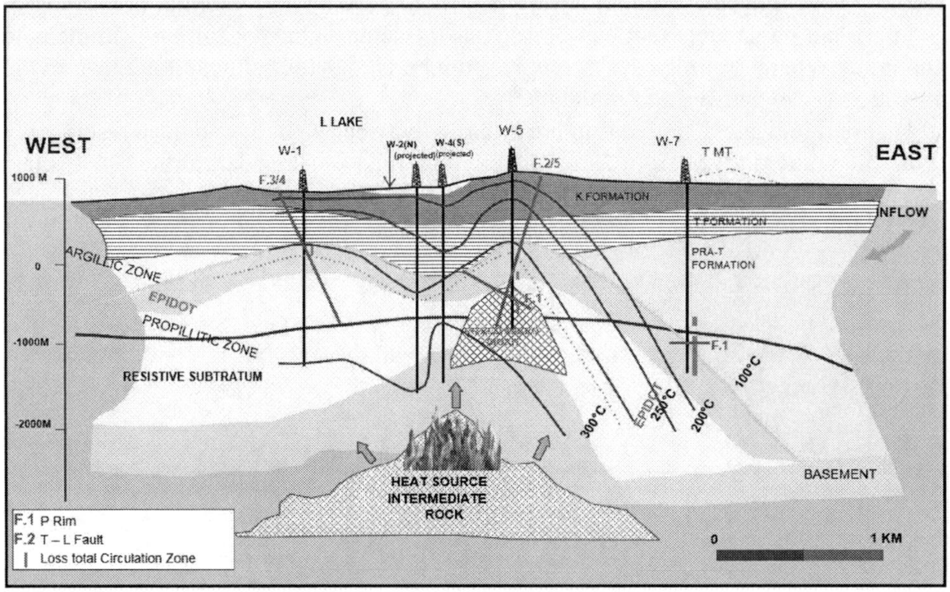

Figure 3. The A geothermal field model, modified from [8].

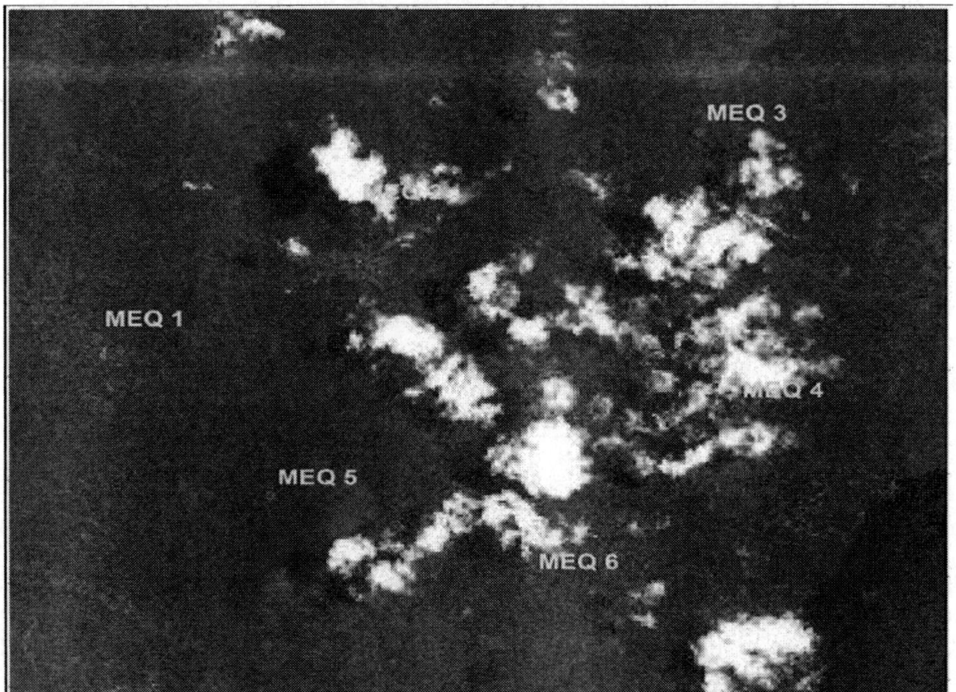

Figure 4. Distribution of seismograph [8]. Coordinate is hidden.

Generally in geothermal Field, the typical microseismic, follow: duration under 50 second, the depths under 5 km, and the P and S arrival (S-P) time differences under 3 second. Therefore, a simple assumption that model is a horizontal layered velocity structure, assuming constant, isotropic velocity within one layer, is reasonable enough to get a quick interpretation as preliminary result [2].

A small error in picking of the arrival time and the inaccuracies of velocity model can produce a large error in determining the hypocenter and epicenter location. Therefore, both factors must be controlled, carefully [2].

Microseismic activity observed continue and real time by Pertamina Geothermal Energy (PGE) used 6 digital seismograph (Fig. 4) with GPS to synchronize time and position. The data on this research area microseismic data on September – October 2010 periods. This research used 14 event microseismic [8].

3. Wadati Diagram

Wadati diagram is important in processing micro-earthquake data for two reasons. The first is that we can control the correct picks of travel time of P wave and S wave (ts and tp) from record data. The second reason is that we can compare the origin time of earthquake (To, if the time of S and P are equal, it means the rock are fractured) with to from a computer program e.g. linier inversion [2].

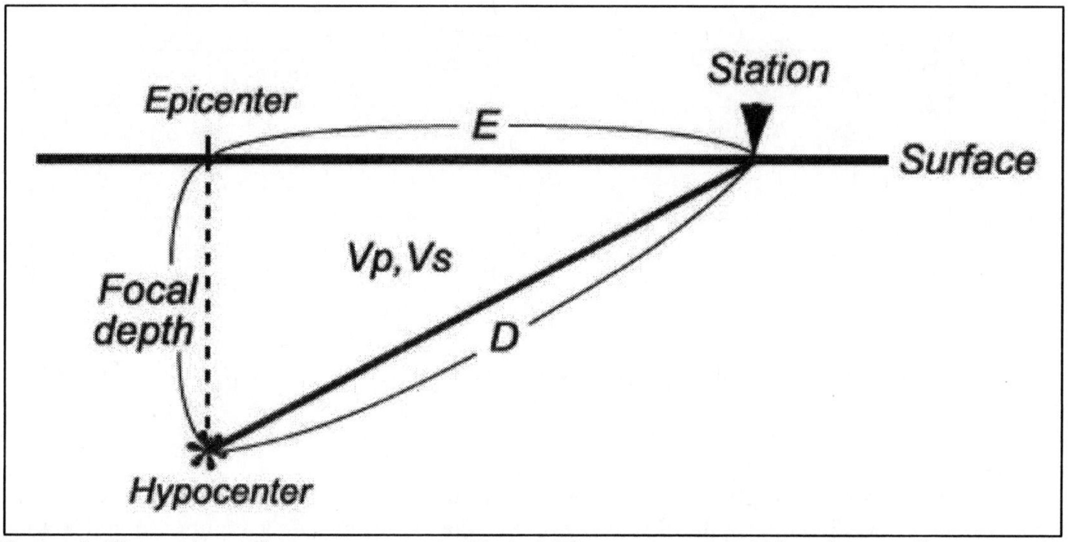

Figure 5. Illustration of hypocenter to get Wadati Diagram concept. E is epicenter distance and D is hypocenter distance [9].

The principle of the Wadati diagram is as follows, assuming that the medium is homogeneous. A hypocentral distance is represented by P-and S-wave travel times and velocities as follows:

$$D = (Tp - To) * Vp = Tp * Vp \tag{1}$$

$$D = (Ts - To) * Vs = \{(Ts - Tp) + (Tp - To)\} * Vs = \{(Ts - Tp) + Tp\} * Vs \tag{2}$$

From equations (1) and (2)

$$Tp * Vp = \{(Ts - Tp) + Tp\} * Vs \tag{3}$$

$$Tp * (Vp - Vs) = (Ts - Tp) * Vs \tag{4}$$

Therefore

$$(Ts - Tp) = \left(\frac{Vp}{Vs} - 1\right) * Tp \tag{5}$$

This is also valid in a layered media with a constant Vp/Vs ratio [9].

The Tp and Ts picked data from one microseismic event as shown in Tabel-1 as example to all microseismic event. The next step is constructing a Wadati diagram by plotting all Ts – Tp vs Tp for all stations, as shown in Fig. 6.

IOP Conf. Series: Earth and Environmental Science **279** (2019) 012025 doi:10.1088/1755-1315/279/1/012025

Table 1. Tp and Ts picked data from one microseismic event

No	Station		Tp (s)	Ts (s)	Ts – Tp (s)
1	MEQ2	22	23.45	25.5	2.1
2	MEQ3	22	22.74	24.2	1.5
3	MEQ4	22	22.74	23.8	1.1
4	MEQ5	22	23.65	25.5	1.8

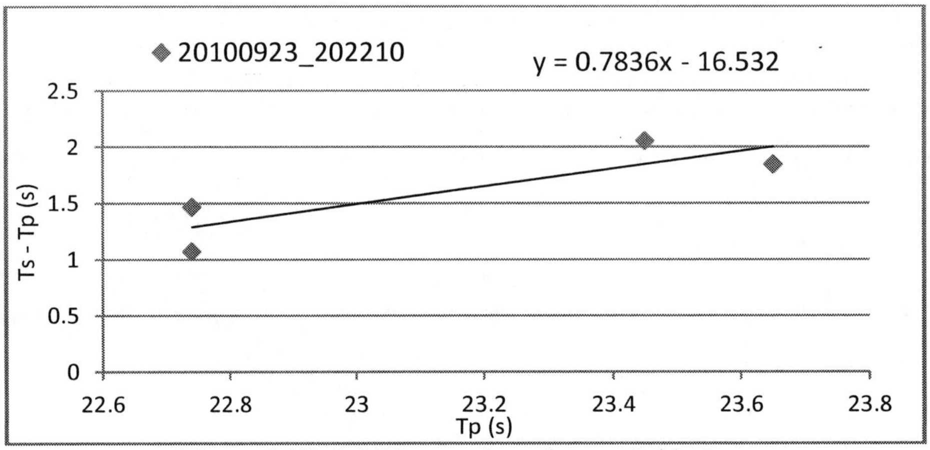

Figure 6. Wadati Diagram from data on Table 1.

Look linear regression equation of Fig. 6. Vp/Vs is the gradient + 1. So, from this Wadati Diagram got the Vp/Vs, is 1.7836.

4. Poisson's Ratio

Basically Poisson's ratio is defined by the ratio of the strain perpendicular to either type of deforming force to that in the direction of the force itself. In other way, by using compressional velocity and transversal velocity or the comparison of both velocity from Wadati diagram the Poisson's ratio is defined as follow [2]:

$$\sigma = \frac{(Vp^2 - 2Vs)}{2(Vp^2 - Vs^2)} \tag{6}$$

here σ , Vp dan Vs are Poisson's ratio, the compressional wave velocity (longitudinal) and the shear wave velocity (transversal) respectively [2].

If we assume that the ray-paths are confined to homogeneous geology, experimental and theoretical results indicate that additional fracturing of a fluid-filled rock will cause an increase in the Poisson's ratio. The P-wave velocity decreases slightly and the S-wave velocity significantly reduced [2].

The study from [2] got that in the steam dominated system, Vp decreases faster than Vs. The two parameters provide poisson ratio lower than 0.25. In the water dominated system at the higher porosity pressure, the Poisson's ratio is bigger than 0.25. The Vs decreases faster than Vp [2].

The Poisson's ratio of the A geothermal field is 0.11197 – 0.468463. The results indicate a differences between shallower and deeper zone. In the shallower zone, the Poisson's ratio dominated 0.2 up to 0.46, the anomalies gave on 0.125 – 0.144. In the deeper zone, the Poisson's ratio varies from 0.137 to 0.291, please look Fig. 7.

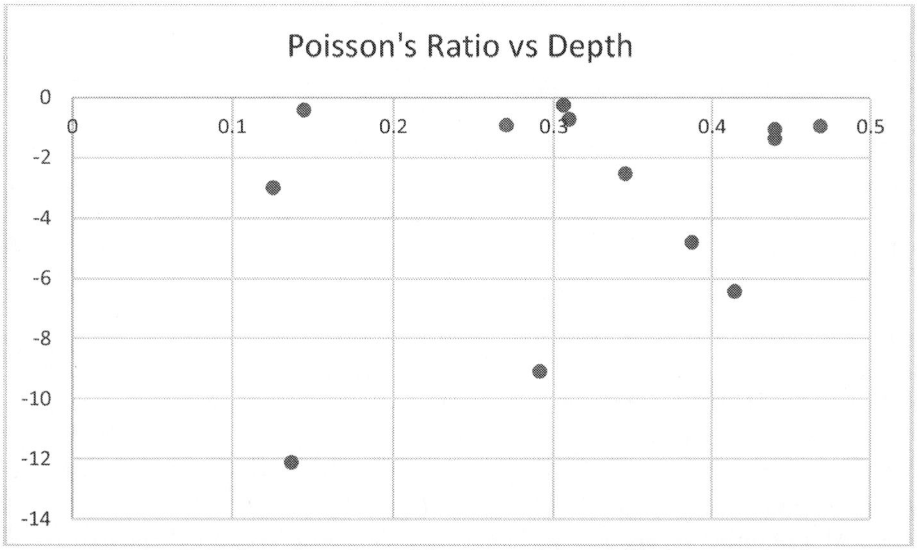

Figure 7. Poisson's ration of the A geothermal filed versus depth.

Figure 8 shows the significant differences between Poisson's ratio in the shallower zone and deeper zone. The value of Poisson's ratio with depth in the A geothermal field can be divided into two zones which correspond to shallower and deeper zone. The shallower zone with interval of Poisson's ratio varies from 0.2 up to 0.46 has depth from surface to 6500 meter depth and the deeper zone with interval of Poisson's ratio varies from 0.137 to 0.291 has depth is deeper than 6500 meter. Based on Fig. 8, the shallower zone is distinguished by the intensively fractured rock, soft and poorly consolidated. For the deeper is hard, rigid and it can interpreted as un-fractured rock.

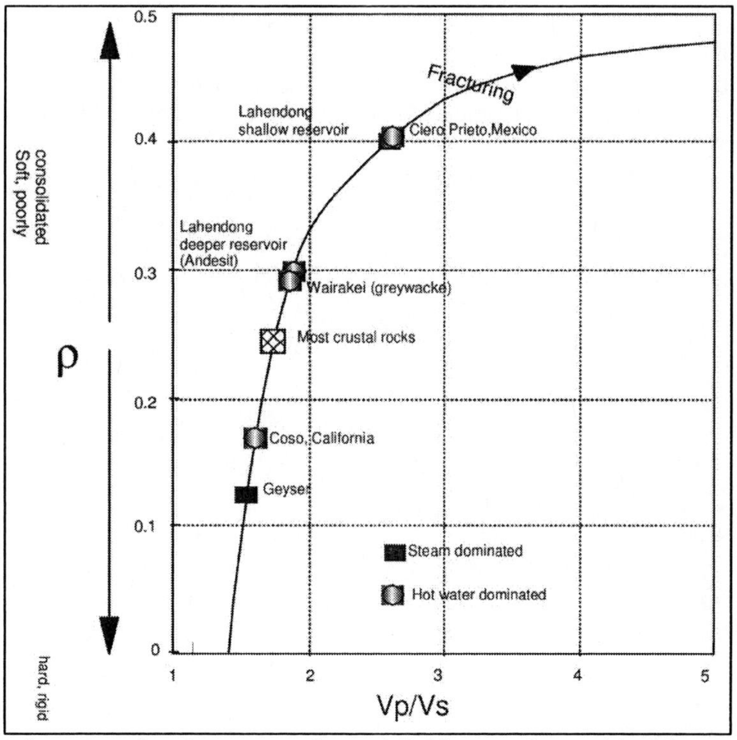

Figure 8. Poisson's ratio of some geothermal fields over the world wide versus Vp/Vs [2].

5. Conclusions

Based on the results, we had been conclude that:

1. Lower Poisson's ratio is associated with shallower zone, is distinguished by the intensively fractured rock, soft and poorly consolidated.
2. Higher Poisson's ratio is associated with deeper zone, is hard, rigid and it can interpreted as un-fractured rock.

In the next step, the study will continue for determining hypocenter to get more accuracy in this result. The result in this study effectively to observe the relation between the Poisson ration and fracture intensity in geothermal field.

6. References

[1] Lees J M & Wu H 2000 *Journal of Volcanology and Geothermal Research* **95** 157 – 173.

[2] Silitonga T H, Siahaan E E, Suroso 2005 *Proc. World Geothermal Congress (Antalya)* p.1 – 10.

[3] Bannister S, Sherburn S, Powell T, Bowyer D 2008 *Geothermal Resources Council – Transactions* **32** 259 – 263.

[4] Sewell S M, Cumming W, Bardsley C J, Winick J, Quinao J, Wallis I C, Sherburn S, Bourguignon S, Bannister S 2015 *Proc. World Geothermal Congress (Melbourne)* p.1 – 10.

[5] Sherburn S, Sewell S M, Bourguignon S, Cumming W, Bannister S, Bardsley C, Winick J, Quinao J, Wallis I C 2015 *Geothermics* **54** 23 – 34.

[6] Okamoto K, Yi L, Asanuma H, Okabe T, Abe Y, Tsuzuki M 2018 *Earth Planets and Space* **70**:15.

[7] Stanêk F, Jechumtálová Z and Eisner L 2015 *Reservoir stress from microseismic source mechanisms* (The Leading Edge: August 2015)

[8] Purwansah B 2012 *Bachelor Thesis* (unpublished, Geophysics Hasanuddin University).

[9] Hurukawa N 2008 *Practical Analyses of Local Earthquakes.* International Institute of Seismology and Earthquake Engineering (IISEE) Building Research Institute, Tsukuba, Japan.

Acknowledgements

We gratefully acknowledge the USGS and BMKG for data. We thank to Geophysics Study Program, Faculty of Mathematics and Natural Science Hasanuddin University (UNHAS) for supporting this research.

Geomorphology study on the Bandar Lampung Capital City for recommendation of development area

Rahmi Mulyasari[1], Hari Wiki Utama[2] and Nandi Haerudin[1]

[1] Geophysics Engineering, Lampung University, Lampung, Indonesia
[2] Geological Engineering, Jambi University, Jambi, Indonesia

rahmi.mulyasari@eng.unila.ac.id, h.wikiutama@unja.ac.id

Abstract. Bandar Lampung area is a capital of Lampung Province that located on the south of Lampung Province. It has highly populated urban area consisting of land and waters with several highlands and mountains lied in Bandar Lampung. Population density supported by high infrastructure development makes Bandar Lampung City become a city with rapid growth. Consequently, spatial planning concept by considering all aspects needed to support sustainable development in a city, one of them is geomorphology aspect. Geomorphology has an important role to give the information and overview on the existing landscape in an area. The aim of this paper is to know geomorphology characteristics of this area that could provide an information to land use, disaster mitigation, and references for infrastructure development area. Analysis of geomorphology aspect was employed in this study, which are morphology aspect (morphography and morphometry), morphogenetics aspect (morphostructure active, passive and dynamic), morphochronology aspect and also consider the morphoconservation aspect. Remote sensing is a method which is carried out to know characteristic of geomorphology aspect become a map. The geomorphologic map is resulted based on analysis of geomorphology aspect. This area could be divided to four geomophologic units, there are fluviatil unit, structural unit, volcanic unit, and denudational unit. In addition, based on analysis of geology and geomorphology, the area composed of two degree of geohazard, which are secure zone and dangerous zone. In addition, geomorphology map could be used to designed and development for recommendation area in Bandar Lampung.
Keywords: Bandar Lampung, geomorphology map, recommendation area.

1. Introduction

Bandar Lampung area is a capital of Lampung Province that located on the south of Lampung Province (Figure 1). It has highly populated urban area consisting of land and waters with several highlands and mountains lied in Bandar Lampung [1]. Population density supported by high infrastructure development makes Bandar Lampung City become a city with rapid growth. Consequently, spatial planning concept by considering all aspects needed to support sustainable development in a city, one of them is geomorphology aspect.

Based on Geological Map of Tanjungkarang Sheet [2], geological conditions of Bandar Lampung City are influenced by faults. On the map, there are clearly some faults that crossed Bandar Lampung city. The existence of faults are reflected in morphological conditions hills with steep slope landforms. Dominated lithology in research area is former of coastal and river deposits that are scattered around Lampung Bay and around Tanjung Karang. It is dominated by terrain resulting from young volcanic activity from Lampung Formation which is generally consist of tuff rock. This geological condition causes geological hazard in this area.

Geomorphology has an important role to give information and overview on the existing landscape in an area. Geomorphological map can be considered as graphical inventories of a landscape depicting landforms and surface as well as subsurface materials. This map can act as a preliminary tool for land management, geomorphological and geological risk management, as well as providing baseline data for other applied sectors of environmental research [3].

The aim of this paper is to know geomorphology characteristics of this area that could provide information to land use, disaster mitigation, and references for infrastructure development area. The geomorphologic aspects would be analyzed using remote sensing analysis.

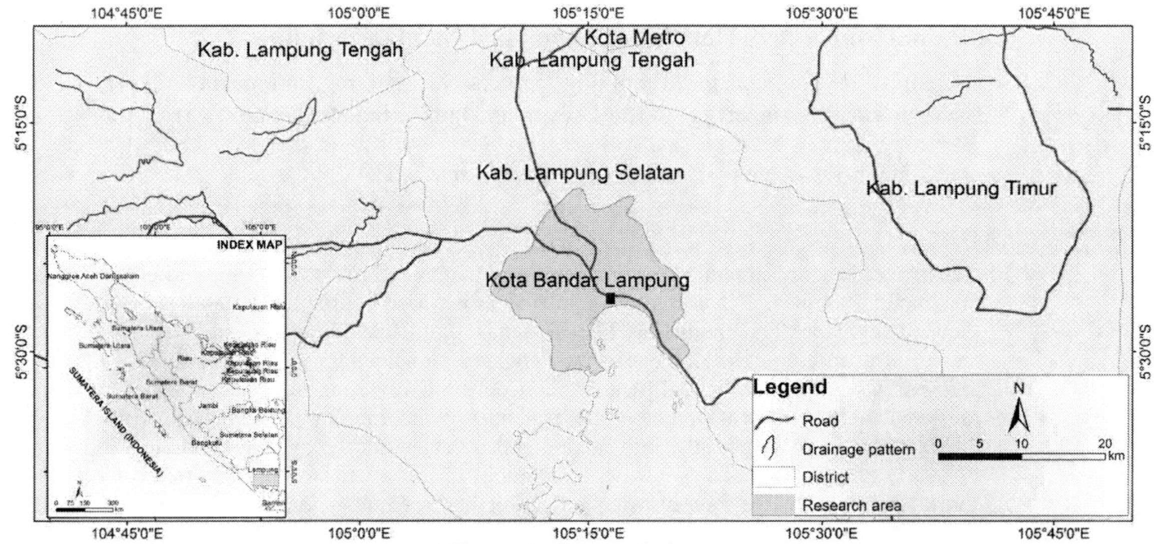

Figure 1. Research Area.

2. Regional Geology

Sumatera Island was resulted from subduction of Indian Ocean plate boundary beneath the Eurasian Continent plate in the Cenozoic Period which is thought to cause a clockwise rotation of Sumatera Island [4, 5, 6]. This subduction made changing position of Sumatera Island, which initially directed west – east to northwest – southeast [7]. Deformation changes in Sumatra Island occurred in Oligo – Miocene Period [8]. This deformation caused movement of Sumatran fault which became active at that time. According to [8] the geological structure formed on the island of Sumatra is generally in the form of a horizontal fault which forms a fault segment.

2.1 Physiography

Physiography of Sumatra Island is divided into several physiographic zones: Bukit Barisan Zone, Semangko Fault Zone (Sumatera Fault Zone), Highland and Hills Zone, Bukit Tiga Puluh Zone, Outer Arc Zone, and Sunda Exposure Zone [9]. Bandar Lampung area is part of Bukit Barisan Zone. This physiographic condition has similarities to Barisan Magmatic Arc in Physiography and Mandala Geology [10] (Figure 2).

Bukit Barisan Zone is a hilly zone with a southeast-northwest orientation and has a longitudinal pattern of about 1,650 km with a width of 100 km [9]. Based on morphology side, in general this area can be divided into three morphological units: wavy plain in the east and northeast, rough mountains in the middle and southwest, and a hilly to flat coastal area. The slopes are generally steep with altitudes up to 500-1,680 m above sea level. Coastal areas are diverse in variety and often consist of rough hills, reaching an altitude of 500 m above sea level.

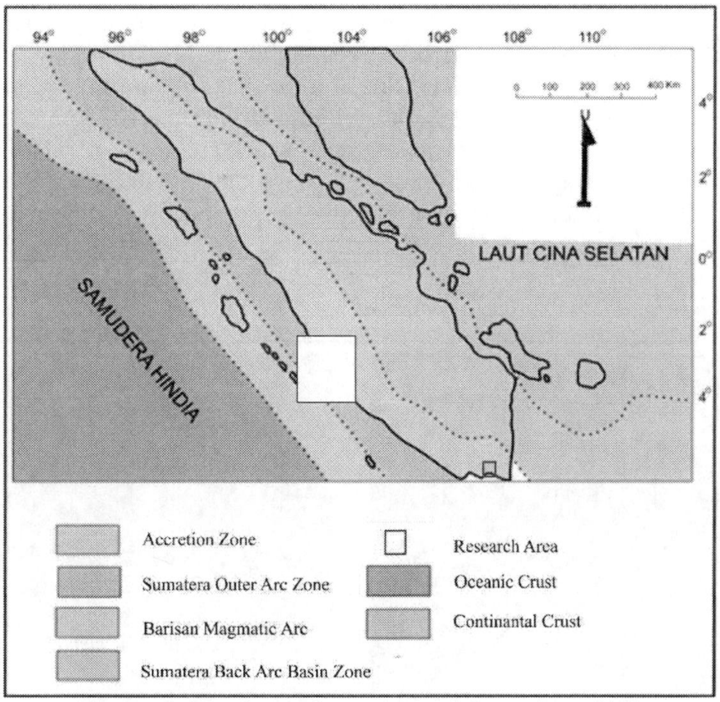

Figure 2. Physiography and mandala geology [10].

2.2 *Stratigraphy*

Based on regional geology map of Tanjungkarang sheet [2], stratigraphy of Bandar Lampung is divided into three parts: Pre-Tertiary, Tertiary, and Quaternary. Quaternary rocks are composed of sedimentary rocks, volcanic products and surface deposits. Tertiary rocks are composed of volcanic and intrusive rock products. Pre-Tertiary Rocks are composed of bedrock of metamorphic rocks.

2.3 *Structural Geology*

Research area was crossed by Lampung-Panjang fault. Lampung-Panjang fault in NE-SW direction was identified in the Tanjungkarang Sheet Geological Map [2] (Figure 4).

3. Data and Method

This research was conducted using some data, Digital Elevation Model National (DEMNAS) image, which has 5 m resolution [11] and regional geological map which scale of 1:250.000 on Tanjungkarang sheet of regional geology [2]. All of data were used to analyze and specify geomorphological aspects.

Research started with literature study to initiate previous publication article and interpretation on remote sensing imagery to detail analysis on digital elevation model of DEMNAS. Second stages than have done processing hill-shade analysis from topo to raster (Figure 3). Hill-shade analysis was conducted to attain an exact figure of relief from DEMNAS, particularly for a steep slope which probably indicated major lineament. Determination lineament and interpreted fault structure geology have been using combined of four azimuth directions. The systematics for to obtained of tentative geologic map. Geologic map to shown of rocks formation to composed of research area, formation boundary, stratigraphy contact, kinds of structure geology.

Determination of drainage pattern is a basic approach for interpreting to morphology units, indicated of structural geology, characteristics of lithology. These several approach might overlay to DEMNAS imagery and so that determination of geomorphic units the research area. The geomorphic aspects were taken into information in geomorphic analysis, which were shapes, processes, and historical geology. Geomorphic analysis was conducted tentative geomorphology map.

In the field were required of data collected, consist of characteristic lithology, confirmed of structural geology, geomorphic shape such as degree of undulation-valley, volcanic controlled, fluviatil processes, and degree of weathering. Specific data taken point to geohazard consist of mass movement related to active structural geology tectonics and volcanic structures. Collecting data have been give factual geologic map and geomorphology map.

Integration of geological map, geomorphology map, and point located of geohazard have done identification related to structural geology active/approximately and understanding tectonic development to control of fault segment in Sumatera, which Semangko fault (western of research area) and Lampung-Panjang fault segment similarly orientation to Semangko fault. Therefore, this step is divided in two group of geohazard zone, consists of secure zone and dangerous zone.

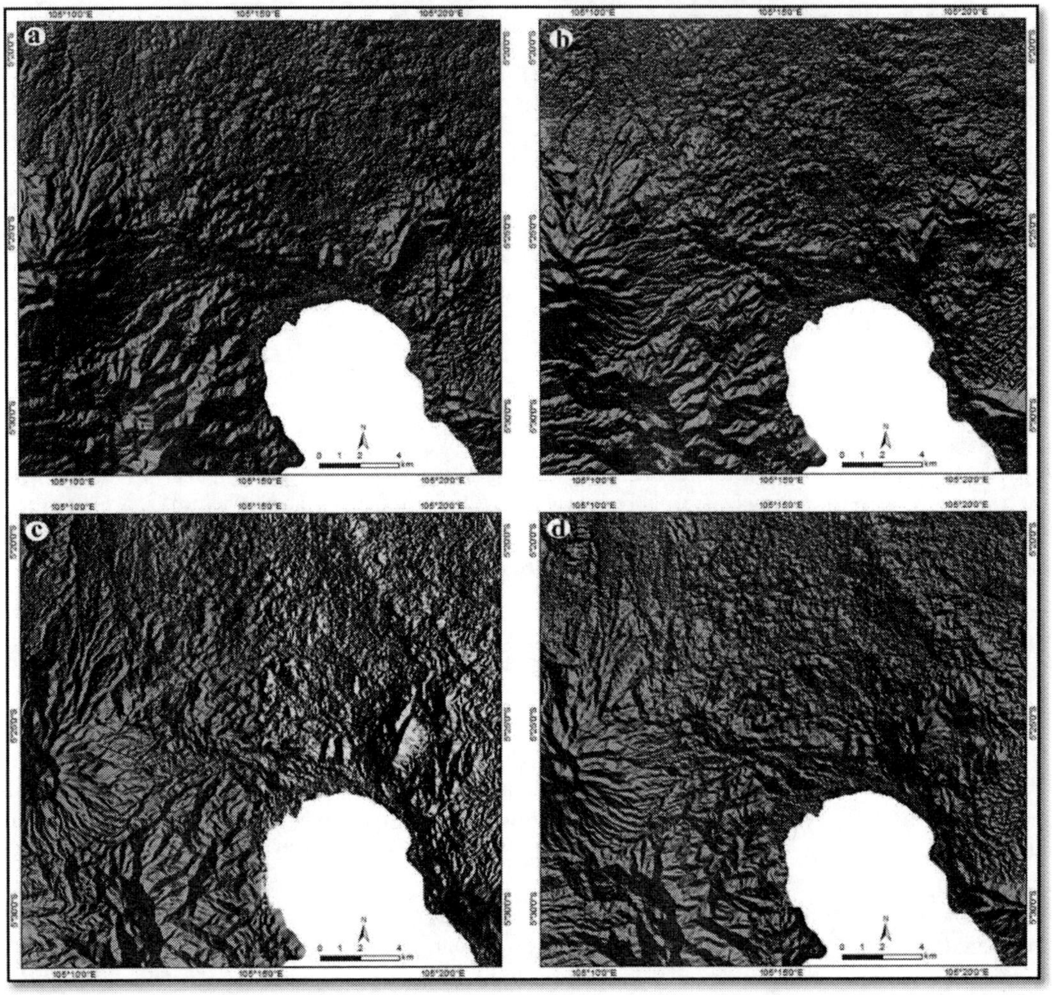

Figure 3. Four azimuth directions from the hill-shade analyze, a) N 315°E, b) N 0°E, c) N 45°E, and d) N 90°E. Four azimuth direction have done combine for produce hill-shale exactly is representative real geological condition surface.

4. Result and Discussion

4.1. Result

4.1.1. Stratigraphy

The research area is composed of Paleozoic metamorphic rocks, Mesozoic intrusive rock overlying to alluvium surficial deposits, Quaternary volcanic product and tertiary sediment mixing to Miocene intrusive rock (Figure 4). Lithology characteristics are divided rocks formation, however distinguishing of formation boundary.

a. Paleozoic Group
   This group is compiled of Schist Way Galih and Quartzite Sidodadi. These formation is depend on orogenesis processes product and related to reverse fault. The existence of rocks formation is basement rock of Bandar Lampung and Sumatra region. This group is distributed of eastern and southeastern of Tanjung Karang (center of research area).
b. Mesozoic Group
   Granodiorite is intrusive igneous rock as feature of magmatisme in Bandar Lampung and distributed to eastern of research area. The existence of granodiorite intruded Paleozoic Sidodadi Quartzite.
c. Tertiary Group
   Tarahan Formation and Campang Formation are sedimentary rock formation composed of research area. Campang Formation in the southeastern part of research area. This formation associated straight of Lampung-Panjang fault, which segmentation fault similarly to Semangko fault (one of big segment fault in Sumatra), whereas Tarahan Formation have width distribution, which is in the southeastern-northeastern and western-southwestern part of research area.
   Tertiary group is also composed of magmatic intrusive igneous rock, which is Miocene Granitoid. This intrusive associated to secondary segment of Lampung-Panjang fault and the existence in the western part of research area.
d. Quaternary Group
   This group can be divide two cluster, which are Alluvium as product Recent geological processes and Quaternary volcanic product. This group is focused to volcanic product. Lampung Formation distributed in the part of northern and southern research area, meanwhile Young Volcanic Deposits (Betung Volcano) distributed in the part of central to western research area. Young Volcanic Deposits of Betung Volcano product is occupied more half than research area extensive.

4.1.2. Geomorphology
In the research area is four geomorphic units that could be identified. Classified of geomorphic unit divided become landscape. There is composed of fluviatil unit, structural unit, volcanic unit, and denudational unit. The geomorphic unit is too the determination based on morphochronology of landforms (Figure 5).

Drainage pattern in the research area compiled of radial, subdendritic, and dendritic. Radial is occupy in Betung Volcano geomorphic unit, subdendritic is covered of structural and denudational geomorphic units, whereas dendritic is only fluvial/alluvial plain unit.

a. Structural Geomorphic
   This landscape have influenced by structural, which was active fault and approximately fault. There were occupied in the central-north-east-southeastern-west-southwestern part of research area. This landscape associated Lampung-Panjang fault with similarly Sumatra Fault System which Semangko fault segmented. This landscape is composed by Way Galih Structural Hills (WGSH), Way Galih Structural Hill (WSH), Sidodadi Structural Hill (SSH), Tarahan Structural Hills (TSH), Campang Structural Hills (CSH), Lampung Structural Hills (LSH), and Lampung Structural Valley (LSV).the existence of fault segment is primary element to formed of landscape.
b. Volcanic Geomorphic
   This landscape controlled by Recent volcanic activity of Betung Volcano and magmatism in Granodiorite Cretaceous and Granitoid Miocene. The existence of active volcano is counted of Barisan Range, meanwhile presence of intrusion magmatism indicated tectonic-magmatism-volcanism the associated with Sumatra Fault Zone. This units geomorphic are composed of Sulan Intrusion (SI) Cretaceous, Panjang Intrusion (PI) Miocene, and active volcano product of Lampung Formation Plio-Plistocene, and Betung Volcano unit (BV) Quaternary.
c. Denudational Geomorphic
   This landscape is only occupied of little distribution geomorphic unit. This geomorphic unit is controlled by eroded, which have characterized as lithology or rock formation on the hill morphology different to lithology in surrounding. Padang Isolated Hill (PIH) is half of Lampung Formation. This indicated that volcanic product such as pyroclastic, lava, and volcanic-sediment.

d. Fluvial Geomorphic
This landscape is product of Recent geological processes. This geomorphic have characteristics as unlitification, sedimentary material and distributed in plain/flat relief, such as Alluvial Plain (AP). Identified of fault segmented Lampung Panjang was dotted where concealed.

4.2. Discussion

4.2.1. Structural Geology and Geohazard

Structural geology is almost covered full in the research area. The existence of research area in between Barisan Range-Magmatic Arc associated Sumatra Fault Zone. Fault segment Lampung-Panjang fault is half Sumatra fault segmented. Lampung-Panjang fault is fault range, it have similarly characteristic which Semangko fault, where fault segments interpreted to be active.

Structural geology development of the research area is resulted of interaction between those two segment Lampung Panjang fault and Tanjung Karang fault (west beside part of Lampung Panjang fault). There is orientation south-southeastern - north-northwestern direction and following orientation the primary fault structure segments of Sumatra Fault System.

Along of segment fault system "tectonic fault" and volcanic fault is indicated to geohazard. Point of mass movement have occurred in fault zone and surrounding (Figure 6). In addition to tectonic compression of Sumatra regional, weight load exerted volcanic edifices. There is also plays important role in creating structural geology controlled. A relatively north-northeastern - south-southwestern compressional stress was generated by dextral fault movement of those fault, there are producing of extensional faults that the distributed of research area.

4.2.2. Geomorphology Study for Recommendation Development area

Geomorphology study could be standard operation procedure for recommendation development area. Integration data and considerately study have made geohazard potential zone which based on degree of geohazard, so that the result could be standard operational procedure. Based on analyzed of DEMNAS image, field datas, geomorphology map, geological map, and previous study, that divided become to 2 zonations (Figure 6).

Base principle of degree geohazard zonation is counted of several aspect geology, between of morphology (morphography), morphogenesis (characteristic lithology, structural geology, volcanism, eroded, fluviatil, and weathering), morphochronology. Characteristic lithology and structural geology is primary factor in determination of degree of geohazard. Active fault system in the research area have conducted change of geomorphic condition. In the research area is also of residence or located capital population, government building, and public transportation.

a. Secure Zone
This zone is category low potential geohazard. This zone occupied of Lampung Structural Hill (LSH) with Lampung Formation is volcanic product, and volcanic-sediment product. This zone secure relatively, meanwhile this zone across fault segment, but approximately segment. Therefore, this zone is very recommendation for development area, such as residence, government building, public transportation, and capital office complex.

b. Dangerous Zone
Dangerous zone is category high - very high risk when will become zone for development area. There is occupied landscape of structural. Denudational, fluvial, and the half of volcanic, which is Betung Volcanic zone. Structural geology is observed in this zone, between of Lampung-Panjang fault (extension stress) dextral fault. Fault orientation in this zone compiled of north-northeastern – south-southeastern direction, southeastern – northwestern direction are typically of dextral/strike slip fault, northeastern – southwestern is typically strike slip fault sinistral with fault observed/measured and approximately, and several normal fault of volcanic structure have east – west direction. Fault volcanic structure to associated of Betung Volcanic. This zone is occupied Paleozoic Group Formation, Mesozoic Group Formation, and Tertiary Group Formation, and Quaternary Alluvium Group. The characterized is very high potential geohazard or mass movement.

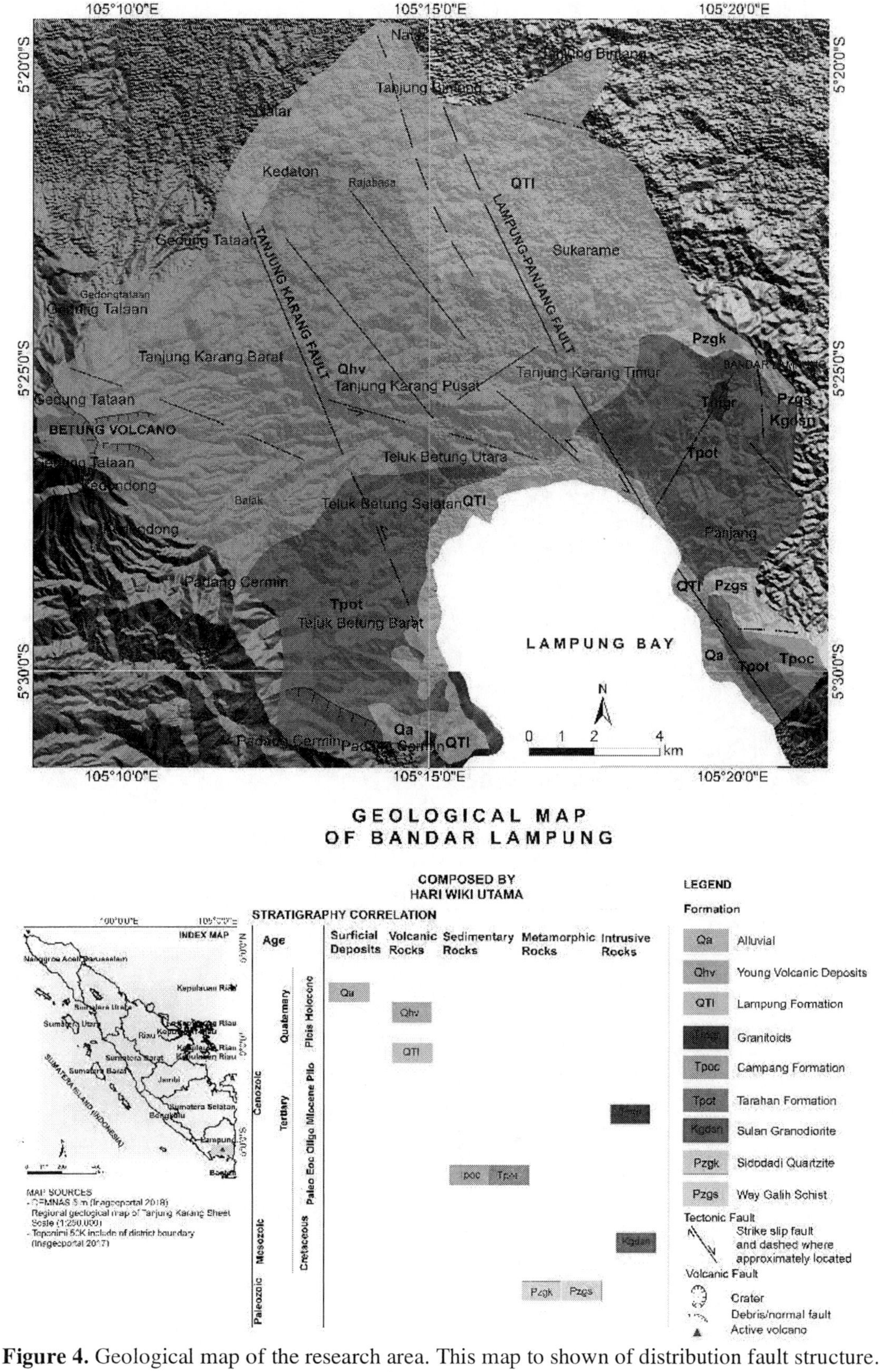

Figure 4. Geological map of the research area. This map to shown of distribution fault structure.

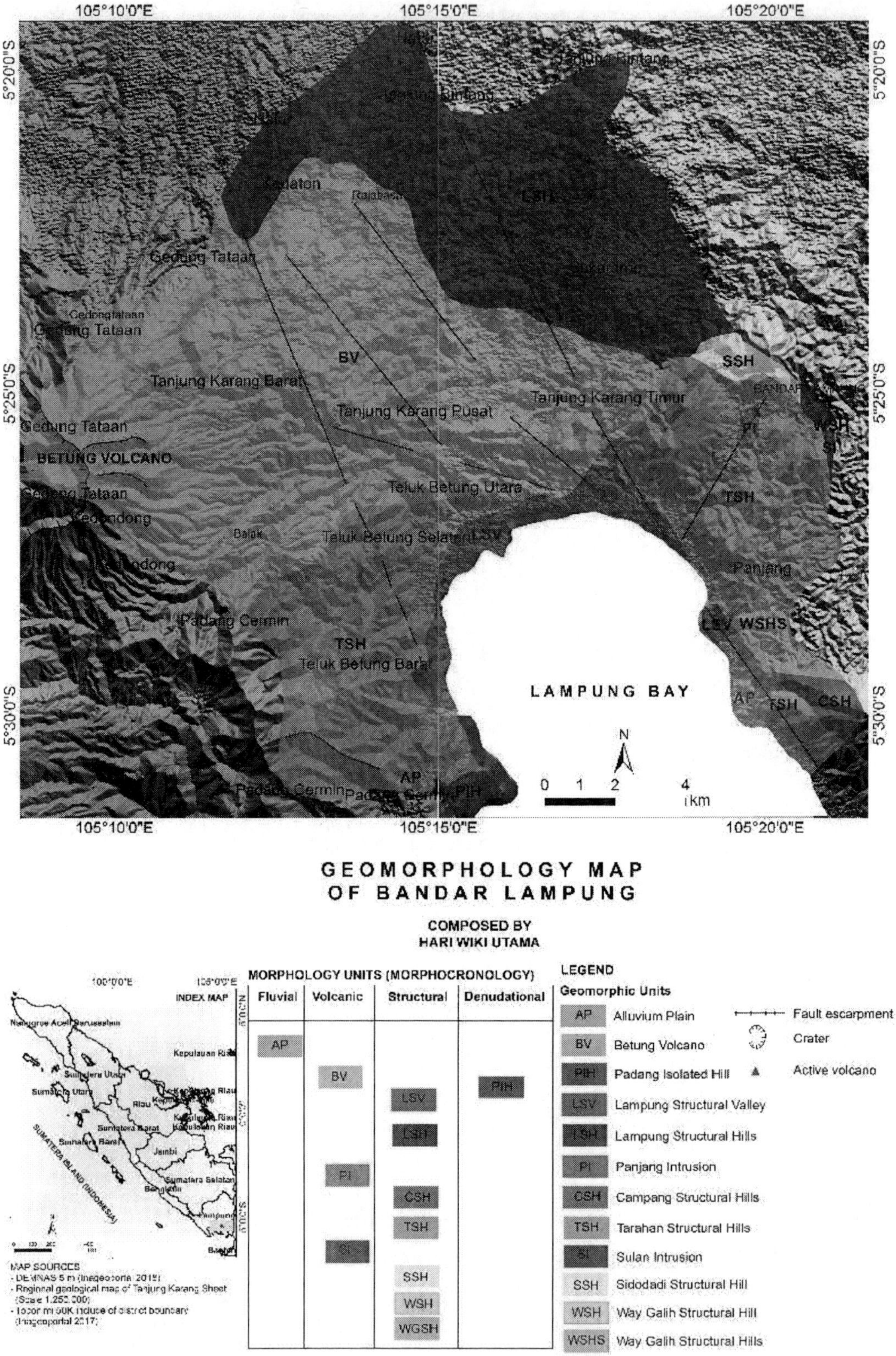

Figure 5. Geomorphology map of the research area. Drainage pattern is indicated morphology unit and control geology, such as volcanic and structural.

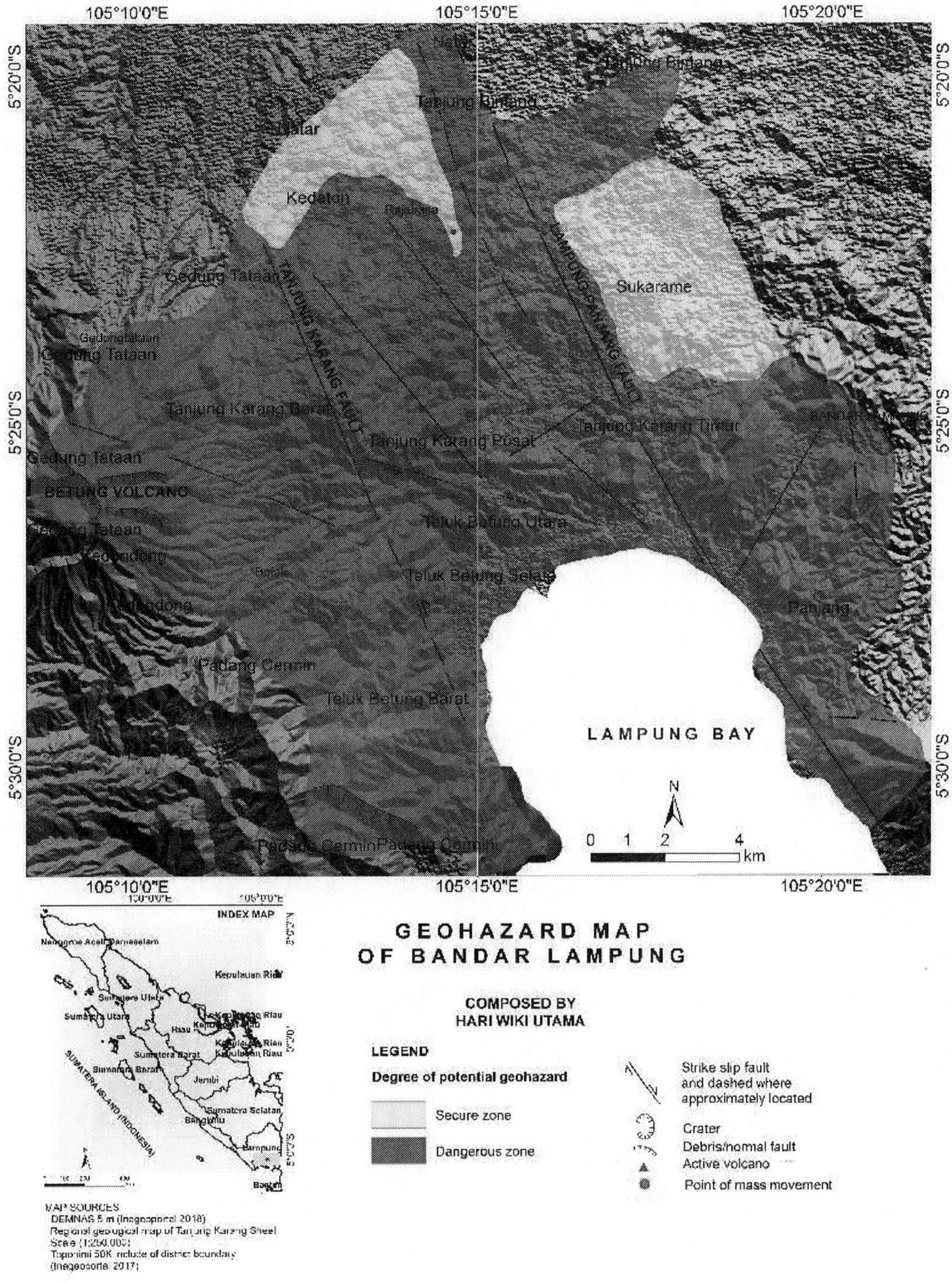

Figure 6. Geohazard map of the research area.

5. Conclusions

Based on structural geology, stratigraphy and geomorphic analysis are control geological setting in research area. Geomorphology map is resulted based on analysis morphology, morphogenesis, and morphochronology aspects. This area could be divided into several geomorphologic units, there are volcanic, structural, denudational, and fluvial morphology. The exist two degree of geohazard, which are secure zone and dangerous zone. In addition, based on geomorphology map, it could be designed and development for recommendation for area in Bandar Lampung.

References

[1] Pemerintah Kota Bandar Lampung 2017 *Sekilas Kota*, https://bandarlampungkota.go.id/sekilas-kota/ [accesed on 16 May 2018].
[2] Mangga S A, Amirudin T, Suwarti S, Gafoer and Sidarto 1993 *Geological Map of Indonesia Quadrangle: Tanjungkarang, Sumatra Scale 1: 250.000* (Bandung: Geological Research and Development Centre).
[3] Otto J C and Smith M J 2013 Geomorphological mapping *Geomorphological Techniques, Chap. 2, Sec. 6.*
[4] Hall R 1997 *Geological Society of London, Special Publication* **126** pp 11-23.
[5] Hall R 2002 *Journal of Asian Earth Sciences* **20** pp 353-356.
[6] Hall R 2014 Indonesia Tectonics: Subduction, Extention, Provenance, and More, Indonesian Petroleum Association *Proceedings 38th Annual Exhibition and Convention, Jakarta, Indonesia IPA 14 G* p 360.
[7] Hamilton W B 1979 Tectonic of the Indonesian Region *Professional Paper 1078* US Geological Survey Washington DC p 345.
[8] Barber A J Crow M J Milsom J S 2005 *Geology, Resources and Tectonics Evolution* (London: Geological Society Memoir) p 290.
[9] van Bemmelen R W 1949 *The Geology of Indonesia Volume 1A* (The Hague, Netherlands: Government Printing Office) p 732.
[10] Kusnama Pardede R, Mangga S A and Sidarto 1992 *Geological Map of Indonesia Quadrangle: Sungai Penuh and Ketahun Scale 1: 250.000* (Bandung: Geological Research and Development Centre).
[11] Geospatial Information Agency 2018 *Digital Elevation Model National (DEMNAS) image* http://tides.big.go.id/DEMNAS/

Geology and Genesis of the Tanco Isolated Hill on the Kerinci Lake, Jambi

F Nabella[1], H W Utama[1], Y M Said[1]

Department of Geological Engineering, Faculty of Science and Technology, Universitas Jambi, Jambi-Ma. Bulian KM 15, Muaro Jambi, 36122, Indonesia

h.wikiutama@unja.ac.id, nabellafiqhy@gmail.com

Abstract. Kerinci area is a small part of the Barisan Range where the Magmatic Arc Zones and a valley extending from the northwest to the southeast flanked by hills with steep slopes on both of sides that flank it. Isolated hill which is on the Kerinci Lake is the main focus to be studied because its existence is located in the middle of a part of the depression Kerinci Lake. In additions, in the vacinity of the research site there has never been research conducted both of in general and specifically. The research was conducted by interpreting the geological controlled by mechanism on isolated hill morphology forming based on geomorphoogical approach and petrographic analysis. Kerinci Lake on the south side of the research site is indicated as volcano - tectonic lake. The isolated hill is thought to be related to the formation of Kerinci Lake. In saveral location along the path of the Sumatran fault saveral lakes were found which were controlled by this fault. Sumatran fault formation is influenced by strike slip, while the formation of the lake is influenced by extensional stress. Based on it is existence, understanding the location of the research could be done analyzedg of volcanism and structural geology.

Keywords: isolated hill, Kerinci Lake, volcano - tectonic, Sumatran Fault.

1. Introduction

The latest Sumatran tectonics are a result of the convergence of the Indian Ocean Plate and Eurasian Continent Plate, thus giving a deformation effect on rocks and are thought to have caused rotation from Sumatra Island, [1, 2, 3, 4, 5]. Changes in the position of the island of Sumatra, which initially directed west-east to northwest - southeast [5]. Deformation changes in Sumatra Island which experienced rotation occurred in Oligo-Miocene Period [6, 8]. This geological condition is also complemented by an active fault movement from the Sumatra Fault System (Siulak Segment) which is extention stress, resulting in a valley of the "Kerinci Valley" which is part of the "pull a part basin", [7].

The research area located in Kerinci Valley is a small part of a series of Magmatic Arc Zones of the Barisan Hills and the Sumatran Fault System in the form of a valley extending from the northwest towards the southeast flanked by hills with steep slopes on both sides flanking it. The Kerinci Valley is then filled with material from the surrounding volcanic eruptions and is part of the siulak and terban fault segments (graben), [8, 9, 10]. In [11, 12] stated that the graben valley is developed in the Quaternary Quarters with a horizontal movement of the ascending (dextral).

General, the location of the study was on a volcanic alignment track controlled by the Sumatra Fault System located on the north side of Lake Kerinci. The study area is an isolated hill that is surrounded by alluvial deposits with alluvial terrain morphology. Based on regional geological data the isolated hill constituent rocks in the Pengasih Formation consist of sedimentary rocks such as claystone, siltstone, floating rocky sandstone with lignite insertion and polymic conglomerates with a thickness of 135 m [12]. Based on the appearance of aerial photographs showing the north side of Kerinci Lake in the form of a hill that is interpreted as isolated hill (Figure 1).

Figure 1. Research area "Isolated Hill" is in the north-northwest part of Kerinci Lake (Google Earth 3D, 2018)

2. Regional Geology

2.1. Tectonic Framework

The island of Sumatra is located along the southwestern edge of the Sunda Plain with an expansion of the Eurasian Plate in the form of the plains of Southeast Asia and is part of the Sunda Arc. The accretion zones located along the west coast of Sumatra, where oceanic crust underlies the Indian Ocean and parts of the Indo-Australian Plate form a sloping subduction along the Sunda trench, then Sumatra Island has complex tectonic characteristics, [5, 13].

Subducted in Sumatra during the Early Tertiary to Recent have caused extention associated of magmatic arcs in the Barisan Range, meanwhile irregularly subducted had been occured since Late Perem [14, 15], although the position of the bow and trench as it is now likely to occur in the Miocene. Compression stress rising of sloping at regular intervals has been released through to dextral fault of the edge continental margin, which resulted in the occurrence of the Sumatran fault system stretching along the island and limiting from the magmatic arc from west to east [16]. [12] Sumatra can be divided into four Mandala Geology zone: Accretion Zone, Outer Arc Zone, Barisan Magmatic Arc Zone, and Back Arc Basin Zone. Geology This sheet includes Pre-Tertiary bedrock, a number of Neogene plutons and sequences of volcanic rock layers and Tertiary and Quaternary sediments which cover it (Figure 2).

Deformation have been occurred on the island of Sumatra caused the movement of the Sumatra fault which began to be active at that time. This plate convergence is estimated to experience movement of 5 - 7 cm / year, in [8]. Sumatra Island is composed of suture-sutures or continental blocks that form a unit of continental crust, including Block Woyla Nappe, Block West Sumatra, Block Sibumasu (East Sumatra), Quartzite-Granite Terrain, Block Indochina [8].

2.2. Physiography

The physiology of Sumatra Island is very reflected in the condition of the elevation model or regional geological conditions, there are at least six physiographic zones, including the Low Hills and Corrugated Hills Physiographic Zones, the Bukit Tigapuluh Zone, Barisan Hills Series Zone, Sumatra Fault Zone, Outer Bow Islands Zone, and Exposure Zones. Sundanese, in [10]. This physiographic condition has similarities to the Geology Mandala which is the Barisan Magmatic Arc, but the research area is precisely located in the valley of the "Kerinci Valley" between the Sumatra Fault System and the Barisan Hills (Figure 3).

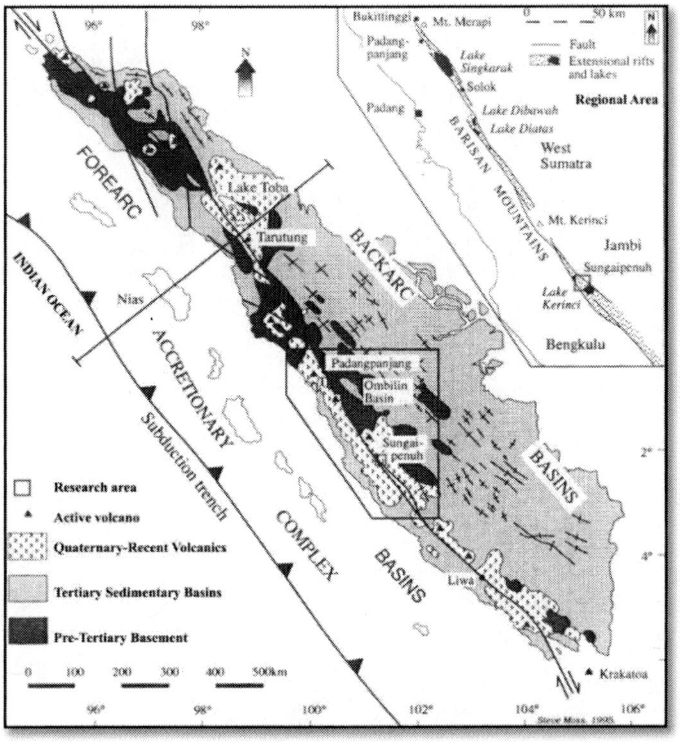

Figure 2. Tectonic setting in Sumatra. Research area associated of Sumatran Fault Zone which characterized extention stress to fault segmented and had gived pull a part basin on the Kerinci Valley, modified from [4]

Kerinci Valley is recognized by many experts as a result of past tectonic volcanic activities [10]. Whereas [17], interpreting Kerinci Valley as one of the active tectonic segments lies in a line stretching from Semangko Bay (southern Sumatra) to Blangkajeren (Aceh - North Sumatra). According to [18], based on the Indonesian Geomorphological Map from the results of remote sensing, the research area is part of a series of volcanic geomorphology units bordering the geomorphology unit formed from the origin of the western Sumatra Island.

2.3. Geological Structure

Geological structure have orientation of north-northwest - south-southeast include Siulak and Dikit faults (Kerinci and surroundings), [19, 20]. Fault segemented generally has movement away from each other (extension) and generally pull apart basin, but some fault characteristics of the study area are formed by compressional stress, thus forming a height (pop up), [8, 12, 20].

The history of the Sheet Riverpenuh and Ketahun structures is dominated by the Jura - Resen tectonic event. Elements of the main structure of enlargement and splitting [12]. According to [9, 12, 19, 20] the study area was traversed by the Siulak fault which is two parallel faults with a length of 37 km, the two faults as a barrier between the Kerinci Valley landscape and lane mountains that flank it (Figure 4).

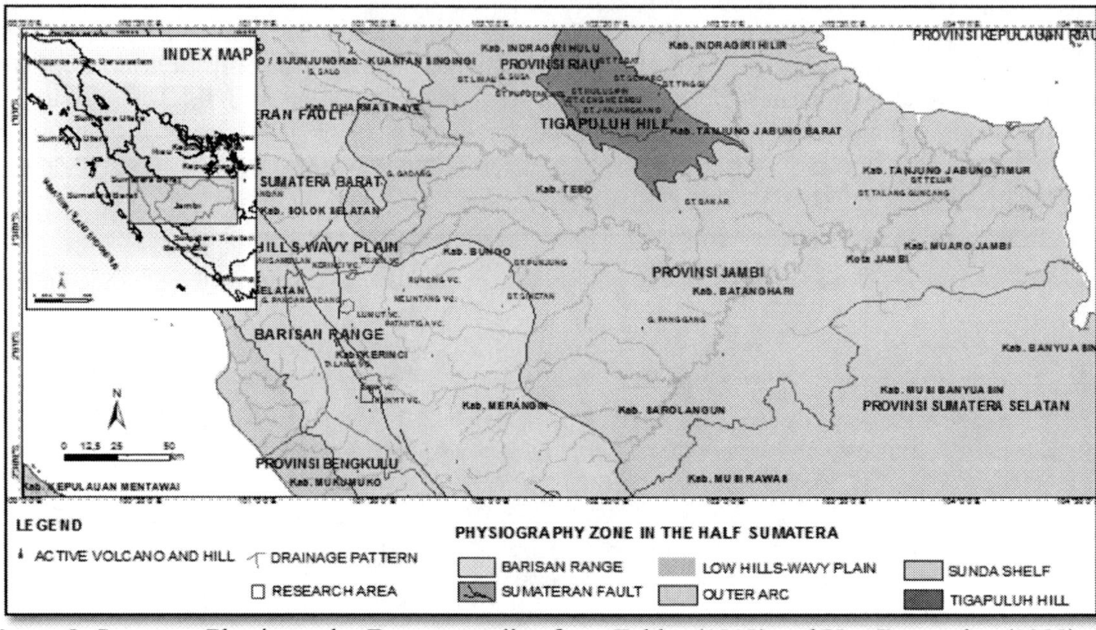

Figure 3. Sumatran Physiography Zone, according from Tobler (1913) and Van Bemmelen (1939), modified from [10]. Isolated Hill is half of Barisan Range - Sumatran Fault Zone

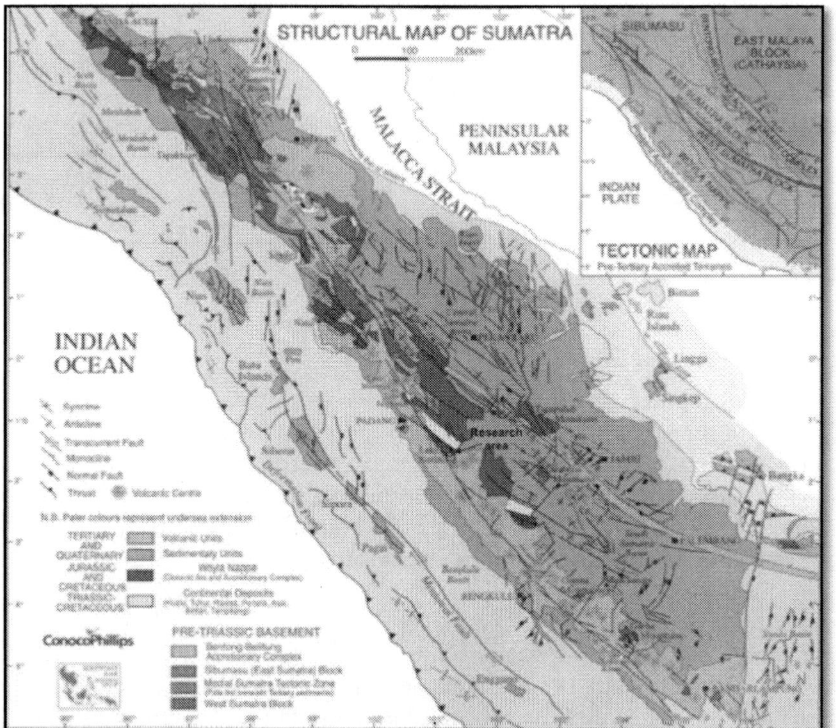

Figure 4. Simplified geological map of Sumatra, comprise of rocks formation and structural geology, [8]. Isolated hill and surrounding is half of volcanic unit product, sedimentary unit, and Woyla Nappe Block

3. Data and Method

Data available in this study include of Digital Earth Map 1: 50,000 from the Badan Informasi Geospasial [21], regional geological maps of Sungaipenuh and Ketaun with a scale of 1: 250.000, and

IOP Conf. Series: Earth and Environmental Science **279** (2019) 012027 doi:10.1088/1755-1315/279/1/012027

ASTER-GDEM with a resolution of 30 m from [22]. Map processing uses ArcGIS 10.1 and Global Mapper 16 with 12.5 m interval digital contours, as to produce a digital elevation model (DEM).

Method used in this study include of indirect mapping, such as using remote sensing data analyzed from ASTER GDEM 30 m and topographic contours from RBI. Based on this map, a fault structure delineation will be carried out, the alignment of the ridges overlapping the regional geological map, so that the combination of the map produces a temporary geological map. From the previous geological map, interpretation of research objects was carried out using geological concepts and also based on preliminary studies with related references, so that from this data will be integrated into field research.

4. Result and Discussion

4.1 Petrologi Volcanic of Isolated Hill

The result of petrographic analysis is the basis of determining the lithological characteristics of isolated hill compilers. In this study several petrographic rock analyzes were carried out for different types of rocks, thin slices a rock consisting of igneous rock, pyroclastic and sediment. Petrographic analysis of rocks in this study is a benchmark in determining the genesis of each rock sample and isolated hill genesis. Through petrographic analysis can be studied about the origin, availability and history of rocks. The important thing from petrographic analysis is the identification of mineral composition and rock texture. Grouping or classification of rocks is based on the observation of the main mineralogical texture and composition.

4.1.1 Ignimbrite. Petrographic analysis that has been known that andesite rocks are igneous rocks that have a crystal size (0.1 - 2 mm). Composed by phenocrysts in the form of plagioclase (0.1 - 2 mm) with An 44 (andesine) and ampibol values (1-2 mm), biotite (0.5 - 1.5 mm), piroxen (1-2 mm), quartz (0.1 - 0.8 mm). The base period consists of glass and microcrystalline with a relative size of 0.05 - 0.1 mm. The main texture of this rock is holocrystalline but with a feneric size. There are several clinopiroxen minerals that stand out as phenocrysts reaching 0.4 mm (fine generic) based on the price of the microcrystalline plagioclase with andesine type and the presence of little hornblande identifies that these volcanic igneous rocks undergo imperfect crystallization processes. Based on laboratory results and on field observation data can be concluded that these rocks come from magma that freezes very quickly (outside the body of the volcano). Included in volcanic igneous rocks. In the field, the presence of these andesite rocks is as a fragment from the ignimbrite associated with pumice.

Through petrographic observations it can be seen that the thin incisions of rock observed contain a lot of microcrocrystalline. This indicates that the rock is formed outside the body of the volcano where the cooling process takes place quickly, causing the formation of small (micro) minerals. It can be concluded that the igneous rock fragments found in the study area are lava (not intrusion) formed during volcanic eruptions (Figure 5).

4.1.2 Pumice. Observation under the microscope showing that pumice from the study area is colorless (tends to be brownish) with a welded structure, and has a clastic texture. There are fragments in the pumice that are observed, namely pyroxene, k-feldspar, plagioclase, and hornblande with a base period of glass juveniles that experience welding and volcanic ash as large as. Pumice fragment is not the main material in rocks, but are formed first before the pumice is deposited. The degree of hypocrystalline crystallization, angular-sub angular grain shape, disaggregated poorly, with open containers (between fragments).

IOP Conf. Series: Earth and Environmental Science **279** (2019) 012027 doi:10.1088/1755-1315/279/1/012027

Figure 5. Ignimbrite is indicated volcanic product, which effusive-explosive eruption volcanic. Ignimbrite have been formed to mixing of pyroclastic flow and pyroclastic airfall. a) outcrop ignimbrite which is nroth-westnorthern part of Kerinci Lake, b) rock chip of ignimbrite fragment, c) micrography cross polarized nicol and d) parallel polarized of ignimbrite fagment. Clinopyroxene (Cpx), potassium feldspar (Kfs), hornblende (Hbl), plagioclase (Pl), microcrystalline (M-Cr)

Results of the petrographic analysis that has been done, it is known that the pumice originating from the study area is pyroclastic rock in the form of deposit flow with piroxen fragments (0.2 - 3 mm), k-felsfar (0.5 - 1 mm), plagioclase (0.7 - 1.5 mm) with values of AN 51 (labradorite), and hornblande (0.3 - 3 mm). With the basic form of glass juveniles that experience welding and volcanic ash. Based on laboratory analyzed and supporte datas if the field observation data can be concluded that the pumice rocks are formed from acidic rock explosive volcanic eruptions that release the material into the air, then transfortasi horizontally and accumulate as pyroclastic rocks. The cavity that forms on the rock comes from the expansion of foam froth contained in it. The fragments in it come from external materials that are mixed during the formation and during transportation. Generally produced by large or strong eruptions of a volcano.

Pumice is a group of pyroclastic rocks originating from magma foam, formed at the time of volcanic eruption. Pumice is the character of an acid volcano. This is reinforced by the pumice found to have a white color indicating that the pumice consists of silica (acid) derived from acid volcanoes - intermediates. Based on the availability in the field of pumice in the study area, the shape tends to be angular, indicating that the rock is not transported far from its source (close to the source). Besides that, the pumice found in the field tends to be flat, which means that the pumice is deposited in hot conditions (immediately after a volcanic eruption occurs) (Figure 6).

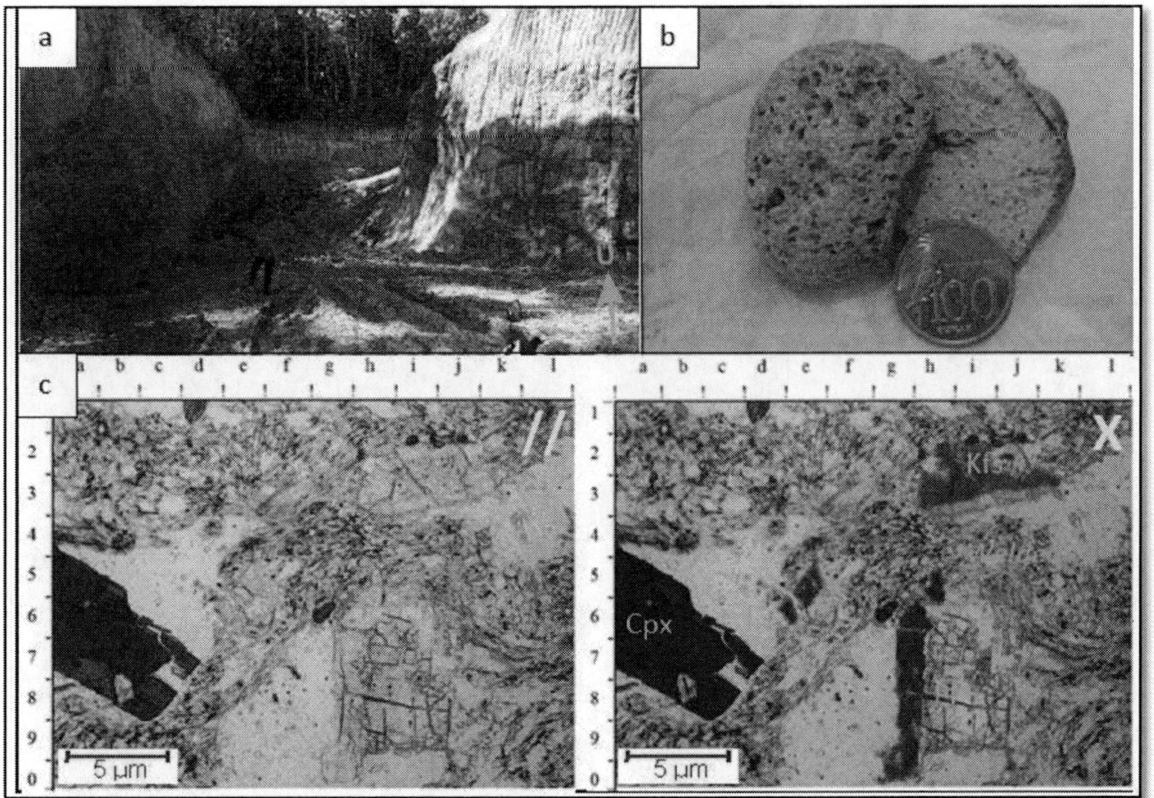

Figure 6. a) outcrop of pyroclastic flow which is pumice fragment, b) rock chip of pumice, c) micrography cross polarized nicol and d) paallel polarized of pumice. Pyoroclastic flow was volcanic product which explosive erupted of volcanic activity. Clinopyroxene (Cpx), potassium feldspar (Kfs), plagioclase (Pl)

4.1.3 Tuff. There are 3 types of tuffs in the field, pure tuff which consists of glass, tuff pasiran, and clay tuff. Petrographic analysis that has been carried out is known that the tuff rocks are pyroclastic rocks in the form of surge deposits with a matrix / base which are all in the form of glass juveniles that experience welding and volcanic ash (<0.004 mm), well sorted, closed container. This rock is formed due to the blowing process. It is a volcanic material that results from explosive eruptions, then consolidates and undergoes rocking. At the time of the formation process the material is still hot and changes in shape so that the rock undergoes a welding process. These tuff rocks are formed due to explosive eruptions of acid volcanoes (Figure 7 and Figure 8).

Result analyzed of laboratory and supported datas from field observation data can be concluded that these rocks are volcanic rocks, produced by explosive eruptions. This rock has undergone consolidation and formation. Composed of rock and mineral and lytic fragments embedded in the base of the glass. Based on the degree of non-uniform sorting and angular (angular - subangular) indicates that this rock does not undergo a complex sedimentation process (originating directly from the source).

The presence of tuffs with large and thick massive structures indicates that the study area is close to the source of volcanic eruptions. Tuff is a type of rock which consists of consolidating volcanic ash released from ventilation holes during volcanic eruptions. Tuff found in the study area has a degree of crystallization of holohyalin where the composition of the majority of minerals is glass. Based on petrographic tuff analysis from the study area has a welded structure. Where this structure can be formed if the eruption material deposited on the surface is still hot and then freezes into welded tuff. This is also evidence that the research location is close to the center of volcanic eruptions.

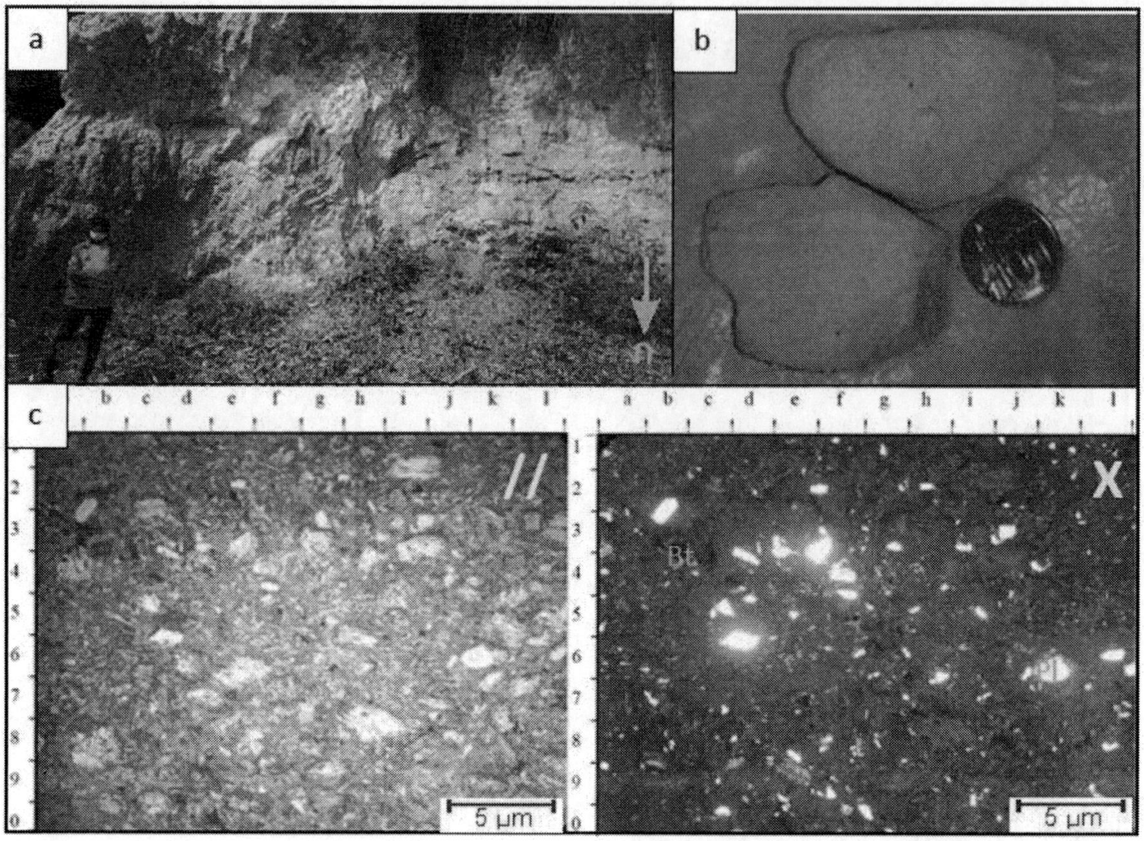

Figure 7. Pyroclastic flow product is indicated not far from source (volcanic activity erupted). a) outcrop of tuff which ticknes realize 4 m, b) hand specement of tuff, c) cross polarized nicol and d) parallel polarized of crystal tuff. Biotite (Bt), plagioclase (Pl), Quartz (Qz)

4.2 Structural Geology of Isolated Hill

Faults that are formed in the observing area are a continuation of the force on the fracture (formed first). Faults were formed due to semi-brittle deformation resulting in a fault where there are moving blocks (Figure 9). From the datas and the results of the analysis that has been done, it can be seen that minor faults which work in the observing area are generally relatively moving to the right (dextrals). This is similar to the movement of the Large Sumatran Fault and the Siulak Fault segment that is a fault that has the direction of horizontal movement to the right. So, it can be concluded that the formation of isolated hill, other than due to volcanic activity, is also influenced by the large segment faults.

Based on this existence, it can also be seen that fractures in the Pengasih ignimbre unit is older (Plio-Plistocene) compared to faults in the Pengasih sandy tuff unit (Quarter) formed in the final phase. Sandy tuff indicated product of volcanic sediment. This product have been controlled by Siulak fault and another fault (secondary fault).

Figure 8. Tuff volcanic product. a) outcrop of tuff, b) hand specement of tuff, c) photomicograph cross polarized nicol and d) parallel polarized of Welded Vitric Tuff. This product id dominantly glass volcanic

4.2.1 Genesis of Isolated Hill

Data and some analyzed were completed, an overview of the g enes of isolated hill (Bukit Tanco) was obtained, which was located in 2 (two) districts, namely Sitinjau Laut District and Danau Kerinci District. Based on its location, the object of the study was in the path of the Great Sumatra Fault just beside the north of Lake Kerinci. The Isolated Hill is the same as Sumatra Island, namely Southeast - Northwest. Based on the initial hypothesis, it is stated that Lake Kerinci is currently the remnant of ancient ancient lakes that have undergone silting. The presence of weathering and erosion causes silting of Lake Kerinci.

The beginning of Lake Kerinci which is currently available is an ancient volcano of acid that has experienced explosive eruption so that it is cut off and leaves a large hole (caldera). The caldera is gradually filled by meteoric water to form a puddle in the form of a lake. As previously explained, the existence of Lake Kerinci which is right in the path of the Sumatran Fault has a direction of horizontal movement, causing the caldera to be drawn so that it is elongated. Lake Kerinci has almost the same shape as Lake Singkarak. The faults that worked in the Lake Kerinci withdrawal process were Siulak Faults on both sides flanking the lake. Siulak Fault itself is a segment of the Sumatran Fault which has a length of 37 km towards the southeast - northwest.

The existing Kerinci Lake when compared to Lake Singkarak and Lake Toba has different forms when observed through aerial portraits. Lake Singkarak and Lake Toba have longitudinal tendencies while Lake Kerinci currently has a rounded shape. This is due to the silting process of the ancient lake. Lake siltation occurs due to weathering and erosion that occurs in the Kumun hills and flanking both sides of the study area. Besides that weathering also comes from isolated hill itself (Figure 10).

Figure 9. Fault outcrops on the *Isolated Hill* that both of them formed downward fault block, the east side of the Tanco Hill Koto Salak Village (left photo) and southwest side of Tanco Hill Ujung Pasir Village (right photo)

The silting of Kerinci Lake can be seen from the evidence left behind in many locations in isolated hill. The figure shows the water tuff outcrop whose settling is affected by water. From this picture it can be seen that the material from the isolated hill tuffs at the bottom indicates that the tuff pasiran has a heavier period compared to the pumice above it because it has a lighter period.

Isolated hill in the study area does not only come from a volcano eruption. This can be observed from the phenomena in isolated hill. The main eruption produces an overlap in thickness and a large amount, then followed by smaller eruptions that produce pyroclastic deposits which are not too thick. The next eruption occurred after the formation of an ancient lake. Indicated when a volcanic eruption occurs, water has already inundated the location. So that when the tuff sediment deposition process occurs it will be precipitated while the pumice material still floats on the water before the entire cavity is filled with water and then deposited. Because of the influence of strong currents, the precipitate formed is not deposited horizontally (wave). In addition, evidence of ancient lake shrinkage can also be observed at this time. When rain occurs with high intensity, the surface of the water will rise towards the north of the lake reaching ± 1.5 km from the lakeside (horizontally) so that it covers the isolated hill and inundates the surrounding plains.

Isolated hill in the study area is not only covered by the eruption of the Mountain that forms Lake Kerinci at this time, but also comes from several volcanic eruptions that are around it. This is known from the thickness of the layers at the isolated hill peak parts which tend to be thin (± 10 cm) compared to the very thick layer below (± 15 m). Sand tuff pasiran and clay tuff as isolated hill wing come from weathering and erotion. Based on its position, sand tuff and clay tuff from the Kumun hills are associated with Pengasih ignimbrit on isolated hill.

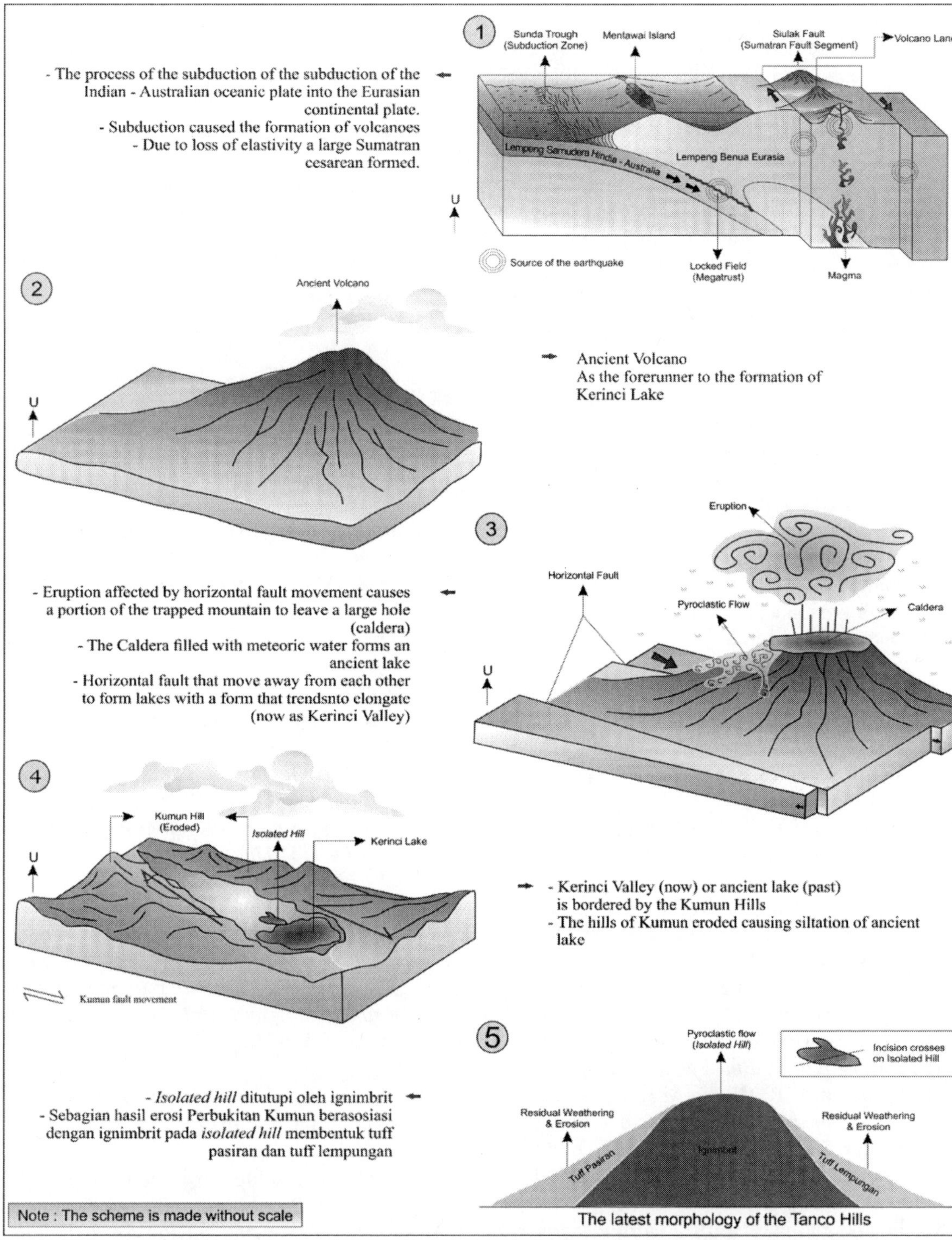

Figure 10. Simple skatch of genesis of isolated hill. 1) Old volcanic, but age of Quarternary Volcanic, 2) erupted of volcano was triggered by volcanic activity and tectonic of Sumatran Fault Zone which Siulak fault segmented, 3) fault segmented of extention is formed of pull a part basin whic is valley "Kerinci Lake", but still set aside volcanic product from erupted "ignimbrite which is pumice fragment" and tuff have been become core of isolated hill, 4) present of isolated hill, this flank of isolated hill was product volcanic-sediment

5. Conclusions

The research area is an isolated hill whose existence is located in the middle of the alluvial plain. Right on the isolated hill south side directly adjacent to Lake Kerinci. Kerinci Lake is an ancient lake formed by volcanic and tectonic activity.

Isolated hill in the study area is part of the Pengasih Formation, consisting entirely of volcanic rock consisting of 3 (three) rock units from old age, namely the ignimbrite Pengasih unit, sand tuff Pengasih unit, and clay tuff Pengasih unit. Based on the petrographic analysis of isolated hill volcanic rocks from the explosive eruption of an acidic volcano which is characterized by the discovery of ignimbrit outcrops in large and thick quantities. Genesis of isolated hill is related to the process of the formation of Lake Kerinci which was once an ancient volcano. The morphological shape is influenced by the volcanic activity of the ancient volcano followed by tectonic activity from the Great Sumatran Fault.

References

[1] Hall R 1997 *Geological Society of London Special Publication* **126** pp. 11-23.
[2] Hall R 2002 *Journal of Asian Earth Sciences* **20** pp. 353-356.
[3] Hall R 2014 Indonesia Tectonics: Subduction, Extention, Provenance, and More *Indonesian Petroleum Association Proceedings 38th Annual Exhibition and Convention* Jakarta, Indonesia (IPA), 14 G, pp. 2-8.
[4] Simandjuntak T O, Barber A J 1996 *Tectonic Evolution of Southeast Asia, Geology Society Special Publication* **106** pp. 185-201.
[5] Hamilton W 1979 *Tectonics of the Indonesian Region*, United States Geological Survey. Profesional Paper, 1078 pp.
[6] Darman H and Sidi F H 2000 *An Outline of the Geology of Indonesia* Indonesian Association of Geologists, Jakarta, 192 pp.
[7] Sieh K and Natawidjaja D H 2000 *Journal of Geophysical Research* **105** pp. 28 295-28 326.
[8] Barber A J, Crow M J, Milsom J S (eds) 2005 *Geological Society, London, Memoirs* **31** 192 pp.
[9] Tjia H D 1977 *Geologi Indonesia* **4** (1) pp. 13-27.
[10] Van Bemmelen R W 1949 *The Geology of Indonesia Vol. I*, Goverment Printing Office, The Hague, 732 pp.
[11] Rosidi H M D, Tjokrosaputro S, Pendowo B 1976 *Geological Map of the Painan and Eastern part of the Muarasibert Quadrangles, Sumatra, scale 1:250.000* Geological Survey of Indonesia, Bandung.
[12] Kusnama, Pardede R, Andi Mangga S, Sidarto 1993 *The Geology of the Sungaipenuh dan Ketaun Quadrangle, Sumatra* Report Geological Research and Development Centre, Bandung in Press.
[13] Curray J R, Emmel F J, Moore D G, Railt R W 1971 *Structure, Tectonics and Geological History of the Northeastern Indian Ocean. in: Ocean Basins and Margins* **6**, The Indian Ocean (Eds. A.E.M. Nairn & F. Stehli), Plenum Press New York, pp. 399-450.
[14] Katili J A 1969 *Bull Volcano* **33** pp 530-540.
[15] Cameron N R, Clarke M C G, Aldiss D T, Aspden J A, Djunuddin A 1980 *Proceedings of Indonesian Petroleum Association, 9th Annual Convention* pp 149-187.
[16] Fitch F J 1972 *J Geophys Res* **77** pp. 4432 60.
[17] Muraoka H, Takashi T, Sundhoro H, Dwipa S, Soeda Y, Momita M, Shimada K 2010 *Proceedings World Geothermal Congress* pp 3-8.
[18] Poedjopradjitno S 2012 *Morphotectonic and Natural Hazard Potential in the Kerinci Valley West Sumatra, based on Aerial Photographinterpretation* Geological Survey Centre, Geological Agency, Bandung, **22** (2) pp 101-113.
[19] Natawidjaja D H, Triyoso W 2007 *Journal of Earthquake and Tsunami* **1** (1) pp 21-47.
[20] Sieh K and Natawidjaja D 2000 *Journal of Geophysical Research* **1 05** (B 12) pp 28,295-28,326.
[21] Badan Informasi Geospasial (BIG) 2018 *Peta Tofografi Kabupaten Kerinci, Provinsi Jambi.*

[22] United State Geological Survey (USGS) *Shuttle Radar Topography Mission (SRTM) 90 m and Global DEM ASTER (*Spaceborne Thermal Emission and Reflection Radiometer*)* http://earthexplorer.usgs.gov/

Acknowledgments
This research is based on Fiqhy Nabela final assignment "in the Universitas Jambi". The authors acknowledge to Allah SWT, some people for their discussion and motivation Raisa, Ghea, and Ina. Our gratitudes to the Department of Geological Engineering of Universitas Jambi. Thank you to The 1st International Conference on Geoscience (ICoGeS) 2018 on the Hasanuddin University Committee for their acceptence of this paper.

Heavy metal cadmium (Cd) content in the medaka fish *Oryzeas javanicus* Bleeker

Magdalena Litaay[1*], Andi Iin Fadliah[1], Irma Andriani[1], Dahlang Tahir[2]

[1]Dept of Biology Faculty Mathematics and Natural Sciences Hasanuddin University, Makassar Indonesia 90245
[2]Dept of Physic Faculty Mathematics and Natural Sciences HasanuddinUniversity, Makassar Indonesis 90245

*mlitaay@fmipa.unhas.ac.id

Abstract. The medaka oryzeas fish *Oryzias javanicus* Bleeker has been widely used as bioindicator animal. The present study aims to assess a lethal dose 50% (LD_{50}) of heavy metal Cadmiun (Cd) towards test animal oryzeas fish. Fish was collected from the estuary of Jeneberang river of Makassar. Samples were acclimatized in a laboratory before being a test. Randomized Complete Design with four treatments and four replicates was applied in this study. Test fish was treated without (control) and with three different levels of Cd concentration as follows 0 mg / L; 0.008 mg / L; 0.01 mg / L and 0.03 mg/L. An observation was done during seven days experiment. Observed parameters included a change in movement, morphological abnormalities (peeled skin) as Cd content in medaka fish. Least Significance Different (LSD) method was applied to see different significant effect amongst treatments. At the end of the experiment, Cd content in test fish was determined by using an Atomic Absorption Spectrophotometer (AAS). The result indicates that medaka fish Oryzeas javanicus showed disorientation movement since day 5 at Cd concentration 0.03 mg / L and increased to the end of the test. Peeled skin happend at day seven at all three treatments (Cd 0.008; 0.01 dan 0.03 mg/L). Cd content in fish was low at concentration of 0.008 mg/L (average 1.50 mg/kg) and the highest at Cd 0.03 mg/L (average 4.19 mg/kg), respectively. Statistical analysis reveals that Cd concentration 0.03 mg/L differs from other treatments. Level of Cd accumulation in *O. Javanicus* Bleeker increased with an increase in metal concentration in water.

1. Introduction

Water pollution is the inclusion of living creatures, substances, energy or other components into water bodies causing water quality polluted. Pollutants may include heavy metal. Metal can accumulate in aquatic organisms, including fish and persist in water and sediment. Nowadays, contamination by heavy metals in waters such as cadmium (Cd), mercury (Hg) and lead (Pb) become a serious threat. Generally, heavy metal at room temperature is not always solid, but in the form of a liquid element, for example, mercury (Hg), lead (Pb), and Cadmium (Cd). Mercury, cadmium, copper, lead and zinc are of the most important pollutants in the aquatic environment and fish [1]. Cadmium is recognized as white silvery soft metal that is easily oxidized [2]. This metal can be found from zinc mining waste, electroplating flush, waste sediment, and metal industries. Widely used of this metal can also lead to Cd pollution into the environment.

Measurement of the levels of contamination, or the detection of pollutants in water, generally using chemical techniques. These techniques tend to be expensive and wastes can pose new pollutants to the environment. Currently, developing techniques for early detection of pollution substances or chemicals in the water using biological indicators. In this study we want to know the type ofbiological indicators for water pollution decent and easy to obtain. One that meets the requirements of aquatic organisms as biological indicators are medaka fish. Oryzeas fish can be one of the biological indicators that because they can live in the pressure of heavy metals in waters, have the high reproductive ability, spawning time is relatively short, the level of adaptation high that can live in freshwater and seawater, settled on a body of water so easily made sample, can be easy developed in the laboratory.The biologic impact of toxic pollutants on fish is an important area of study in ecotoxicology. Fish models, such as zebrafish (*Danio rerio*), tilapia (*Oreochromis niloticus*), and rainbow trout (*Oncorhynchus mykiss*), medaka fish, have been widely used for ecotoxicological studies in the freshwater and marine environment [3][4][5]. Zebra fish and medaka fish has been used as a model organism for modern biomedical research [6] and also transgenic zebra fish can be served as bioindicators of environmental toxicants at the cellular level [7]. Fish used in this study is medaka javanicus *Oryzias javanicus* as a local endemic fish which are found in South Sulawesi and is also easily to be cultivated. Therefore, research on the use of this fish as biological indicators in polluted waters is needed.

2. Material and Method

2.1. Materials

Equipment used in laboratory included glass jar measuring 10 x 10 x 10 cm, aerators, measuring devices, digital cameras, sample boxes, petridish, scales, horn spoons, scissors, gill net and AAS. Materials used in this study included medaka javanicus *Oryzias javanicus* fish, Cadmium (Cd) solution with a concentration of 0.01, 0.03, and 0.05 mg/L, glass glue, water, pellets, board marker, and rolls tissue.

2.2. Sample collection.

This study was conducted in May-August 2016. Sampling was conducted at the mouth of the river otters Je'ne Berang, Makassar of South Sulawesi province. Acclimatization and experiment were done at Medaka Study Center, the 5th Floor Research Center Building Hasanuddin University, while Cd content was analyzed Department Chemistry Faculty of Mathematics and Natural Sciences Hasanuddin University Makassar. Prior to sampling, an observation was done in order to get a general idea of the condition of estuary waters Je'ne Berang, Makassar, South Sulawesi. Sampling was done randomly using a gill net. Collected fish was photographed for documentation and samples were taken to the laboratory. Fish was sorted accordingly, with a size of 2 ± 0.1 / 0.2 cm.

2.3. Acclimatization

Fourthly animal test that had been sorted, adapted to the physical conditions in the laboratory (testing environment) for 10 days. Acclimation for at least 10 days because if within 48 hours more than 3% of the population of test animals die, the population of test animals deemed ineligible for testing.

2.4. Experimental design

LD_{50} test was conducted prior to application of different concentration of Cd solution as the main task of this study. The lethal doses of Cd with varying concentrations of 0.75, 1, 1.25 and 1.5 mg/L were administered towards test fishes. Mortality (%) of test animals was recorded after 24.

Experiment on heavy metals Cd content was done using a completely randomized design (RAL), which consists of 4 treatments and 4 repetitions as follows:

Treatment (tr) 1: No additional Cd (0 mg / L) (control).
Treatment (tr) 2: 0.008 mg / L Cd (below the threshold).
Treatment (tr) 3: 0.01 mg / L Cd (threshold).
Treatment (tr) 4: 0.03 mg / L Cd (above the threshold).

Cd content in test animals was measured using Atomic Absorbance Spectrophotometer (AAS) at the wave length of 248.3 nm. At the end of the experiment, test animal was crushed, homogeneous and analyses according to the AAS method elsewhere.

2.5. *Data Analysis*

Data analysis consists of Cd content in the test fish as abnormalities in fish movement. Parameters in fish movement such as disorientation characterized by weakened fish movement and fish body becomes tilted position, as well as abnormalities characterized by morphological changes in fish especially change in gills colour into red and flaky fish scales.

3. Result and Discussion

3.1 Lethal Dose 50% (LD_{50})

The result of the LD_{50} test for oryzeas fish is shown in Figure 1. As seen in the Figure. At the concentration of 1.25 mgL Cd, 50% of the fish population died after 24h. This reveals that LD_{50} for *Oryzias javanicus* Bleeker, respectively.

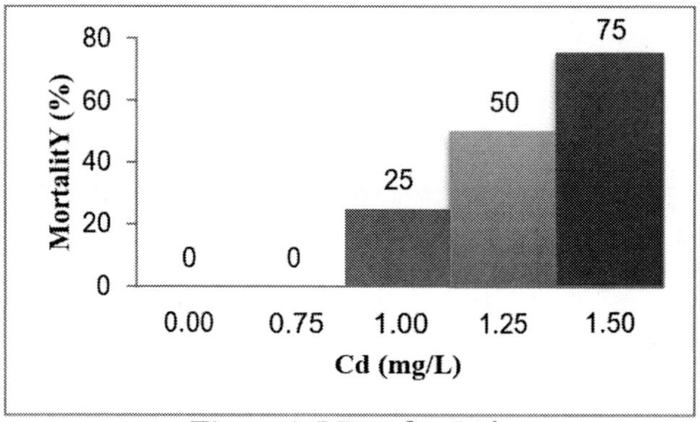

Figure 1. LD_{50} after 24 h

LD_{50} is an initial test to observe acute toxicity potential, toxicity symptom, toxicity spectrum and dying mechanism from a toxic agent. LD_{50} aims to detect toxicity of material, target organ and its sensitivity, to obtain data on the effect on the acute chemical, as to initial information on target dose [8].

3.2 Effect on Cadmium on Oryzeas javanicus Bleeker

The effect on the different concentration of Cadmium in the water on fish *O. Javanicus* Bleeke abnormalities (movement and skin) is described in Table 1.

Table 1. The relationship between time and number of fish that show abnormalities [disoriented movement (DM) and peeled skin (PS)]

Time (h)	Control (n = 12)		Cd 0.008 mg/L (n = 12)		Cd 0.01 mg/L (n = 12)		Cd 0.03 mg/L (n = 12)	
	DM	LS	DM	PS	DM	LS	DM	PS
0	-	-	-	-	-	-	-	-
24	-	-	-	-	-	-	-	-
48	-	-	-	-	-	-	-	-
72	-	-	-	-	-	-	-	-
96	-	-	-	-	-	-	-	-
120	-	-	-	-	-	-	+	-
144	-	-	+	-	+ (2)	-	+ (2)	-
165	-	-	+ (2)	+	+ (3)	+ (2)	+ (5)	+ (3)

Percent (%)	0	0	17	8	25	16	41	25

Note:
n = number of fish
(-) = no abnormalities
(+) = number of abnormal fish

As shown in Table 1, no abnormalities are shown by test animal at all concentration before 120h after the experiment started. One fish starts to show disoriented move at day 5 (120h) at the highest Cd concentration. At the highest concentration, the number of fish show disorientation movements increase towards the end of the experiment. Overall, percentage of disorientated fish increases with an increase in Cd concentration, that of Cd 0.008 mg/L (17%); Cd 0.01 mg/L (25%) and Cd 0.03 mg/L (41%). As also shown in Table 1, scale abnormality (Figure 2) happened at last day of the experiment at the lowest Cd concentration (0.008 mg/L), and a number of fish that shows similar abnormality increases as an increase in Cd concentration.

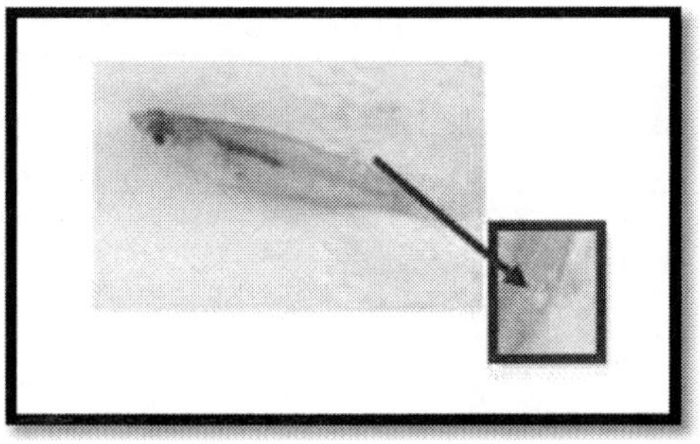

Figure 2. Scale abnormality (peeled skin) on the test animal

Heavy metal enters organisms through food, air respiration, as also through the skin [2]. A present of heavy metal in water or sediment, hence the ability of organisms to accumulate it inside their body, will affect existing of organisms [9].

Pathophisiolocial effect on fish can be shown by a change in fish behaviours such as movement, body balance, mucus production and color change. The present of Cd in fish media in a small amount can affect biological activity that can lead to the dead of organisms [10]. Accumulation of pollutant for example any heavy metal in fish is allowed when one is exposed to such metal. However, the effect of any heavy metal pollutant to organisms is depended on the toxicity of that heavy metal and a successful detoxification and excretion. An increased level of heavy metal inside fish would disturb organs function, and when it over a threshold, the fish organ will be damaged.

Study done by [11], on the accumulation of Cd content in fresh water shell, concluded that, biotransformation and bioaccumulation happened inside water's organisms. Cd enters organisms through food or water, being accumulated in gills, digestive tract, even in a kidney. A bio-magnification process can also allow an amount of such metal is higher inside organ compared to the surrounding water. Carcinogen Cd gives a slowly toxic effect, when heavy metal content over the threshold level, this can lead to a dead. Toxic effect is depended on the level of contaminant, for example, moderate level results in physic weakness, loose appetite, un-normal breath, and un-control swimming behavior. Fish is categories as bio-indicator of chemical pollutant as in the certain level of pollutant concentrations, as can show a change in movement, unusual growth or dead.

3.3 Cd content in test fish

The result of Cd content analysis in test fish is given in Table 2, while the result of statistical analysis amongst treatments is shown in Table 3. As shown in Table 2, Cadmium content in test fish increase with an increase of Cd concentration in media. Average of Cd content in different Cd concentration as follow 0.50 mg/kg (tr 1); 0.61 mg/kg (tr 2) and 1.39 mg/kg (tr 3) (Table 2). Table 3 indicates that there is no significant different between control with tr 1 (Cd 0.008 mg/L) and tr 2 (Cd 0.01 mg/L) but differs to treatment 3 (Cd 0.03 mg/L).

Table 2. Cadmium content in *Oryzias javanicus* Bleeker

Treatment (tr) Cd (mg/L)	Cd content (mg/kg)				Average Cd content
	I	II	III	IV	(mg/kg)/fish
Control	0.22	0.67	0.00	0.00	0.07
0.008	0.81	1.32	3.05	0.83	0.50
0.01	0.92	1.49	3.28	1.63	0.61
0.03	7.06	2.23	3.99	3.49	1.39

Table 3. The result of Least Significance Different (LSD) analysis for Cd accumulation in Oryzias javanicus Bleeker

Treatment (tr)	Average	
Control	0.22^{ns}	
Cd 0.008 mg/L	1.50^{ns}	LSD value at 5 % =1.93
Cd 0.01 mg/L	1.83^{ns}	
Cd 0.03 mg/L	4.19*	

Note: ns = not significant * = Significant

Heavy metal can be accumulated inside fish, in which this will form a complex compound with organic material from fish itself. Hence, heavy metal will be fixed and cannot be excreted and being accumulated inside a fish body [12]. Heavy metal accumulated inside a body, to some extent higher compared its level in the environment. As shown in ANOVA analysis and LSD, a higher level of Cd content inside media, the higher amount is being accumulated inside fish. The relationship between metal absorption in an organism is in accordance with its level in media, in which an increase of metal content in tissues follows an increased level in the environment.

Naturally, heavy metals are present in freshwater, brackish water, and marine environment. However, human activities have also contributed to heavy metal pollution in a different ecosystem. Heavy metals pollutant comes from industrial wastes as domestic wastes [13]. This study has shown that oryzeas fish can be used as bio-indicator of heavy metal Cd in an environment.

4. Conclusion

1. Medaka Javanicus *O. Javanicus* Bleeker show sub-lethal sensitivity towards heavy metal Cadmium (Cd) in form of disoriented move at day-5 and increase with an increase of exposure time as scale abnormality at day-7.
2. Cd accumulation in Medaka fish Javanicus *O. Javanicus* Bleeker in accordance with concentration in media.

References

[1] Authman MMN, Zaki MS, Khallaf EA, Abbas HH 2015 *J Aquac Res Development* **6**(4). http://dx.doi.org/10.4172/2155-9546.1000328
[2] Palar H 2008 *Pencemaran dan Toksikologi Logam Berat* Penerbit Rineka Cipta Jakarta.
[3] Dong S, Kang M, Wu X, Ye T 2014 *Development of a promising fish model (Oryzias melastigma) for Assessing Multiple Responses to Stresses in the Marine Environment.*

Hindawi Publishing Corporation. BioMed Research International Volume 2014, Article ID 563131, 17 pages. http://dx.doi.org/10.1155/2014/563131

[4] Ishikawa Y 2000 Medakafish as a Model System For Vertebrate Development Genetics. *Bioessays* **22**(5):487-495

[5] Andriani I, Litaay M, Sartika, Tahir D 2019 *Medaka Fish Oryzias javanicus* Bleeker *as bio-indicator of Lead (Pb) content in waters*. Paper presented at The 3rd ICOS 26-27 July 2019. Makassar.

[6] Lin CY, Chiang CY, Tsai HJ 2016 *Journal of Biomedical Science* **23**(19) DOI 10.1186/s12929-016-0236-5.

[7] Tsai HJ 2017 Transgenic ZebraFish can serve as Bioindicator of Enviromental Toxicants at the Cellular Level. *Ann Mar Biol Res* **4**(2): 1022.

[8] Ibrahim M 2012 *Dose 50 % (LD50) of poliherbal (*Curcuma xanthorriza, Kheinhovia hospita, Nigella sativa, Arcangelisia flava *and Ophiocephalus* striatus*) on heparmin Mus Musculus.* Research And Development. PT Royal Medicalink Pharmalab. 4-5.

[9] Jagfar A, Abdul M 2014 Deteksi logam Timbal (Pb) pada ikan Nila sepanjang Sungai Kali Mas. Jurnal Ilmiah Perikanan dan Kelautan **6**(1): 42-48.

[10] Dewi NK 2004 *Decrease toxicity level of cadmium in Chanos chanos Forskal using Eichornia Crassipes (Mart.) Solms and its transport phenomena*. Thesis. Universitas Diponegoro, Semarang.

[11] Sunarto 2007 *Bioindicator pollutant cadmium (Cd) using microanatomi structure analysis, gills function efisiensi, morphology and shell condition of fresh water shell* Anodonta woodiana *Lea)*. Dissertation. Universitas Airlangga. Surabaya.

[12] Ansari TM, Marr I L, Tarig N 2004 Heavy metals in Marine Pollutan Perspective Mini Review. *Journal Applied Science* **4**(1):1-20.

[13] Casas JS, Sordo J 2006 *Lead, Chemistry, Analytical Aspects, Environmental Impac and Health Effects*. Departamento de Quimica Inorganica Facultad de Farmacia, Universidad de Santiago Compostela, Galicia, Spain.

IOP Conf. Series: Earth and Environmental Science **279** (2019) 012029 doi:10.1088/1755-1315/279/1/012029

Field Geological Characteristic of Mesozoic Paremba Sandstone, Bantimala Complex, South Sulawesi

A Maulana*, A Jaya, U R Irfan, M Farida

Departement of Geology, Hasanuddin University

*adi-maulana@unhas.ac.id

Abstract. We conducted preliminary survey to study field geological characteristic of Mesozoic Paremba Sandstone, Bantimala Complex, South Sulawesi in order to investigate the field occurrence and the petrographic characteristic of some samples from the sandstone unit. The samples were collected from measuring section campaign in three rivers, namely Mallapie, Ceppaga and Bontorio. The field investigation shows that The Paremba sandstone occurs as a tectonic slice within mélange complex which is sandwiched by schist and melanges. The sandstones range from fine to coarse-grained sandstone forming a layer of 0.5 up to 50 cm in thickness interlayered with siltstone and shale, forming a sequence set. The sequences are intercalated with other lithologies such as limestone, mudstone, conglomerate, breccia and chert, forming a distinctive unit. In certain place, the units are intruded by basalt dike. Generally, the sandstones are grey to blackish in colour, sometimes show brown to yellowish colour if weathered. Some sandstone is calcareous as shown by strong reaction with acid and the occurrence of carbonate mineral but other samples relatively show less carbonate minerals. In all three rivers, the sandstone units have been strongly tectonized as seen by irregular fold and fault from the strike and dip of the unit layer. Petrographic analyses show that the studied sandstones consist of arkosic sandstone, siltstone and calcareous sandstone which are composed of quartz, orthoclase, plagioclase and calcite in the groundmass with opaque minerals occur as main accessory. Some samples contain muscovite, especially calcareous sample. The sandstone unit shows a typical of coarsening- and thickening updward trend with flaser, lenticular and parallel lamination sedimentary structure. It is interpreted that the sandstone were deposited in shallow marine or shelf region.

1. Introduction

Sulawesi is located in the central part of Indonesian Archipelago which has experienced numerous geological processes such as collision of microcontinent, high-pressure (HP) metamorphism, emplacement of oceanic crust segment and intensive magmatic and volcanic activity during Mesozoic to Tertiary time [1]. These processes produce a complex geological and tectonic setting of the island. The island can be simply divided into 4 provinces based on geological terrane, namely 1) South Arm; 2) North Arm; 3) Southeast Arm; and 4) East Arm. South Arm terrane of Sulawesi has been considered as part of Kalimantan hence eastern margin of Sundaland before spreading of Makassar Strait occurred in Miocene time. As a fragment of Sundaland, the arm has been an object of research by many researchers, especially the occurrence of Mesozoic basement rock in Bantimala and Barru area.

The basement rocks complex consists of various lithology including metamorphic, metasedimentary, sedimentary rocks, block of high-pressure rocks and series of ophiolite rocks forming northwest – southeast trend. There are some distinctive characteristic in this basement

IOP Conf. Series: Earth and Environmental Science **279** (2019) 012029 doi:10.1088/1755-1315/279/1/012029

complex, especially the occurrence of mélange, high-pressure metamorphic rocks and un-metamorphosed sedimentary rocks. Previous studies have reported detail discussion on metamorphic rocks [2]; [3]; [4]; [5]; [6] as well as ultramafic rocks [7]. However, report on the occurrence of un-metamorphosed sedimentary has never been conducted. The main purpose of this study is to investigate the field geological characteristic of the Jurassic Paremba Sandstone in Bantimala Complex, South Arm Sulawesi based on measuring section and petrographic observation.

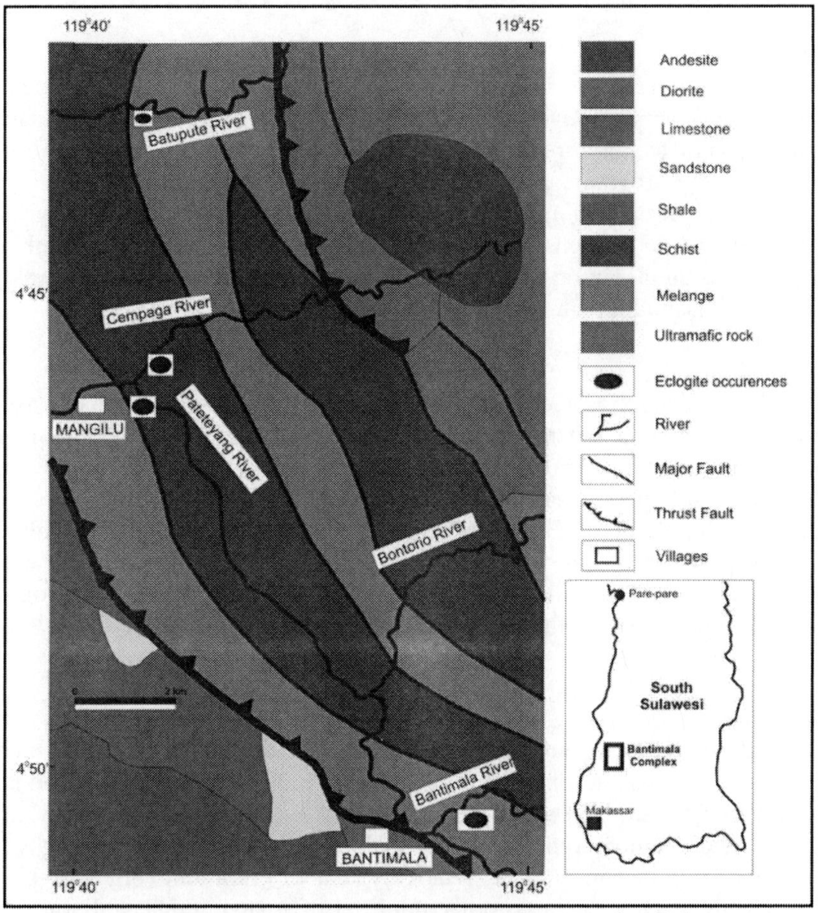

Fig. 1. Geological Map of Bantimala Complex [3]

2. Tectonic Setting and Regional Geology

Paremba sandstone occurs as a tectonic slab within Bantimala Melange Complex. Stratigraphy of the Bantimala Melange Complex was discussed in detail by [5] and [3] (Fig. 1). The complex consists of tectonic slices forming a northeast-dipping trend, which included sedimentary, metamorphic, meta-sedimentary and ultramafic rocks, locally intruded by dacitic to andesitic rocks in composition. The metamorphic rocks show an increasing grade from southwest to the northeast (greenschist to amphibolite facies). HP metamorphic rocks including glaucophane schists and eclogite found as tectonic block within metamorphic terrane. Sandstone, shale, conglomerate, radiolarian chert, siliceous shale and mélange formed a sedimentary and meta-sedimentary sequence within the complex. The peridotite showed moderately to highly serpentinization process, with local chromite lenses and basalt layers. The mélange consists of sandstone, shale, siliceous shale, chert, basalt, schist and felsic igneous rock in a sheared matrix of shale, as well as schist breccia [2]. The Bantimala Complex is sandwiched by the Balangbaru Formation in the south and north whereas bounded by Tonasa Limestone to the southwest, and by the Camba Formation to the east. The similarities of radiolarians between the

Balangbaru Formation and chert in the Bantimala Complex suggested that the Balangbaru Formation is part of Bantimala Complex.

K-Ar ages of 111 Ma were reported by Obradovich in [8] and 113 – 115 ±6 Ma from muscovite in schist by [5] and [6], whereas [7] reported an age of 111± 3 Ma from microfossils in the Balangbaru Formation. The radiolarian in the chert suggested upper Albian to lower Cenomanian in age [5], i.e. from the central part of the interval 112 Ma – 93.6 Ma. Some sedimentary rocks from Bantimala Complex appear to be older than Cretaceous as suggested by [9] and [10].

3. Methodology

The study was based on field work activity using measuring section method in three river sections namely Mallapie River, Ceppaga River and Bontorio River which are perpendicularly cross cut the Bantimala Complex major tectonic trend. We performed petrographic analyses in order to determine the field relationship and petrograpic characteristic of the Paremba Sandstone. 20 samples rocks were systematically collected from various localities in three rivers section. Petrographic study was conducted by examining thin sections to determine the rock types, mineral assemblages, fabric and textural relations by using a Nikon petrographic microscope in Optical Mineralogy Laboratory, Geology Department, Hasanuddin University, Makassar.

4. Results

The Paremba sandstone occurs as tectonic slab within mélange complex. It consists of alternating layer of thinly sandstone with siltstone and shale and sometimes with conglomerate forming a distinctive unit. The sandstone range from fine to coarse-grained sandstone intercalated with limestone, mudstone, conglomeratic sandstone, breccia, chert and shale. In certain place, the sandstone units are intruded by basalt dike. Generally, the sandstones unit are grey to blackish in colour, sometimes show brown to yellowish color if weathered. Most sandstones units are calcareous, shown by strong reaction with acid. In all three sections, the sandstone units have been strongly tectonized as seen by irregular fold and fault from the strike and dip of the unit layer.

4.1 Mallapie Section

Total measured length of Paremba Sandstone in Mallapie Section is 800 meters which is the thickest compare to other two sections. It consists of sandstone which is tectonically overlaid by schist and chert unit and ophiolitic basalt. The section is dominated by interlayering of fine- to coarse-grained sandstone; silt; shale and conglomeratic sandstone with thickness of sandstone ranges from 0.5 – 1 cm for fine-grained unit and up to 1 m thick for relatively coarser grain unit (Fig. 2. a–b). The sandstones are calcareous, showing dark grey in colour composed of quartz, mica and some rock fragments. Pararel lamination, flaser and lenticular are common sedimentary structure in sandstone layer. Cross bedding structure also found but very rare. The unit generally shows coarsening and thickening upward trend. The occurrence of thick layer chert (up to 65 m) and limestone intercalation (up to 40 m) and basalt dikes in certain part are distinctive to this section. The sedimentary environment is interpreted as shelf. In thin section, coarse-grained sandstone show moderate sortation with sub-rounded to angular quartz (Qtz) grains, orthoclase (Or) and Plagioclase (Pl), and grains of opaque (Op).

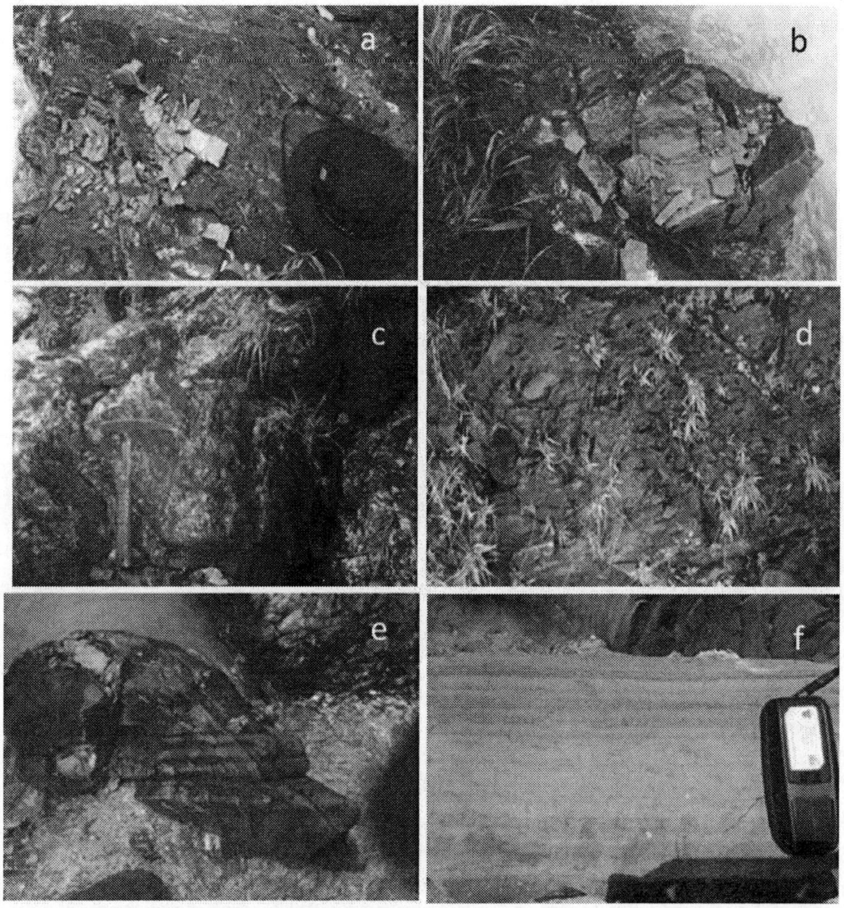

Fig. 2. Field occurrence of Paremba Sandstone. (a) fine-grained sandstone interlayering with silt and shale in Mallapie Section, (b) grey to blackhish medium- to coarse-grained calcareous sandstone in Mallapie Section. (c) Coarse-grained sandstone layer in Ceppaga Section (d) Interlayering of fine- to medium- grained sandstone and silt showing parallel lamination structure in Ceppaga Section. (e) Paremba Sandstone outcrops in Bontorio Section showing a medium to fine grained texture and (f) Fine- to medium-grained sandstone showing typical of flaser and lamination sedimentary structure in Bontorio Section.

4.2 Ceppaga Section

Ceppaga Section is located in the middle region of the studied area, extending about 370 m in length. The section is tectonically sandwiched by schist and dominated by interlayering of calcareous fine- to coarse-grained sandstone and grey to blackish conglomerate. Sandstone layer range from 10 cm – 50 cm but in some layer it reach up to 1 m thick (Fig. 2. c-d). Thin layer of shale is also found associated with scaly clay, breccia and chert. The unit is interpreted to be deposited in shelf environment. The limestone is classified as wackestone. Photomicrograph of siltstone layer in Paremba Sandstone from Ceppaga Section shows a fine-grained texture consists of quartz, feldspar and calcite.

4.3 Bontorio Section

Bontorio Section is located in the most southern area with total thickness about 350 m in the main river and another 190 m in its river branch. The total thickness of Paremba Sandstone is less than 500 m occur as tectonic slices sandwiched by schist unit. Generally it is dominated by interlayering of fine- to coarse-grained sandstone with intercalation of limestone and breccia. The sandstone unit shows a typical of coarsening- and thickening updward trend. Flaser and pararel laminations are

IOP Conf. Series: Earth and Environmental Science **279** (2019) 012029 doi:10.1088/1755-1315/279/1/012029

common. Photomicrograph of fine-grained calcareous sandstone from Bontorio Section consists of quartz, feldspar and calcite in groundmass.

5. Conclusions

(1) Paremba sandstone found as a tectonic slab included in mélange complex. The sandstones show grey to blackish color, calcareous and are sandwiched by schist. The Paremba sandstone composed of alternating of fine to coarse-grained sandstone unit with various in thickness (range from 0.5 – 50 cm) intercalated with limestone, mudstone, conglomeratic sandstone, breccia, chert, shale and intruded by basalt dike. The sandstone has been strongly tectonized and folded.
(2) Petrographically, the studied sandstones consist of arkosic sandstone, siltstone and calcareous sandstone which are composed of quartz, orthoclase, plagioclase and calcite in the groundmass with opaque minerals occur as main accessory. Some samples contain muscovite.
(3) The sandstone unit shows a typical of coarsening- and thickening updward trend show flaser, lenticular and parallel lamination sedimentary structure. It is interpreted that the sandstone were deposited in a shallow marine area or shelf.

Acknowledgments
RUNAS UNHAS Research Grant 2018 is highly acknowledged for permit and financial support to conduct this field work. This work is part of collaboration research with Pertamina EP in Mesozoic Geology of South Sulawesi research topic which is highly acknowledged. We are grateful to local people, especially Head of Village in Bontorio Village, Bone for their kind support during field work.

References
[1] Maulana A, Imai A, van Leeuwen T, Koichiro W, Yonezu K, Takanori N, Boyce A, Page L, Schersten A 2016 *Journal of Asian Earth Sciences* **124** 102-25,
[2] Sukamto R 1982 *The geology of the Pangkajene and Western part of Watampone, South Sulawesi, scale 1 : 250.000* Geological Research and Development Centre, Bandung.
[3] Maulana A, Christy A, Ellis D, Imai A, Watanabe K 2013 *The Island Arc* **22**(4) 427-52.
[4] Parkinson C D 1998 *The Island Arc* **7** 231-245.
[5] Wakita K, Sopaheluwakan J, Miyazaki K, Munasri 1996 *Geological Society (London) Special Publication* **106** pp. 353-64.
[6] Miyazaki K, Zulkarnain I, Sopaheluwakan J, Wakita K 1996 *Journal of Metamorphic Geology* **14**, 549-63.
[7] Maulana A. Christy A, Ellis D, 2015 *Chemie der Erde* **75** (1), 73 – 87.
[8] Hasan, K 1991 *Proceeding Indonesia Petroleum Association 20th annual convention*, 183-208.
[9] Haile N S 1978 *Tectonophysics* **46** 77-85.
[10] Sasajima S, Nishimura S, Hirooka K, Otofuji Y, van Leeuwen T, Hehuwat F 1980 *Tectonophysics* **64** 163-72.

IOP Conf. Series: Earth and Environmental Science **279** (2019) 012030 doi:10.1088/1755-1315/279/1/012030

Lithogenic and Antropogenic in Surface Sediment From Outlet of Tempe Lake, South Sulawesi

A. M. Imran[1], Ratna Husain L[1], Safruddim[1], Baso Rezki Maulana[2]

[1)]Geological Department, Hasanuddin University, Makassar Indonesia
[2)]Graduate Student of Geology Engineering, Hasanuddin University, Makassar, Indonesia

Corresponding author: mudhimran@gmail.com

Abstract. Prehistory of Central of South Sulawesi was a marine environment connecting Bone Bay and Makassar Strait through Tempe Lake. The influence of marine can be traced nowadays in outlet of Tempe Lake (Cenrana River). The concentration of SiO_2 decreases toward downstream, on contrary Fe_2O_3 and CaO tend to increase downstream. Heavy mineral (Zr and TiO2) provides an increasing trend downstream . Most of the sediments originate from the rock covering the area and precipitation from the seawater. The latter can be found as far as 34 km from the coast of Bone Bay to the west.

1. Introduction

Lake Tempe locates in South Sulawesi in which several rivers as a inlet and only Cenrana River as an outlet of the lake. The lake faces serious degradation due to sedimentation from several rivers and population pressure surround the lake. Sedimentation rates reach 1 – 3 cm/ year [1]. Formerly, Lake Tempe had been connecting between Bone Bay in the east and Makassar Strait in the west in the form of depression as stated in the La Galigo legend, (Figure 1) [2]. Furthermore Gremmen [3] revealed that Lake Tempe had been inundated by seawater around 7,100 to 2,600 years ago based on analysis of pollen analysis. Furthermore Whitten et al. [4] detected sea levels of the southwestern peninsula of Sulawesi by 5 and 2.5 m higher than now in 4500 and 1600 years ago and Sulawesi separated between north and south by the Tempe Depression. Caldwell & Lillie [5] explained that in large inundation conditions, Lake Tempe became a link between Bone Bay in the east and the Makassar Strait in the west around 4,500 years where the sea level is a maximum of 6.55 m above the present surface.

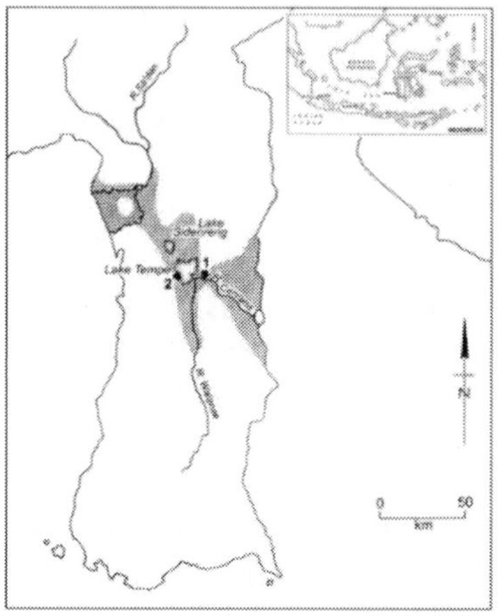

Figure 1. The estimated boundaries of formerly was sea water connecting Bone Bay in the east and Makassar Strait in the west [5].

An important outlet of Tempe Lake is Cenrana River in which the water flows to the Bone Bay. Assuming that the sediment influenced by marine and fresh water along the river, sediment deposited in the river gives characteristic showing marine and/or terrestrial minerals by studying geochemical characteristics of sediment.

2. Paleoenvironment of South Sulawesi

Tempe Lake was a part of small strait correponding to the sea level fluctuation in South Sulawesi during late Pleistocene – early Holocene age. Sea level fall was as high as 5 meters above sea level today at 4.500 years ago recorded from the coral reefs of the Spermeonde Islands [6]. A decrease in sea level has also been studied on the Quaternary coral reef at the southern tip of South Sulawesi [7]. They concluded that sea level had reached as high as 20 m above sea level today during Pleistocene time. Sea level fall at Panambungan Island and Barrang Lompo Island occurred since 6000 years ago [8]. He founds that sea level was at altitude of 0.5 m Panambungan Island and 0.8 m Barrang Lompo Island around 5600 cal. yr BP of sea level today. Based on the data above, most likely the depression forming the Lake Tempe was once a strait.

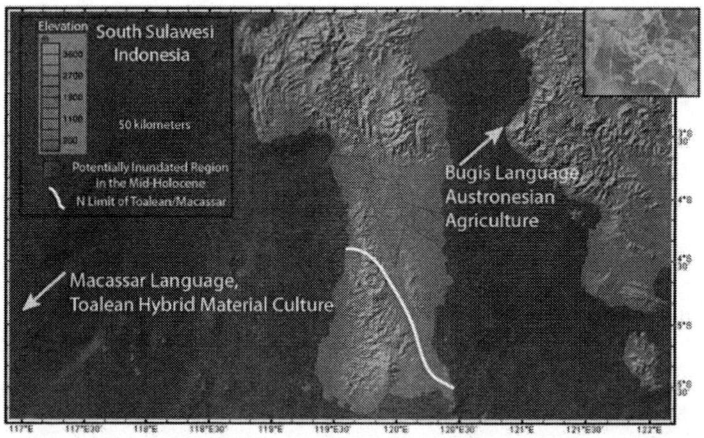

Figure 2. Physiography of South Sulawesi showing depression trending southeast through Tempe Lake. (from: Bulbeck et al., [9]).

Concentration of major element such as SiO_2, Al_2O_3, Fe_2O_3, CaO and MgO is an important of depositional environment. SiO_2 tends to decrease toward downstream, on the otherhand Al_2O_3, Fe_2O_3, CaO and MgO tend to increase downstream [10]. Kasim et al. [11] explain that surface sediments in

Lake Tempe originate from terrestrial (land) which is characterized by the presence of heavy minerals such as zircons, tourmaline, rutile (ZTR) even present less stable minerals such as epidote, olivine , hypersten, biotite, magnetite and iron oxide.

3. Material and Method

Surface sediment was carried out along the Cenrana River in 9 stations collected by percussion core. Sampling locations were choosed in straight stream in order to get laminar water flow. Lithogenic and antropogenic minerals in surface sediment of Tempe Lake outlet is identified from grain size, mineral composition and geochemical analyses. Selected samples were analyzed by petrographic analisis, grain size siever and X-ray diffraction (XRD) at Geological Department, Hasanuddin University. Due to the sediments are loose, they were impregnated for sand size before cutting and analysed under microscope.

4. Result and Discussion

Grain size analysis shows a distribution along the stream dominated by fine sand – clay. Two zones of sediment distribution based on its grain size (fig 3): fine sand – silt which accumulated in upstream (Padduppi to Kampiri) and silt – clay deposited in down stream (Pincengpute to Kampoti). The grains size distribution implies that river current from the Tempe Lake dominantly influences sediment deposit in the river rather than sea current.

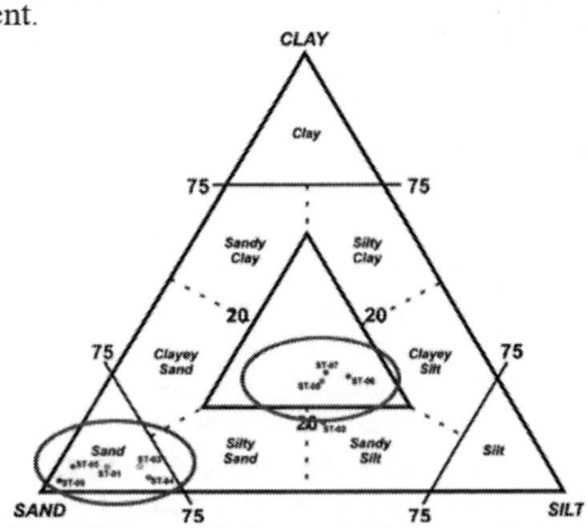

Figure 3. Sediment distribution of Tempe Lake *outlet base on its texture plotted in* Shephard's diagram.

4.1. Mineral Characteristics

Petrographic analysis on sediment shows vary in mineral type and even rock fragments. Most common mineral deposited along the stream are pyroxine. Others are rock fragments, quartz, plagioclase, biotite, and opac minerals (Table 1). There are three type of presence of mineral i.e a) undulating present such as rock fragments (max35%), Quartz (25%), Quartz (max 25%), b) other mineral increasingly occurs toward downstream such as pyroxene and opac mineral, and c) plagioclase decreases toward downstream.

XRD analysis on clay mineral shows a presence of marine influence toward downstream. Upstream of Cenrana River is dominantly composed of silica minerals (quartz, clay minerals and plagioclase), furthermore downstream consists dominantly of heavy minerals, salts (gypsum and halite), and sulfur. Silica minerals such as quartz, clay and plagioclase tends to decrease toward downstream, on the other hand halite, gypsum, sulfur minerals increase toward downstream (Table 2). Carbonate minerals are highly deposited in the ST05 (15,3%) due to intruding of sediment input from small river (such as Salo Lampoe) covered by carbonate rock formation. The statement is confirmed by decreasing volume even neglected toward coastal area of Bone Bay. Heavy minerals is mostly supplied from the basic rocks which covered the northern area of South Sulawesi. The minerals are transported through Bila River to Tempe Lake [11].

Table 1. Distribution of main components on surface sediment in outlet of Tempe Lake

IOP Conf. Series: Earth and Environmental Science **279** (2019) 012030 doi:10.1088/1755-1315/279/1/012030

Component	Mineral Composition (%)				
	ST-01	ST-03	ST-04	ST-05	ST-09
Rock Fragment	35,00	35,00	35,00	15,00	35,00
Quartz	10,00	5,00	25,00	25,00	5,00
Plagioclase	30,00	30,00	5,00	10,00	5,00
Pyroxene	15,00	25,00	10,00	30,00	30,00
Hornblende	0,00	0,00	20,00	0,00	0,00
Biotite	0,00	0,00	0,00	0,00	10,00
Opaque Mineral	10,00	5,00	20,00	20,00	15,00

Silicate group (quartz, clay, plagioclase and volcanic glass) shows decreasing trend downstream. On contrary saline minerals (halite and gypsum) begin to be present at ST-05 and increase toward downstream (Table 2). The concentration of mineral elements show a decreasing trend of SiO_2 toward downstream, on contrary Fe_2O_3 and CaO tend to increase downstream (Fig. 4). The trend of SiO_2, Fe_2O_3 and CaO support the trend line proposed by Verma [9] in which the concentrations of SiO_2 tend to decrease to downstream reversely of Fe_2O_3, CaO tends to increase downstream. Heavy mineral (Zr and TiO2) provides an increasing trend downstream.

4.2 Sediment origin

Heavy mineral occurs without trend showing probably influence of sediment input from surrounding rivers. Lithology of the area is dominated by sandstone of Walanae Formation as main contribution of sediments in outlet of Tempe Lake (Fig.5). The limited present (ST-05 and ST-08) and decreasing trend of carbonate mineral on the stream indicates also limited origin of material. This case figures out lithogenic sediment origin which suppose from carbonate rock of Taccipi Formation in the southern part (Fig. 5).

Petrographic and geochemical data indicate sediment characteristics of outlet of Tempe Lake is partly influenced by marine and mostly by lithogenic origin. The data mentioned above suggests that the effect of sea water in Cenrana River (±67 km long) as an outlet of Tempe Lake reaches Kampiri village (ST-06), Bone regency as part as ± 35 km northwest. The mineral characteristics inconsistently occur in outlet of Cenrana River. The phenomena suggest that the occurrence of mineral at the coastal area surround Cencarana River under pressure of man activities and sea water current. The factors allow fine sediments (clay) to reveal and be transported to the sea. The condition gives no chance to halite and gypsum minerals to develop in the coastal area.

The presence of SiO2, Zr, TiO2 and CaO indicates that most of the sediment origin deposited in Cenrana River come from the rock composing the area. Others are from sea water precipitation.

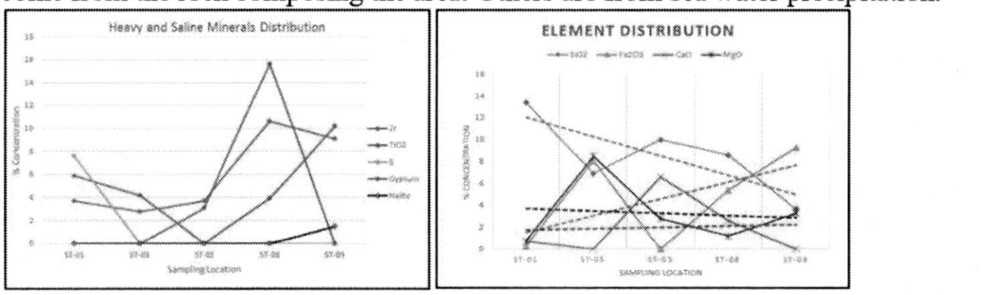

Figure 4. Concentration of element and saline minerals along the Cenrana River.

5. Conclusion

Minerals deposited in Cenrana River as an outlet of Tempe Lake originating from several condition: a) sediment transported from Tempe Lake which mainly fulfilled by Bila River and Walanae River. The type of sediments are mainly in small size grain and heavy minerals; b) lithogenic sediments transported through small stream from surround area covered by sandstone of Walanae Formation and carbonate rock of Taccipi Formation such as silicate mineral group and heavy mineral, and c) precipitation mineral

IOP Conf. Series: Earth and Environmental Science **279** (2019) 012030 doi:10.1088/1755-1315/279/1/012030

influenced by sea water of Bone Bay such as gypsum and halite. Beside that there is also influence from sea current and man activity. This influence can be traced by the irregularly present of mineral that should be increase or decrease trend.

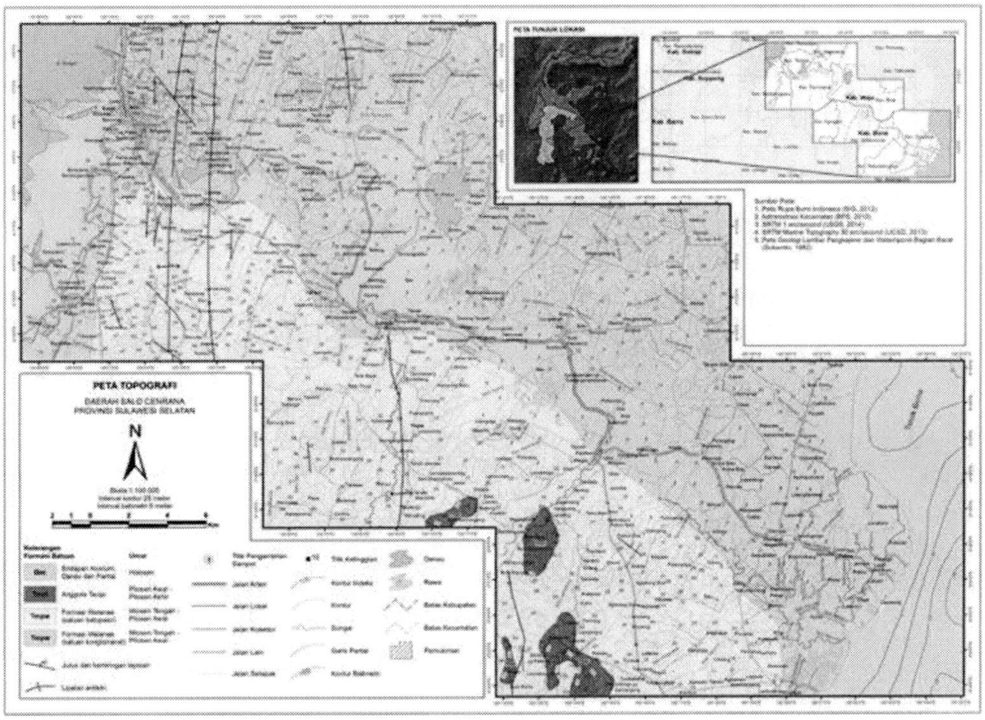

Figure 5. Geografical dan lithological map of Cenrana River.

References

[1] Kementerian Lingkungan Hidup 2014 *Gerakan Penyelamatan Danau (GERMADAN) Tempe*, Kementerian Lingkungan Hidup. Jakarta.

[2] Pelras C 2006 *The Bugis* Oxford: Blackwell. Indonesian version.

[3] Gremmen, W H E 1990 *Modern Quaternary Research in Southeast Asia* **11**: 123-134. Rotterdam: A.A. Balkema Publishers.

[4] Whitten A J, Mustafa M, Henderson G S 1987 *The Ecology of Sulawesi* Yogyakarta: Gajah Mada University Press.

[5] Caldwell I and Lillie M 2004 *Mod. Quaternary Res. SE Asia* **18**: 259-272, A.A. Balkema, Leiden, ISBN 90 5809 674 2.

[6] De Klerk L G 1983 Zeespiegels, *Riffen en Kustvlakten in Zuidwest Sulawesi, Indonesië; een Morfogenetische-Bodemkundige Studie*. PhD thesis. Utrecht: Rijksuniversiteit Utrecht.

[7] Imran A M, Farida M, Arifin M F, Husain R, Hafidz A 2016 *Indonesian Journal on Geoscience* **3** (1) p. 53-66.

[8] Mann T, Rovere A, Schöne T, Klicpera A, Stocchi P, Lukman M, Westphal H 2016 Geomorphology **257**, 155–163. doi:10.1016/j.geomorph.2015.12.023.

[9] Bulbeck D, Pasqua M, Di Lello A 2000 *Asian Perspect* **39** 71–108. doi:10.1353/asi.2000.0004.

[10] Verma M, Singh B P, Srivastava A and Mishra M 2012 *Geology* **33** 1-14.

[11] Kasim M, Imran A M, Hamzah A, Busthan, Pachri H 2016 *IJEScA* **3** (1) May 2016 p. 81-90.

IOP Conf. Series: Earth and Environmental Science **279** (2019) 012031 doi:10.1088/1755-1315/279/1/012031

High Efficiency Dye Sensitized Solar Cell from *Tectona Grandis* and red *Crysanthemum Morifolium* as a Sensitizer

Dahlang Tahir*, Ruzaini H. Abidin, Paulus Lobo Gareso

Department of Physics,
Hasanuddin University, Tamalanrea, Makassar, 90245 Indonesia

*Email: dtahir@fmipa.unhas.ac.id

Abstract. *Tectona Grandis* leaves and red *Crysanthemum Morifolium* have been used as sensitizers for dye sensitized solar cell (DSSC) using organic solvents by ethanol, citric acid, and distilled water in respectively with the ratio of 5 : 1 : 4. TiO_2 paste was spin-coated on top Indium Tin Oxide (ITO) conductive glass and calcinated at temperature of 450°C for 10 min. The chemical properties were analysis by using Fourier transform infrared spectroscopy (FTIR) indicated the presence of anthocyanin from *Tectona Grandis* leaves and red *Crysanthemum Morifolium* petal. The physical properties were analysis by using UV-Vis spectrophotometer was performed to measure the absorbance of dye solution in the visible spectrum range and X-Ray Diffractometer (XRD) was used to determine the structure of TiO_2/ITO and Dye/TiO_2/ITO. Among the three dyes studied, the mixed dye solution from *Tectona Grandis* leaves and red *Crysanthemum Morifolium* was shows the best photosensitization effects with the conversion efficiency is 0.834 %.

1. Introduction

Research activities in the field of energy especially for green sources which are renewable, low-cost, and eco-friendly increase every year [1], due to the limitation of energy sources from fossil fuel which is non-renewable. Ref. [2] described in detail about the availability of amount of the fossil fuel in the form of coal, natural gas, and crude oil are 1.139 billion tonnes, 187 trillion cubic meters, 1.707 billion barrels, respectively [2]. The energy crisis occurring revealed that the total energy demand was much higher than that the supply energy available from fossil fuel. To overcome these problems, solar energy is regarded as the most alternative energy source. Because it is a promising alternative energy sources which are available in every country around the world. In addition, solar energy is the most powerful sources of energy providing a continuous stream with solar radiation with approximately of $3x10^{24}$J per year [1, 3]. One example of solar energy technology is photovoltaic. Photovoltaic solar energy is being widely studied as one of the efficient technology due to its ability to convert sunlight directly into electricity [3].

DSSC is kind of photoelectrochemical solar cell which is the third generation and working on the principle of plant photosynthesis [1, 4]. DSSC appeared immediately as a promising low-cost and eco-friendly photovoltaic technology since 1991 [1, 5]. Unlike the conventional systems, based on DSSCs function principle in both: light absorption and charge transport performed separately [4, 6]. The dye absorbs the photon energy, the electrons in the dye become exited and then injected to the conduction band of semiconductor. As a result, the dye becomes oxidized and then the electrolyte donates electron to the initial state of dye. The electrons move through the semiconductor to a current collector and

IOP Conf. Series: Earth and Environmental Science **279** (2019) 012031 doi:10.1088/1755-1315/279/1/012031

electrolyte restores its initial state by accepting electrons from the external circuit and the process is cyclic [1, 4].

Several previous works have been published in DSSC by utilizing natural dyes as sources of natural sensitizer [5]. Natural dyes were used in DSSCs because of their availability and easy extraction, nontoxicity, complete biodegradability, and temperature compatibility [1, 5]. Several of natural dyes have been extensively investigated as a sensitizer for DSSC such as anthocyanin, tannin, carotene, betalain, and chlorophyll [5], but only anthocyanin is responsible for cyanic colors [1]. Many studies utilized anthocyanin as sensitizer for DSSC, Abdou et al. (2013) used anthocyanin from *Hibiscus Sabdariffa L.* and they reported that the conversion efficiency was 0.27 % ; Ayalew & Ayele (2016) also used anthocyanin from *Acanthus sennii chiovenda* flower as a sensitizer with the conversion efficiency of 0.150 % [7, 8].

Tectona Grandis leaves and red *Chrysanthemum morifolium* contain an anthocyanin that can be utilized as a sensitizer. Recently, the utilization of the *Tectona Grandis* leaves waste in everyday life is not maximized so that it would be wasted. The purpose of the present study is fabrication DSSCs were prepared with natural dyes extracted from *Tectona Grandis* leaves. Red *Chrysanthemum morifolium* petal is another natural dye that used in this study. Furthermore, *Tectona Grandis* leaves and red *Chrysanthemum morifolium* petal has been selected due to its widely available in Indonesia.

2. Experimental

2.1 Preparation of Dye Extraction

Tectona Grandis leaves and red *Chrysanthemum morifolium* petal were washed with distilled water and cut into small pieces about 2 cm. They were dried in the open air for three days then ovened them at 40°C for an hour and crushed into fine powder by using a blender. Infrared spectroscopy was carried out on IRPrestige-21 FT-IR spectrometer (Shimadzu Corp, Japan) equipped with a bright ceramic light source, a KBR beamsplitter, and deuterated triglycine sulfate doped with L-alanine (DLATGS) detector. The measurements of the sample were collected over the range of 4000-500 cm^{-1} and 16 co-added scans. All spectra were in Transmittance units.

Organic solvents was made by ethanol, citric acid, and distilled water in the ratio 5 : 1 : 4. 10 gr dye powders were mixed with 35 ml organic solvents for *Tectona Grandis* and 20 ml for red *Chrysanthemum Morifolium* using a magnetic stirrer for 30 minutes. The solution was stored in the dark room and covered with aluminum foil for 24 hours then filtered. Each solution with the ratio of 50 : 50 were mixed using a magnetic stirrer for 30 minutes used as mixed dye. The absorbance of the dyes in this study was determined by UV-2600 Series (Shimadzu Corp, Japan).

2.2 Preparation of Working Electrode

1 gr TiO_2 powders were mixed with 8 ml ethanol in a beaker and stirred using a magnetic stirrer for 30 minutes to form TiO_2 paste.

The conductive side of ITO glass was determined by a multimeter. A size of 2.1x2.1 cm^2 area on the conductive glass was made by using scotch tape to cover the four-sided border. Applying a few drops of the TiO_2 paste until the conductive side of the ITO glass surface was covered evenly by spin coating method at 3000 rpm for 30 seconds, there processed was repeated in three times. Then sample was calcined in low oxygen environment at temperature of 450°C for 10 minutes to solidify TiO_2. The X-ray Diffraction (XRD) spectra was collected (Shimadzu 7000) with Cu Kα radiation (λ =1.5405 Å) over the anguler range $15° \leq 2\theta \leq 80°$, operating at 30 kV and 10 mA. It was performed to examine the crystallinity of the TiO_2/ITO.

The TiO_2/ITO was directly immersed in the each dye solution extracted for 24 hours in the dark environment at room temperature for good adsorption of the dye on the TiO_2 surface. The Dye/TiO_2/ ITO was rinsed with distilled water to remove non-adsorbed dyes and excess water from porous TiO_2 and then dried for several minutes in the open air. The sample was characterized by XRD again to examine the crystallinity of Dye/TiO_2/ ITO.

The electrolyte solution was prepared by mixing 0.8 gr Potassium Iodide (KI) and 10 ml Polyethylene Glycol (PEG) in a beaker and added by 0.127gr Iodine (I2) then stored in a sealed bottle.

2.3 DSSC Assembly

Another ITO glass was coated with graphite on the conductive side. Apply electrolyte solution on the graphite/ITO side until surface of the sample covered. Then DSSC was assembled to form a sandwich by placing the working electrode face to face with the counter electrode. The two pieces of ITO glass were clamped on the opposite side. The DSSC were tested under natural sunlight and the intensity of sunlight was measured by LIGHTMETER CA813 AEMC.

3. Result And Discussion

3.1 XRD Characterization

Figure 1 shows the XRD pattern of TiO_2 nanomaterial which has been coated on top of cunductive side of ITO glass and calcined at 450°C for 10 minutes. This result indicates high intensity of the main peak due to the absorption of the anthocyanin dye on the TiO_2 surface. While figure 1 (b), (c), and (d) represent the *Tectona Grandis,* red *Chrysanthemum morifolium,* and mixed dye were absorbed on the TiO_2 nanomaterial surface. Table 1 shows intensity of the main peak after adsoption of dye on the TiO_2 nanomaterial surface compared with peak of pure TiO_2. For more clear the difference of pure TiO_2 and TiO_2 which has been immersed into dye as shown in the Table 1. All peaks of TiO_2 are anatase structure as in Figure 1 (indicated by A). The intensity and the crystal orientation of the main peaks of TiO_2 shows increased after absoption of dye solution compared with pure TiO_2. This result is comparable with the previous research which is reported by Al-Alwani et al. (2017) [9].

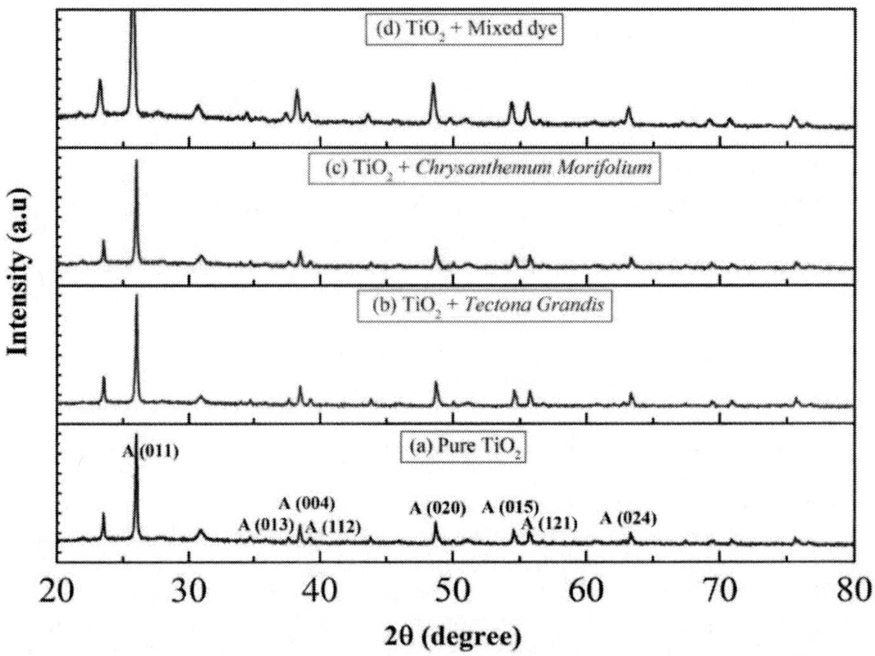

Figure 1. XRD of (a) Pure TiO_2, (b) TiO_2 + *Tectona Grandis,* (c) TiO_2 + Red *Chrysanthemum morifolium,* and (d) TiO_2 + Mixed dye. We have added symbol A indicated the anatase structure and crystal orientations for pure TiO_2.

3.2 FTIR Characterization

Figure 2 shows the FTIR spectrum was recorded in the range from 500 – 4500 cm^{-1}. Spectra of TiO_2 confirms the –OH group stretching appears at wavenumber of 3419.79 cm^{-1} with low signal. Thus C-H bonds of the aromatic ring at wavenumber of 2926.01 cm^{-1} with moderate intensity. Then, the C=C bonds of an alkene compound type seen at wavenumber 1645.28 cm^{-1}, while at 1514.21 cm^{-1} – 1541.12 cm^{-1} indicate C=C bonds of the aromatic ring compound type. The existence of O-Ti-O bond whic is shown by a broad wavenumber from 490 cm^{-1} to 800 cm^{-1}.

Spectra of *Tectona Grandis* and red *Chrysanthemum morifolium* exhibit a wide and strong band at 3414 cm^{-1} which is related to the –OH stretching. At wavenumber 2920.23 cm^{-1} was shows the sharp peak and at 2852.72 cm^{-1} shows small shoulder which is related to the C-H bonds stretching vibration. C=O bonds stretching vibration at wavenumber of 1735.93 cm^{-1} and 1732.08 cm^{-1} while C=C bonds at 1647.21 cm^{-1} and 1668.43 cm^{-1}. C-H bonds of an alkene compound type at 1361.74 cm^{-1} – 1452.40 cm^{-1} (bending vibration). At wavenumber from 1098 cm^{-1} to 1246 cm^{-1} were found the stretching vibration of C-O due to presence of some esters [10]. The band at 775.38 cm^{-1} and 763.81 cm^{-1} indicated the bending vibration of C-H. The IR spectra of extracts are shows existence of C=O, C-H, and C-O bands that contribute to coloring components in *Tectona Grandis* and red *Chrysanthemum morifolium* extracts which is indicated the existence of the anthocyanin [11].

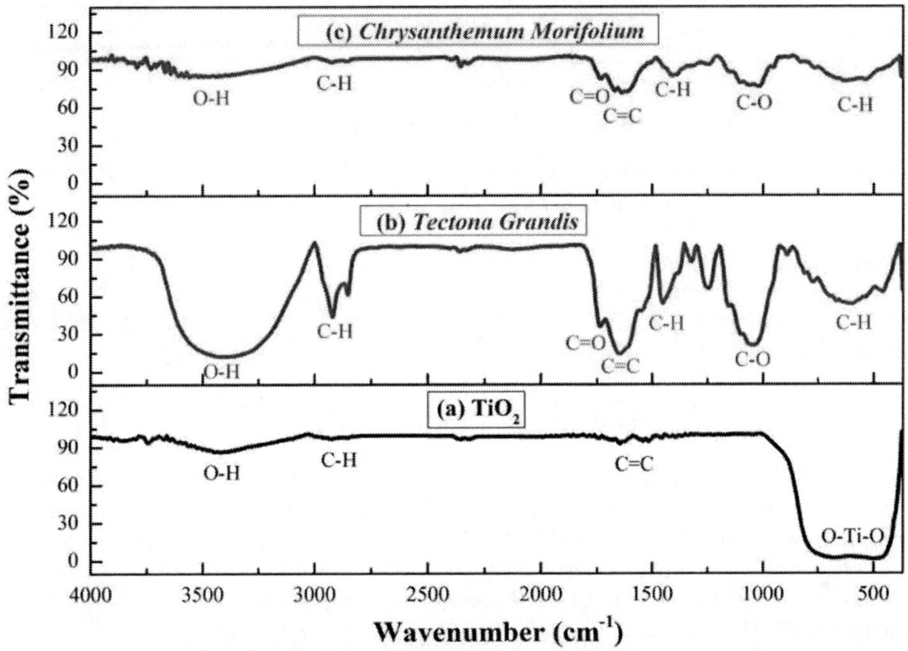

Figure 2. FTIR spectra of (a) TiO_2, (b) *Tectona Grandis,* and (c) Red *Chrysanthemum morifolium*. The anthocyanin structure exist in C=O, C-H, and C-O bond.

3.3 UV-Vis Characterization

Figure 3 compares the absorption spectra of natural dye solutions extracted from *Tectona Grandis* leaves, red *Chrysanthemum morifolium*, and mixed dye. The results exhibited the absorption at the visible region of light spectrum. To be more specific, broad absorption band of *Tectona Grandis* leaves in the range 400-480 nm with an absorption peak at 420 nm, broad absorption band of red *Chrysanthemum morifolium* in the range 400-490 nm with an absorption peak at 440 nm, and mixed dye showing a sharp absorption at 460 nm with the broad absorption band in the range 500-580 nm with an absorption peak at 532 nm . The differences in the absorption spectra of dyes may due to the different colors of the pigments present in them [10]. Common peak of anthocyanin pigments at 500–600 nm which is the main characteristics [8].

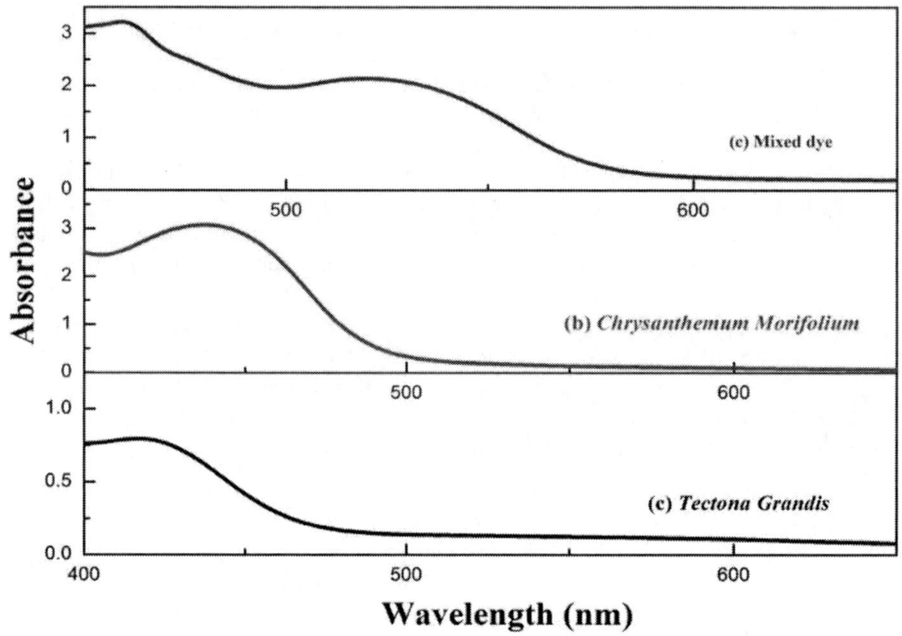

Figure 3. UV-Vis absorption spectra of dye solution (a) *Tectona Grandis*, (b) red *Chrysanthemum morifolium*, and (c) Mixed dye

3.4 Conversion Efficiencies

Table 2 shows the cell output and conversion efficiencies of DSSCs sensitized by three kinds of natural dyes. The power convertion efficiency (η) from the data in Table 2 was calculated by using the following expression.

$$\eta = \frac{Pm\ (\mathrm{mW}/cm^2)}{Light\ intensity\ (mW/cm^2)} \text{x } 100\ \%$$

From the table can be concluded that the extract from mixed dye performed slighty better than the two other dyes. Mixed dye was found to have the highest efficiency of 0.834 % while *Tectona Grandis* and *Chrysanthemum morifolium* is found to be 18.92 x 10^{-4} % and 37.99 x 10^{-4} % respectively. Mixed dye shows higher efficiency than previous research used *Hibiscus sabdariffa L.* (0.27 %) and *Acanthus sennii chiovenda* flower (0.150 %) which are reported by Abdou et al. (2013) and Ayalew & Ayele (2016), respectively [7, 8].

4. Conclusion

In this research, fabrication and characterization DSSCs have been successfully prepared from TiO_2 nanomaterial and extract *Tectona Grandis* leaves and red *Chrysanthemum morifolium* as a dye. The extracted dyes contain anthocyanin as the major coloring. Among the three dyes studied, the extract obtained from mixed dye has shown the best photosensitization I this study with the highest efficiency of 0.834 %.

References

[1] Mehmood U, Rahman S, Harrabi K, Hussein I A, Reddy B V 2014 *Advances in Materials Science and Engineering* http://dx.doi.org/10.1155/2014/974782.
[2] Knoema B P 2017 *World Reserves of Fossil Fuels* https://knoema.com/smsfgud/bp-world-reserves-of-fossil-fuels accessed 28 July 2017.
[3] Mescia, Luciano & Massaro, Alessandro 2014 *Advances in Materials Science and Engineering* http://dx.doi.org/10.1155/2014/252879.
[4] Pablo C, Vela C, Enrique R, Rocha José A R G, Enrique M P, Juan L H, Eddie N A M 2016 *Materials Today Proceedings* **3** 194-200.
[5] Obotowo I N, Obot I B, Ekpe U J, 2016 *Journal of Molecular Structure* **1122** 80-87.

[6] Abdou E M, Hafez H S, Bakir E, Abdel-Mottaleb M S A 2013 *Spectrochimica Acta Part A: Molecular and Biomolecular Spectroscopy* **115** 202–207.
[7] Cavallo C, Pascasio F D, Latini A, Bonomo M, Dini D 2017 *Journal of Nanomaterials* https://doi.org/10.1155/2017/5323164.
[8] Ayalew, Andargie W, Ayele D Worku 2016 *Journal of Science: Advanced Materials and Devices* **1** 488-494.
[9] Al-Alwani, Mahmoud A M, Mohamad A Bakar, Kadhum, Abd Amir H, Ludin, Norasikin A, Safie N E, Razali M Z, Ismail M, Sopian Kamaruzzaman 2017 *Int J Electrochem Sci* **12** 747-761.
[10] Kushwaha R, Srivastava P, Bahadur L 2013 *Journal of Energy* http://dx.doi.org/10.1155/2013/654953.
[11] Pereira Jr V, Aniceto, Arruda, Queiroz de I N, Stefani Ricardo 2015 *Food Hydrocolloids* **43** 180-188.

Characteristic of Coal Deposits in the East Kolaka, Southeast Sulawesi Province

Deniyatno[1], A Armid[2], Harisma[1], E Anshari[1] and E S Hasan[1]

[1] Faculty of Earth Sciences and Technology, Halu Oleo University, Kampus Hijau Bumi Tridharma Anduonohu Jl. HEA Mokodompit Kendari, Sulawesi Tenggara, Indonesia
[2] Faculty of Mathematics and Natural Sciences, Halu Oleo University, Kampus Hijau Bumi Tridharma Anduonohu Jl. HEA Mokodompit Kendari, Sulawesi Tenggara, Indonesia

E-mail: harisma@uho.ac.id

Abstract. Coal deposits in Southeast Sulawesi may not be large, one of the occurrence and deposit has been recorded in the Tawanga Village, East Kolaka Regency, Southeast Sulawesi Province. The study area is part of Southeast Arm of Sulawesi and the occurrence of coal is indicated from Meluhu Formation (TRm) which uncomformably overlying with schist of the Mekongga Complex (Pzm). In this study, field survey and proximate analysis of coal had been conducted to determine the distribution and quality of coal. Six coal samples from different area were subjected to the proximate analysis and the comparison of different properties including moisture content, volatile matter, fixed carbon, ash content and calorific values has been studied. The study is aimed to characterize of coal in the area through proximate analysis method. Field survey showed that the occurrence of coal as thin layers or lenses set in shale with thickness is vary from 5 to 20 cm. The analytical evaluation of properties showed the value of fixed carbon content 5.2 to 12.4%, the ash content 8.4 to 35.7%, the total moisture content 47.1 to 73.0% and volatile matter 7.0 to 10.7%. The analytical results suggest the existing of coal in the study area is classified as low rank calorie with the calorific value ranges between 1246.7 to 3644.6 cal/gr (adb).

Keywords: Coal, Proximate Analysis, Low Rank Calorie

1. Introduction

The occurrence of coal deposits in Sulawesi is not large. Directorate General of Mineral and Coal [1] reported the deposits of coal in the Sulawesi region came from South Sulawesi and Central Sulawesi with total resources at 231.12 and 1.98 million tons respectively. However, Southeast Sulawesi is also recorded the occurrences and deposits of coal. Surono [10] reported two coal seams have been found in Kendari city from Toronipa Member of the Meluhu Formation. The coal is characterized black, sheared due to the fault, abundant of pyrite, and rank of coal is between sub-bituminous and high volatile bituminous.

The coal deposits in Southeast Sulawesi are distributed in the Tawanga Village, East Kolaka Regency. This area is located in the eastern part of Kendari, a capital city of Southeast Sulawesi with geographic coordinates between $3^0 46'30"$ - $3^0 48'30"$ latitude and between $121^0 41'00"$ - $121^0 43'15"$

longitude. Based on geological map of thc Lasusua - Kendari Sheet, the study area is consisted of metamorphic rock of Mekongga Complex in age of Paleozoic and Alluvial deposits (Figure 1). This thing is different with another deposits of coal in others part of Sulawesi. In South Sulawesi, coal has been identified in the Eocene stratigraphic units of Malawa Formation and Toraja Formation while in the East Sulawesi, coal has been identified in the Triassic (Winto Formation) and Jurassic units (Nanaka Formation), as explained in Ref. [3].

Numerous previous studies had been conducted in this area. Ngkoimani et al [4] used petrography study to identification of coal in this area and resulted the rank of coal is lignite. In 2015, Ngkoimani et al [5] were back to study estimation of coal distribution in this area by using DC Resistivity Method and resulted that Tawanga area is potential for coal mining.

In this current study, it will be conducted field geological survey in the Tawanga Village to identification the distribution of coal and determine the characteristics and quality of coal by proximate analysis method.

2. Geological setting

Geologically, Sulawesi Island and its surrounding area are complex regions. The complexity was caused by convergence between three lithospheric plates: the nortward-moving Autralian Plate, the westward-moving Pacific Plate, and the south-southeast-moving Eurasian Plate [11]. Because of that, regional structures were developed especially in the Southeast Arm of Sulawesi (Figure 1) and its surrounding area.

The Southeast Sulawesi continental terrane which trends NW-SE is bounded by two major strike-slip fault systems. The northern-most fault is referred to as the Lawanopo Fault System, and the southern-most fault is named the Kolaka Fault [9]. The Lawanopo Fault System and Kolaka Fault are active sinistral faults [8]. Both of these faults are considered to extend from onshore the southeast arm of Sulawesi into Bone Bay. These faults are considered to have been active during the Plio-Pleistocene [8] according to cross-cutting relationships and deformation of sediments within Bone Bay.

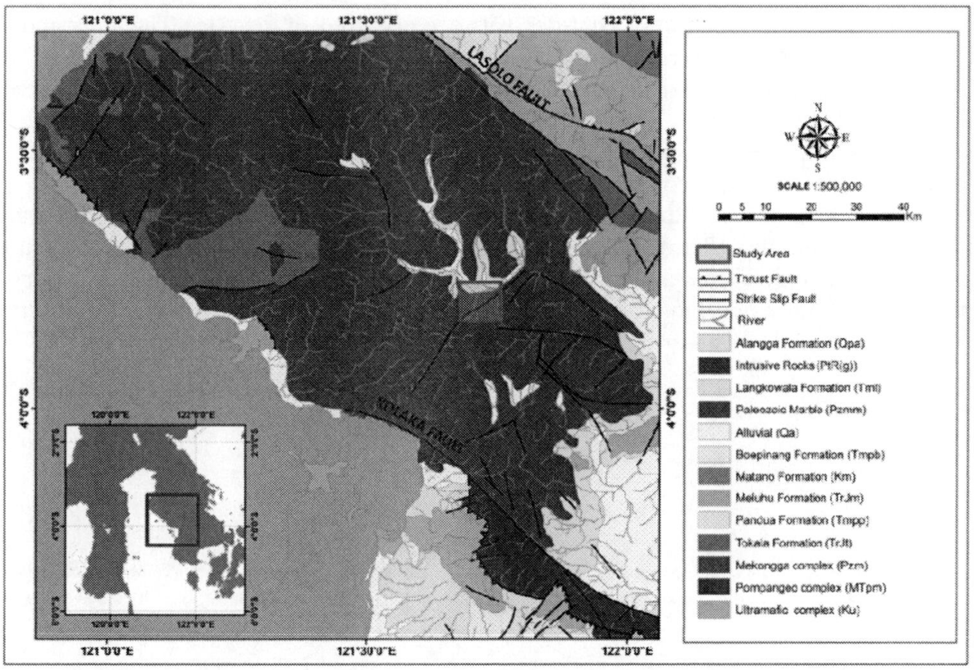

Figure 1. Geological map of the East Kolaka Regency, Southeast Sulawesi
(Part of the Geological Map of the Kolaka Sheet, scale 1:250.000, Simandjuntak et al, 1993 and Geological Map of the Lasusua-Kendari Quadrangles, scale 1:250.000, Rusmana et al, 1993)

IOP Conf. Series: Earth and Environmental Science **279** (2019) 012032 doi:10.1088/1755-1315/279/1/012032

Stratigraphy of the Southeast Arm of Sulawesi has detailed explained by Surono [8]. The oldest rocks in the Southeast Sulawesi continental terrane are metamorphic rocks which were intruded by granitic rocks. These are unconformably, overlain by clastic-dominated sediments of the Triassic Meluhu Formation, which is unconformably covered by a carbonate sequence of the Paleogene Tampakura Formation. Collision between the Southeast Sulawesi and Buton terranes occurred in the Latest Oligocene and developed the thrusts. The ophiolite which was located between these micro-continents, was thrust over the margins of the Southeast Sulawesi and Buton terrane during the collision. Thrusting was followed by uplift and erosion of the ophiolite and micro-continent with deposition in local fluvial basins which were part of the Early Miocene syn-orogenic clastic sediments of the Sulawesi Molasse. The Sulawesi Molasse is dominated by clastic units with local limestone. The clastic sequence consists of conglomerate of the Langkowala and Pandua Formations, sandy marl to claystone of the Boepinang Formation and sandstone with local coral reefs of the Buara Formation. After thrusting ended, collapse from block faulting caused trangression in the region and deposition of carbonate of the molasse.

3. Methodology

The methodology in this study by collected coal samples from six locations during field survey. Field sampling procedure was conducted both using chip and channel sampling methods. Most of coal samples were taken from Alaa Tawanga River and Palembua River which located in the Tawanga Village, East Kolaka Regency, Southeast Sulawesi Province.

Determination of coal quality was carried out by proximate analysis. All samples from six locations were taken and sent to Pusat Studi Pangan dan Gizi Universitas Gadja Mada. Proximate analysis is consisted of moisture value, ash value, volatile matter value, fixed carbon value, and calorie value.

4. Results

4.1. Field survey

Based on field survey, distribution of coal deposits are occurred in the Alaa Tawanga and Palembua Rivers. There are six stations were found of coal deposits (ST.6, ST.21, ST.22, ST.23, ST.24, and ST.25) and all are located in the river with fresh to weathered condition. Distribution of coal deposits in the study area relatively NE-SW to N-S. The occurrence of coal is mostly in thin layers or lenses set in shale with the thickness up to ±20 cm. The deposits are commonly associated with shale and sandstone. The contact between coal deposits with associated rocks are conformable.

The geological map of the study area (Figure 2) shows the major distribution is schist metamorphic rock of Mekongga Complex (Pzm) with distribution is very wide around 70% of the study area. While the remains distribution are coal and alluvial deposits (Qa). The distribution of coal is very minor or the smallest in the study and mostly as thin layers or lenses in shale and it can be observed in the northern and eastern parts of study area or near with the main river of Konaweha River.

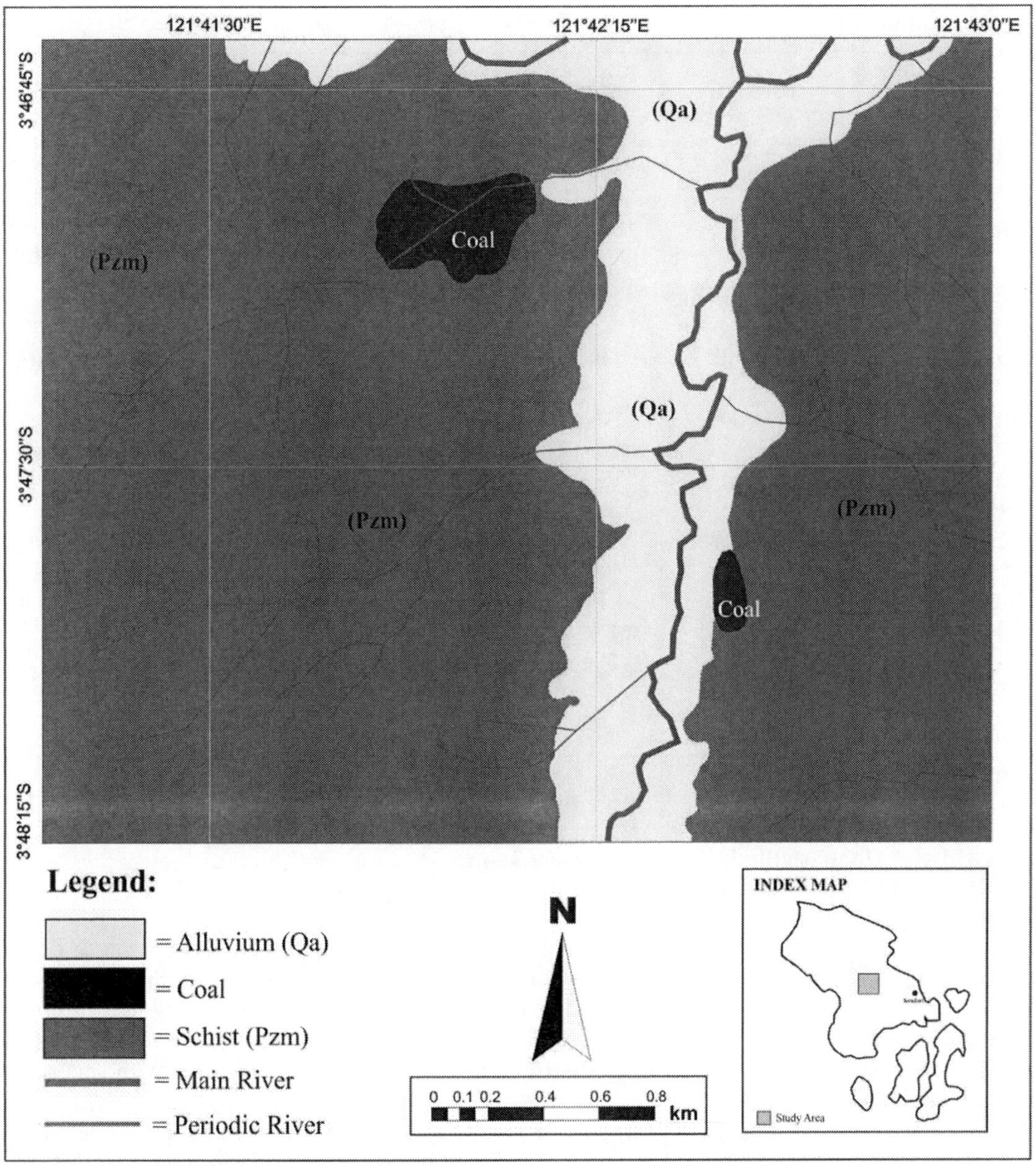

Figure 2. Geological map of the study area based on field survey.

Macroscopic appearance of coal (Table 1) in the Alaa Tawanga (ST.6) showed brownish to black, brittle, and laminated structure. The orientation of coal in this area relatively N-S with strike and dip value at N166°E/90°. The thickness of coal is from 6-13 cm. While coal exposure in the Palembua River (ST.21, ST.22, ST.23, ST.24, and ST.25) showed black, mostly weathered, very brittle, very wet, and coal deposits exhibited cleats or fractures. All strike and dip values show the orientation of coal in this area relatively NE-SW and its different with the orientation in the Alaa Tawanga River. A mass of plant debris is present in the rock texture. The thickness of coal is vary from 5 to 20 cm.

IOP Conf. Series: Earth and Environmental Science **279** (2019) 012032 doi:10.1088/1755-1315/279/1/012032

Table 1. The result of field survey for six stations of coal in the Tawanga Village

Description \ Station	ST.6	ST.21	ST.22	ST.23	ST.24	ST. 25
Location	AT River	PB River	PB River	PB River	PB River	PB River
Strike/Dip	N166°E /90°	N240°E /27°	N232°E /42°	N232°E /42°	N220°E /74°	N256°E /45°
Colour	Bwn Black	Black	Black	Black	Black	Black
Luster	Dull	DMB	DMB	Dull	Dull	Dull
Streak	Brown	Black	Black	Black	Black	Black
Hardness	Brt	Vr Brt	Vr Brt	Vr Brt	Vr Brt	Vr Brt
Sorting	Poor	Poor	Poor	Poor	Poor	Poor
Impurities	-	PR	PR	PR	PR	PR
Structure	-	Cleat	Cleat	Cleat	-	-
Thickness	±6-13 cm	±18 cm	±20 cm	±13 cm	-	±5 cm

Explanation:
AT River = Alaa Tawanga River
PB River = Palembua River
Bwn Black = Brownish Black
DMB = Dull with Minor Bright
Brt = Brittle
Vr Brt = Very Brittle
PR = Plant Roots

Coal deposits in all stations showed the contact with sediments rock. The station six (ST.6) at the Alaa Tawanga River showed as thin layer of coal in shale. The coal observed with thickness is vary from 6-13 cm while shale observed with thickness up to 2.4 m. The uppert part of shale are overlain with sandstone and conglomerate with thickness is 80 cm and 1.5 m respectively with bedding pattern is coarsening upward (Figure 3A). At the bottom part of shale is also interbedded with sandstone and the thickness up to 2 m.

In the Palembua River (ST.21, ST.22, ST.23, ST.24, and ST.25), the occurrence of coal is showed by thin layers or lenses set in shale. The thickness of coal is vary from 5 to 20 cm (Figure 3B, 3C, 3D, 3E and 3F).

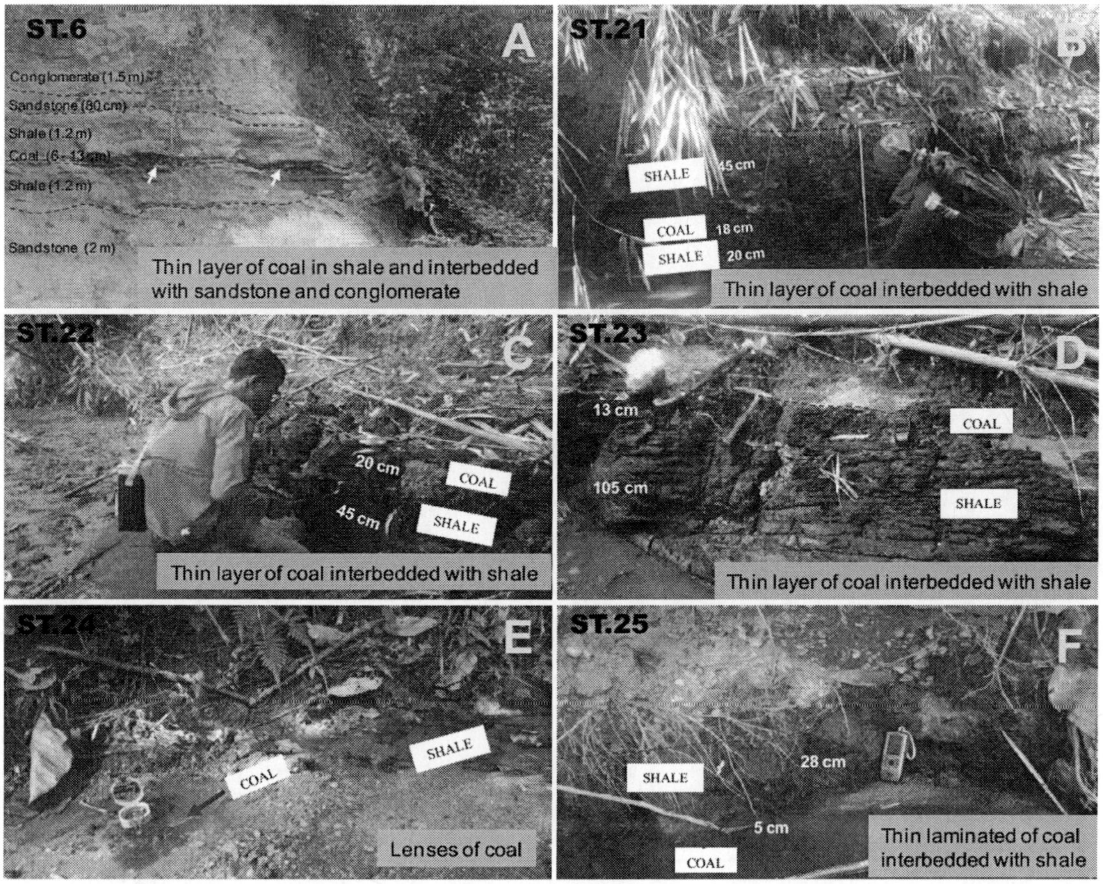

Figure 3. Photographs of the occurrence of coal observed in the Tawanga Village

4.2. *Quality of coal*

Coal proximate analysis is performed in Laboratorium Studi Pangan dan Gizi Universitas Gadja Mada to define ash content, moisture content, volatile matter, fixed carbon, and calorie value from coal samples from Tawanga Village, East Kolaka Regency, Southeast Sulawesi Province.

The result of proximate analysis from coal sample in six stations is showed on figure 4.

4.2.1. Ash content. Ash content is an impurity that will not burn. Ash will reduce burning capacity. Based on information, general coal properties of Indonesia for ash content is characterized by typical value at <10% as explained in Ref. [2], while ash contents range in the study area from 8.4 to 35.7% (Table 2). It is considered that ash contents in the study area are higher than normal value.

4.2.2. Moisture content. Moisture in coal must be transported, handled and stored. Since it replaces combustible matter, it decreases the heat content per kg of coal. General coal properties of Indonesia for moisture content is characterized by typical range is 10 to 45% as explained in Ref. [2], while moisture contents range in the study area between 47.1 to 73.5% (Table 2) and it is over of normal range value. This range indicate the typical of the low rank of coal with very high moisture.

4.2.3. Volatile matter. Volatile matter is an index of the gaseous (methane, hydrocarbons, hydrogen and carbon monoxide, and incombustible gases like carbon dioxide and nitrogen found in coal) fuels present. Proportionately, volatile matter increases flame length, and helps in easier ignition of coal. General coal properties of Indonesia for volatile matter is characterized by typical range is 25 to 45%

as explained in Ref. [2]. The volatile matter contents in the study area are relatively lower, ranging between 7 to 10.7% (Table 2). This range value indicates the typical of the low rank of coal.

4.2.4. Fixed carbon. Fixed carbon gives a rough estimate of heating value of coal. The increasing of fixed carbon content means the calorie value of coal will be higher. General coal properties of Indonesia for fixed carbon is characterized by typical range is 30 to 50% as explained in Ref. [2]. The Fixed carbon contents in the study area are relatively lower, ranging between 5.2 to 12.4% (Table 2). This range value indicates low calorie value.

4.2.5. Calorie value. Calorie value of coal samples in the study area are vary from the lowest value is 1246.7 cal/gr to the higher value is 3644.6 cal/gr (Table 2). According to the Goverment Regulation No.13/2000 and No. 45/2004, coal quality is classified into 4 type based on calorific value: 1) Low Calorie: <5100 cal/gr; 2) Medium Calorie: 5100-6100 cal/gr; 3) High Calorie: 6100-7100 cal/gr; and 4) Very High Calorie: >7100 cal/gr, as explained in Ref. [2]. Based on that classification, calorie values in the study area are classified into low rank calorie.

Table 2. The result of proximate analysis for six coal samples from Tawanga Village

Samples No.	Proximate Analysis (wt %)				CV (cal/gr)	Rank of Calorie
	AC	MC	VM	FC		
ST. 6	35.7	47.1	7.0	10.1	1246.7	Low Calorie
ST. 21	32.7	46.4	8.8	11.8	1816.6	Low Calorie
ST. 22	8.4	73.0	8.1	10.3	3644.6	Low Calorie
ST. 23	32.1	55.1	7.4	5.2	1876.9	Low Calorie
ST. 24	17.3	60.2	10.2	12.1	2833.8	Low Calorie
ST. 25	9.9	66.8	10.7	12.4	3190.4	Low Calorie
Mean:	22.7	58.1	8.7	10.3	2434.8	Low Calorie

Explanation:
AC : Ash Contents (wt %) FC : Fixed Carbon (wt %); CV : Calorie Value (cal/gr)
VM : Volatile Matter (wt %); MC : Moisture Contents (wt %);

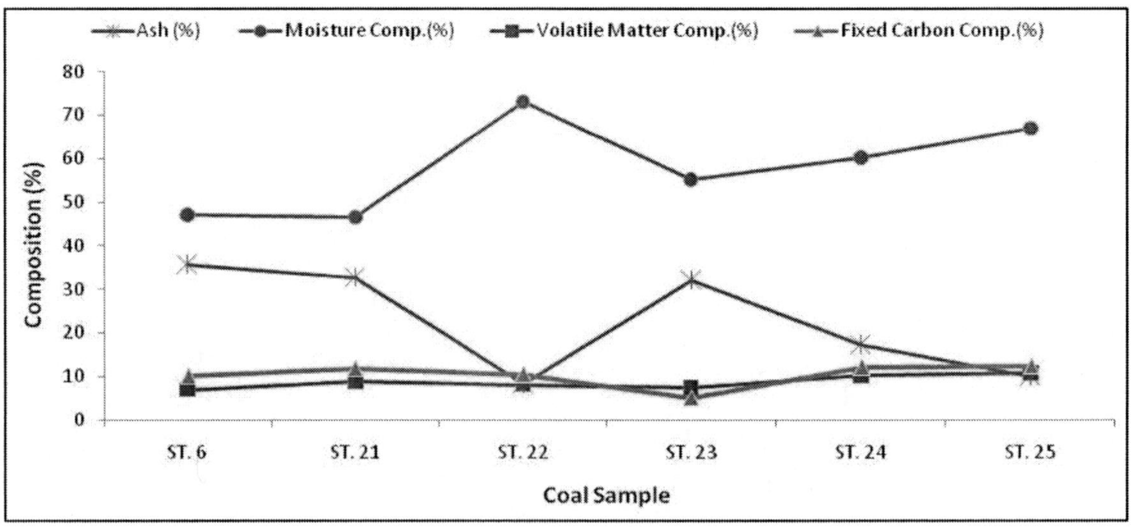

Figure 4. The result of proximate analysis from coal sample in six stations

5. Discussion

Based on field survey in the study area, typical of coal exposures display as thin layers or lenses set in shale. Shale in the study area showed interbedded with sandstone and conglomerate with sequence pattern is coarsening upward. Characteristic of the lithologies in study area can indicate that in the study area is developed of Meluhu Formation. This thing is based on identification of litology and texture of rocks which have similar with litology from Meluhu Formation. So, we postulate that the occurence of coal in the study area may be came from Meluhu Formation.

Proximate analysis is also showed that the value of ash content, moisture content, volatile matter and fixed carbon from six samples of coal indicates that typical of coal is tend to low rank of coal. This is also confirmed by calorie value which indicates the range of calorie is classified into low rank calorie. So, we indicate that typical coal in the study area is lignite coal with low rank calorie.

6. Conclusion

Based on field survey and proximate analysis of coal samples in the Tawanga Area, some conclusions can be drawn as follow:

1. Coal deposits in the study area are found in the six stations (ST.6, ST.21, ST.22, ST.23, ST.24, and ST.25). The coal display as thin layers or lenses set in shale with thickness is vary from 5 to 20 cm. The occurence of coal in study area may be carried by Meluhu Formation.
2. Based on proximate analysis, coal deposits in the study area is classified as low rank calorie with range between 1246.7 to 3644.6 cal/gr (adb).

7. References

[1] Directorate General of Mineral and Coal 2015 *Indonesia Mineral and Coal Information* (Jakarta: Directorate General of Mineral and Coal Ministry of energy and Mineral Resources) p 10 and 44

[2] Geological Agency 2006 *Indonesia Lower Rank Coal Resources Development Policy* (Jakarta: Geological Agency Ministry of Energy and Mineral Resources) p 21

[3] Matasak T 2011 Coal deposits in sulawesi *Proceedings of the Sulawesi Mineral Resources Seminar MGEI-IAGI* p 332

[4] Ngkoimani L O, Makkawaru A and Jahidin 2014 Coal petrographic study of southeast sulawesi *The 1st International Conference Geoscience for Energy, Mineral Resources and Environment Applied* p 6

[5] Ngkoimani L O, Isa L S, Jahidin, Syamsurizal, Asfar S and Makkawaru A 2015 Estimation of coal distribution in tawanga village, east kolaka regency, southeast sulawesi by using dc resistivity method *J. Phys.: Conf. Ser.* **846** p 5

[6] Rusmana E, Sukido, Sukarna D, Haryono E and Simandjuntak T O 1993. Peta geologi lembar lasusua-kendari, sulawesi skala 1: 250.000 *Pusat Penelitian dan Pengembangan Geologi*

[7] Simandjuntak T O, Surono and Sukido 1993. Peta geologi lembar kolaka, sulawesi skala 1: 250.000 *Pusat Penelitian dan Pengembangan Geologi*

[8] Surono 1994 Stratigraphy of the southeast sulawesi continental terrane, eastern indonesia *Journal of Geology and Mineral Resources* **IV** No.31 p 8 and 10

[9] Surono 1998 Geology and origin of the southeast sulawesi continental terrane, indonesia *Media Teknik 3* **XX** p 14 and 23

[10] Surono 1999 An organic petrology study on coal and carbonaceous rocks from the triassic meluhu formation, southeast arm of sulawesi *Majalah Geologi Indonesia* **15 No.1-2** p 70

[11] Surono 2013 *Geologi Lengan Tenggara Sulawesi* (Bandung: Badan Geologi Kementerian Energi dan Sumber Daya Mineral) p 1 and 32

IOP Conf. Series: Earth and Environmental Science **279** (2019) 012033 doi:10.1088/1755-1315/279/1/012033

Limited off-set consideration in de-multiple process of seismic Lake Towuti

Ryan Bobby Andika*[1], Wahyu Triyoso[1], Satria Bijaksana[1], Abd. Hafidz[1], James M. Russell[2], Nigel Watruss[3], Hendrik Vogel[4],

[1]Faculty of Mining and Petroleum Engineering, Institut Teknologi Bandung, Jalan Ganesha 10, Bandung 40132, Indonesia;
[2]Department of Earth, Environmental, and Planetary Sciences, Brown University, 324 Brook St., Providence, RI, 02912, USA
[3] Large Lakes Observatory, University of Minnesota Duluth, Duluth, MN, 55812, USA
[4] Institute of Geological Science & Oeschger Center for Climate Change Research, University of Bern, Baltzerstrasse 1+3, 3012 Bern, Switzerland

Abstract. Multiple attenuation effect study in Lake Towuti seismic data is based on limited offset recording. The goal of this study is to better understand the sediments structure of Lake Towuti that layered above East Sulawesi ophiolitic and metamorphic rocks for stratigraphic interpretation. Surface-related multiple and internal multiple are a common noise signal that recorded in this seismic data, alongside the primary signal from the sediments layer. In order to get a representative subsurface image of the lake, a combination of some de-multiple method to suppress this noise signal has been applied in previous study. However, the line seismic that used in this seismic recording is less than 100 m. From our perspective, this limited offset line fact have to be considered carefully. It's because some application of de-multiple method in this case will not effectively suppress the noise. In this study, we give an alternative combination method in function to put out the surface-related multiple of the case. First, we enhance the S/N ratio and velocity analysis of the data by Common Reflection Surface (CRS) method. From this enhancement, CRS can give a direct advantage when applying some de-multiple method, especially in Predictive Deconvolution and SRME method. Moreover, we combine the application of de-multiple theory with the strict arrangement of Predictive Deconvolution, Surface-Related Multiple Elimination (SRME), and the last is F-K filter. Those combination methods precisely give a better imaging of Lake Towuti sediments layer in this seismic processing study.

1. Background

Lake Towuti is one of the lakes in South Sulawesi that are in the Lake Malili system. The sediment layer that coats the basin of the lake naturally is able to store uniqueness in reconstructing the ancient environment included in the center of the Indo-Pacific heat pool (IPWP) (Russell et al., 2016). So that in determining the drilling point, 2D sea seismic survey activities were carried out with the total length of the trajectory to reach the meter's height (Russell et al., 2016).

The seismic survey work carried out at Lake Towuti has a data recording configuration with an offset of less than 100 m. This is what makes the authors feel the need for consideration of limited offset (limited offset) in processing data. With the main objective of eliminating multiple signals, a combination of de-multiple methods such as Predictive Deconvolution, Surface-Related Multiple Elimination (SRME), and F-K filters are specifically combined with the use of Common Reflection Surface (CRS) methods. This CRS

IOP Conf. Series: Earth and Environmental Science **279** (2019) 012033 doi:10.1088/1755-1315/279/1/012033

method is projected to enhance those de-multiple methods in order to eliminate the multiple signals that exist in Lake Towuti seismic data for line 04.

2. Methodology

2.1 Towuti *Drilling Project* (TDP)

Towuti Drilling Project (TDP) is a scientific project that has been carried out since 2012. This project was held with the aim to find out information about the long-term changes in the geomicrobiology of this system, including changes in climate, microbial geochemical cycling, carbon storage, metal deposition, and microbial processed operating at a depth of the lake sediment layer (Russell et Al., 2016). From the results of the data acquisition, it was determined that the two main areas became drilling targets in this project. The selection of the two drilling locations, in short, is based on the subsurface image of the seismic recording results. The results show that the sediment layer in both locations has a relatively high thickness.

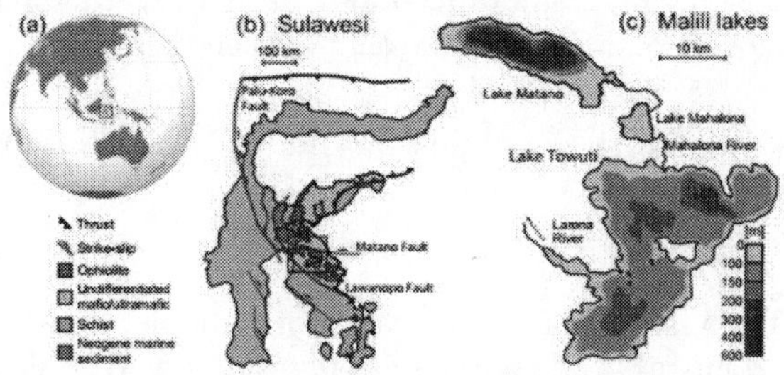

Figure 1. The appearance of the TDP study area map showing (a) the location of Sulawesi Island, (b) regional geology around the study area, (c) Lake Malili system complex (Russell et al., 2016)

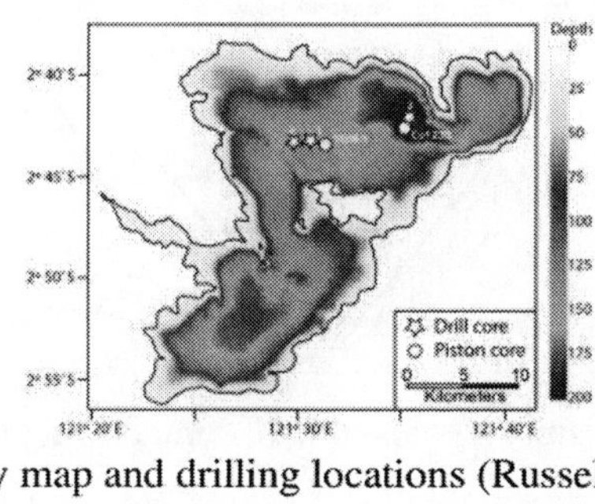

Figure 2. Lake Towuti bathymetry map and drilling locations (Russell et al., 2016)

2.2 *Predictive Deconvolution*

Predictive deconvolution is a method for estimating the value of a series of signal amplitude values at a certain lag. In its application to the concept of the de-multiple method, Predictive deconvolution can be used to predict multiple signals. This is because multiple signals, on the results of recording, are generally described as having a certain period of occurrence.

$$\begin{bmatrix} a_0 & \cdots & a_{n-1} \\ \vdots & \ddots & \vdots \\ a_{n-1} & \cdots & a_0 \end{bmatrix} \begin{bmatrix} f_0 \\ \vdots \\ f_{n-1} \end{bmatrix} = \begin{bmatrix} a_a \\ \vdots \\ f_{a+n-1} \end{bmatrix} \tag{1}$$

IOP Conf. Series: Earth and Environmental Science **279** (2019) 012033 doi:10.1088/1755-1315/279/1/012033

2.3 *Common Reflection Surface* (CRS)

Common reflection surface (CRS) is a method for approximating traveltime from a reflection response point (Xm) around the midpoint (Xo) point on the seismic line (Hocht et al., 1999). In principle, CRS is a stacking method that is carried out on reflection data, similar to stacking on Conventional CMP. But in the CRS method, the stack process and results use more traces than the CMP gather (Mann, 2007). This is what makes the CRS stack able to increase the S / N Ratio compared to the conventional stack (Mann, 2007).

$$t^2_{hyperbol}(x_m, h) = ((t_0 + \frac{2sin\alpha}{v_0}(x_m - x_0))^2 + 2t_0\frac{cos^2\alpha}{v_0}(\frac{(x_m-x_0)^2}{R_N} + \frac{h^2}{R_{NIP}}) \quad (2)$$

2.4 *Surface-Related Multiple Elemination* (SRME)

The Surface-related multiple elemination method is a method that is specifically applied to emphasize the presence of multiple surface-related multiple types. The method proposed by D.J. Verschuur is simply made to predict multiple waves on the results of recording. The prediction is done by convolution of each trace from a gather, with the whole gather the result of recording in an area. After that, we just subtract the input data with the multiple prediction by SRME to get the data with no multiple interference.

$$X_0(z_0, z_0) = X(z_0, z_0) - r_0\{X_0(z_0, z_0)\}^2 + r_0^2\{X_0(z_0, z_0)R^-(z_0)\}^3 - r_0^3\{X_0(z_0, z_0)R^-(z_0)\}^4 + \cdots \quad (3)$$

2.5 F-K *Filter*

In its orientation to reduce the noise signals in the form of multiple, transformation into the frequency domain - wave number (F-K domain) is one of the right things to do. With this domain change from X-T, and applying NMO correction to the data, the primary signal is in the form of a reflector field and multiple noise can be distinguished based on the wave number and frequency in more contrast.

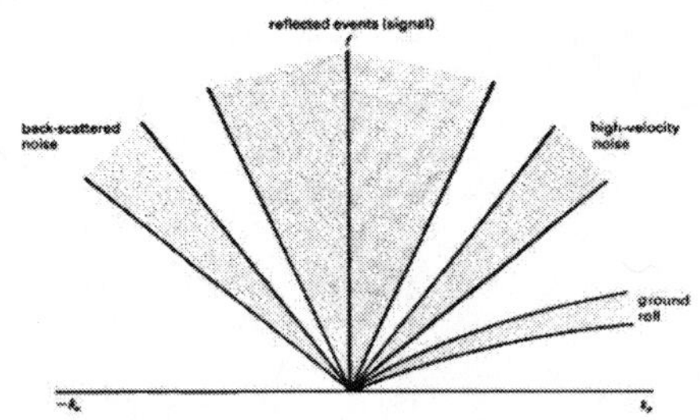

Figure 3. The response of seismic signals to F-K domains theoretically (geosci.xyz, 2017)

III.1 Data

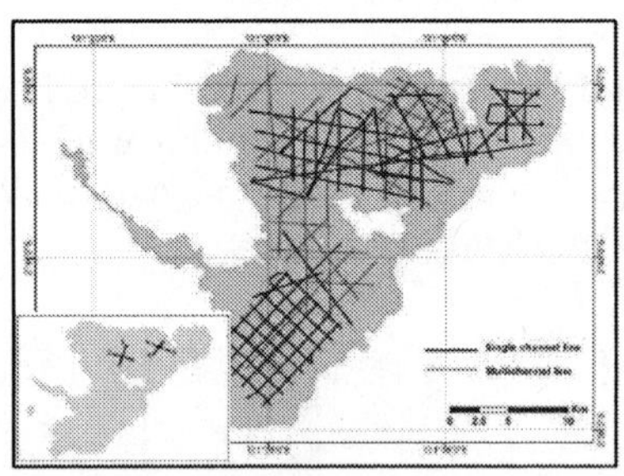

Figure 4. Seismic trajectory map at Lake Towuti (Hafidz, 2017)

IOP Conf. Series: Earth and Environmental Science **279** (2019) 012033 doi:10.1088/1755-1315/279/1/012033

In this study, the authorss specifically carried out the treatment of Lake Towuti seismic data on the line-04 line, with the initial subsurface image and data recording parameters as follows:

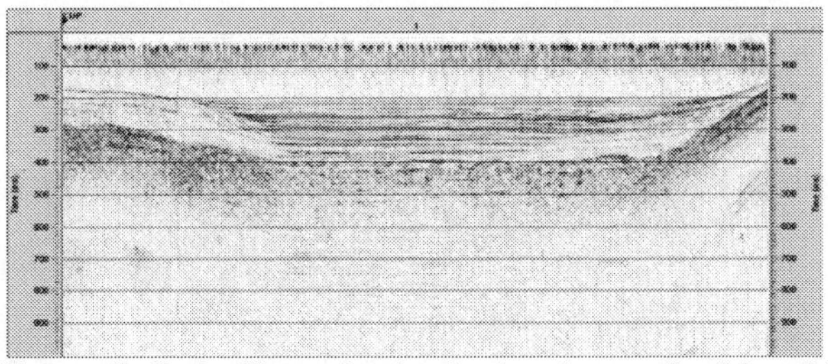

Figure 5. Lake Towuti seismic data image for Line-04

Table 1. Data recording configuration parameters on Lake Towuti for Line-04

Parameter	Keterangan
Total Receiver	24
Total Source	1163
Shot / Receiver Spacing	3.5 : 3.125 (meter)
Near Offset	25 m
Far Offset	96.875 m
Configuration	Off – End

3. Data Processing

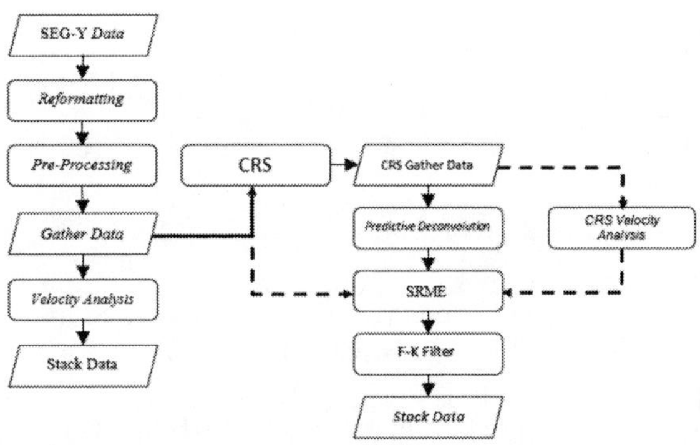

Figure 6. Research flow chart.

In general, data processing deals with the application of de-multiple methods such as F-K Filter, Surface-Related Multiple Elimination, and Predictive deconvolution. But before de-multiple processing is done, the Common reflection surface method is applied to seismic data. It aims to increase the S / N Ratio first from the data to be processed so that the use of multiple-method will be more effective, especially on the application of Predictive deconvolution and Surface-Related Multiple Elimination.

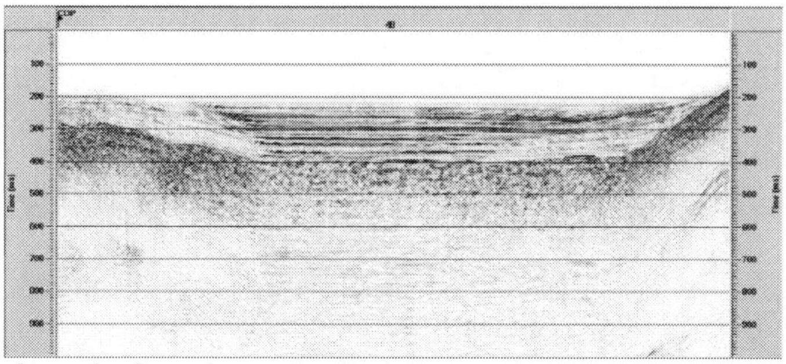

Figure 7. Data stack results from the application of pre-processing using conventional background velocity.

The authors projects that the use of the CRS method as the kernel for the use of the Predictive deconvolution method will produce a representative output in predicting multiple signals. Furthermore, in this study, multiple signals that have been predicted by the Predictive deconvolution equation are not directly used as a component of the extraction of the initial data or input. With relatively thin sediment layering conditions for seismic surveys, as well as limited seismic line facts (limited-offset), the authorss suspect that the predicted signals may also be present along with the primary signal. To reduce this possibility, further predictive data will be taken to the multiple prediction process using the Surface-Related Multiple Elimination (SRME) method named SRME Macro.

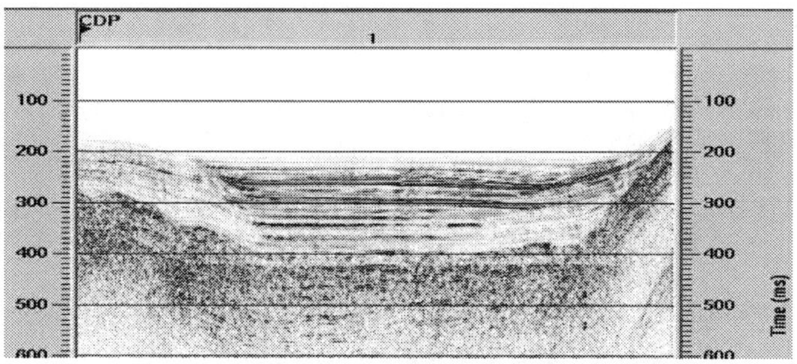

Figure 8. Results of stack application of Predictive deconvolution and SRME on Lake Towuti line-04 seismic data.

After obtaining the output data from the prediction process (SRME Macro), the SRME Match-Filter process of the data is carried out with conventional data gather which is not applied by the Common reflection surface (CRS) method. It aims to represent multiple predictive data on its original purpose, maintain data authenticity, and the subtraction process in the Adaptive Subtraction process can produce good output and be free from multiple signals. After that, authors apply F-K filter to attenuate the multiple signals.

IOP Conf. Series: Earth and Environmental Science **279** (2019) 012033 doi:10.1088/1755-1315/279/1/012033

4. Analysis and Discussion

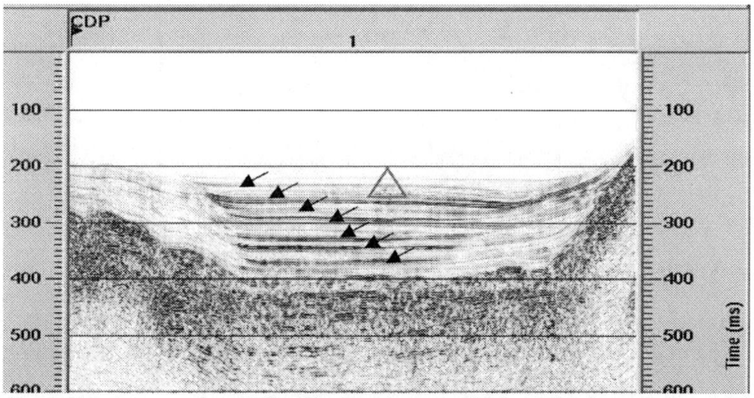

Figure 9. Data stack application results of Predictive deconvolution, Surface-Related Multiple Elimination (SRME), and F-K Filter on Lake Towuti line-04 seismic data.

In general, the authorss interpreted the presence of seven markers with a slump material. To strengthen the accuracy of the processing results in reducing multiple of seismic data and its interpretation, the writer analyzes the data from the initial and after data as shown below.

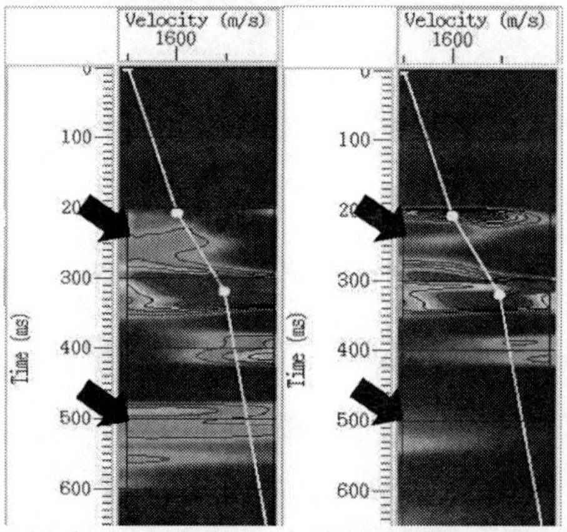

Figure 10. Semblance CDP 201 for before (left) and after (right) processing seismic data of Lake Towuti line-04.

From Figure 10, it can be shown that a reduction in the value of Semblance in the area marked by a black arrow after data processing is performed. These removals, based on images, are entirely in areas with low velocity and TWT in which indicate the removal of multiple signals. Based on theory, the presence of semblance which has a decrease in velocity value, and when the TWT value increases, is a common feature of the presence of multiple signals in seismic data. This is in accordance with the following theoretical

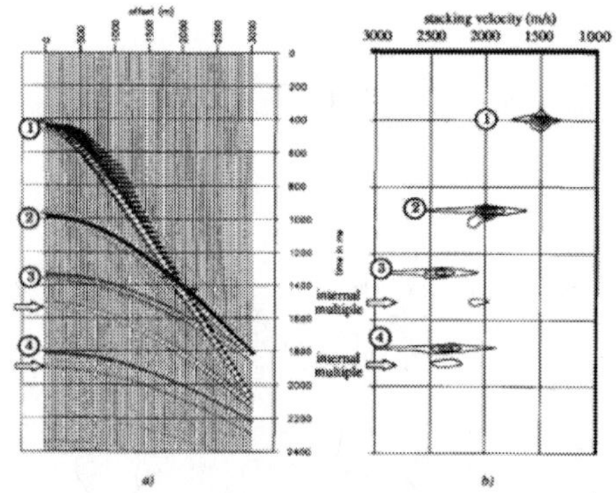

Figure 11. *Semblance* CDP *gather* dimana panah nomor 4 menunjukan kehadiran *multiple* pada data seismik (Verschuur,1991).

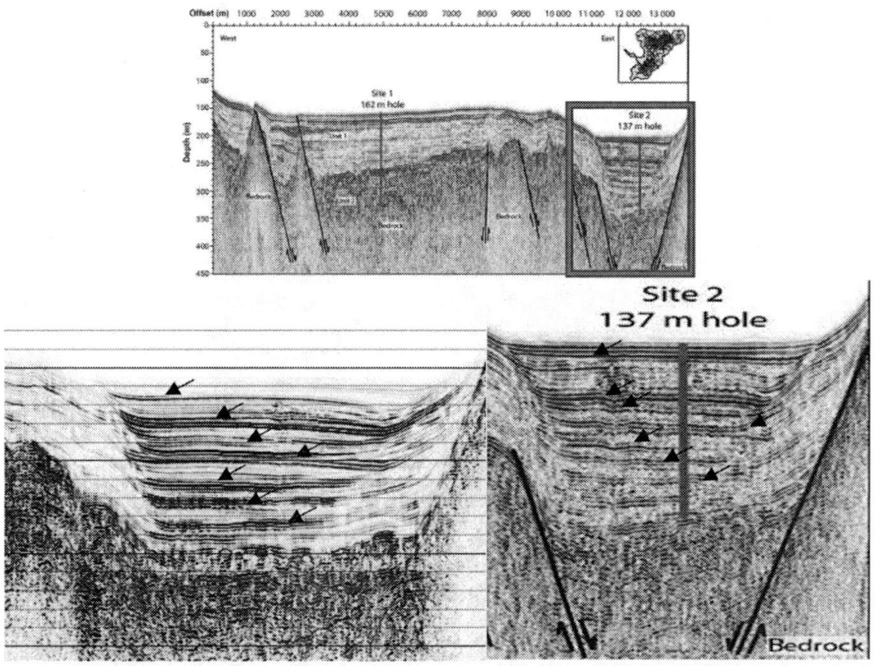

Figure 12. Comparison between the authors's data processing line (left) and data processing by Russell et al. (right) with different azimuths (Russell et al. 2016).

Furthermore, the authors also compares the results of data processing that has been done with data processing that has been carried out in previous studies by Russell et al. (2006). Data processing that has been done previously has the same target area as the authors, but in the direction of a different recording azimuth. It can be seen that there are 7 markers on the target sediment layer.

For the last, the authors re-autocorrelated the Lake Towuti data stack after de-multiple processing was carried out. From the processing data stack, it can be seen in the picture below that the seismic peak from water bottom lake Towuti (0 ms lag) is relatively weak, tends to disappear, at 37 ms lag as in Figure 13. This proves that the processing carried out by the authors has succeeded in reducing multiple signals in the Lake Towuti seismic data.

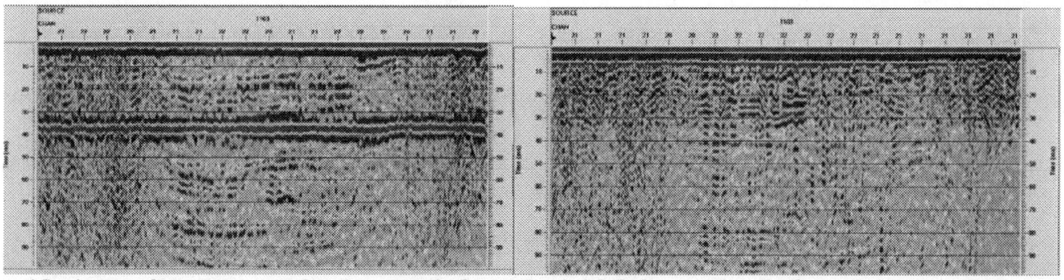

Figure 13. Autocolerogram on the data stack from the initial data gather before (left) and after (right) the application of Preprocessing and conventional velocity background which shows a lag of 37 ms.

5. Conclusion

Based on the results of the processing of seismic data from Lake Towuti for line 04, the following conclusion points are obtained:

1. The results of seismic data processing by the authors to eliminate multiple signals (increase S / N Ratio), and delineate markers for Lake Towuti on line 04 are relatively good and representative.

2. Using a combination of Common reflection surface application methods on seismic data with de-multiple methods in the form of, sequentially, Predictive Deconvolution, SRME, and, F-K filters were able to image sediment markers for Lake Towuti on line 04 well.

6. Acknowledgement

Token of appreciation and gratitude goes to Institut Teknologi Bandung, Pertamina Hulu Energi, Parents of first Author, Abdul Hafidz, and Haritsari Dewi who helped a lot during the making of this project.

7. References

[1] Höcht, G., de Bazelaire, E., Majer, P., dan Hubral, P. (1999): Seismics and optics: hyperbolae and curvatures, Journal of Applied Geophysics, 42, 261-281.
[2] Mann J., Schleicher J., dan Hertweck T., (2007): CRS stacking – a simplified explanation, EAGE 69th Conference & Technical Exhibition, London.
[3] Russel J.M, et al., (2016): The towuti drilling project: paleoenvironments, biological evolution, and geomicrobiology of a tropical pacific lake, SCI. Dril., 21, 29-40.
[4] Verschuur D.J., (1991): Surface-Related Multiple Elimination, an inversion approach, Delft University of Technology, Belanda.
[5] Image Response of seismic signals on the F-K domain theoretically, images obtained through the internet site: https://gpg.geosci.xyz/content/seismic/seismic_reflection_filtering.html. May 10 2018.

Integrated of Gravity and Magnetic Data to Modeling Structure Subsurface In Bora Geothermal Field Central Sulawesi

Rustan Efendi[1], Abdullah[1], Erfan Syamsuddin[2], M. Rusydi[1], Nirmayanti[1], Asep Sugianto[3], Wiwid Joni [3]

[1]Physics Department Tadulako University Palu Indonesia
[2]Geophysics Department Hasanuddin University Indonesia
[3]Centre for Geological Resources - Indonesian Geological Agency

Email: rstefendi67@gmail.com

Abstract. This field research is located in a geothermal area expressed by the existence of hot springs. A complex geological structure has shown that the area was in the Palu-Koro fault. The purpose of the study is to achieve a model of subsurface of the area using data collected over gravity data and magnetic data. The two Data were compared in order to validate the models. Furthermore 2-dimensional modeling was performed using two gravity and magnetic lines by means of geo-sys software from Geosoft. The results of the models have a bit difference response by means of the existence of residual indicator that expected as a geothermal reservoir due to fault activity

Keywords: *gravity, magnetic, fault and geothermal reservoir*

1. Introduction

Bora village, district of Sigi is an area that has a complex geological structure and is an active fault line of Koro-Palu so that this area is a zone of depression. These conditions make this area a prospect for a geothermal field. In this area also found hot springs with temperatures around 90.1 ° C, hot ground (100.6 ° C) and altered rocks. The rock structure of this area consists of Quaternary sediment, Tertiary plutonic intrusive rocks and Pre-Tertiary metamorphic rocks. The terrain is characterized by extensive faulting forming almost the N-S trending ridges and fault scarps [1].

The application of geomagnetic and gravity methods has been carried out by various researchers, for example [1] conducting exploration research using gravity and geomagnetic methods that model volcanic magma in the Brazilian region. [2] have conducted research by combining magnetic and gravity data to describe subsurface structures in sedimentary basins in the Sinai Peninsula region. [3] conducted a numerical study on the interpretation of magnetic and gravity anomalies in the Thuringian Basin, Germany. [4] using magnetic and gravity data to describe the characteristics of subsurface structures in the complex basin in Egyptian Central Sinai. [5] conducted a study using magnetic and gravity data to draw geological structures in Cairo Egypt

IOP Conf. Series: Earth and Environmental Science **279** (2019) 012034 doi:10.1088/1755-1315/279/1/012034

Particularly in the study area by [1] integrated measurement of geophysics (gravity, gravity, magneto telluric) and the results in the form of qualitative interpretations which indicate the existence of gravity anomalies that are positive and negative and geomagnetic measurements indicate the presence of Bora depression zone with negative anomaly characteristics while positive magnetic anomaly indicates intrusion. The magneto telluric results show that the Bora geothermal field has a variation of resistivity from 0 to 200 ohms to a depth of 1000 meters. In this study modeling is done using magnetic and gravity data that has been measured by [1]. The aim is to describe the structure model under the surface in the study area using integrated magnetic and gravity data.

2. Geology setting

Geothermal manifestations in Bora Village appear to be caused by the Palu-Koro Fault. This fault is found in the South Arm of Sulawesi trending north-west to south-southeast. This fault is called the Palu-Koro Fault because it passes through Palu City and Koro River. On land, this fault is characterized by the presence of a fault canyon that is flat at its base, with a width reaching 5 km around Palu, and its wall reaches a height of 1,500-2,000 m above the valley floor, while in the sea it is characterized by lineaments of the bath, namely basic slope straightness the sea is steep and ends at the Poso normal fault.

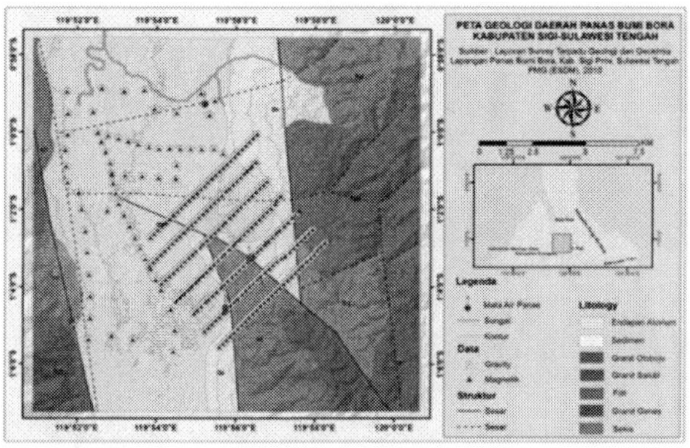

Figure 1. Geological maps and gravity and magnetic measurement points [1]

Stratigraphy of constituent rocks in the research location of Bora Village and its surroundings are sequentially from the youngest to the oldest rocks consisting of: Alluvium (Qal), Sediment (Qs), Oloboju Granite (Tgo), Granite salubi (Tgs), Filite (Kf), Granite genes (Trg) and Secis (Trs) as shown in Figure 1.

3. Geophysical Data

In this study the gravity and magnetic data used were obtained from the Bandung Geology Agency (Center for Geology Resource Indonesia Geology Agency). The results of the corrections made to the gravity data are Bouguer anomalies and corrections to the magnetic data are total magnetic intensity as shown in Figures (2) and (3) below.

IOP Conf. Series: Earth and Environmental Science **279** (2019) 012034 doi:10.1088/1755-1315/279/1/012034

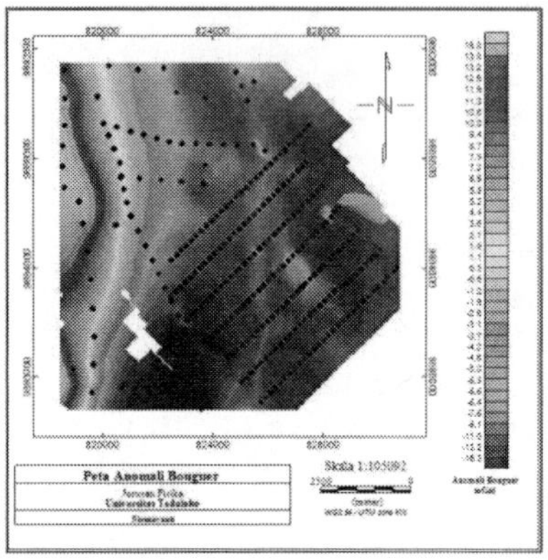

Figure 2. Map of Bouguer Anomaly distribution and gravity data measurement point

It can be seen in Figure (2) that the distribution of Bouguer anomalies in the study location is the response of the subsurface density distribution. The distribution of Bouguer anomalies below the surface varies between -17 mGal to 17 mGal. This anomaly is divided into low and high anomalies. Low anomalies as shown in dark blue to light green ranging from -17 mGal to 0.3 mGal. High anomalies as shown in yellow to pink range from 1.1 mGal to 17 mGal.

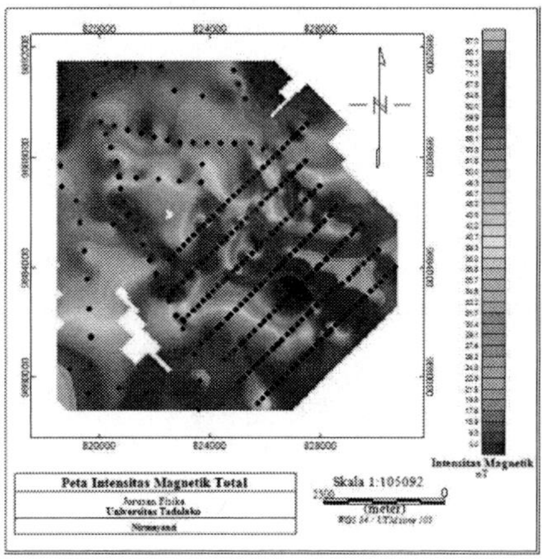

Figure 3. Map of total magnetic intensity distribution and gravity data measurement point

Figure 3 shows that the distribution of magnetic intensity values ranges from 4.0 nT to 90 nT. The value of magnetic intensity varies greatly due to the unevenness of the rock below the surface, resulting in a low and high magnetic intensity value. Relatively low magnetic intensity is in the north and magnetic intensity is high relative to the south. Based on magnetic intensity distribution (Figure 3), there is a low magnetic intensity whose magnetic intensity around it has a relatively higher magnetic intensity. This anomaly is located in the location of Bora hot water manifestations which can be seen on the geological map (Figure 1). This low anomaly is suspected because the heated rock will have a low level of magnetization and indicated manifestations that appear on the surface perpendicular to the reservoir.

4. Gravity and Magnetic modeling

In this study modeling was carried out using GM-SYS 2D software from Geosoft. The model is built by integrating gravity data and magnetic data to draw subsurface structures carrying surfaces. Furthermore, to obtain a subsurface picture of the study site, two trajectories were made on the gravity and magnetic contour maps (Figure 4). One track is made across the springs (BB track) with relatively North-South direction / N 45 ° E. The length of the track between B-point to B' point is 7.44 km. The

IOP Conf. Series: Earth and Environmental Science **279** (2019) 012034 doi:10.1088/1755-1315/279/1/012034

second track is the FF path 'relatively North-South direction N 45 ° E. The distance from point F to point F' is 8.33 km

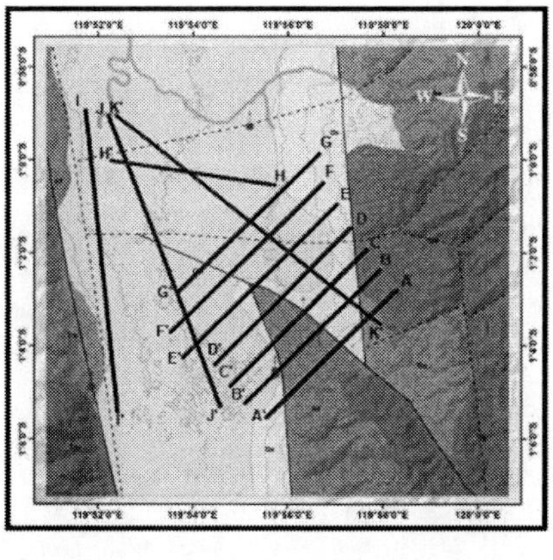

Figure 4. Profile of overpass trajectory with geological maps [1]

Modeling is done through several iterations until a match between the gravity and magnetic data model matches with the response of the measurement data with presentation errors of less than 5%. Two-dimensional model (2D) rock coating distribution built up to 5 km. The modeling results on the BB 'and DD' paths can be seen in Figures 5 and 6. The BB trajectory modeling results' obtained by 3 layers of rock distribution. The surface layer has a susceptibility (k) of 0.0002 SI and a density (ρ) of 1.960 kg/m3 with a thickness of 0.8 km extending from positions 5.79 to 7.44 km allegedly alluvium rocks. In the middle rock layer obtained susceptibility (k) of 0.0088 SI and density (ρ) of 2,500 kg/m3. This layer has a model that is suspected of sedimentary rock. The third layer shown in pink has susceptibility (k) of 0.0015 SI and density (ρ) of 2.770 kg/m3 extends to 7.44 km. These rocks are found in depths varying from 0.18 to 3.60 km with different layer thicknesses from 0.18 to 3.28 km thought to be phyllite rocks.

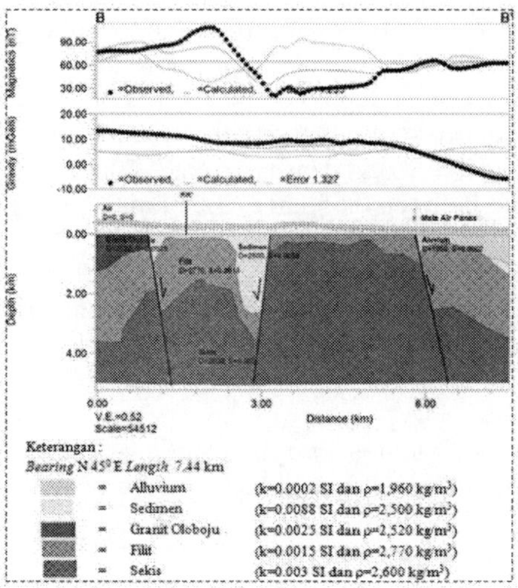

Figure 5. Model of the track BB 'subsurface structure.

IOP Conf. Series: Earth and Environmental Science **279** (2019) 012034 doi:10.1088/1755-1315/279/1/012034

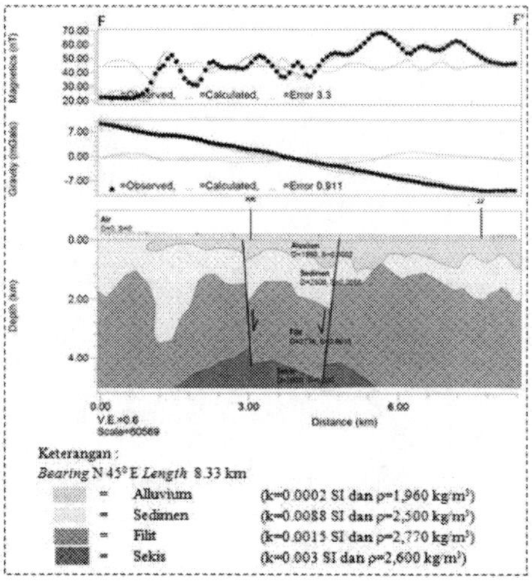

Figure 6. Model of track FF' subsurface structure

Figure 6 shows a rock distribution model that has varying density and susceptibility values. The first layer is shown in blue with the susceptibility value (k) of 0.0002 SI and the density (ρ) of 1.960 kg/m3. The rock has a depth of 0.12 to 0.98 km and extends to 8.33 km and is thought to be alluvium. The second layer is shown in yellow with susceptibility (k) of 0.0088 SI and density (ρ) of 2,500 kg/m3 with depths varying between 1.11 to 3.49 km extending to 8.33 km and thought to be sedimentary rocks. The third layer is indicated by color pink has a susceptibility (k) of 0.0015 SI and density (ρ) of 2.770 kg/m3. Extending to 8.33 km with a depth between 1.03 and 5 km is thought to be phyllite rock. The fourth layer shown in green is interpreted with a susceptibility of (k) 0.003 SI and a density of (ρ) 2.600 kg/m3, extending from 1.50 to 5.56 km at depths of up to 5 km and thicknesses varying from 0.24 to 1.34 km and thought to be schist.

The geological structure in the form of a fault appears on the two tracks that cut rock layers. There are 3 faults which are thought to be normal faults. The appearance of the water which is a manifestation of geothermal energy at the study site is likely to be caused by these faults. The two rock layer distribution models carrying surfaces show that geothermal reservoirs are interpreted in schist rocks with susceptibility values (k) of 0.003 SI and density (ρ) of 2.600 kg / m3 which averages from 0.48 to 7.61 km with thicknesses varying between 0.65 km to 3.55 km.

5. Conclusion

The purpose of this study is to model the subsurface structure by integrating gravity data and magnetic data simultaneously in the area of the geothermal field. Some important conclusions reached in this study are.

- Modeling of rock structures carrying surfaces using data gravity integration with magnetic data simultaneously can reduce ambiguity.
- The subsurface model obtained shows that there are several faults that cross the study area.
- Rocks that behave as reservoirs are metamorphic (schist) rocks that have undergone fractures.

References

[1] Alanda IDRAL, W.R.M., *Integrated Geophysical Studies of Palu-Koro Depression Zone, Indonesia: Implications for Geothermal Resources in Bora Central Sulawesi.* Proceedings World Geothermal Congress, 2015. Melbourne, Australia, 19-25 April 2015.

[2] Juliana G. Damaceno, D.L.d.C., Samir N. Valcácio, Zorano S. Souza, *Magnetic and gravity modeling of a Paleogene diabase plug in Northeast Brazil.* Journal of Applied Geophysics, 2017: p. 219-30.

[3] Selim, E.S.I., *The integration of gravity, magnetic and seismic data in delineating the sedimentary basins of northern Sinai and deducing their structural controls.* Journal of Asian Earth Sciences, 2015.

[4] Ilya Prutkin, P.V., Thomas Jahr, Florian Bleibinhaus, Pavel Nov´ak, and R. Tenzer, *Interpretation of gravity and magnetic data with geological constraints for 3D structure of the Thuringian Basin, Germany.* Journal of Applied Geophysics, 2016.

[5] Sultan Awad Sultan Araffa, M.E.-b., Abou Heleika M., Mahmoud Mekkawi, E. Ismail , Ahmed Khalil, Enas M. Abd EL-Razek, *Implementation of magnetic and gravity methods to delineate the subsurface structural features of the basement complex in central Sinai area, Egypt.* NRIAG Journal of Astronomy and Geophysics, 2018. **7**.

[6] Sultan Awad Sultan Araffa, A.S.H., Ashraf Khozium, and S.A.S. Amir M.S. Lala, Noha M. Hassan, *Delineating groundwater and subsurface structuresby using 2D resistivity, gravity and 3D magnetic data interpretation around Cairo–Belbies Desert road, Egypt.* NRIAG Journal of Astronomy and Geophysics, 2015.

[7] Wibowo, A.E.A., Hermawan, D., dan Widodo, S. , *Penyelidikan Terpadu Geologi Dan Geokimia Daerah Panas Bumi Bora Kabupaten Sigi Provinsi Sulawesi Tengah.* Prosiding Hasil Kegiatan Pusat Sumber Daya Geologi, 2010.

Derivative using complex variable conjugate approach for analytic signal of magnetic field anomaly due to 2D finite prism

A Manan[1*], L Hamimu[1] and R Chahyani[2]

[1] Department of Geophysical Engineering, Universitas Halu Oleo, Kampus Hijau Bumi Tridharma, Anduonohu, Kota Kendari 93232, Indonesia
[2] LPPM IAIN Kendari, Jln. Sultan Qaimuddin No. 17, Baruga, Kota Kendari 93116, Indonesia

[*]corresponding e-mail : amanan.geophysics@gmail.com

Abstract. The complex variable conjugate approach has been derived analyticaly for derivative computation. Computational results are then used in calculating the amplitude of analytic signal. It is the square root of the square of the total magnetic field anomaly derivative. The total magnetic fields are generated by the upper and lower parts of a 2D finite prism, and subtraction of both parts yields the magnetic field anomaly. While the approach is obtained by truncating the Taylor series expansion of the total magnetic field function in argument of the complex conjugate in the h^2 order-term. Truncating the series does not significantly affect the computational results. This is because when step-size h get smaller, and at $h \leq 10^{-2}$, then the errors due to the series truncation became 0 ($K_p \rightarrow 0$). For derivative computation, the approach has precision on the order of 10^{-17} to 10^{-12} towards analytical settlement. On the order of $h \leq 10^{-2}$, the approach is insensitive to the selection of step-size h for small numbers, so it can be done arbitrarily without any particular treatment or requires a complicated combination of numbers. The computational results of the analytic signal amplitude show that the positive and negative polarity on the magnetic profile is transformed into a positive profile only. This can facilitate the interpretation of actual magnetic data, especially in determining the causative source position of anomalies.

Keywords: Derivative, Complex Variable Conjugate, Analytic Signal, 2D Finite Prism

1. Introduction

In geophysics, magnetic method is one method that utilizes potential field data that is measured on the surface. This method is widely used in mineral exploration, preliminary surveys of oil and gas prospect areas, archeology and environmental surveys because it is relatively simple and easy to do, and is also relatively economically inexpensive.

In general, magnetic data show positive and negative polarities which are due to the dipolar nature of the anomalous source magnetic field [1], so that it often creates difficulties in data interpretation. The problem of magnetic data polarity can be solved by using analytic signal. Analytic signal transforms magnetic anomalies into positive anomalies only. Signal analytic is a complex function with the real and imaginary components being the horizontal and vertical derivatives respectively which can be proven that the imaginary component is a Hilbert transform of the real one [2,3]. So, it is the sum between the original signal and its Hilbert transform. Analytic signals for interpretation of magnetic data were first applied by [2], and are currently generally used in the form of amplitude. The amplitude is the absolute value of the analytic signal.

In this study, derivative in computational analytic signal utilizes conjugate of complex number or variable. This approach is relatively new because so far the completion of derivatives is not in the form of a conjugate. A conjugate of complex variable is a complex variable containing the imaginary part in the negative form. The use of complex variables for derivative estimate was first introduced by [4,5]. Furthermore [6] developed a simple form of Complex Variable Approximation by using the theory of previous researchers for functions that are real, continuous and analytic in complex plane.

Estimates of derivative using the complex variable conjugate approach, and hereinafter called the Complex Conjugate, were arranged in Matlab® programming language using simple codes. This approach was carried out by first truncating the Taylor series of the function in argument of the complex conjugate in the h^2 order-term. The codes that had been made were then used in computing the analytic signal of the magnetic field anomaly generated by a causative source in shape of 2D finite prism.

2. Analytic Signal

The analytic signal $a(x)$ of the function $f(x)$ is a complex variable defined by equation [3]

$$a(x) = f(x) - iH[f(x)] \tag{1}$$

where $H(x)$ is the Hilbert transform of $f(x)$. $H[f(x)]$ itself is

$$H[f(x)] = -\frac{1}{\pi}\int_{-\infty}^{\infty} \frac{f(x')}{x - x'} dx' \tag{2}$$

and for the inverse

$$f(x') = \frac{1}{\pi}\int_{-\infty}^{\infty} \frac{H[f(x)]}{x' - x} dx \tag{3}$$

Analytic signals have been widely applied, including by [7-10] for two-dimensional analytic signal. For a potential field M generated by a 2D source aligned parallel to the y-axis, and measured along the x-axis, then equation (1) can be written [2]

$$a(x,z) = \frac{\partial M}{\partial x} + i\frac{\partial M}{\partial z} \tag{4}$$

which satisfies the Cauchy-Riemann conditions [3]. Equations (1) and (4) apply to relationships

$$f(x) = \frac{\partial M}{\partial x} \quad \text{and} \quad H[f(x)] = -\frac{\partial M}{\partial z} \tag{5}$$

The absolute value of equation (4) yields the analytic signal amplitude as follows

$$|a(x,z)| = \sqrt{\left(\frac{\partial M}{\partial x}\right)^2 + \left(\frac{\partial M}{\partial z}\right)^2} \tag{6}$$

For magnetic fields generated by a causative object whose position does not change vertically or are at a certain height, then $\frac{\partial M}{\partial z} = 0$. Therefore, the amplitude in equation (6) can be written as

$$|a(x, z = \text{konstan})| = \sqrt{\left(\frac{\partial M}{\partial x}\right)^2} \tag{7}$$

In equation (7), the amplitude is only determined by a magnetic field M at the measurement points x on the surface.

3. Derivative Using Complex Conjugate

The approach to the derivative of a function, $f'(x)$, using complex conjugate is conducted by truncating the Taylor series expansion of the function $f(x)$ in argument of $z^*=x-ih$, that is $f(z^*)$. If f is a real and analytic function, then the series expansion can be expressed

$$f(z^*) = f(x-ih) = f(x) - ih\frac{f'(x)}{1!} - h^2\frac{f''(x)}{2!} + ih^3\frac{f'''(x)}{3!} + h^4\frac{f^4(x)}{4!} - ih^5\frac{f^5(x)}{5!} - \ldots \tag{8}$$

If what is taken in equation (8) is the imaginary part and multiplies it by $1/h$, then the complex conjugate approach for the first derivative is obtained as follows

$$f'(x) = -\frac{\mathrm{Im}[f(x-ih)]}{h} + h^2\frac{f'''(x)}{3!} - \dots \quad (9)$$

Whereas if taking only the real one of both sides, then yields the complex conjugate approach for the value of the function

$$f(x) = \mathrm{Re}[f(x-ih)] + h^2\frac{f''(x)}{2!} - \dots \quad (10)$$

The h^2 and higher order terms can be ignored since the step-size h can be selected arbitrarily for sufficiently small numbers, so that the completion of complex conjugate in equation (9) and (10) becomes

$$f'(x) = -\frac{\mathrm{Im}[f(x-ih)]}{h} + E_T \quad (11)$$

$$f(x) = \mathrm{Re}[f(x-ih)] + E_T \quad (12)$$

with E_T is the truncation error. Equations (11) and (12) are analogous to the original form of the complex variable approach that has been formulated by [6].

Function in argument of the complex conjugate $f(z^*)$ can also be derived from function having argument of the complex variable $f(z)$. Equation (8) is rearranged into

$$f(z^*) = \left\{f(x) - h^2\frac{f''(x)}{2!} + h^4\frac{f^4(x)}{4!} - \dots\right\} - i\left\{h\frac{f'(x)}{1!} - h^3\frac{f'''(x)}{3!} + \dots\right\}$$

$$f(z^*) = \mathrm{Re}\{f(z)\} - i\,\mathrm{Im}\{f(z)\} \quad (13)$$

which shows the relationship between the complex conjugate (z^*) and the complex variable (z). The complex variable approach has been widely used for estimates of the derivative, among which can be seen in [4,5,6,11].

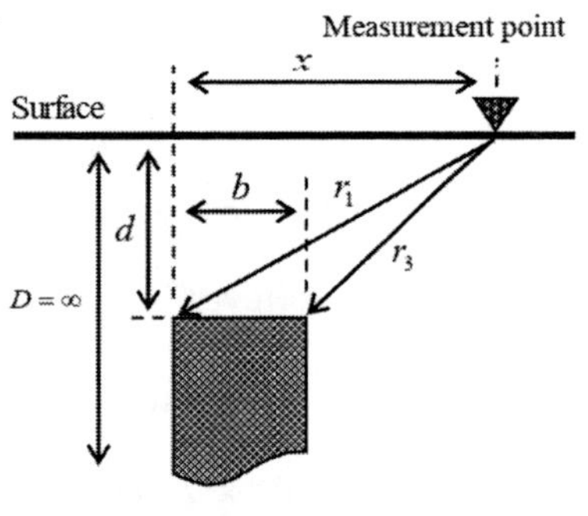

Figure 1. $2\frac{1}{2}$D vertical prism with infinite cross section

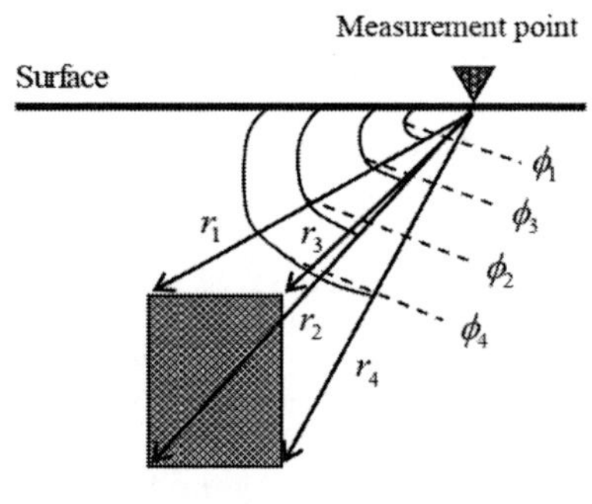

Figure 2. 2D vertical prism with finite cross section

4. Causative Object of Magnetic Field

Model $2\frac{1}{2}$D for the causative source of a total magnetic field in the form of a vertical prism with an infinite cross section is shown in Figure 1. The magnitude of the causative total magnetic field is [12]

$$F = kF_e\Big\langle \sin 2I \sin\beta\Big[\ln\Big(\big(r_1^2+L^2\big)^{\frac{1}{2}}+L\Big) - \ln\Big(\big(r_1^2+L^2\big)^{\frac{1}{2}}-L\Big) + \ln\Big(\big(r_3^2+L^2\big)^{\frac{1}{2}}-L\Big) - \ln\Big(\big(r_3^2+L^2\big)^{\frac{1}{2}}+L\Big)\Big]$$
$$-\big(\cos^2 I\sin\beta - \sin^2 I\big)\Big[\tan^{-1}\Big(\frac{L}{x}\Big) - \tan^{-1}\Big(\frac{L}{x-b}\Big) - \tan^{-1}\Big(\frac{Ld}{x\big(r_1^2+L^2\big)^{\frac{1}{2}}}\Big) + \tan^{-1}\Big(\frac{Ld}{(x-b)\big(r_3^2+L^2\big)^{\frac{1}{2}}}\Big)\Big]\Big\rangle \quad (14)$$

Whereas for prism with finite cross section such as Figure 2

$$F = 2kF_e\left[\ln\left(r_2 r_3 / r_4 r_1\right)\sin(2I)\sin\beta + \left(\cos^2 I\sin^2\beta - \sin^2 I\right)\left(\phi_1 - \phi_2 - \phi_3 + \phi_4\right)\right] \quad (15)$$

where F is the measured total field on the surface, k the magnetic susceptibility, F_e the intensity of the earth's magnetic field, I the inclination of the earth's magnetic field, β the prism strike angle relative to the magnetic north, b the prism width, d the prism depth, L the half-strike length of the prism, r_i and ϕ_i with i=1,2,3 and 4 are distances and angles that establish the prism geometry.

In general, a magnetic object can be considered as 2D object when its strike length is at least 10 times its width ($2L \geq 10b$) [12], and the strike perpendicular to the profile [13]. Therefore, a 2D prism model with a finite cross-section can be approximated from $2\frac{1}{2}$D model by following the definition [13], and assuming that the prism composed of many very small square prisms.

A single square prism is illustrated as Figure 2. This prism generates a magnetic field anomaly that can be calculated by first calculating each of the total magnetic fields caused by the Upper Prism and the Lower Prism by using equation (14), then subtracting the results.

Determination of distance r and angle ϕ in Figure 2 is done by considering equation (15). The technique of subtracting the total magnetic field caused by the Upper with the Lower of the Prism was proposed and applied first by [14], and was also further applied by [13] for modeling and inversion of the magnetic data.

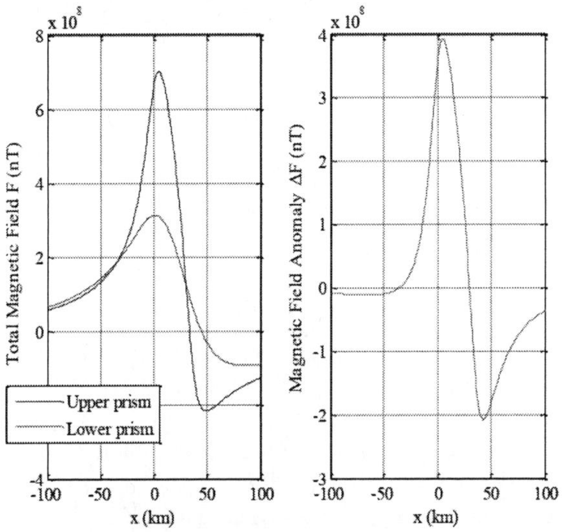

Figure 3. Profiles of (a) the total magnetic field due to the Upper and Lower parts of the Prism, and (b) the magnetic field anomaly

5. Synthetic Magnetic Field Source

The total magnetic field measured on the surface at the measurement points ($-100 \leq x \leq 100$) km with spaces 0.5 km was designed because it was generated by a causative object in shape of 2D finite prism using equation (14) and following [12-14]. The used parameter values are F_e=45.000 nT, k=7x10^3 SI (chromite), I=70°, β=30°, b=35 km, L=8b km, and d=10 m for Upper Prism and 30 m for Lower Prism. Whereas the magnitude of r_i and ϕ_i can be seen in Table 1 in the form of parametric equations with the dip angle ξ=90°.

Table 1. Parametric equations of distance and angle

Index *i*	**Distance r_i**	**Angle ϕ_i**
1	$d^2 + (x + d\cot\xi)^2$	$\tan^{-1}\{d/(x + d\cot\xi)\}$
2	$D^2 + (x + D\cot\xi)^2$	$\tan^{-1}\{D/(x + D\cot\xi)\}$
3	$d^2 + (x + d\cot\xi - b)^2$	$\tan^{-1}\{d/(x + d\cot\xi - b)\}$
4	$D^2 + (x + D\cot\xi - b)^2$	$\tan^{-1}\{D/(x + D\cot\xi - b)\}$

6. Result and Discussion

6.1. Magnetic field anomaly due to a 2D finite prism

Figure 3a displays the total magnetic field profile F caused by the Upper and Lower parts of the Prism which is calculated analyticaly. While the magnetic field anomaly ΔF which is calculated by subtracting the total magnetic fields of the Upper and Lower parts of the Prism shown in Figure 3b. Magnetic profile in the direction of N-S with strike in the direction of E 30° W.

The profiles in Figure 3 shows the presence of positive and negative polarities. These polarities are present because of the dipolar nature of the magnetic field of the 2D finite prism. In case of the field data, polarity can cause difficulties in interpreting magnetic data, especially in determining the location and depth of the source.

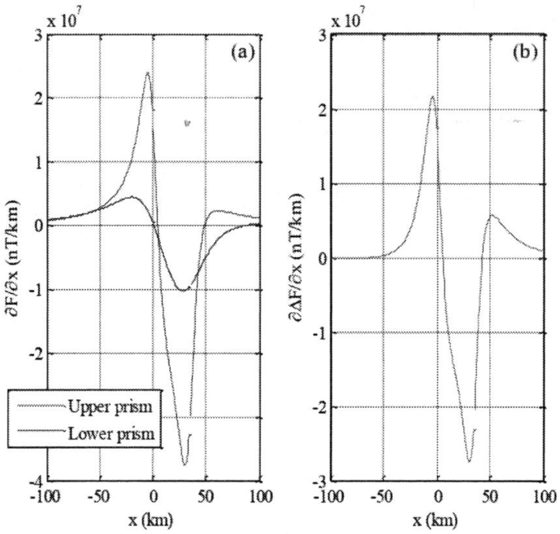

Figure 4. Analytical derivative profiles of (a) the total magnetic field due to the Upper and Lower parts of the prism, and (b) the magnetic field anomaly

6.2. Analytical derivative of magnetic field anomaly

The results of the derivative of the total magnetic field due to the Upper and Lower parts of the Prism are analyticaly shown in Figure 4a. As for the derivative of the magnetic field anomaly shown in Figure 4b.

Profile resulting from the total magnetic field derivatives $\partial F/\partial x$ in Figure 4a for each of the Upper and Lower parts of the Prism still shows polarity like the total field profile F. Similarly for the magnetic field anomaly derivative $\partial \Delta F/\partial x$ in the Figure 4b also shows the same thing as ΔF.

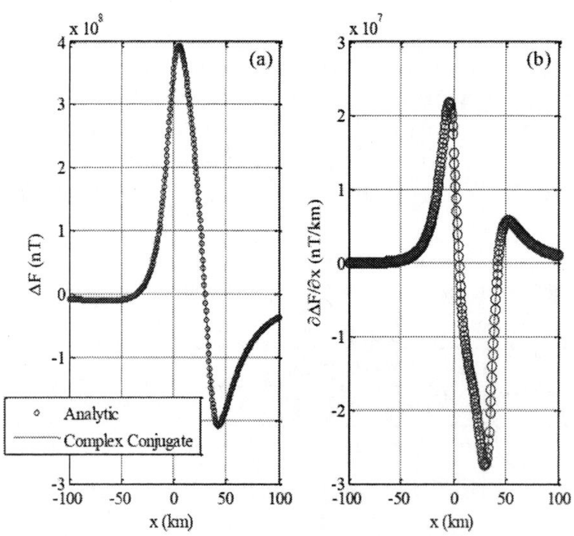

Figure 5. The complex Conjugate of (a) the magnetic field anomaly, and (b) The derivative of the magnetic field anomaly

6.3. The complex conjugate

Derivative of the magnetic field anomaly using the complex conjugate is calculated by deriving subtraction results of the total magnetic field caused by the Upper and the Lower parts of the Prism (ΔF) with respect to measurement points (x) where each total magnetic field is also obtained using the complex conjugate. Furthermore compared to the analytical solution obtained in Section 6.2, and its results can be seen in Figure 5. The complex conjugate approach (red line) and the analytical solution (blue line) are almost exact for all measurement points ($-100 \leq x \leq 100$). The red line coincides with the blue line. To find out the precision of computational works, relative error of the complex conjugate E_r is calculated towards the analytical solution using formula

$$E_r = \frac{\Delta F_{\text{complex conjugate}} - \Delta F_{\text{analytic}}}{\Delta F_{\text{analytic}}} \tag{16}$$

for the magnetic field anomaly, and

$$E_r = \frac{\partial \Delta F_{\text{complex conjugate}} / \partial x - \partial \Delta F_{\text{analytic}} / \partial x}{\partial \Delta F_{\text{analytic}} / \partial x} \tag{17}$$

for approach to the derivative. Figure 6 shows the E_r values for both approaches in equations (16) and (17).

Based on Figure 6, the use of the complex conjugate in estimating a function and derivative exhibits highly accuracy. Accuracy can be identified from a very small Relative Error value. Figure 6a displays E_r for estimates of the magnetic field anomaly function ΔF, which it is between $7.163413336035101x10^{-20}$ and $1.185821058201096x10^{-13}$. While Figure 6b displays E_r for the approach to the derivative of the magnetic field anomaly $\partial \Delta F / \partial x$ which E_r is obtained between $3.159607284251312x10^{-17}$ and $7.079603533496899x10^{-12}$.

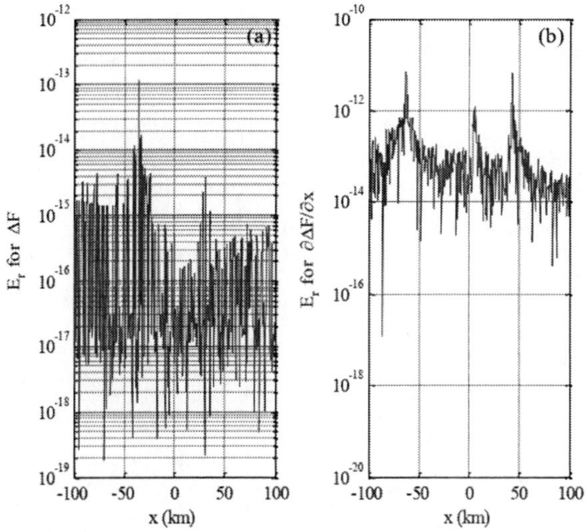

Figure 6. Relative errors of the approach of the complex conjugate towards the analytical solution for (a) the magnetic field anomaly, and (b) the magnetic field anomaly derivative

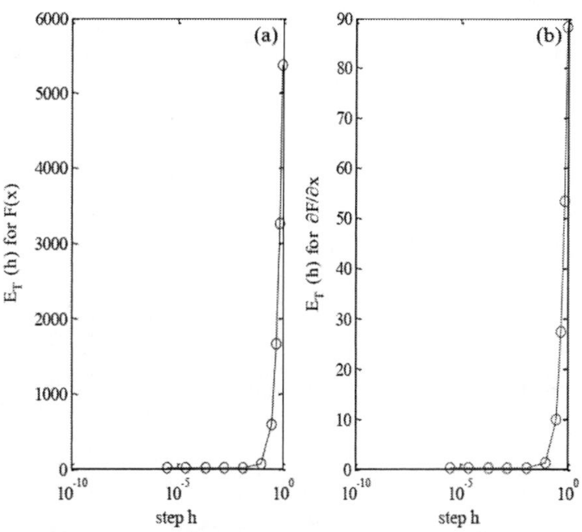

Figure 7. Truncation errors as a function of step-size h (x = 100 km) (a) the total magnetic field function $F(x)$, and (b) the total magnetic field derivative $\partial F/\partial x$

If the values E_r of the approach of the complex conjugate in Figure 6 are compared with machine epsilon which is $2.220446049250313x10^{-16}$, then it can be said that the approach for the magnetic field anomaly and its derivative has high precision. This is because the truncation of the series in the h^2 order-term in equation (9) and (10) does not have any significant effect on the computational results. The magnitude of the error due to the series truncation in equation (10) can be written

$$E_T = h^2 \frac{f''(x)}{2!} = \frac{h^2}{2!} \frac{\partial^2 F_{analytic}}{\partial x^2} \quad (18)$$

for estimates of the total magnetic field function, and

$$E_T = h^2 \frac{f'''(x)}{3!} = \frac{h^2}{3!} \frac{\partial^3 F_{analytic}}{\partial x^3} \quad (19)$$

for approach to the derivative in equation (9). Both of these equations are similar to the equations that had been used by [11] to tested how precise the codes were made in Fortran® for derivative computation.

Truncation of the Taylor series of the total magnetic field function $F(x)$ in argument of the complex conjugate parsed like equation (8) in the h^2 order-term does not have any significant effect on the computational results. Figure 7 shows the errors due to the series truncation is getting smaller with increasing the small step-size h ($10^0 \rightarrow 10^{-10}$). At $h \leq 10^{-2}$, the E_T value has become and is stable at 0. Therefore, to reduce or eliminate the truncation effect of the complex conjugate series on the computational results, it can be done by minimizing the step-size h. The choice of a small h will result in minimal truncation errors. The choice of h can be done in such a manner without requiring certain treatment.

In its implementation, derivative computation using the complex conjugate approach does not present a possible subtraction of two possible numbers which are the same as seen in the mathematical form of equation (9), thus eliminating the possibility of cancelation errors. Its implementation in the Matlab® programming language is relatively easy to do just by creating simple codes.

6.4. Analytic signal amplitude

The amplitude of the analytic signal from the magnetic field anomaly in Figure 8b is made by implementing equation (7) in the Matlab® language codes with derivative using the complex conjugate resulted in Figure 5b.

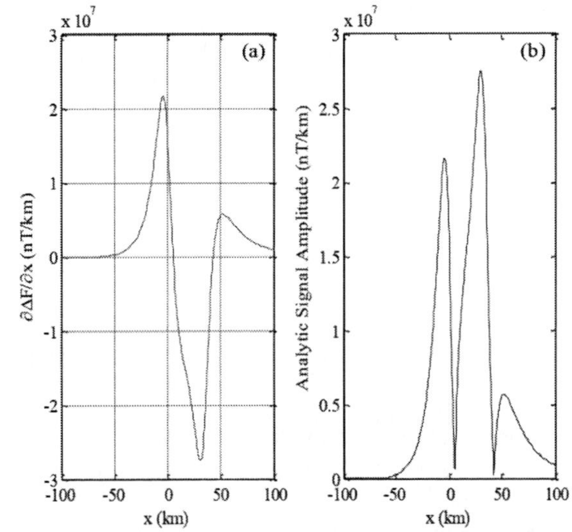

Figure 8. Profiles of (a) The magnetic field anomaly derivative using the complex conjugate, and (b) Analytic Signal Amplitude

The computational result in Figure 8b shows that the Analytic Signal Amplitude transforms the profile of the magnetic field anomaly derivative with positive and negative polarities being positive anomaly without further data processing. This is because it does not depend on other parameters, and only on the magnetic field anomaly or 2D finite prism total magnetic field is measured on the surface ($-100 \leq x \leq 100$ km). In Nabighian's paper [2] explained that the amplitude was invariant with respect to object magnetization vector direction and local geomagnetic field. Besides that, it's also independent to induced field vector [7]. Therefore, in interpretation of actual magnetic data, this transformation will be very useful and makes it easy especially to determine the location of the causative source of anomalous objects.

7. Conclusion

The approach of the complex conjugate to estimating derivatives is extremely easy to implement in the Matlab® programming language by only designing simple codes, easy to run and highly accurate. Accurate precision is on the order of 10^{-17} to 10^{-12} for derivative computation of the the total magnetic field or magnetic field anomaly. The effect of the Taylor series truncation of the total magnetic field function mathematically in the complex conjugate argument in the h^2 order-term does not affect the computational results because for a small step-size $h \leq 10^{-2}$, then the truncation errors become 0. Because of the complex conjugate is insensitive to the selection of h of the complex conjugate argument for $\leq 10^{-2}$, then the selection of h for small numbers can be done arbitrarily without requiring certain treatments. Besides that, the complex conjugate equations does not allow the presence of a subtraction in the same two numbers, so that it is free from the cancelation operation. The resulting derivatives are then used in computing the Analytic Signal Amplitude of the magnetic field anomaly of the 2D finite prism-shaped object. Computational results show the loss of negative polarities from the magnetic profiles. So, the amplitude is able to transform negative and positive polarities on magnetic data into positive polarity only without further data processing.

References

[1] Young C T 2004 Basic magnetic processing and display in MATLAB *http://pages.mtu.edu/~ctyoung/somepdfs/sageep2004.pdf* accessed on August 02, 2018
[2] Nabighian M N 1972 The analytic signal of two-dimensional magnetic bodies with polygonal cross-section: its properties and use for automated anomaly interpretation *Geophysics* **37** (3) 507-517
[3] Blakely R J 1995 *Potential Theory in Gravity and Magnetic Applications* (New York: Cambridge University Press) pp 350-352
[4] Lyness J N and Moler C B 1967 Numerical differentiation of analytic functions *SIAM J NumAnal* **4** 202-210
[5] Lyness J N 1967 Numerical algorithms based on the theory of complex variables *Proc. of the 22nd Nat. Conf. ACM* **1** 124-134
[6] Squire W and Trapp G 1998 Using complex variables to estimate derivatives of real functions *SIAM Rev* **40** 110-112
[7] Bournas N and Baker H A 2001 Interpretation of magnetic anomalies using the horizontal gradient analytic signal *ANNALI DI GEOFISICA* **44** (3) 505-526
[8] Pilkington M and Keating P 2006 Short Note: The relationship between local wavenumber and analytic signal in magnetic interpretation *Geophysics* **71** (1) L1-L3
[9] Tuma S L and Mendonça C A 2007 Stepped inversion of magnetic data *Geophysics* **73** (3) L21-L30
[10] Ansari A H and Alamdar K 2009 Reduction to the pole of magnetic anomalies using analytic signal *World Applied Sciences Journal* **7** (4) 405-409
[11] Martins J R R A Sturdza P and Alonso J 2003 The complex derivative approximation *ACM Trans. Math. Softw.* **29** 245-262
[12] Telford W M Gedart L P and Sheriff R E 2001 *Applied Geophysics 2nd Ed* (New York: Cambridge University Press) pp 92-95
[13] Stocco S Godi A and Sambuelli L 2009 Modelling and compact inversion of magnetic data: A Matlab code *Computers & Geosciences* **35** 2111-2118
[14] Bhattacharyya B K 1964 Magnetic anomalies due to prism-shaped bodies with arbitrary polarization *Geophysics* **29** (4) 517-531

IOP Conf. Series: Earth and Environmental Science **279** (2019) 012036 doi:10.1088/1755-1315/279/1/012036

Two-Dimensional Inversion Magnetotelluric Investigation of the Pariangan Geothermal Field, Indonesia

Wahyuni[1*], Sabrianto Aswad[1], Maria[1], Wiwid Joni[2], Jamaluddin[3]

[1]Geophysics Departement, Faculty Of Mathematics and Natural Sciences, Hasanuddin University, Perintis Kemerdekaan Street, Km. 10, Makassar City, South Sulawesi Province, Pos Code 90245.

[2] Resource Center of Mineral, Geothermal and Coal, Soekarno Hatta Street, Number 444, Pasirluyu, Regol, Bandung City, West Java, Pos Code 40254, Indonesia.

[3] Geological Engineering Departement, School Of Geosciences, China University of Petroleum, Qingdao, China

*wahyuni.dulen10@gmail.com

Abstract. Pariangan geothermal field is located in Tanah Datar District, West Sumatera Province, Indonesia. The structure developed in this region is influenced by a large fault that extends through Sumatera from Banda Aceh to the Gulf Semangko in Lampung and also the pattern of radial faults that follow the development of Complex Marapi Volcanism. This study is aimed at delineating the shallow as well as deeper conductive anomalies related to the geothermal system in the Pariangan area. Dimensionality analysis showed a two-dimensional (2D) behavior for the data, and therefore 2D inversion was performed. The result of magnetotelluric (MT) data processing indicates a low resistivity (<15 Ω.m) distribution at depths of 1300-3000 meter, which is interpreted as the caprock. Reservoir rock is indicated by the medium resistivity (15-93 Ω.m) at depths of 1200-3200 m. Below the reservoir the deep reaching conductive anomaly is interpreted as circulating hot fluids which heat up the fluids in the reservoir due to the subduction process. Type of alteration developed in research area are dominated by argillic type characterized by the presence of illite, kaolinite, montmorillonite, dickite, and halloysite; propylitic type characterized by the presence of chlorite, epidote and, quartz; and phyllic type characterized by the presence of sericite, quartz, and chlorite. Basement rock zone is indicated by the high resistivity (93-1000 Ω.m) at the depths of 1500-6000 m. MT in this study shows a powerful geothermal exploration tools because of its sensitivity to image hot saline fluids as zones of high conductivity.
Keyword: Basement rock, Caprock, Complex Marapi Volcanism, Magnetotelluric, Reservoir Rock, Two-dimensional.

1. Introduction

Tectonic processes that occur in Sumatera are controlled by three major fault systems. The first fault system is thrust fault (TF) due to subduction zone, the second fault system is Sumatera fault zone (SFZ) and the third fault system is Mentawai fault (MF). SFZ is a sliding fault with a rightward movement associated with a series of valleys along the row of hills. SFZ are characterized by volcanism and the flow of high-level hydrothermal solutions associated with extension and compressional regimes. And this fault system is interpreted as the magmatic path of the Sumatera tectonic. This has led to the emergence of geothermal manifestations along the path of the SFZ [1]. Geothermal of Pariangan which is administratively located in the Tanah Datar district, West Sumatera Province (Figure 1) was located in the SFZ and active volcanoes, which range from Sabang in the north to Semangko Bay in the south [2].

The parameters measured of magnetotelluric (MT) from natural elecromagnetic (EM) signals are magnetic field and electric field in orthogonal directions at the Earth's surface. While the parameters analyzed are apparent resistivity and phase.

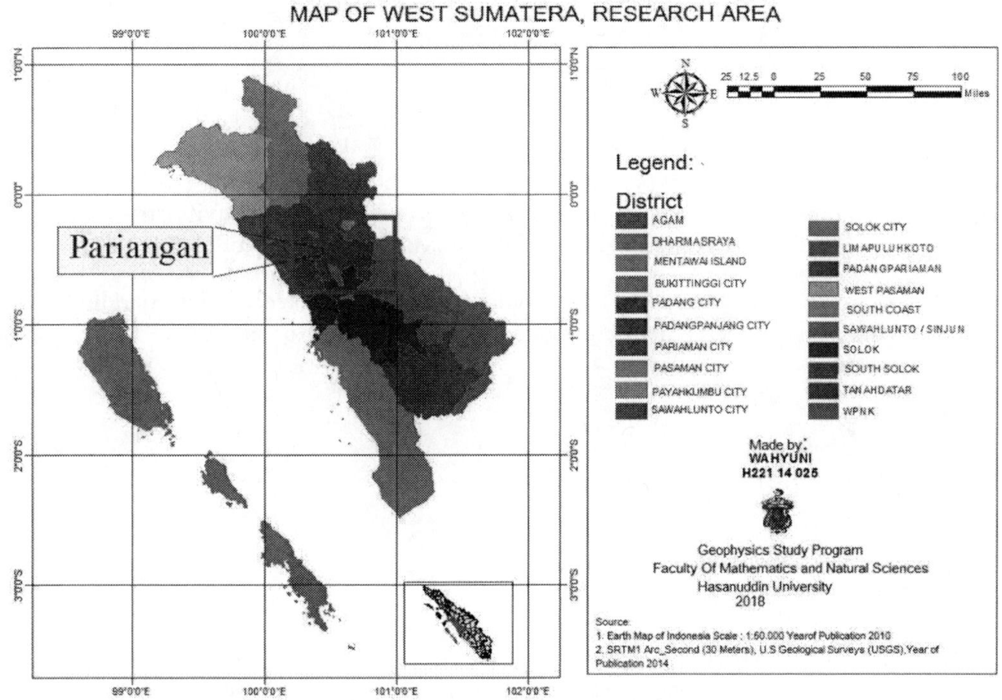

Figure 1. Map of West Sumatera, Pariangan is area in this study indicated by red square

2. Geology Structure

The geological structure that developed in this area was greatly influenced by the tectonic activitiy of the SFZ and also the radial fault pattern following the development of the complex Volcanism of Marapi. The main pattern of fractures and fault is northwest (NW)-southeast (SE). The second order is the alignment of the southwest (SW)-northeast (NE) and the main fault antitetic. Third order in the form of fractures which are directed almost North and South [2].

3. Data Analysis and Methods

MT method is used to investigate subsurface based on rock conduciviy value depends on the amount of frequency size [3]. The EM wave fluctuations range frequencies of 10^{-3}-10^{5} Hertz depending on their origin [4]. This can be determined based on how much penetration of EM wave penetrates the subsurface which is formulated as follows [5]:

$$p(T) = \left(\frac{T}{\pi\bar{\sigma}\mu}\right)^{1/2} \quad (1)$$

Equation can be approximated as:

$$p(T) = 500\sqrt{\rho_a T} \quad (2)$$

with ρ_a a is apparent resistivity, $\bar{\sigma}$ is rock conductivity, and μ is permeability and $p(T)$ is penetration EM wave.

The impedance (Z) is a measure characteristic that are analyzed as a function of the MT responses from EM wave which can be written in the following equation [5]:

$$\begin{pmatrix} E_x \\ E_y \end{pmatrix} = \begin{pmatrix} Z_{xx} & Z_{xy} \\ Z_{yx} & Z_{yy} \end{pmatrix} \begin{pmatrix} H_x \\ H_y \end{pmatrix} \quad (3)$$

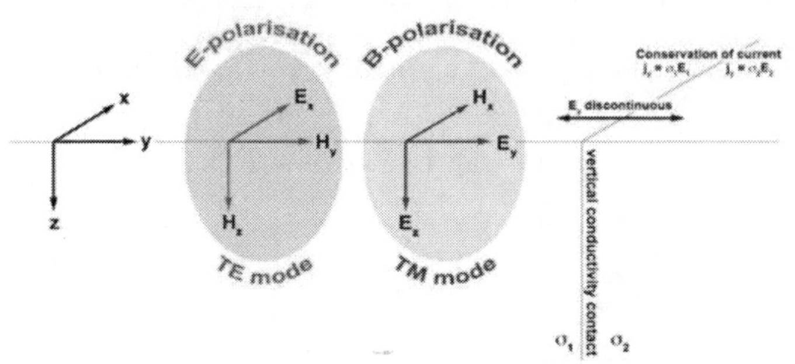

Figure 2. Polarizations in MT with two zones of different vertical conductivity structure by Tranverse Electric (TE mode) and Tranverse Magnetic (TM mode) [6].

Based on Figure 2, TE mode is a EM fiel component which is electric field parallel to the electric strike depends only on magnetic field components perpendicular to the strike. The equation can be written as follows [6]:

$$\frac{\partial Ex}{\partial y} = \frac{\partial Bz}{\partial t} = i\omega \mathbf{B}_z \quad (4)$$

$$\frac{\partial E_x}{\partial z} = \frac{\partial By}{\partial t} = -i\omega \mathbf{B}_y \quad (5)$$

$$\frac{\partial B_z}{\partial y} - \frac{\partial By}{\partial z} = \mu\sigma \boldsymbol{E}_x \quad (6)$$

TM mode is a EM field component which is electric field perpendicular to strike depends only on magnetic fields parallel to the electric strike. The equation can be writen as follows [6]:

$$\frac{\partial Ex}{\partial y} = \frac{\partial Bz}{\partial t} = i\omega \mathbf{B}_z \quad (7)$$

$$\frac{\partial E_x}{\partial z} = \frac{\partial By}{\partial t} = -i\omega \mathbf{B}_y \quad (8)$$

$$\frac{\partial B_z}{\partial y} - \frac{\partial By}{\partial z} = \mu\sigma \boldsymbol{E}_x \quad (9)$$

Root mean square error (nRMS) is obtained static shift correction and two-dimensional inversion process from the response of the MT model with $\tilde{r}$ is the residual between calculated and observed data, can be writen [8]:

$$nRMS = \sqrt{N^{-1}\sum_{i=1}^{N}\tilde{r}_i^2} \quad (10)$$

In two-dimensional inversion process, general form of relationship between data and model parameters are searched can be written with [9]:

$$\boldsymbol{d} = F(\boldsymbol{m}) + e \quad (11)$$

With **F** is forward function which theoretically predicts d (data) for give **m** (model), **λ** is regularization parameter and **e** is error function. A suitable model is usually defined as one minimizing the objective function $\boldsymbol{\Psi(m)}$ measured by the difference between **d** and **F(m)**. Damped least square requires specifications regularization paramaters. This parameters minimize data mismatch respect forward model and stability functions **Ω**(m).

$$\boldsymbol{\Psi(m)} = \left(\boldsymbol{d} - \boldsymbol{F(m)}\right)^{\boldsymbol{T}} \boldsymbol{V}^{-1}\left(\boldsymbol{d} - \boldsymbol{F(m)}\right) + +\boldsymbol{\lambda m^T L^T L m} \quad (12)$$

With **V** is vector covariance matrix, **L** is linear operator (smoothness).

Data measured in the Geothermal Field which are directed SW-NE were 9 measurements points, cutting (perpendicular) to the regional structure of the study area with distances between measuring points around 1200-1500 m [2]. Processing of MT data is done from raw data in the form of time series. The initial principle processing the data is to change the time domain data into frequency domain by FFT (Fast Fourier Transform) method. The transformation process to frequency domain is done because physical parameters such as impedance, apparent resistivity, and phase are frequency functions. After that, robust processing to damped the noise and make the smooth curves by editing process. We can calculate the value of impedance, apparent resistivity and phase. Then a static process and processing two-dimensional inversion modelling for the subsurface model of the measurement results (calculate and observation data).

4. Results and Discussion

The editing process aims to eliminate the noise signal and get the smooth curves. Curves the match the trendline during observation will describe the subsurface conditions. This process result by FFT in robust processing form high frequency (.MTH) and low frequency (.MTL) shown in Figure 3.

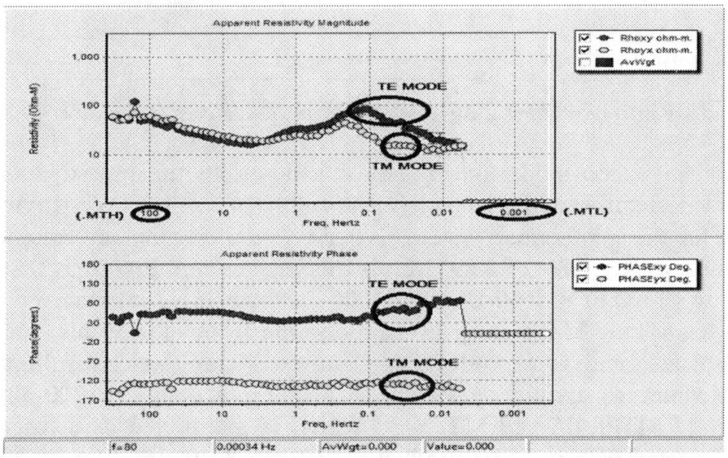

Figure 3. Cross Power Selection of measurement points MTPR-33 Line 4

Static Shift Correction can be done with the smoothing curve process. The data entered is the editing result data apparent resistivity and phase respect to frequency that is already in the format *.edi. This correction aims to obtain smooth data apparent resistivity and phase respect to period and damped the amount of noise due to influence of subsurface topography. Based in the Figure 4, showed TE+TM calculate dan observation data coincide with each other indicating the value of noise is getting smaller with rms 1.4374.

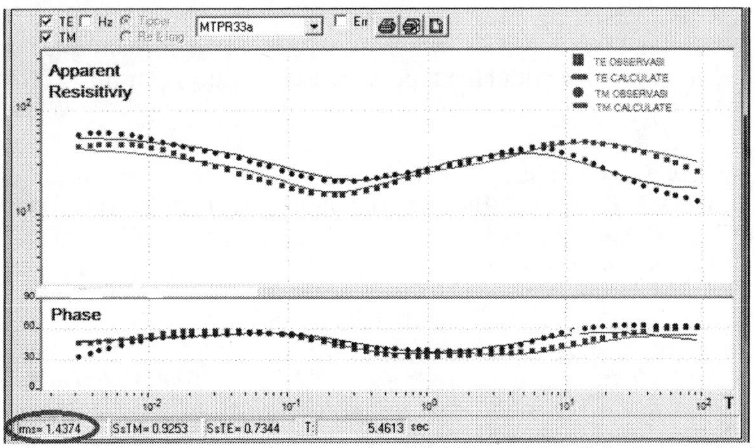

Figure 4. Comparison Static Correction Results Of TE and TM Calculate and Observation of measurement points MTPR-33 on Track 4

IOP Conf. Series: Earth and Environmental Science **279** (2019) 012036 doi:10.1088/1755-1315/279/1/012036

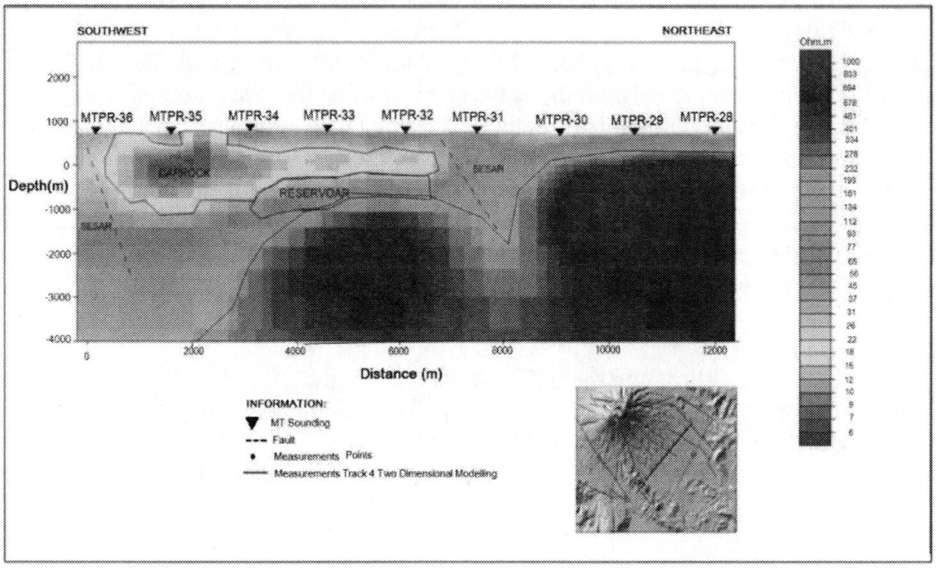

Figure 5. Modelling Two-Dimensional inversion Track 4

Two-dimensional inversion model is a model that describes the investigation of the subsurface structure conductivies laterally and horizontally. The color represents the resistivity value and the type of rock concerned in the geothermal system. After two-dimensional inversion modeling obtained results from inversion with a value of 2.0951 error rms as shown in Figure 5. Distribution of Track 4, shows low resistivity <15 Ω.m at depth of 1300-3000 meters below the measurement point MTPR-32, MTPR-33, MTPR-34, MTPR-35 indicated as Caprock zone. The value of this type of low resistivity is estimated to be associated with rock products of Complex Marapi Volcanism in the form of lava and pyroclastic falls. Medium resistivity ranges from 15-112 Ω.m at depth of 1200-3200 meters below the measurement point of MTPR-35 and MTPR-34 which is interpreted as a reservoir zone which is estimated to be associated with volcanic product rocks from Complex Marapi Active Volcano. High resistivity ranges from 112-1000 Ω.m at depth of 1500-6000 meters which is interpreted Basement rock. High resistivity is thought to be associated with metamorphic rocks.

5. Conclusions

Low resistivity values <22 Ω.m are interpreted as caprock zones (red to yellow). The caprock zone has a low resistivity type (conductive) because in this zone there are impermeable rocks, so that the fluid is trapped in this stamp both physically and minerals. Medium resistivity ranging from 22-112 Ω.m (green color) is interpreted as a reservoir zone. The reservoir zone has permeable rock because the fluid in this zone evaporates as a result of its position closer to the heat source. While the blue color is interpreted as impermeable zone which is suspected as metamorphic rock which acts as basement rock from the study area with high type of resistivity ranging from 112-1000 Ω.m.

References

[1] Sieh K and Natawidjaja D 2000 *J Geophysics Res* **105** 28.295 – 28.326.
[2] Resource Center of Geology 2014 *Pariangan Geothermal Survey Report* West Sumatera (Internal Report) Not Published.
[3] Cagniard L 1953 *Geophysics* **18** 605-635.
[4] Vozoff K 1991 *The Magnetotelluric Method in Electromagnetic methods in Applied Geophysics* Vol 2 Application, M N Nabighian (ed) SEG Publishing.
[5] Simpson F and Bahr K 2005 *Practical Magnettelluric* Cambridge University Press: London.
[6] Beyene A T 2012 *Multidimensional Inversion of MT Data from Alid Geothermal Area, Eritrea; Comparison with Geological Structures and Identification of A Geothermal Reservoir* Thesis Magister Scientiarium degree in Geophysics. University of Iceland.

[7] Chave, Alan D, Alan G J 2012 *The Magnetotelluric Method* Cambridge University Press: Cambridge.
[8] Robert Delhaye, Volker Rath, Alan G Jones, Mark R Muller, Derek Reay 2017 *Correcting for Static Shift of Magnetotelluric data with Airborne Electromagnetic Measurements: A Case Study from Rathlin Basin, Northern Ireland. Ireland* : Geophysics Section, School of Cosmic Physics, Dublin Institute for Advanced Studies (DIAS), 5 Merrion Square, Dublin 2.
[9] Grandis H 2009 *Introduction to Geophysical Inversion Modelling* Bandung Institute of Technology: Bandung.

Acknowledgments

First, praises and thanks for the presence of the God, the Almighty, the Lord of the Worlds for the abundance of His mercy and guidance, the author can be complete the research successfully. Gratefully acknowledge the Resources Center for Mineral, Geothermal and Coal which is located in Bandung, West Java for the opportunity given to publish this paper at once to Mr. Wiwid Joni, S.Si, MT as mentor. Especially to Mr. Sabrianto Aswad, S.Si, MT and Mrs. Dra. Maria, M.Si as a supervisor provide knowledge and motivation to the author. And most importantly, both parents the writer who always gives prayers and endless motivation for the author.

Analysis of Regional Anomaly on Magnetic Data Using the Upward Continuation Method

Ilapadila[1], Bambang Harimei[2], Maria[3],
[1,2,3)]Solid Geophysic Laboratory, Geophysics Program Study, Faculty of Mathematic and Natural Science, Hasanuddin University, Indonesia

Email: ilapadila8@gmail.com

Abstract. Separation of regional anomalies and residual anomalies are important steps in processing magnetic data. The better separation of the two anomalies resulting data which can be easily interpreting. The most commonly used in the separation of regional-residual anomalies is the upward continuation method. In this study, we used synthetic models I (consists of four blocks with different depths) and II (consists of four blocks with different susceptibility) to determine the height of uplift in the upward continuation method. The results obtained by the appointment that illustrate the regional anomaly in the model I are 50 to 90 meters, while model II at the lift height of 20 to 60 meters. The model I shows the lifting results of 50 to 90 meters almost close to the regional anomaly response which has a range of anomalous value of -0.703-2.408 nT. Among 50-90 meters, the lifting height of 90 m is closest to the contour and range of regional component anomalies. Model II shows the lift results that are close to the regional anomaly response at 20 to 60 meters. Model II shows the lift results that approach regional anomalies (-7,780-18,012 nT) at 20 to 60 meters, which most closely approximates the regional component anomaly range is the lifting of 40 meters with -6,152-16,629 nT. The correlation between these two models is very useful for analyzing the earth's magnetic field signal.

Keywords: Geomagnetic, Regional Anomaly, Synthetic models, Upward Continuation

1. Introduction

Magnetic method (geomagnetism) is done based on the measurement of magnetic anomalies caused by differences in susceptibility contrast. If there is a field value obtained has a contrast value from the value of the surrounding magnetic field, commonly called magnetic anomalies [1]. The magnetic field value measured in the field consists of two superposing components, regional anomalies, and residual anomalies. Components of regional anomalies that have low frequencies provide information about the source of anomalous objects at great depths. The residual anomaly component provides the opposite information [2].

Separation of regional-residual anomalies becomes a very important stage. Therefore, a data filter is needed that can be appropriate to obtain representative data results so that the results of the interpretation become better. There are several methods used to separate the two anomalies. But the most commonly used is the upward continuation method. Previous research using upward continuation like Purnomo, et al., 2013 [4], Hiskiawan, 2016 [5], Nurdin, et al., 2017 [6] and Satiawan, 2008 [2]. The separation between regional anomalies and local anomalies in processing using upward continuation will result in a more accurate interpretation of the existence of magnetic anomalies [3]. But there is one problem in this method is that there is no parameter to determine the results obtained is appropriate or not.

One of a way to find out the upward continuation results is that it is suitable to create a synthetic model that is almost the same as in the field. So that, the results obtained using synthetic models can be applied to the data obtained from field measurements. The model used in previous research [2] only uses depth parameters between regional anomalies and residual anomalies. Some blocks of residual anomalies have the same depth and block regional anomalies are only one block which cannot describe the situation in the field.

In this study the models are made by using the UBC-GIF Mag3D software. It is processed by using the upward continuation method. The resulting contour was analyzed using the kriging method. The purpose of this study is to determine the height of upward continuation lift in accordance with synthetic models.

2. Methodology

In this study using two synthetic models, the model I and model II (figure 1) consisting of two regional anomaly blocks and two residual anomaly blocks based on table 1. The distance between the anomaly residual and regional anomaly approximately 100 meters, the model was made in the area of 190 X 200 meters. The model is created using software UBC-GIF MAG3D based on table 1. The equation used to calculate the total magnetic field caused by the block is [7]

$$\Delta T = C_m M \left[\frac{\alpha_{23}}{2} \log\left(\frac{r-x'}{r+x'}\right) + \frac{\alpha_{13}}{2} log\left(\frac{y-x'}{y+x'}\right) - \alpha_{12} \log(r+z_1) - \hat{M}_x \hat{F}_x arctan\left(\frac{x'y'}{x'^2+rz_1+z_1^2}\right) - \hat{M}_y \hat{F}_y arctan\left(\frac{x'y'}{r^2+rz_1-x'^2}\right) + \hat{M}_z \hat{F}_z arctan\left(\frac{x'y'}{rz_1}\right) \right] \Bigg|_{x'=x1}^{x'=x2} \Bigg|_{y'=y1}^{y'=y2} \quad (1)$$

Table 1. Coordinates of model I and model II

Model	Block	Coordinates (m) X	Y	Z	Susceptibility $(Wb.A^{-1}.m^{-1})$	Inc (°)	Dec (°)
Model I	1	60,80	80,100	150,180	0,1	45	0
	2	100,120	140,160	120,150	0,1		
	3	40,80	160,180	50,70	0,1		
	4	140,160	120,140	30,40	0,1		
Model II	1	60,100	100,120	150,180	0,3	45	0
	2	140,160	60,80	150,180	1,4		
	3	80,100	100,120	50,70	0,3		
	4	140,180	40,60	50,70	1,4		

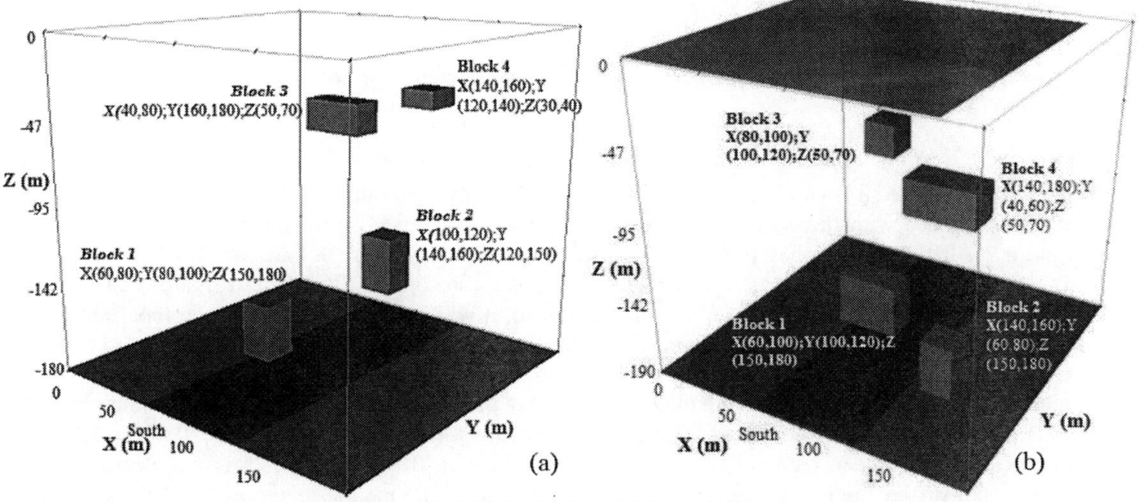

Figure 1. (a) Model I and (b) Model II

The difference between model I (figure 1a) and model II (figure 1b) is the depth and value of block susceptibility. Before being processed using the upward continuation method, the two models are transformed Fourier (Figure 2). To simplify the analysis, the data in the space domain is changed to frequency dominance using Fourier transformation analysis. The results of Fourier transform still contain two regional and residual anomalies. Both components are comparable to determining the upward continuation results in accordance with the model. Because it is difficult to determine the results of upward continuation that match the model without comparison. So the comparison is necessary in this case the regional component of the model I. The response of regional anomaly is obtained by removing two blocks of residual anomalies in the model (figure 3) as a comparison of the results of the upward continuation.This study will focus on comparing the results of regional anomalies.

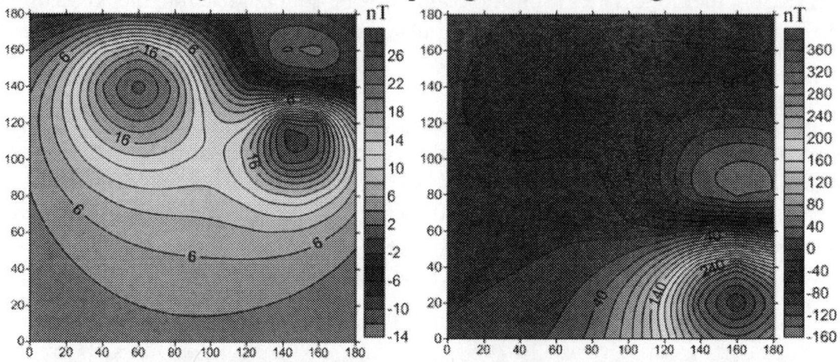

Figure 2 the model that has been transformed (left) model 1 (right) model 2

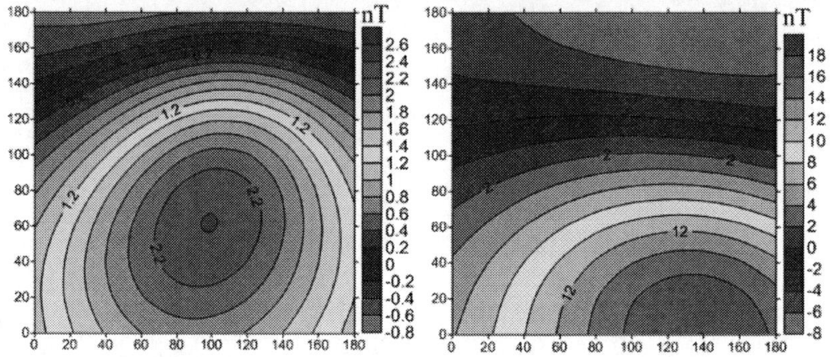

Figure 3 komponen/respon anomali regional (left) model I (right) model II

3. Results and Discussion

3.1 Model I

To calculate the potential field generated in the upward continuation field by using the following equation [8]

$$P(x,y,z_0-\Delta z)=\frac{\Delta z}{2\pi}\int_{-\infty}^{\infty}\int_{-\infty}^{\infty}\frac{Q(x',y',z_0)}{((x-x')^2+(y-y')^2+\Delta z^2)}dx'\,dy' \quad (2)$$

The model I is lifted from a height of 20 to 90 meters. Then the contours of the regional anomaly upward results compared with the contours of regional anomaly response. Of all the contours produced, the contour at a lifting height of 80 meters (figure 4b) and 90 meters (figure 4c) are the closer to the contour of the regional anomaly response (figure 4a). Based on the results of Satiawan's research, 2009 [2] the more similar the contours produced by the results of upward to the contours of regional anomalies response, then the results are considered in accordance with the model made. In addition to contours produced, the range of anomalous values also determines the height of the right lift. When the range of anomalous values obtained is getting closer to the range of regional component values, the height of the lift is correct. at a lifting height of 90 meters anomalous value range is obtained -0.482 - 2.554 nT approaching the range of regional anomaly response values of -0.703-2408 nT. for more detail see table 2. The color difference in the contour shows the variation of the magnetic anomaly. The orange color on the contour shows a high anomalous field (2,408 nT). In this model, the red color indicates the location of the anomaly that has been made.

Table 2 Range values of regional anomaly upward results and regional anomalous responses model I

Lifting height (m)		80	90
Range Anomaly value (nT)	Regional anomaly upward result	-0,658 to 3,002	-0,482 to 2,554
	Regional anomaly response	-0,703 to 2,408	

The RMSE is used to determine the lifting height that is closest to the regional component between 80 and 90 meters slice is carried out the same coordinates (figure 4) so that the magnetic anomaly value curve appears (Figure 5). By using the RMSE equation (Root Mean Square Error) an error value is obtained between the upward lift height and the regional component. N is the amount of data, x is the upward result, x 'is a regional component.

$$RMSE = \sqrt{\frac{1}{N}\sum_{i=1}^{N}(x - x')^2} \quad (3)$$

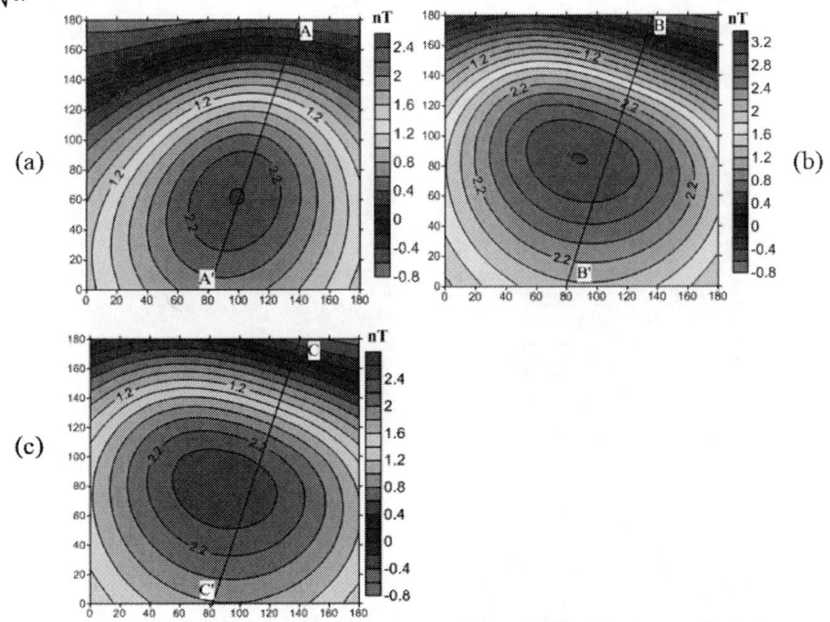

Figure 4. The track slice (a) contour response regional anomalies (b) the results of upward 80 meter contour (c) the results of 90 m contour

The error value generated in the 80-meter lift is 0.11198% while the 90-meter lift is 0.05394%. So that the lifting height that is closest to the regional component is 90 meters. The blue curve shows the regional component of model I while the orange color shows upward results.

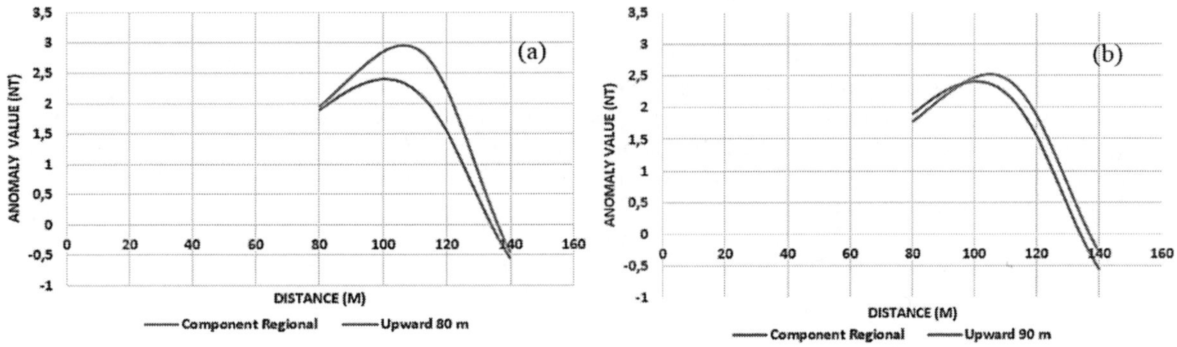

Figure 5. comparison of regional component of model I curves with (a) 80 m and (b) 90 m

3.2 Model II

The most difference between model II models I apart from the model depth is the susceptibility value which is very different between blocks (see table 1). The difference in susceptibility values which is quite large between 0.3 and 1.4 has an impact on the range of anomalies that are greater than that of model I. The model II is lifted to a height of 20 to 60 meters as model I.

IOP Conf. Series: Earth and Environmental Science **279** (2019) 012037 doi:10.1088/1755-1315/279/1/012037

The resulting contour of 20-60 meters compared to the contour of the regional component of model II. As a result, lifting height of 30 meters (figure 6b) and 40 meters (figure 6c) have the contour that is most similar to a regional component (figure 6a). For more details see table 3. The orange color on the contour shows the location of the regional block anomaly (especially in blocks with 1.4 susceptibility values).

Table 3 Range values of regional anomaly upward results and regional anomalous responses model II

Lifting height (m)		30	40
Range Anomaly value (nT)	Regional anomaly upward result	-7,249 to 20,171	-6,152 to 16,629
	Regional anomaly response	-7,780 to 18,012	

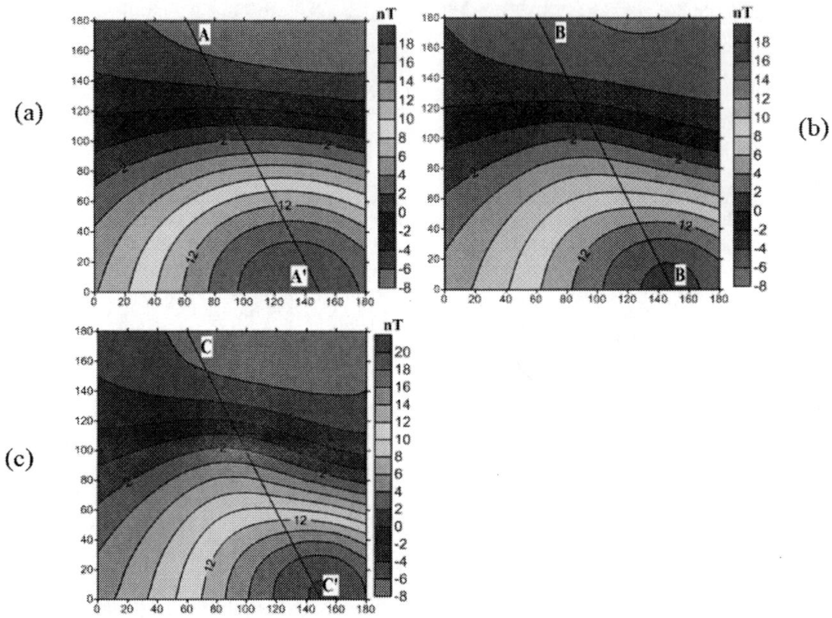

Figure 6. The track slice (a) contour response regional anomalies (b) contour results upward to 30 metersz (c) the results of 40 m contour

Furthermore, the lifting height is determined between 30 meters and 40 meters by slice the contour at the same coordinates (figure 6) so that the curve appears (figure 7). From the comparison of regional component curves with upward results, the error value can be calculated using RMSE. The error value on the 30 meters lift is 0.36724% while the 40 meters lift is 0.46236%.

From the results of the RMSE, the lifting height can be determined which is closest to the regional component of model II, which is 30 meters.

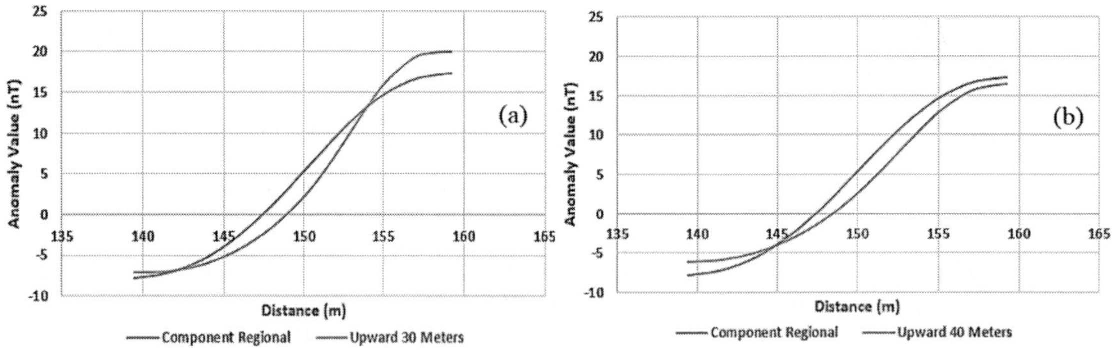

Figure 7. comparison of regional component of model II curves with (a) 30 m and (b) 40 m

In model II the distance between the top of the regional anomaly block and the bottom of the residual anomaly block is 80 meters. in contrast to the model I where upward results are closest when the two block residual anomalies are completely lost, in model II the lifting of 30 meters or 1/3 from the distance of residual and regional anomalies results in value contours and ranges that are closest to the regional component model II.

4. Conclusions

The model I lifting height which is closest to the regional component is 90 meters with the RMSE value of 0.05394% while the model II is 30 meters with RMSE 0.36724%. a large susceptibility value of 1.4 in model II produces a greater magnetic field value (-6.152 to 16.692 nT) compared to the model I whose susceptibility is only 0.1 (-0.482 to 2.554 nT).

References

[1] Broto, S. & Putranto, T. T., 2011. *Aplikasi Metode Geomagnet Dalam Eksploasi Panasbumi.* TEKNIK, 32(1), pp. 79-87.

[2] Satiawan, S., 2009. *Aplikasi Kontinuasi Ke Atas dan Filter Panjang Gelombang untuk Pemisahan Anomali Regional-Residual Pada Data Geomagnetik,* Bandung: Institut Teknologi Bandung.

[3] Nugraha, Y. A., Hiskiawan, P. & S., 2015. *Kontinuasi ke Atas Anomali Bawah Permukaan Memanfaatkan Data Magnetik di DAS Bedadung Wilayah Kota Jember. Jurnal Ilmu Dasar,* pp. 69-74.

[4] Purnomo, J., Koesuma, S. & Yunianto, M., 2013. *Pemisahan Anomali Regional-Residual Pada Metode Gravitasi Menggunakan Metode Moving Average, Polinomial, dan Inversi. Indonesian Journal of Applied Physics,* pp. 1-10.

[5] Hisikawan, P., 2016. *Pengaruh Pola Kontur Hasil Kontinuasi Ke Atas Pada Data Geomagnetik Interpretasi Reduksi Ke Kutub. Saintifika,* Juny, 18(1), pp. 18-26.

[6] Nurdin, N. H., Massinai, M. A. & Aswad, S., 2017. Identifikasi Pola Sebaran Intrusi Batuan Bawah Permukaan Menggunakan Metode Geomagnet di Sungai Jenelata Kabupaten Gowa. *Jurnal Geocelebes,* 1(1), pp. 17-22.

[7] Bhattacharyya, B. K., 1964. *Magnetic Anomalies Due to Prism-Bodies with Arbitary Polarization. Geophysic,* XXIX (4), pp. 517-531.

[8] Blakely, R. J., 1995. *Potential Theory in Gravity and Magnetic Applications.* Cambridge: Cambridge University Press.

IOP Conf. Series: Earth and Environmental Science **279** (2019) 012038 doi:10.1088/1755-1315/279/1/012038

Application of Transient Electromagnetic (TEM) Method for Delineation of Mineralized Fracture Zones

Douglas Law*, Khairul Arifin Mohd Noh and Abdul Ghani Md Rafek

Department of Geosciences, Universiti Teknologi PETRONAS, 32610, Bandar Seri Iskandar, Perak, Malaysia

*douglas_g03711@utp.edu.my

Abstract. Fracture zones have always been areas of interest in mineral exploration. These geological structures are often studied and mapped as prospective zones for mineralization through the process of hydrothermal alteration. The transient electromagnetic (TEM) survey has been utilized to delineate the possible mineralized fracture zone of eastern Kuantan, Pahang. 35 single loop TEM soundings were acquired in a scattered manner over a region of high relief terrain, encompassing an area of approximately 5 km^2. Data acquisition was conducted using the TerraTEM system by Monex Geoscope, with a 30 m x 30 m wire loop. The wire loop functions both as the transmitter coil and the receiver coil. The TEM responses indicate the presence of 3 electrically conductive zones (A, B and C) which are attributed to possible mineralized fracture regions within the country rock. All three regions record high conductivity readings with values of more than 15 mS/m and have a directional trend of NW-SE orientation. Target A is observed to be 300 m in width (E-W) and extends to 100 m in depth while B and C both have larger dimensions, exceeding 400 m in length (E-W) and up to depths of 150 m. The 3-D model generated from the 1-D soundings of the TEM dataset demonstrates the anomaly patterns that could be caused by the electrically conductive fracture zones within a resistive metasediment country rock. Geoelectrical properties of targets A, B and C may suggest the location of possible mineralized fracture zones within the study area.

1. Introduction

Electromagnetic (EM) surveys have been utilized in the sector of geological exploration for more than seventy years. Several uses of the EM method includes the exploration for iron ore deposits [1], detection of buried valleys [2], delineation of freshwater and saltwater boundary [3], reservoir explorations [4] and many more. The method is selected due to its simplicity of usage and ability to produce diagnostic data within a short period of time. Most EM systems are conducted in the frequency-domain and involve constant transmission at a predetermined frequency. The more current development of time-domain electromagnetic systems, also known as transient electromagnetic (TEM) methods, offer considerable advantages over the conventional electromagnetic systems. In the TEM system, signals are transmitted, and the transient decay of the resultant secondary field is recorded after the primary field has been switched off. Many of the TEM systems generate a train of primary pulses with measurements made in the interval between each pulse.

The concept of utilizing TEM signals in the field of exploration for electrically conductive ore bodies was constructed upon the theoretical grounds back in year 1951 by Wait [5]. Advancing from there, geoscientists from Russia have also been active in the progress and application of the method [6]. In the

IOP Conf. Series: Earth and Environmental Science **279** (2019) 012038 doi:10.1088/1755-1315/279/1/012038

west, Barringer Research assembled an airborne version of the geophysical survey method and acquisitions were successfully conducted [7]. The TEM surveys have been conducted for multiple purposes and yielded many successful results. It is one of the most reliable geological prospecting methods and therefore was selected to map the study area of eastern Kuantan, Pahang.

Fracture zones are described as regions where cracks in rocks and minerals reduced the tensile strength of the overall material. It is one of the most common geological structures due to many contributing factors such as movements of the tectonic plates, compressional and extensional forces, and even plutonic uprising activities. Displacements and void spaces generated by the Mode I fractures are termed as joints. The void spaces within the joints serve as conduits for meteoric fluid and gasses. Formation of hydrothermal veins within these fracture zones are favorable locations for mineralization process and therefore are highly sought after by exploration geoscientists. Geophysical prospecting has significantly reduced the time and cost of mineral exploration and therefore is often used for initial studies when exploring a new study area.

Sections of the subsurface with high concentration of sulphides, graphite and magnetite are areas of mineralization potentials that would cause anomalies in the EM readings. The objective of this study is to explore and locate the possible regions of mineralization that is recorded to be confined within certain structural features such as fracture zones. This survey focuses on the electrical characterization of the subsurface sections within the high relief regions of the study area in Sungai Lembing, Pahang. The TEM soundings were acquired using the single loop configuration with a 30 m x 30 m loop size.

1.1. The electromagnetic method

All TEM systems function on the principle of inducing eddy currents into the subsurface and examining the decay time of the resultant signal to provide information of electrically conducting materials in the ground. The acquisition process begins when the alternating current in the transmitting coil (Tx) generates a magnetic field in the nearby location which in turn, induces an electrical field in any proximal electrical conductors, thus forming eddy currents. The induced electrical fields generated depend on the conductivity, shape and size of the conductor material and its location with respect to the acquisition coil. After the initial signal generation, the eddy currents in the conductor material tend to diffuse inwards towards the center of the body, gradually dissipated by resistive heat losses to the surrounding. In the case of highly conductive body, the current tends to circulate within the boundary of the material and decays at a slower rate.

The eddy currents, which are essentially the secondary magnetic field, decay within a relatively longer period depending on the electrical and geometrical properties of the conductive material. The receiver coil (Rx) records an output voltage which is relative to the time derivative, also known as the rate of decay, of the vertical component from the resultant secondary field. In general, the EM properties of mineralization within the subsurface produce transient decays lasting from a fraction of a millisecond to approximately 20 milliseconds. Figure 1 briefly explains the process of a general EM surveying method. Firstly, an EM field is generated by the transmission of alternating current through a large loop of ungrounded wire at Tx. Next, the primary EM field travels from Tx to Rx via paths, both above and below the surface. Electrically conductive bodies in the subsurface generate eddy current through the magnetic component of the primary EM field penetrating the ground. The eddy currents in return generate a secondary EM field that is detected by the Rx and the difference between Tx and Rx fields indicate the presence and properties of conductive bodies.

EM methods measure the electrical conductivity of materials. The apparent conductivity of a quantified body is given by

$$\sigma_a = \frac{4}{\omega \mu_0 s^2}\left(\frac{H_S}{H_P}\right) \qquad (1)$$

where σ_a is the value of the apparent conductivity of the material (Siemen/m)

ω = angular velocity (rad/s)

= $2\pi f$ where f is the frequency of the EM wave (f in Hz)

μ_0 = permeability of free space

IOP Conf. Series: Earth and Environmental Science **279** (2019) 012038 doi:10.1088/1755-1315/279/1/012038

s = distance between Tx and Rx (meters)
H_s = secondary magnetic field at Rx
H_p = primary magnetic field at Rx

Permeability in electromagnetism is defined as the measure of a material's ability to support the formation of a magnetic field within itself. It is the degree of magnetization that a material obtains in response to an applied magnetic field. Ore-bearing rocks have very high magnetic permeabilities and are able to impact the measured responses of EM surveys.

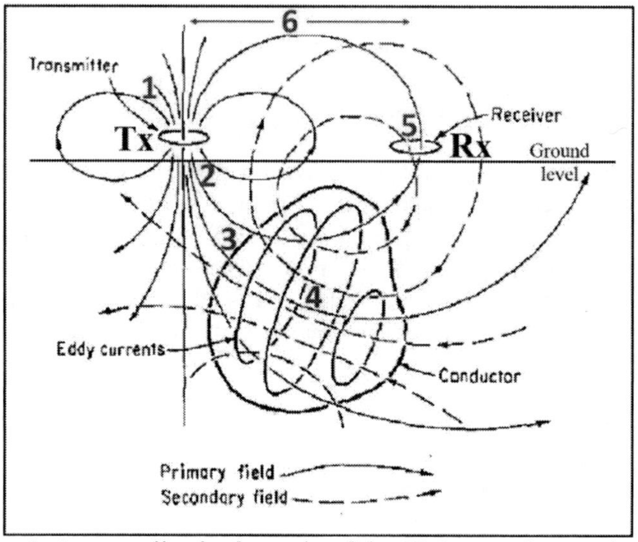

Figure 1. A generalized schematic of the EM surveying method [8].

1.2. Geology of study area

Sungai Lembing is located approximately 30 km NW from Kuantan, Pahang. The study was conducted in the hilly regions of a secondary forest near the town, with elevations ranging from 20 m to 270 m above sea level. The majority part of the study area is covered by dense vegetation. Slate and phyllite are the two main lithologies recorded within the study area. The strike directions of the beddings observed were NE-SW and NW-SE. Two others minor lithologies that were found within the survey area were shale and quartz. Weathering profiles of the surface outcrops ranged from slightly weathered to highly weathered. Structural features such as quartz veins and fractures were also observed and documented to be mainly trending E-W and NE-SW within the study area. It is recorded that the faulting and folding of the argillaceous beds of Sungai Lembing are the effects of the granite intrusions [9]. Mineralization of the hydrothermal vein deposits are also genetically supported by these plutonic uplifts. Geological studies within the area of Sungai Lembing revealed that the hydrothermal veins were found to be situated within the sedimentary rocks following the azimuthal trend of E-W, NW-SE, SE-NW and NE-SW. Paragenesis of the minerals studied within the same area are composed of cassiterite, followed by arsenopyrite, pyrite, pyrrhotite, sphalerite, chalcopyrite and galena. Several other minerals such as limonite, malachite and dendritic manganese oxides were also found within the study area due to a high degree of weathering.

Other studies recorded that mineralization of the deposits were also observed to be confined within the contact zones and the intruded country rocks. Alteration and deposition of minerals via hot ascending waters of uncertain origin [10] through the fracture zones of the country rocks have contributed to the formation of the mineralized zones. Sungai Lembing is interpreted to be constrained locally by steeply dipping faults that range from 70° to 90° in angle. Vein swarms that were observed to occur in large amounts throughout the entire study area forms potential mineralized regions through the process of hydrothermal vein deposition. These deposits can be explored by using appropriate geophysical techniques and therefore, the TEM survey is selected to locate these potential mineralized zones.

2. Methods and materials

35 single loop TEM soundings were acquired in a scattered manner over the high relief regions of the study area. The geoelectrical survey encompasses an area of approximately 5 km^2. Figure 2 illustrates the layout of the 35 TEM sounding points over a topography map of the study area. Each survey point had a minimum separation distance of 200 m from another survey point. This was to ensure proper distribution of the TEM data. The 1-D TEM soundings were conducted using the TerraTEM system by Monex Geoscope that was connected to a single square loop setup that functions both as the Tx and the Rx. The dimension of the square loop was 30 m x 30 m and the value of the electrical current was set to 10 amperes. Maximum depth of penetration for each TEM sounding point was approximately 150 m. GPS location for each survey points were obtained by using a handheld Garmin GPS at the center of the loop after each reading was acquired. The setup of the acquisition system is demonstrated in Figure 3.

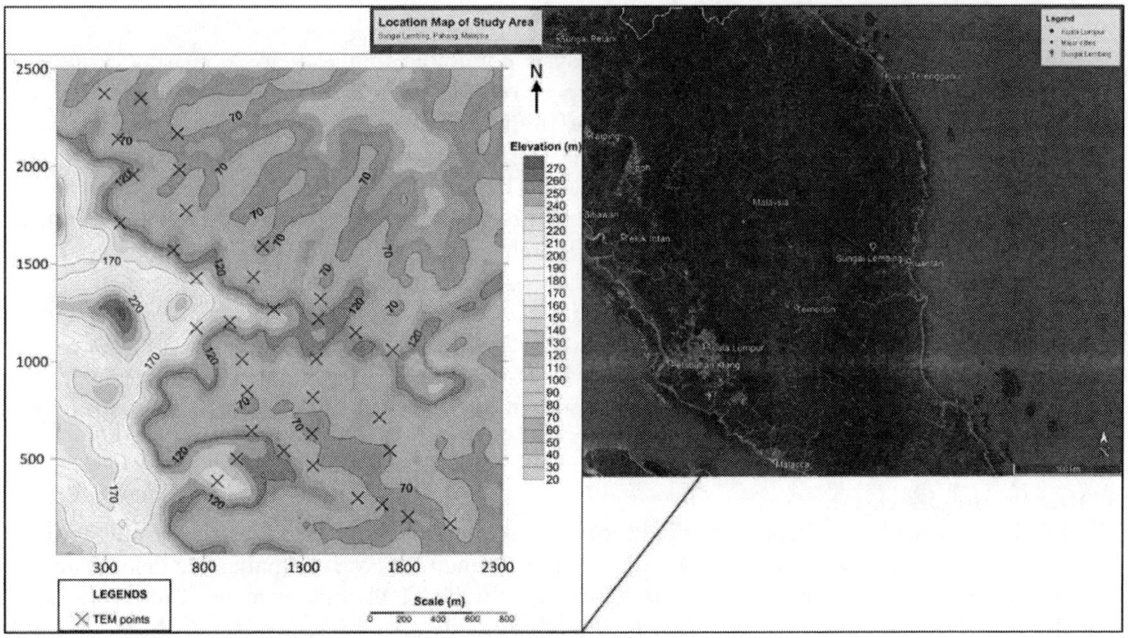

Figure 2. Location map and layout of the 35 TEM sounding points over the high relief region of Sungai Lembing, Pahang.

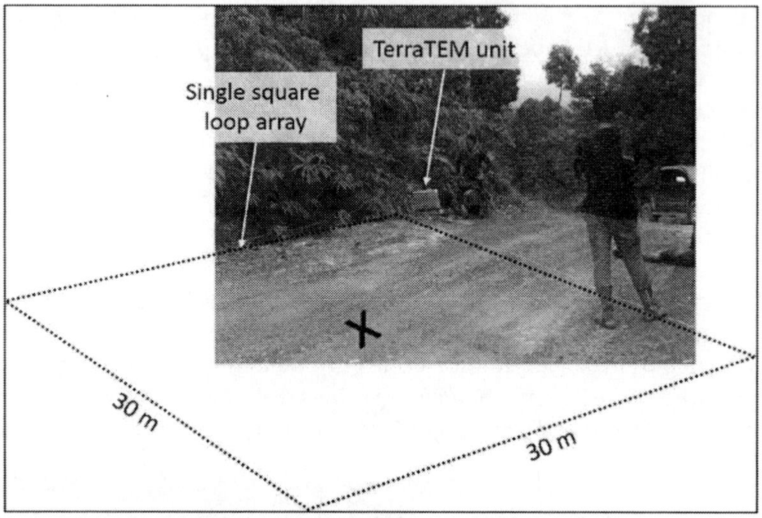

Figure 3. Setup of the TerraTEM acquisition system with the single square loop array. Scale of the loop is not drawn to size but for illustration purpose only. X is the midpoint of the square loop that indicates the location of the TEM sounding.

During the model construction procedure, all 35 TEM sounding data were processed and compiled to obtain a 3-D conductivity model of the subsurface. The decay time of the EM field that was acquired is then transformed into images using the TEMPlot software. This data alteration involves the inversion of time series into space domain using the SPIKER algorithm [11]. During the inversion process, all physical characteristics of the recorded TEM decay has been taken into deliberation such as the value of the original voltage and the rate of decay. In the space domain, the 1-D TEM sounding data is imaged as a depth-conductivity plot. 2-D contour maps (depth slices) are extracted from the 3-D electrical conductivity model produced by using the Oasis Montaj software. The contour maps were then overlaid onto the Digital Elevation Model (DEM) to present the vertical and lateral continuation of the geoelectrical responses measured.

3. Results and discussion

The TEM survey have been conducted by discrete sampling procedure. Based on the results, three high electrical conductivity zones can be observed: Zone A, B and C. These anomalies are distinct and have an elongated pattern that trends E-W when observed from the TEM contour maps. The results suggest that all three anomalies may indicate regions of possibly mineralized fracture zones. Based on the 3-D model produced, the vertical and lateral continuation of the target zones can be observed and studied. Four contour maps from different depths were extracted and presented in Figure 5. The electrical conductivity response of the study is presented in milli Siemens per meter (mS/m) with the highest recorded value at target C with readings above 1000 mS/m.

The possible mineralized fracture zone of target A is located at the NW section of the study area. Target A is observed to be 300 m in width and extends to a depth of 100 m into the subsurface before the response fades off. Target B and C are imaged at the central and SE part of the study area respectively. Both targets are considerably larger than A, with widths over 400 m in length and having positive anomaly response up to a depth of 150 m. All three target zones produced significant anomaly responses in contrast to the electrically resistive country rocks. A cross-section of Target B with respect to depth is obtained from the 3-D conductivity model and displayed in Figure 4. The high electrical conductivity target of zone B is observed at the central region of the cross-section and is labeled.

All TEM contour maps were overlain onto the DEM of the survey area to study the possible relationship between the high relief regions and recorded geoelectrical responses. Figure 6 illustrates the contour map at depths of -50 m that was overlain onto the topographic structures of the study area. It is observed that the trend of the electrical conductivity targets is NW-SE, along the ridges at all three zones instead of EW as initially interpreted from the contour maps. Zone A, B and C are independent, elongated bodies that are identified as the possible mineralized fracture zones.

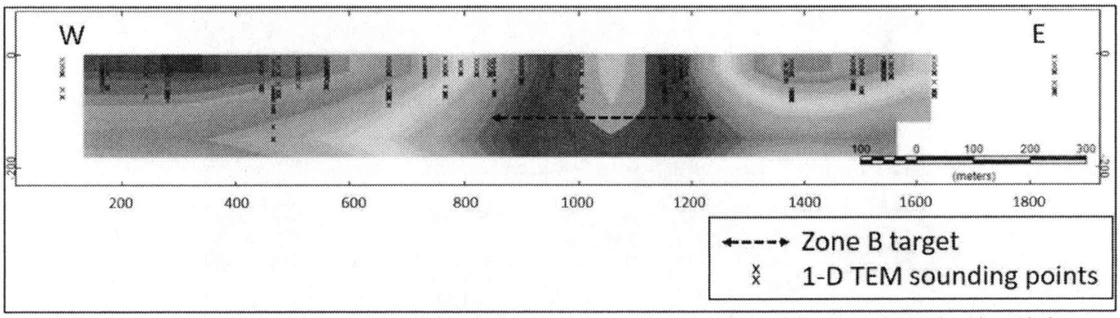

Figure 4. Cross-section of Target B and TEM sounding points with respect to depth. High electrically conductive zone measures up to 450 m in width.

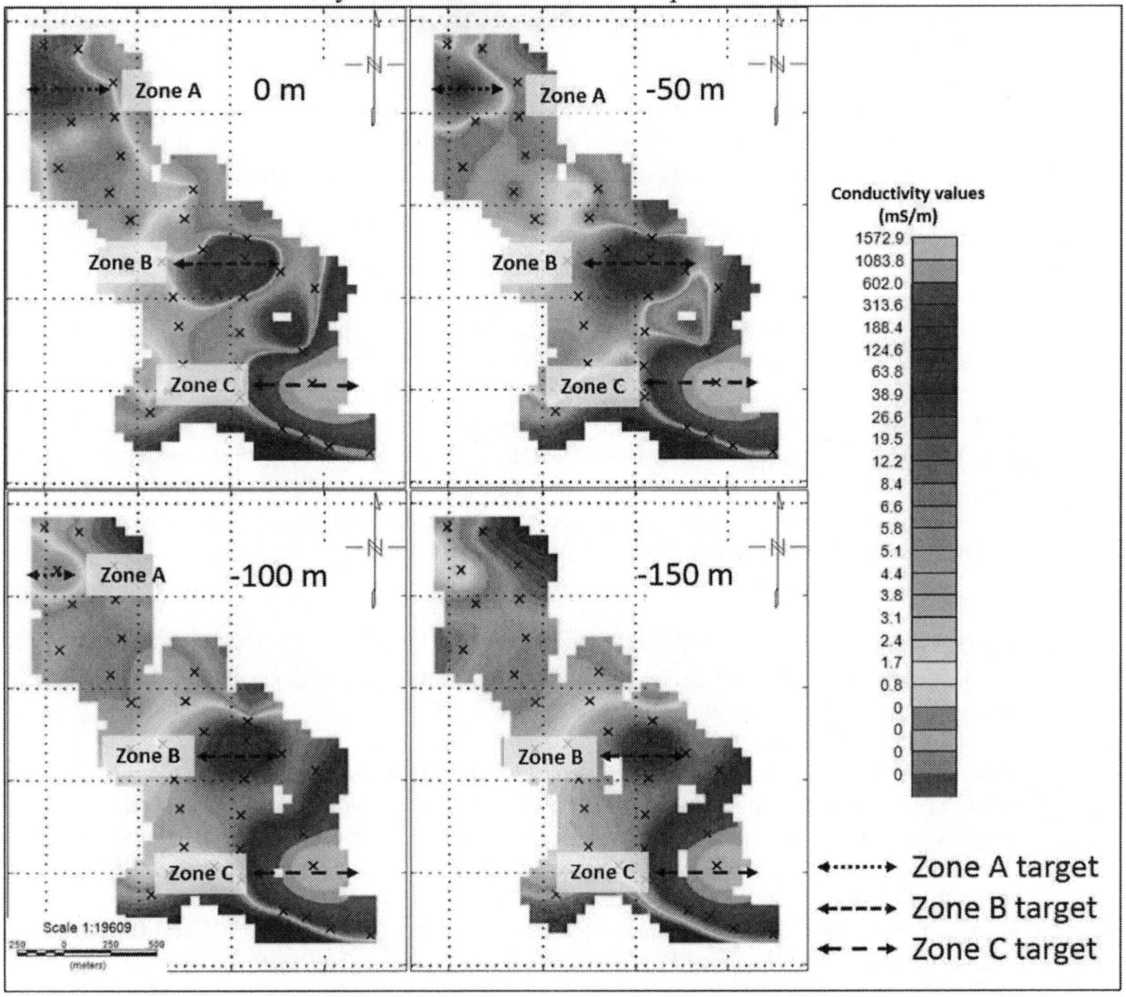

Figure 5. Contour maps of various depths to study the vertical and lateral continuity of the geoelectrical responses. Targets A, B and C have high electrical conductivity responses and are interpreted as possible mineralized fracture zones.

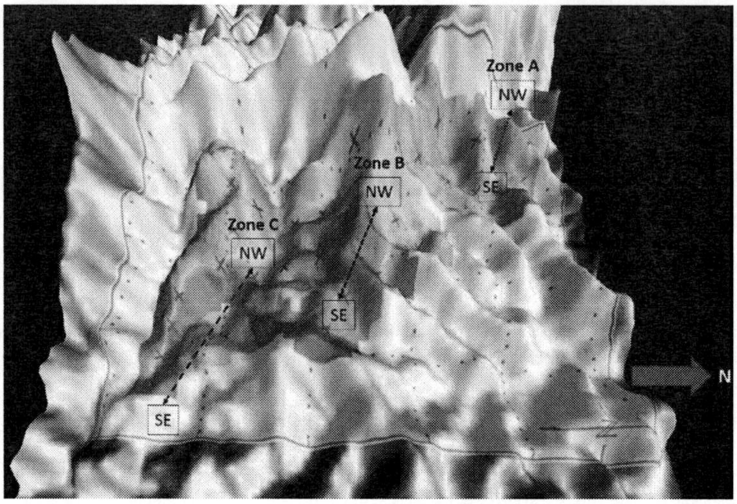

Figure 6. TEM contour map of -50 m overlaid onto the DEM of the study area. Three independent, elongated bodies of high electrical conductivity response may suggest possible mineralized fracture zones along the ridges with directional trends of NW-SE.

4. Conclusion and recommendations

The TEM survey have successfully been utilized to study the geoelectrical characteristics of the hilly region of Sungai Lembing, Pahang. A 3-D electrical conductivity model of the study area have been constructed based on 35 single loop TEM soundings that have been acquired. Contour maps in the form of depth slices have been extracted from the model to study the vertical and lateral continuity of the geoelectrical data. Three high electrically conductive anomalies have been determined and are labelled as target A, B and C. These targets are observed to have a directional trend of NW-SE along the ridges based on the contour maps that are placed over the DEM of the survey area. Targets A, B and C are identified as possible mineralized fracture zones and further investigations such as drilling and other geophysical studies should be carried out to support this initial finding. The TEM method has proven successful as an initial geoelectrical mapping and prospecting method.

5. References

[1] Xue G Q, Gelius L J, Sakyi P A, Zhou N N, Chen W Y, Su B C, Li H, Zhong H S and Su Y P 2014 *Discovery of a hidden BIF deposit in Anhui province, China by integrated geological and geophysical investigations* **63** 470-77
[2] Masrom S N and Samsudin A R 2012 *Application of airborne transient electromagnetic (SkyTEM) technique for buried valley detection in part of Hadsten, Aarhus County, Denmark*
[3] Mohamad I C, Samsudin A R, and Rafek A G M 2001 *Penggunaan bersama data geofizik dan geologi dalam kajian akuifer aluvium pantai, kawasan Pekan-Rompin, Pahang (IN MALAY), Joint use of geophysical and geological data in the study of coastal alluvial aquifer, Pekan-Rompin, Pahang*
[4] Agafonov Y A, Salleh M M, Kompaniets S V, Markovceva O V, Egorov I V and Bouddo I V 2011 *Efficiency of TDEM and EM-IP Methods Application for Reservoirs Exploration in South East Asia* in *PGCE 2011*
[5] Wait J R 1951 *A conducting sphere in a time varying magnetic field* **16** 666-672
[6] Velikin A B and Bulgakov Y I *Transient method of electrical prospecting (one loop version)* in *International Seminar on Geophysical Methods, Moscow, USSR*
[7] Boniwell J B 1967 *Some recent results with INPUT airborne EM system* **60** 325
[8] Grant F S and West G F 1965 *Interpretation theory in applied geophysics*

[9] Hutchison C S and Tan D N K 2009 *Geology of peninsular Malaysia* 479
[10] Edwards R and Atkinson K 1986 *Hydrothermal vein deposits* in *Ore Deposit Geology and its influence on mineral exploration* 143-174
[11] Smith R S, Edwards R N and Buselli G 1994 *An automatic technique for presentation of coincident-loop, impulse-response, transient, electromagnetic data* **59** 1542-50

Acknowledgement
The author would like to thank all parties that have been involved in making this TEM study a successful one, especially to the acquisition team that have braved through the challenging terrains of the tropical jungles of Malaysia. Special thanks to all the other supporting parties and their contribution towards the success of this project as well.

Reconstructing past landslide models based on subsurface resistivity profiles: case study from Majene West Sulawesi

Ardy Arsyad[1], Sabrianto Aswad[2], Erfan Syamsuddin[2]

[1]Department of Civil Engineering, Faculty of Engineering, Hasanuddin University Makassar Indonesia
[2]Study Program of Geophysics, Faculty of Mathematics and Natural Science, Hasanuddin University, Makassar, Indonesia

ardy.arsyad@unhas.ac.id

Abstract. This paper presents a case study of reconstructing past landslide failure models in a napal geological formation by undertaking a series of geoelectrical tomographies and geotechnical bor-SPT surveys. Study area is located in the National Artery Road 135 Kilometres, Majene – Mamuju, West Sulawesi. Geological formation found in the area is Napal Pambauang Formation. Massive ground movement is obviously detected in the area, and several landslide events have been occured in the last 10 years. In order to derive an effective technnique of mitigating the landslide in the area, past landslide failure models are needed to reconstruct. Therefore, 2D geoelectric tomographies survey combined several geotechnical bors were undertaken. A Framework of geophysics and geotechnical investigations was developed and implemented. The results show that slope failure models can be mapped, following unique patterns in the subsurface resistivity models. The finding slope failure models are well agreement with successive scarp patterns founded in the slope area. Past and recent landslides and future landslide are derived and this can be employed to design effective landslide mitigation strategies.

Keywords: past landslide, electrical resistivity tomography, slip failure plane, reconstructing

IOP Conf. Series: Earth and Environmental Science **279** (2019) 012039 doi:10.1088/1755-1315/279/1/012039

1. Introduction

Landslides have become the second of the common natural disaster in Indonesia after flood disaster. Landslides and floods events comprise 18% and 32% of the total natural disaster events, recorded since 1815 [1]. In 2018, landslide events have been accounted for at least 1142 events throughout Indonesia. Therefore, recent landslide disaster mitigation techniques have been developed since significant economic losses and even fatalities caused by this disasters. Current studies on detection and investigation techniques of old landslide have been conducted due to most landslide events in many regions are in some cases as a result from reactivation of old landslides. Bell et al. [2] performed old landslide study for the case of Swabian Alb Germany. New and old landslide body structures were identified by using 2D electrical resistivity tomography (ERT), while their moisture distributions were monitored. The use of 2D ERT is widely applied, how to apply this geophysical method for geomorphology can be explored in Schrott et al. [3]. Other method such ground penetrating radar (GPR) is just suitable for shallow small scale landslide, due to its drawback of damping by soft sediment and reflecting by woody material [4]. By considering that conventional techniques such as drillings (Bor-SPT) can provide point information of slope stratigraphy, while 2D ERT can derive continuous profile of slope stratigraphy [5] [6], it is necessary to combine those techniques to obtain the most reliable of landslide models. In the case of Swabian Alb [2] [4], the sliding plane and lateral boundary can be detected clearly. Therefore, this paper aims to present the use of geophysical and geotechnical techniques in investigation old landslide body structures in the case of Majene Tanjung Rangas West Sulawesi. Geophysical survey of ERT and geotechnical drilling and penetrating tests were undertaken on that slope area.

2. Site Characteristics

2.1. Geological setting

The landslide area is situated on Tanjung Rangas Majene, National Artery Road Km. 135, West Sulawesi (Figure 1). Geologically, the area (Figure 2) is on the Pleistocene Napal Pambauang which deposits Tuffaceous marl, marly shales, tuffaceous sandstone, and lenses of conglomerate [7]. The Pambuang Marl is about 300 m thick, formed in a shallow marine environment. This formation crops out in the South-West corner of Majene Regency. General stratigraphy of this area is represented by a marly shales and tuffaceous sandstone, which is overlain by Tuffaceous marl (Figure 3). Marl is calcium carbonate rock which is dominated by clay and silt. Originally, marl was formed as mixtured between clay and calcium carbonate. The marl is found to containing nodules.

Figure 1. Location of the landslide area.

Figure 2. Geological Map of the landslide area (Qpps = Napal Pambuang, Tmm = Formation of Mandar, Tmpm = Formation of Mapi).

Figure 3. Stratigraphy of the rock formation in the landslide area.

2.2. Borehole Data

Geotechnical investigations were undertaken in the landslide area. Four boreholes were drilled over the landslide materials and the surrounding areas (Figure 4). The depth range of the boreholes is about 40 m, except for BH 2, which is just 30 m. Figure 5 show the depth of groundwater is at 15 m below the ground of road level, whereas the thickness of landslide material varies from 5 m to 7 m. This can be seen in the stratigraphy based on BH 1 - BH 2 (Figure 5) and BH 3 – BH 1 – BH 4 (Figure 6). The landslide material was underlaid by stiff clay and napal. It is obvious that the soft clay and stiff clay indicate the weathering of napal. At 5 m below the ground, the napal is heavily weathered due to the fluctuating groundwater level in this depth.

Figure 4. Layout of the boreholes location in the landslide area and the environment.

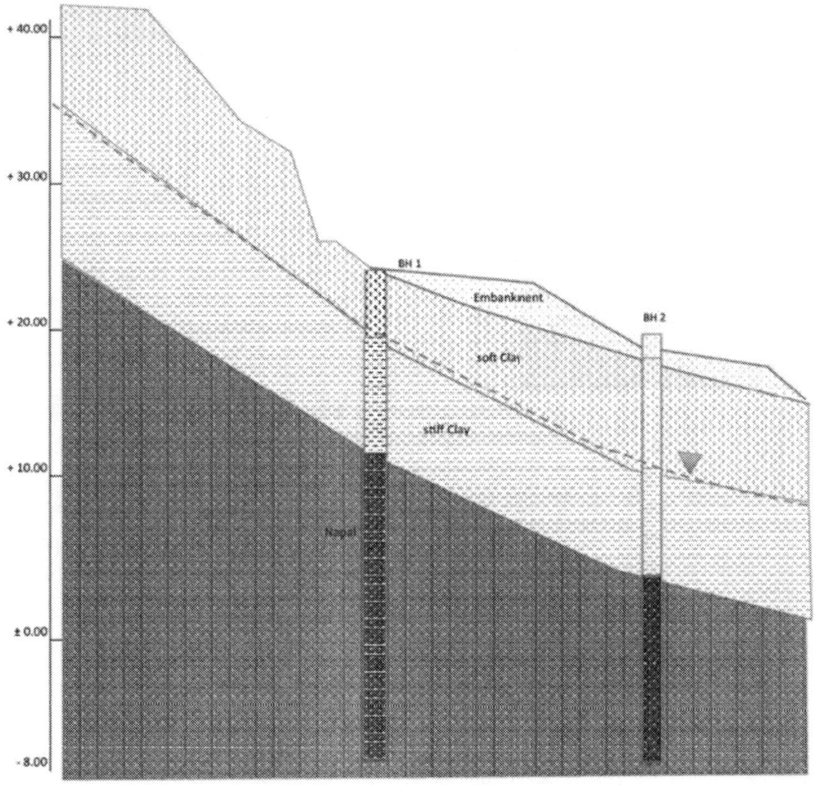

Figure 5. Estimated stratigraphy derived from BH 1 – BH 2, where are situated parallel to landslide movement.

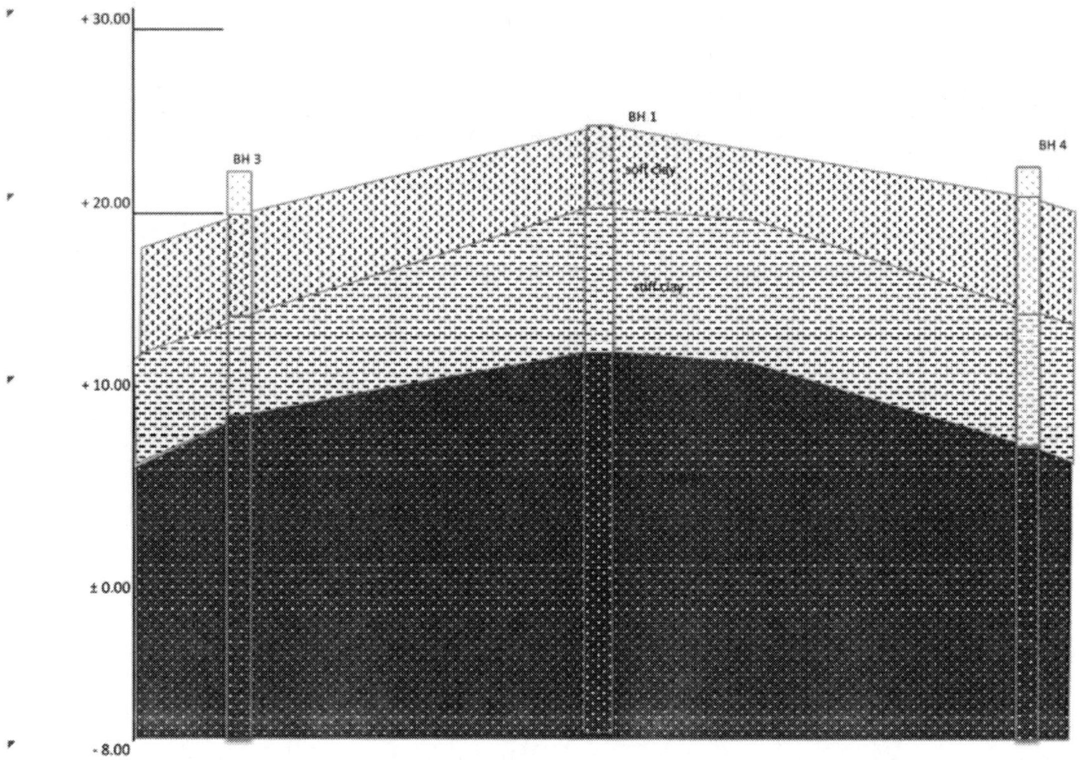

Figure 6. Estimated stratigraphy derived from BH 3 – BH 1 – BH 4, where are situated perpendicular to landslide movement.

3. 2D Electrical Resistivity Tomography Survey

A 2D ERT survey was undertaken in order to measure resistivity distribution in below the ground of the landslide area. The survey measured the electrical potential between a pair of electrodes due to direct injection between another pair of electrodes. After that, apparent resistivity is measured. In this study, two-dimensional ERT was performed within a multi electrode cable, arranged in three lines including Line-1, Line-2 and Line-3 (Figure 7). Data were acquired by a Wenner-Schlumberger configuration using a 60-electrode cable 5 m apart. RES2DINV software was employed to process the data for a 2D tomographic inversion technique based on the smoothness-constrained least squares inversion of pseudosection data [8], and the calculated pseudosection based on finite difference method [9].

4. Results

Figure 8, 9 and 10 show the results of electrical resistivity tomography with the simplified logs of the boreholes. The data along Lines-1 was measured in parallel, while Line -2 and -3 were measured in perpendicular directions to the landslide movement. The resulting tomogram along this profile

IOP Conf. Series: Earth and Environmental Science **279** (2019) 012039 doi:10.1088/1755-1315/279/1/012039

suggests contrast resistivity values from low to high resistivity. Figure 8 illustrates that there are four distinct contrast resistivity zones along the line, indicating four slip planes in the landslide areas. In addition, landslide material such as full saturated soft clay, stiff clay and napal observed in the boreholes are in good agreement with the result of ERT. For instance, Figure 9 and 10 show low resistivity area which has similar profile to soil layers shown in the stratigraphy of BH 3 – BH 1 – BH4 as illustrated in Figure 6. Profound results in this study based on ERT profiles reveal several events of landslide have been occurred, from the oldest to the youngest landslide. The mechanism of such grand landslide can be seen in the oldest landslide in Figure 8. After the movement of this oldest landslide, small landslides occurred sequentially from old to very young landslide. Future landslide can be predicted at the potential slip failure between the oldest landslide and the very young landslide. The ERT profile is well agreement with the existing landslide scarps found the field (Figure 11).

Figure 7. layout of ERT Line measurements.

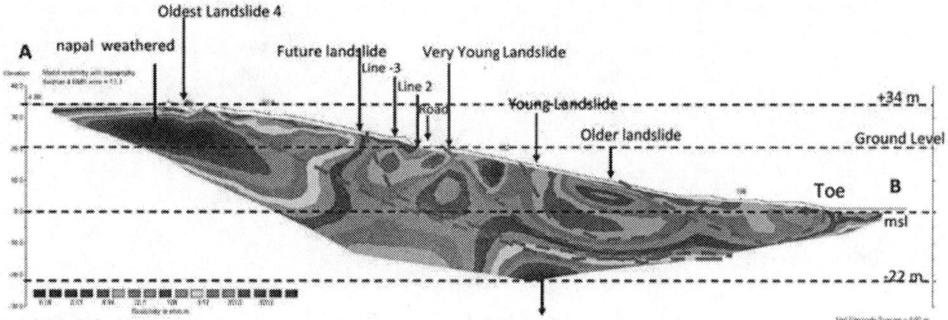

Figure 8. Resistivity modelling of the landslide area through Line 1.

IOP Conf. Series: Earth and Environmental Science **279** (2019) 012039 doi:10.1088/1755-1315/279/1/012039

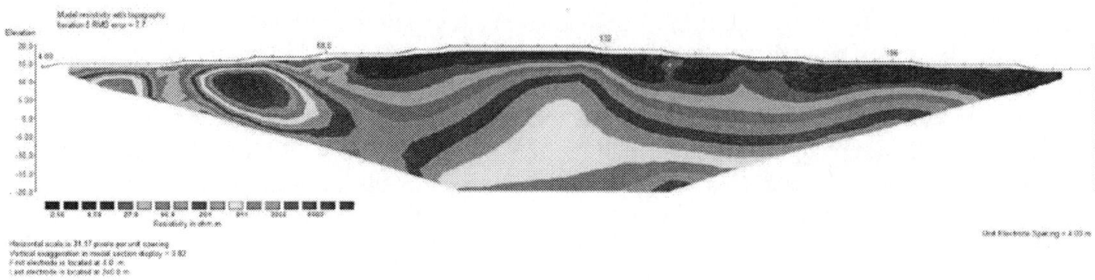

Figure 9. Resistivity modelling of the landslide area through Line 2.

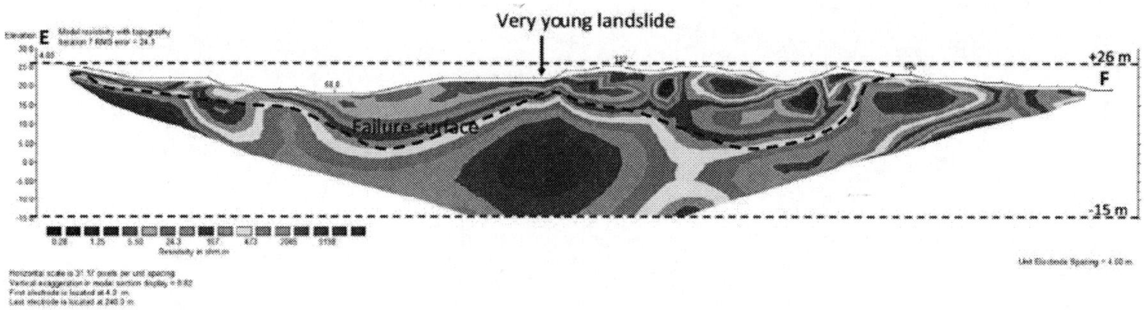

Figure 10. Resistivity modelling of the landslide area through Line 3.

Figure 11. Existing landslide scarps in the landslide areas indicates periodical events of landslide have occurred.

5. Conclusions

This study presents case of study of reconstructing the mechanism of sequential landslide models that took place in the Tanjung Rangas Majene West Sulawesi. The results suggested that the electrical resistivity tomography is a useful technique for investigation of past landslides. The ERT could be employed to determine geometry and water saturated zones in landslide areas, as well as mapping the location of slip failure planes, which can be used to reconstruct the sequential of landslide events occurred in the past. The ERT profiles have been confirmed well with the boreholes. For the case study, it was found that landslide materials are around 5 m to 7 m thick below the ground and at least five slip failure planes. The results suggested that landslides have been taking place in many times for five periods. A future landslide located in between the oldest landslide and the recent landslide would predictably occur if the prevention measures could not be immediately undertaken.

6. References

[1] BNPB 2018, Data Kejadian Bencana Longsor, Badan Nasional Penanganggulangan Bencana, Jakarta.

[2] Bell, R., Kruse, J-E., Garcia, A., Bonn, Hordt, A., Braucschweig, 2006 Subsurface investigations of landslides using geophysical methods geoelectrical applications in the Swabian Alb (Germany), Subsurface Landslide Investigation, pp. 201-208.

[3] Schrott, L., Hördt, A. & R. Dikau (eds) 2003 Geophysical applications in geomorphology. - Zeitschrift für Geomorphologie, Supplementband 132, Berlin, Stuttgart: Gebrüder Borntraeger.

[4] Sass, O., Bell, R., Glade, T, 2008. Comparison of GPR, 2D-resistivity and traditional techniques for the subsurface exploration of the Öschingen landslide, Swabian Alb (Germany), Geomorphology, 93, pp. 89-103.

[5] Drahor,M., Berge, M., Gorktukler, G., Kurtumulus, T.O., 2006. Application of electrical resistivity tomography technique for investigation of landslides: A case from Turkey. Environmental Geology, vol 50, pp. 147-155.

[6] Perrone, A., Lapenna, V., Piscitelli, S. 2014. Electrical resistivity tomography technique for landslide investigation: A review. Earth Science Review 135, pp. 65-82.

[7] Djuri, Sudjatmiko, Bahri, S., Sukido, 1998. Peta Geologi Lembar Majene dan Lembar Bagian Barat Palopo, Pusat Penelitian dan Pengembangan Geologi, Bandung.

[8] Loke MH, Barker RD 1996 Rapid least- squares inversion of apparent resistivity pseudosections using a quasi-Newton.

[9] Dey A, Morrison HF 1979 Resistivity modelling for arbitrarily shaped two- dimensional structures. Geophys Pros- pect 27:106–136

Identification of eroded sediment layer thickness and depth of Mahakam River at Tenggarong Bridge Area, Kutai Kartanegara East Kalimantan using Ground Penetrating Radar Method

S Rasimeng [1,2*], I Mandang [3] Suharno[1]

[1] Geophysics Engineering Department, University of Lampung, Lampung, Indonesia
[2] Doctoral Programme of Environmental Sciences, University of Lampung, Lampung, Indonesia
[3] Department of Physics, Mulawarman University, Samarinda, Indonesia
*syamsurijal.rasimeng@eng.unila.ac.id

Abstract. GPR measurements of lines are two lines intersect under the bridge of Tenggarong (southwest to northeast) along the 377 meters (segment-1 Line04) and 457 meters (Line03) and the other is Southeast to Northwest along the 533 meters (segment-2 Line04) with a starting point It is located on the side of the Mahakam River (Southern) to the West of Kumala island. and there is one track in this area, which is to the northeast to the southwest along the south side of the Mahakam River (Line02) with the length of line is 780 meters. This research used the AKULA-9000C, 100 MHz antenna frequencies, and in time mode. The purpose of this research is to know the depth of the bottom of the Mahakam River, and the thickness of sediments eroded around the bridge of Tenggarong. GPR data processing using the ReflexW software ver. 8.5 through static correction, subtract-mean (dewow) filter, gain, background removal, band-pass filters, and fk-filter. To convert two-way time of electromagnetic wave propagation data to depth units, we use velocity adaptation methods by selecting velocity value each layer anomaly. Cross sections of the line from line02 and Line03 showed that the average thickness of the layer of sediment erosion over 15 meters and depth variation of the river was 10 to 40 meters. The shallow depth is under the Tenggarong Bridge (10m). In addition, on Line04 shows the deepest depth of the Mahakam River is 35 meters, located in the middle of Line04 or the center of Mahakam river flow.

Keywords: Mahakam, GPR, Tenggarong bridge, riverbed.

1. Introduction

The Mahakam River has a flow system that passes through Kutai Barat, Kutai Kartanegara, and Samarinda City, and the Mahakam upstream in West Kutai flows to the Mahakam delta, with estuaries in the Makassar Straits. Some notes explain that the Mahakam River is the second longest river in Indonesia with a length of 920 km. Mahakam River is widely used by humans as an important transportation route in the area of Samarinda City and Kutai Kartanegara Regency, so the Mahakam River falls into the class of priority river based on the Republic of Indonesia Presidential Decree Number 12 of 2012 about the Determination of River Areas.

The silting of the Mahakam River caused by the soil layer which eroded by rainwater and turned through the tributaries into the Mahakam river flow. In addition, the overhaul of the soil layer by coal mining activities also contributes to sedimentation on the Mahakam river. Both processes are the impact of the high sedimentation process on the Mahakam river. In 2009 to 2013 the Mahakam watershed lost 128,000 hectares of natural forest due to mining activities, HTI, forest concession rights and plantations. Until 2013, the Mahakam watershed left only 4.1 million hectares of natural forest or 50% of total area of the watershed. Conversion of the function forest to non-forest is increasingly uncontrollable. Moreover, there will be more overlapping use of the area between corporations on the same land. East Kalimantan province has lost 112 ha of forest each year [1].

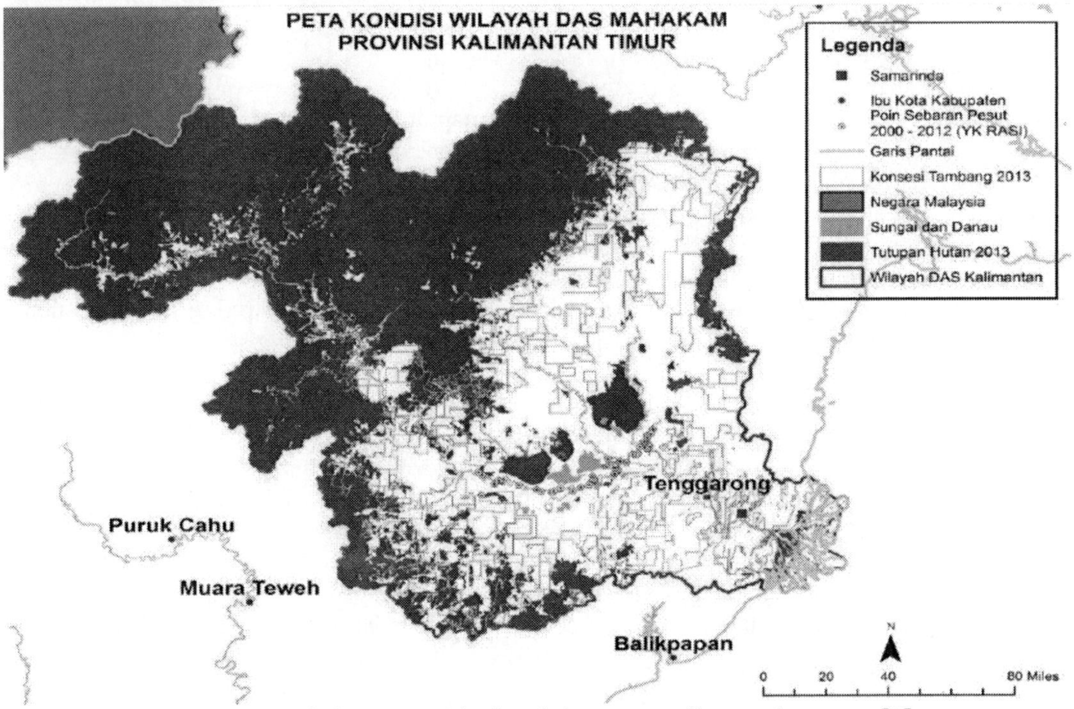

Figure 1. Mahakam watershed and the surrounding environment [1].

Formation of Mahakam delta since Quaternary period was strongly influenced by the mixing of fluvial sedimentation transport processes and influence tidal currents [2, 3, 4] to form fan morphology in transition environment low energy. The stratigraphy of Mahakam delta from Late Plistocene to recent shows continuation of gradation process starting from sea level decline, first rise up to the modern phase/highstand [5]. Interaction between river flow, ocean waves and the slight influence of low to medium energy sequences forming "lobately" of Mahakam delta was drained by river canals and tidal canals. Mahakam delta morphology characterized by absence of alluvial floods, deposits of alluvial levees, fluvial overbank, and splay deposits in delta plain which is a distinguishing feature of the Mahakam Delta with other fluvial-dominated deltas [2].

Vegetation of deltas Mahakam consists of nipa palm and mangrove forests and other primary forests with almost 80% has undergone a conversion process and the remaining only 15 to 20% (22,000 to 30,000 ha). Land use changes have led to an increase rate sedimentation (4 to 10 million tons/year, followed by high abrasion process (139 ha/year) [6]. These evidences indicate the high dissolved soil material along the Mahakam river flow.

These evidences indicate the high dissolved soil material along Mahakam river flow. Sediment analysis is often used as an environmental indicator such as sedimentary texture to find sediment deposition and transport environments [7, 8], sediment grain size to decide sedimentary processes [9], minerals in sediments to determine the origin of sediment sources/rock [9].

The GPR methods are a geophysical technique non-destructive investigation for detecting and identifying subsurface structures to guide decision making of engineer by increasing knowledge about subsurface features, both of natural or artificial. In the civil structures case, the scientific literature only reports several cases of GPR applications for detecting cavities and discontinuities in hydraulic resilience structures such as river embankments and embankment systems [32]. GPR techniques have been used for a century for developing military, detect tunnels and stockpile mines [10]. Currently, GPR applied in various technical surveys for mapping of subsoil features. Some literature reports many applications that documented GPR to at some practical problems like following cases,

- Utility locations: drains, sewer and storm channels, electricity, telephone, cable TV, etc. [11].
- Concrete detection: localization of reinforcement and metal channels, measurement of rebar radius, slab thickness and other property detection [12, 13, 15].
- Monitoring bridges and trains [14].
- Road checks: analysis of pavement structures [16].
- Geological investigations: bedrock profiles, fracture mapping, sedimentology [18, 19].
- Environmental studies and underground hydrogeophysical studies: storage tank locations, soil contamination, mapping water tables, groundwater content [20, 21, 22]

In recent years, investigations of civil structures have improved and supported by national and international standards [32]. However, the literature for application of GPR techniques for river bed sedimentation monitoring is still limited. Identification of sediment layer thickness on the river bed is important information for predicting sediment deposition rates and river shipping safety, so that river water overflows and disruption of vessel traffic can be prevented in river transportation.

2. Methodology

GPR method is one of the geophysical methods for subsurface mapping in shallow layers. Measurement of GPR data can be displayed in real time, so it is possible to assess the quality of data obtained directly in the field. Another advantage of the GPR method is the ability to record data continuously over large areas in a relatively short time. This is helpful because parameter adjustments and quick acquisition arrangements can be made. The basic principle of the GPR method is recording high frequency electromagnetic waves produced by the boundaries of subsurface layers and rocks, based on the reflector response it is possible to detect the dielectric characteristics of existing soil layers [32]. In other words, the response shown contrasts effect of dielectric permittivity on the boundaries of layers and targets buried in subsurface.

The variation of dielectric permittivity contrast value is directly proportional to the contrast of reflectivity coefficient variation. Where the value is related to changes in texture, lithology, porosity and density including water content in the soil layer. Water content also affects the amount of absorption of electromagnetic energy which causes a decrease in wave energy. The difference in soil moisture content can produce a relative dielectric permittivity surge.

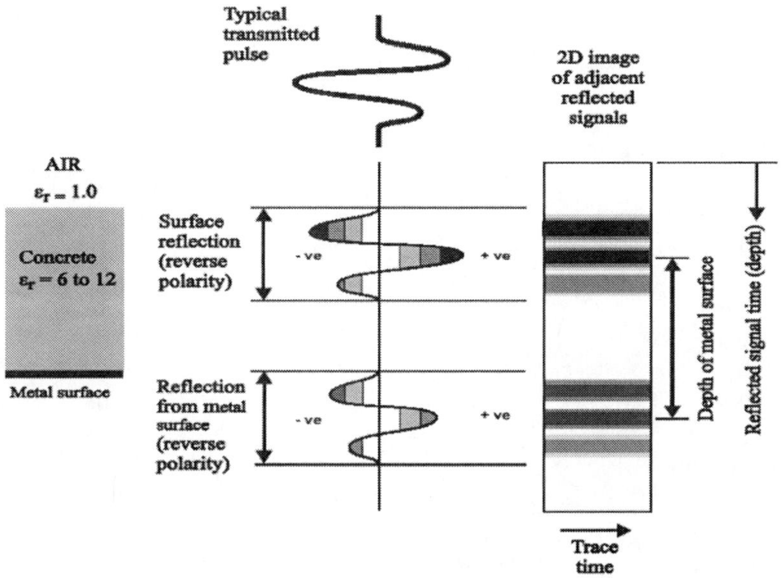

Figure 2. Schematic of radar signal reflection [15].

Source of electromagnetic waves generated from GPR transmitters and transmitted through an antenna, emits pulses with a certain frequency, and spreads in the subsurface have a velocity propagation depending on the medium of the soil layer being passed. When a wave transmitted through a medium that has different dielectric properties, then some of the energy will be reflected and partially transmitted to a deeper layer of soil. Two important dielectric parameters that influence the propagation characteristics of electromagnetic waves in the reflector are conductivity and relative dielectric permittivity [32],

$$\sigma = \frac{1}{\rho} \quad (1)$$

$$\varepsilon_r = \frac{\varepsilon_b}{\varepsilon_o} \quad (2)$$

where σ is conductivity (1/Ωm), ρ is resistivity (Ωm), ε_r is permittivity of material (F/m) and ε_o is permittivity of the vacuum ($\varepsilon_o = 8.85\ 10^{-12}$ F/m) and ε_b is porous media can be considered as the sum of ε_r of different phases,

$$\varepsilon^\alpha = \theta_l \varepsilon_l^\alpha + \theta_a \varepsilon_a^\alpha + \theta_s \varepsilon_s^\alpha \quad (3)$$

where ε_l, ε_a, and ε_s are the permittivity of the dielectric of water, air and solid phase. α is a geometric parameter (usually equal to 0.5) which depends on the position of the ineral particles. θ_l, θ_a, and θ_s are the volumetric fractions of each phase. This relationship is called a dielectric mixture model [23]. Seen in Table (1) water has the highest ε_r value which can be used as an indication to determine the layer with a large water content.

Table 1. Conductivity (σ) relative dielectric permittivity (εr) and velocity (v) reference values for common land and rocks [32].

Material	$\mathcal{E}_r$	σ (1/ohm.m)	v(m/ns)
Air	1	0	0.3
Distilled water	80	0.01	0.033
Water	80	0.5	0.033
Marine water	80	0.003	0.01
Dry sand	3-5	0.01	0.15
Wet sand	20-30	0.1-1	0.06
Limestone	4-8	0.5-2	0.12
Shale	5-15	1-100	0.09
Silt	5-30	1-100	0.07
Clay	5-40	2-1000	0.06
Granite	4-6	0.01-1	0.13
Dry salt	5-6	0.01-1	0.13
Ice	3-4	0.01	0.16
Frozen ground	3-6	-	-
Concrete	4-10	-	-
Metal	1-2	-	-

The equipment that will use in this study is a set of GPR tools with equipment of; (i) AKULA 9000C main unit (ii) GPR transmitter (iii) receiver (iv) antenna 100 MHz and (v) GPS. In addition, we use software GasXp (acquisition software) GPRSoft and ReflexW (processing and modeling software). All equipment, laptops and software used are available at Geophysics Exploration Laboratory and Geophysical Data Processing and Modeling Laboratories, Geophysical Engineering Department, Faculty of Engineering, University of Lampung.

The GPR method or commonly called the geo-radar methods, comes from two words: geo means earth and radar (radio detection and ranging). Literally, it means the earth tracking device use radio waves. The component of GPR method for measuring subsurface conditions usually consists of control unit, transmitter antenna and receiver. The transmitter and receiver antenna modes on the GPR consist of mono-static and bi-static modes. Mono-static mode is when the transmitter and receiver combination in one antenna so that there is no distance between transmitters and receiver, while bi-static mode if the two antennas were separation distance [24].

The transmitter generates EM wave pulses at certain frequencies according to characteristics of the antenna (10 MHz to 1000 MHz). The receiver is set to do a scan that normally manages 32 to 512 scans/second. Each scan result is displayed on the PC (real time) as a time function as two-way time, in time needed for EM wave to propagate from transmitters, target and to the receiver is called radar-gram. Principle of GPR tool is transmitting radar waves into the medium by the antenna and then reflected waves to surface and received by radar receiver from reflection wave in response various kinds of objects and recorded in radar-gram. A mechanism of GPR and the example of radar-gram recording are shown in the following figure (2),

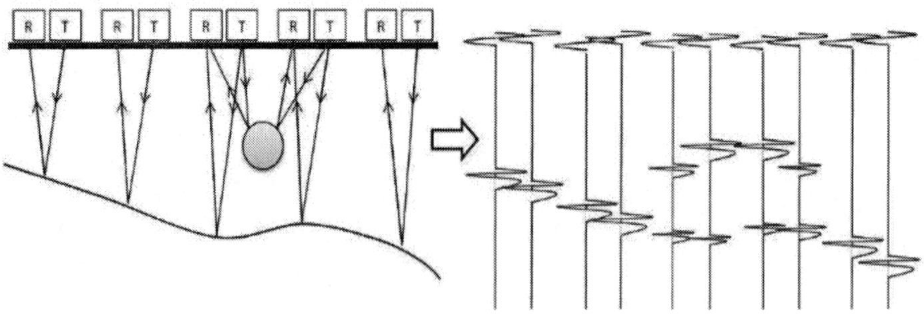

Figure 3. Illustration of how a radargram is generated.

IOP Conf. Series: Earth and Environmental Science **279** (2019) 012040 doi:10.1088/1755-1315/279/1/012040

Transmitter and receiver are transducers that convert electric current to EM waves and the antenna transmit electromagnetic waves to propagate to the subsurface. Currently GPR equipment generally records data digitally, making it easier to process data. Digital control of GPR system consisting of a micro-processor, memory and storage for storing field measurement data. In addition, GPR equipment is also equipped with a microcomputer, a standard operating system and an interface to control the process of measuring and storing data by the operator.

GPR data is taken along the path and simultaneously recorded on the hard drive. When radar waves encounter structural continuity (a drastic difference in material properties), a portion of the wave will be reflected and will be in the form of a secondary impulse. The Impulse is then captured by the receiver's antenna and then recorded as observation data, and if the data is interpreted correctly, then the data will show the subsurface structure of the object/material being observed. When the data is taken continuously in horizontal scale on radar-gram is determined by speed of movement of the antenna. Vertical scale is the depth recording interval which is set at frequency sampling. Recording interval presents maximum two-way travel time recorded. The GPR travel time is then converted into depth by calibrating the known object in it or by performing a common midpoint stack with a bi-static antenna around the flat reflector and separating the transmitter and receiver.

Response of radar system relates to the filter of the transmitter antenna, receiver and target response is related with reflection of subsurface objects. GPR penetration capability depends on signal frequency, antenna radiation efficiency and material dielectric properties. Radar signals with high frequencies will produce higher resolutions with limited depth, whereas low-frequency radar signals will produce a deep penetration depth but the resolution is low [25]. The radar wave frequency emitted can be adjusted by changing the antenna. Antenna dimensions vary with radar wave frequency, for example; 1 GHz antenna measuring 30 cm while 25 MHz antenna has a length of 6 m, frequency used depends on the size of the target, approximation of the depth range and maximum aperture of penetration depth.

GPR investigation consists of measuring wave propagation time based on the reflection of conductive rock layers below the surface. Based on the mechanical wave velocity approach v (m/s) with a path along L (m), the position of the reflected field can be determined [11] based on the equation,

$$v = \frac{2L}{t} \tag{4}$$

where t is the time (s).

Equation (4) describes reflected waves for surveys using bi-static antennas. For GPR signals that are affected by the separation between receivers and transmitter (offset) [26] the speed is,

$$v = \frac{c}{\sqrt{\mu_r \varepsilon_r}} \tag{5}$$

where c is the speed of light in a vacuum of 3.10^8 m/s, $\mu = \mu_r \mu_o$ is magnetic permittivity ($\mu_o = 4\pi.10^{-7}$ V/Am), ε_r is dielectric permittivity relative.

In the GPR survey, the depth of the subsurface layer obtained is not exact, but depends on reducing the wave energy that passes through the subsurface medium. Several factors that influence the weakening of electromagnetic waves in the GPR method include; high conductivity of medium, water content, presence of clay minerals, salt content in soil and antenna frequency. The depth interval ranges from 0.40 to 120 m especially for highly conductive medium, whereas in non-conductive medium it can reach 80 m [27]. The vertical resolution of a GPR cross section is the ability to distinguish two reflectors vertically and is usually called type A scanning [10]. According to Nyquist rule, the distance X in ideal conditions equal to 0.25λ, but in reality X is greater than half of the wavelength. Therefore, the selection of antenna frequency values depends on the target anomaly to be detected by considering the high resolution and depth of investigation [32].

3. Analysis and Discussion

GPR investigation for clay layer sometimes presents its own difficulties due to several processes and double layer polarization or spatial polarization [29, 30, 31]. The polarization effect occurs at different frequencies causes dielectric permittivity, and electrical conductivity will be increased, and energy dissipation and weakening reflection. In addition, clay is usually associated with concentration of dissolved ions through ground water, which causes a large absorption of electromagnetic waves energy [28, 29].

IOP Conf. Series: Earth and Environmental Science **279** (2019) 012040 doi:10.1088/1755-1315/279/1/012040

Field measurement data processing is carried out based on stages; (i) Static Correction, (ii) substract-mean (dewow), (iii). Background removal, (iv) band-pass filter, (v) time to depth conversion. The other stages, FK-filter, automatic gain, resampling trace, equidistance trace, fk-migration, picking and mingration. The following is the final results of the GPR cross section in the distance depth region.
Profile Line-02

Figure 4. Location of profile Line02 (MHK21).

The Line02 trajectory with the Northeast to Southwestern direction is located south of Kumala Island, Tenggarong, with a line length of 780 meters. Interpretation of GPR cross sections based on; (i) the continuity of the dominant reflector on each measurement path, (ii) the difference in reflection patterns and (iii) the difference in amplitude between the reflectors [17].

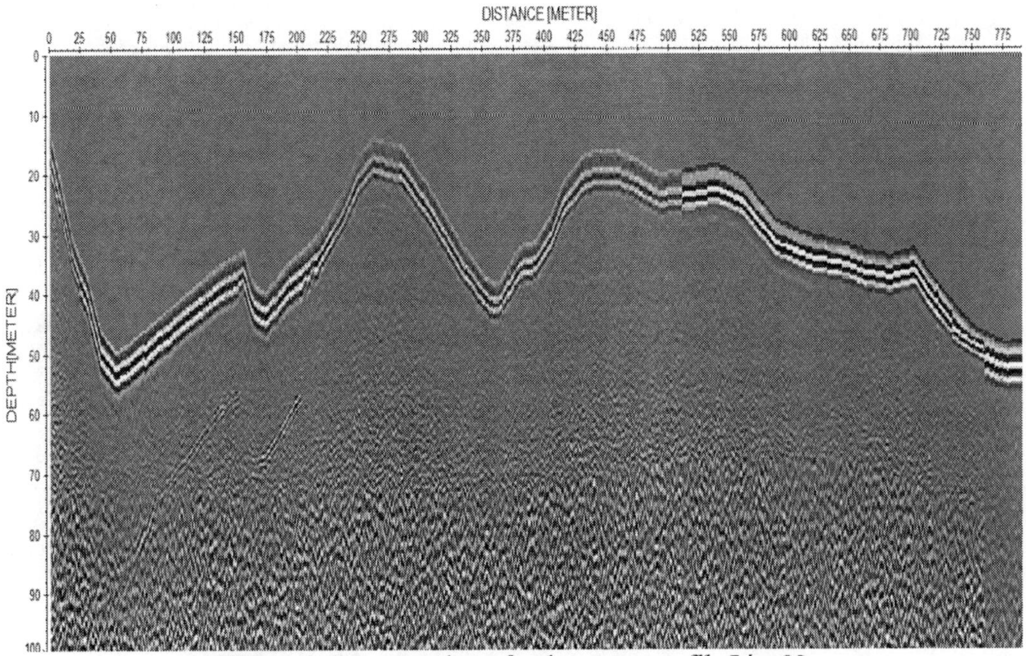

Figure 5. Cross-section of radar-gram profile Line02.

Line02 shows a varied undulation pattern of Mahakam river bed, the depth its 15 to 50 meters with shallow depth at a distance of 275 meters from the starting point of the track. The shallow depth of Line02 is interpreted as a result of the influence of sedimentation from the land. This can be seen from the changes in river banks that are relatively protruding into the river. The other shallow depths are at a distance of 450m which also strongly influenced by the pattern of river banks jutting towards to the river. The thickness of river bed sediment layer around 5 to 10 meters which is a eroded from land and has not compacting, this can be seen from amplitude pattern. Under the sediment layers are estimated as a more compact sediment layer.

Profile Line-03

Figure 6. Location of profile Line03 (P03).

Line03 with relatively direction at Northeast to Southwestern is located on the west of Kumala Island Tenggarong and intersects with Tenggarong Bridge, with a measurement line of 450 meters.

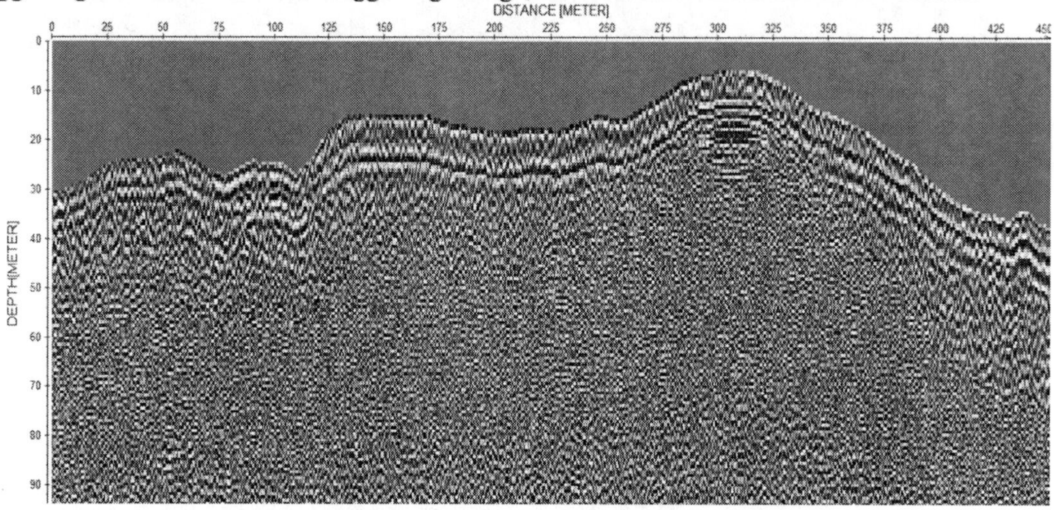

Figure 6. Cross-section of radar-gram profile Line03

The content of organic matter in sediment layer is a sign that there is a limit to the sediment sequence. Whereas non-sedimentary rock layers can be identified based on dielectric contrast values in the GPR cross section. [11]. Line03 shows that the undulation pattern of the Mahakam river bed is relatively stable. The depth of river bed between 5 to 40 meters, with a sludge sediment thickness of more than 20 meters. Location of shallow depth point of Mahakam river bed at 280 to 325 meters from the starting point of line03, exactly below the Tenggarong Bridge. In addition, anomalies around this location are seen in lenses form at radar-gram. The existence of anomaly lens was interpreted as part of the Tenggarong bridge material which collapsed in 2011. The silting of Mahakam river bed around Tenggarong Bridge was caused by the accumulation of sedimentary material due to obstruction of the debris from collapsed bridge. Below the sediment layers are estimated to be a basement rock layer.

Profile Line-04

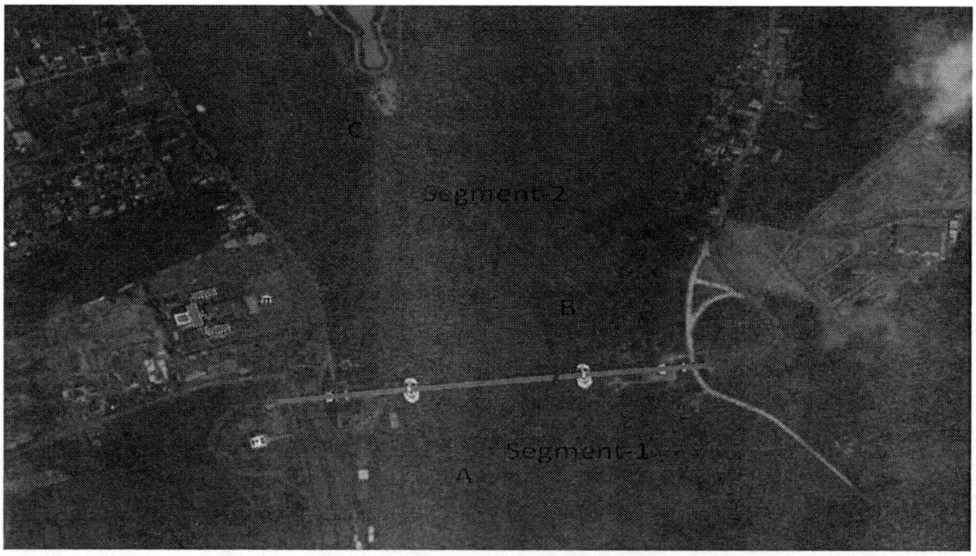

Figure 7. Location of profile Line04.

IOP Conf. Series: Earth and Environmental Science **279** (2019) 012040 doi:10.1088/1755-1315/279/1/012040

Location of Line04; Southwest to Northeast direction (segment-1: 0 to 375 meter) and Southeast to Northwest direction (Segment-2: 375 to 535 meter) so that the total length is 910 meters. This line intersects Tenggarong Bridge in segment-1, and segment-2 intersects of Mahakam River the end at the western edge of Kumala Island.

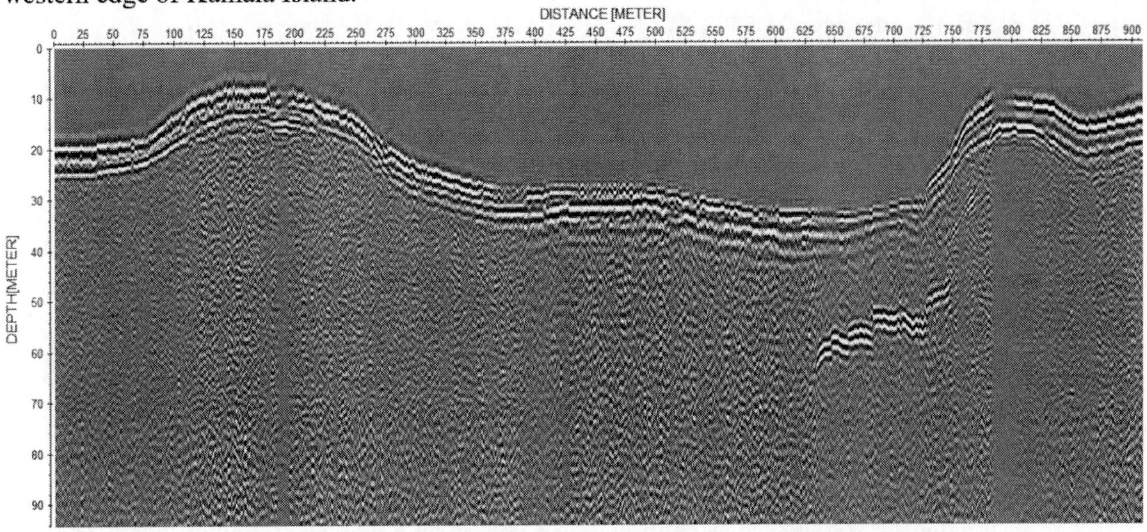

Figure 8. Cross-section of radar-gram profile Line04.

Line04 shows that the undulation pattern of the Mahakam river bed is relatively stable. The depth of river bed between 7 to 30 meters, where shallow depths are at a distance of 130 to 225 meters from starting point of lines. The shallow depth of this line below Tenggarong Bridge. This siltation is probably caused by obstruction of sediment material carried by water flow of the Mahakam river debris, because this path is in line with Mahakam river flow.

The presence of the Mahakam river debris is seen from the Mahakam river bed reflector pattern just below continuous Tenggarong Bridge. Besides that the emergence of anomalous patterns just below the river bed sediment layer. The thickness of sediment layer of river bed is more than 10 meters which is a sediment from soil overhaul and has not undergone compaction. This can be seen from the amplitude pattern. Below the sediment layers are estimated as a more compact sediment layer.

4. Conclusion and Discussion

Based on the results of the GPR cross section analysis, it was concluded that,

a. Undulation pattern of Mahakam river bed that varies with a depth of 15 to 50 meters (Line02); between 7 and 40 meter (Line03); between 7 and 30 meter (Line 04).
b. The shallow depth from 2D GPR section is interpreted due to influence of sedimentation from the land and obstruction of sediment material carried over by the remnants of the Mahakam river debris.
c. The thickness of the river bed sediment layer around 5 to 10 meters which a sediment resulting from errored land and has not compacting.
d. Further search is needed to determine the Mahakam river sedimentation rate.
e. Cross checking of chemical sediment content is needed, to determine the basic sediment materials at the river bed.

References

[1] Forest Watch Indonesia 2016 *Intip Hutan Indonesia: Menyelamatkan Hutan, Menyelematkan Bumi* http://fwi.or.id/ (access on Oct 25th, 2018).

[2] Allen G P, Mercier F 1994 *Proc. of the Indonesian Petroleum Association* p. 261-273.

[3] Lambert B 2003 Micropaleontological investigations in the modern Mahakam delta, East Kalimantan (Indonesia) *Carnets de Géologie/Notebooks on Geology–Article* 2003/02 (CG2003_A02_BL).
[4] Husein S, Lambiase J J 2005 *Proc. of the Indonesian Petroleum Association* p. 367-379.
[5] Roberts H H and Sydow J 1996 *Proc. of the Indonesian Petroleum Association Twenty-Fifth Silver Anniversary Convention* p. 147-161.
[6] Hidayati D, Yogaswara H, Djohan E 2006 *Partisipasi Stakeholders Dalam Konsolidasi Tanah Delta Mahakam: Alternatif Penataan Kawasan* (unpublished).
[7] Sambuelli L, Calzoni C, Pesenti M 2009 *GEOPHYSICS* **74** (4) pp. B95-B104.
[8] McLaren P 1981 *Journal of Sedimentary Petrology* **51** pp. 611-624.
[9] Pettijohn F J 1975 *Sedimentary Rocks*, 3rd ed., Harper and Row, Publishers, Inc. New York, Evanston, San fransisco, and London, pp. 628.
[10] Daniels D J 2004 *Ground Penetrating Radar* 2nd Edition, IET.
[11] Brito G L M, Coutinho A P, Cabral J J S P, Neto S M S, Antonino A C D, Cirilo J A, Braga R A P, Filho S L S 2018 *Revista Brasileira de Recursos Hídricos Brazilian Journal of Water Resources*.
[12] Bungey J H 2004 *Constr Build Mater* **18** pp. 1–8.
[13] Barrile V, Paccinotti R 2005 *NDT&E Int* **38** pp. 596–604.
[14] Chang C W, Lin C H, Lien H S 2009 *Constr Build Mater* **23** pp. 1057–1063.
[15] Sukma A M, Krisningrum N N, Assyiffa T N, Alami F, Rasimeng S, Alhadi A 2018 Ground penetrating radar (GPR) for identifying the depth of spun pile gas station at Batam *The 43rd Annual Scientific Meeting of Indonesian Geophysicists Association* Semarang.
[16] Evans R, Frost M, Stonecliffe-Jones M, Dixon N 2006 *Proc Inst Civ Eng Munic Eng* **159** (2) 105–111.
[17] Emanuel Huber and Peter Huggenberger 2016 *Hydrol Earth Syst Sci* **20** pp. 2035-2046.
[18] Beres M, Haeni F P 1991 *Ground Water* **29** 375–386.
[19] Neal A 2004 *Earth-Sci Rev* **66** 261–330.
[20] Mellet J S 1995 *Appl Geophys* **33** 157–166.
[21] Pyke K, Eyuboglu S, Daniels J J, Vendl M 2008 *J Environ Eng Geophys* **13** 335–342.
[22] Gerhards H, Wollschläger U, Yu Q, Schiwek P, Pan X, Roth K 2008 *Geophysics* **73** (4) J15–J23.
[23] Birchak J R, Gardner C, Hipp J E, Victor J M 1974 Proc. IEEE **62** 93–98.
[24] Akula 2018 *9000 Radar Control Units* www.geoscanners.com. Sweden.
[25] Arcone S A, Finnegan D, Laatsch J E 2006 Bathymetric and subbottom surveying in shallow and conductive water: *Proc. of the 11th International Conference on Ground Penetrating Radar*.
[26] Huisman J A, Hubbard S S, Redman J D, Annan A P 2003 *Vadose Zone J* **2** 476–491.
[27] Sheng Huoo N, Ching-Kuan C, Hong-Ming L 2002 Application of ground penetrating radar on the void-detection in levee. *Proc. Int. J. Offshore Polar Eng.*
[28] Saarenketo T 1998 *J Appl Geophys* **40** 73–88.
[29] Ishida T and Makino T 1999 *J Colloid Interface Sci* 212.
[30] Santamarina J C, Klein K A, Fam M A 2001 *Soils and Waves: Particulate Materials Behavior, Characterization and Process Monitoring* John Wiley and Sons New York.
[31] Bittelli M, Salvatorelli F, Rossi Pisa P 2008 Geoderma 143, pp. 133–142.
[32] Di Prinzio M, Bittelli M, Castellarin A, Pisa P R 2010 *J Appl Geophys* **71** 53–61.

Acknowledgments
We would like to thank for Ministry of Environment, Republic of Indonesia as funding. Thank's to Hydro-Oceanography Team, Department of Physics, Faculty of Mathematics and Natural Science, Mulawarman University for the GPR acquisition team. Thank's to Geophysics Exploration Laboratory and Geophysical Data Processing and Modeling Laboratory, Department of Geophysical Engineering, University of Lampung for equipment and software supporting.

Seismic Hazard Measurement of Earthquake Swarms Activity Based on Horizontal Vertical to Spectral Ratio Analysis (HVSR) in West Halmahera, Indonesia

R W Ningrum[1*], H Fauzi[2], W Suryanto[3] , and E T W Mei[4]

[1] Dept. of Physics Education, Khairun University, Ternate, North Maluku, Indonesia
[2] Dept. of Architectural engineering, Khairun University, Ternate, North Maluku, Indonesia
[3] Dept. of Physics, Universitas Gadjah Mada, Yogyakarta, Indonesia
[4] Dept. of Geographic, Universitas Gadjah Mada, Yogyakarta, Indonesia
*Corresponding Author : ima.geolive@gmail.com

Abstract. The swarm earthquake that occurred in West Halmahera which recorded a lot resulted in damage to houses in several rural areas. Although human casualties are not too significant, they affect the stability and capacity of an area in terms of regional development. Mapping of seismic hazard zone is a solution to minimize the impact that will be caused by the earthquake. The purpose of the study is to initial mapping seismic hazard zoning based on Horizontal Vertical to Spectral Ratio (HVSR) analysis. The estimated values of the fundamental frequency range from 0.0 Hz up to 4.9 Hz increases with decreasing depths of basement rock. Regions with the lowest dominant frequency value are South Ibu sub-District, this shows that the area has a large sediment thickness and is a vulnerable area of high. Amplification value range of 4.7 – 51, where the higher the amplification value, the thickness of sediments. The highest amplification value area is Jailolo sub-District and is an area that has a high risk of damage if shaken by an earthquake. This shows that areas that have high amplification values are areas that have the greatest potential for damage due to earthquakes. Analysis obtained from the results of the data shows that the topographic influence is also very significant with the earthquake waves that occur.

1. Introduction

West Halmahera is one of the areas in North Maluku with the highest and unique earthquake vulnerability. This is because the area is located in the convergence of three main plates, the Eurasian plate, the Pacific plate and the Australian Plate, and small plates such as the Philippine plate, Sangihe plate and Halmahera plate. The intensity of the earthquake is centered in the Jailolo area which is the capital of West Halmahera Regency and the center of community activity. Earthquakes in the Jailolo area are dominated by earthquakes with a type of swarm which has a small magnitude character with a very high frequency of occurrence and lasts a long time and is not accompanied by a strong earthquake as the main earthquake [1].

The earthquakes that occurred in Jailolo, West Halmahera were not only sourced from the Maluku Sea Plate but also local earthquakes in the waters of the Jailolo bay and around the Jailolo Mountain. Vibrations felt around Mount Jailolo include shallow earthquakes and cause damage to regional infrastructure. Earthquake events, especially the Tectonic Earthquake in Jailolo, West Halmahera in the last few years, have resulted in damage to houses in several rural areas. Although human casualties are not too significant,

IOP Conf. Series: Earth and Environmental Science **279** (2019) 012041 doi:10.1088/1755-1315/279/1/012041

they affect the stability and capacity of an area in terms of regional development. It was recorded that the earthquake in 2015, which was 2.385 houses was severely damaged, 1.902 were damaged, and 5.149 were slightly damaged. This is due to the limited level of vulnerability and infrastructure that has a greater earthquake impact [2].

Seismic activity in the Maluku Sea is dominated around the coast of West Halmahera. Significant earthquakes in the area. The distribution of seismic activity in West Halmahera is shown in the figure below:

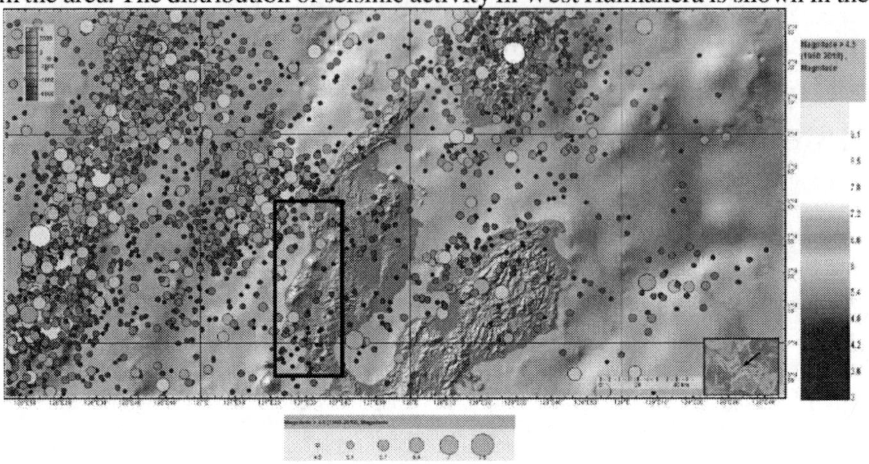

Figure 1. The epicenters of earthquake swarm area of West Halmahera, Indonesia (Data Source: usgs.gov)

Earthquake that has a large and seismic level with an unclear period of time are called earthquake swarm [3]. Earthquakes are expected to come from the pressure of a moving plate and produce a deformation in the rock due to the growing pressure, that's when the earthquake occurs. The earthquake that occurred in Jailolo is influenced by normal faults but is also centered on the Jaiolo mountain which is an old volcano that is inactive and underdeveloped. The cause and trigger of the Jailolo swarm are rocks that burst and propagate laterally. For Jailolo area it is indicated that Jailolo volcano has active magma activity and is expected to carry out research related to volcanic hazards [3] Based on the data obtained that Gunung jailolo is classified as an active volcano although the manifestations and eruptions that occur are not documented in historical records. this volcano can erupt in the Holocene [4].

Mapping of earthquake hazards is a solution to minimize the impact that will be caused by the earthquake. Information on land characteristics and local geology is needed to determine the level of vulnerability of seismic hazard in the West Halmahera area. Microtremor method with Horizontal Vertical Spectral Ratio (HVSR) analysis is a method used to predict subsurface responses based on the dominant period (T) and soil amplification (A) values. This HVSR method is easy to apply and is environmentally friendly to an area with a dense population.

2. Tectonic Setting

Tectonic background in Maluku Sea, one of which is the earthquake in 2014 with a magnitude of 7.1 SR [5]. Figure 1 shows tectonics from the research area [5].

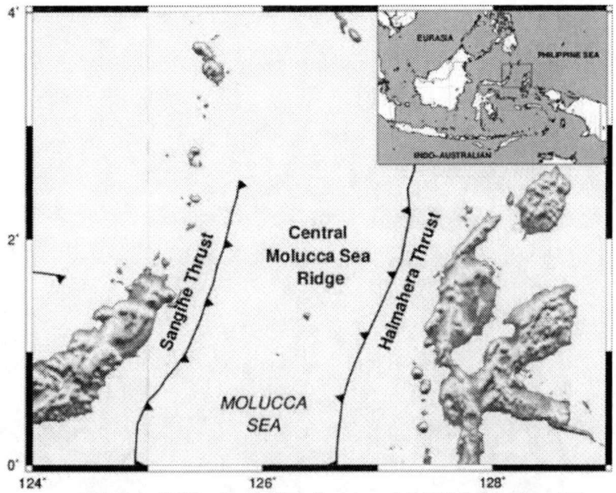

Figure 2. Tectonic Background in Maluku Sea [5]

Maluku Sea is a collision zone with tectonic conditions with the most complex seismicity and is a collision interaction between the Sangihe Archipelago arc that moves eastward with the Halmahera Archipelago arc towards the West [6]. Active volcanoes in the Halmahera arc are produced due to subduction from the Northeast of Maluku [7].

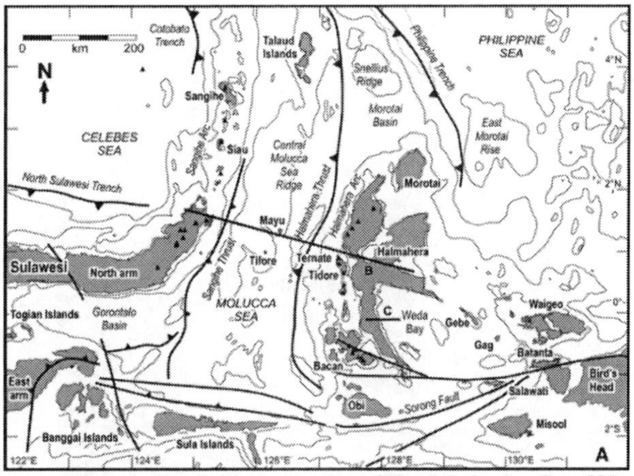

Figure 3. Major tectonic of the Molucca Sea [8]

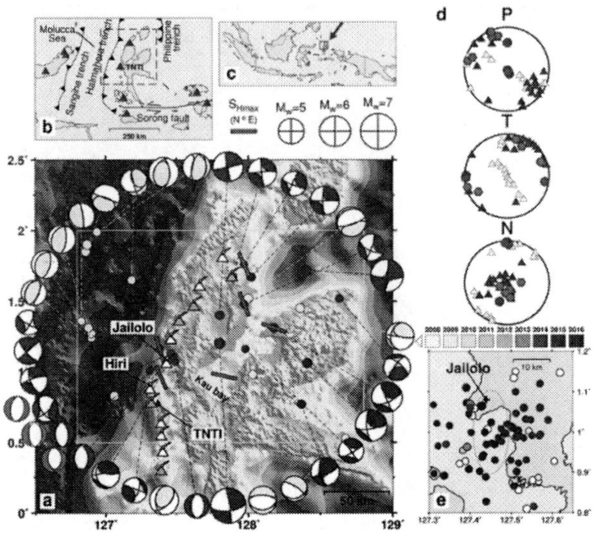

Figure 4. Map of the focal mechanism Halmahera [3]

Jailolo volcano does not have fumarolic, but the emergence of hot springs and hot steamy soil [9]. Earthquake swarm in Jailolo caused severe damage and collapse of houses and buildings in villages located on the south side of Jailolo volcano (Jailolo, Saria and Bobanehena village) and minor damage occurred at Idamdehe caldera. There is a crack along the coastline near Bobanehena and Saria which is indicated as a new hot spring [10]. Figure 5 is a Jailolo mountain located south of the western Halmahera volcanic arc.

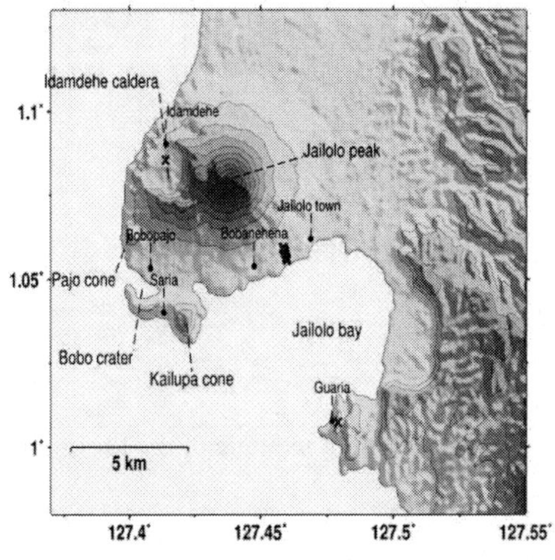

Figure 5. Map of Main Structure in Jailolo Volcano [3]

Swarm earthquakes in the Jailolo volcano occur in a relatively long time, which is more than 2 years starting from 2015 to 2016. It is suspected that there is magma activity, swarm earthquake period, and composition [11, 12].

3. Potential Seismic Hazard

Seen from the tectonic order, the Jailolo area is very vulnerable to earthquake disasters. Earthquake data BMKG records earthquake disasters that occurred in 2015-2016 resulting in damage to infrastructure.

Geological studies show that the Jailolo area is a series of young volcanic mountains. Earthquakes that occur from plate movements, active faults and other types of complex earthquakes. The source of the earthquake due to plate movement includes the Mayu ridge zone (Maluku sea plate), Sangihe earthquake zone and Halmahera-Irian earthquake zone. While the source of the active fault earthquake is the Sorong fault and the complex earthquake source includes the earthquake zone of the Obi fragment and the earthquake source zone of the Banggai-Sula fragment.

Based on data from the North Maluku Disaster Risk Assessment 2016-2020, the West Halmahera region includes a high potential earthquake hazard (BNPB, 2016). The description of the level of earthquake disaster risk in the North Maluku region based on the Disaster Risk Assessment can be seen in the figure below:

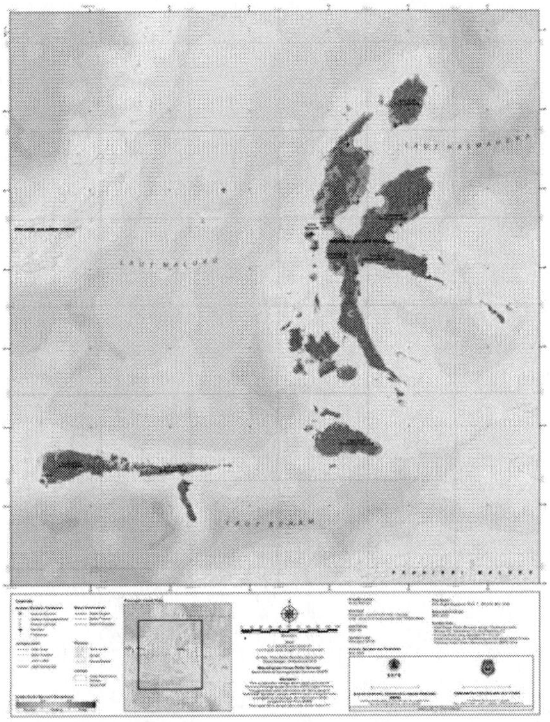

Figure 6. Earthquake Disaster Risk Map in North Maluku

4. Method

The primary data used are seismic wave data (microtremor) as measured by a TDS-303S (Digital Portable Seismograph-three component) seismometer with a sampling frequency of 100Hz. Seismic data obtained were 57 points scattered in West Halmahera (Figure 7). In this research, we use HVSR method or called as the Nakamura technique (1989). The method used in this study is the microtremor method, where this method utilizes ambient waves that appear around the location or also called local earthquakes which are used to determine some characteristic parameters of soil dynamics. Microtremor data processing is assisted with Matlab, surfer, and ArcGIS programs. Data that is processed to produce fo value, Ao.

Figure 7. Distribution of microtremor measurements at West Halmahera

Table 1. Microtremor Measurement Point in West Halmahera

No	Point Code	Date	Latitude	Longitude	Location
1	B1	13 July 2018	1,0538	127,4474	Bobanehena
2	B2	13 July 2018	1,0552	127,4486	Bobanehena
3	B3	13 July 2018	1,0556	127,4458	Bobanehena
4	B4	13 July 2018	1,0531	127,4483	Bobanehena
5	B5	13 July 2018	1,0503	127,4442	Bobanehena
6	C1	14 July 2018	1,0972	127,4172	Idamdehe
7	C2	14 July 2018	1,0874	127,4152	Idamdehe
8	C3	14 July 2018	1,0865	127,4147	Idamdehe
9	C4	14 July 2018	1,0856	127,4152	Idamdehe
10	C5	14 July 2018	1,0886	127,4136	Idamdehe
11	C6	14 July 2018	1,0902	127,4141	Idamdehe
12	C7	14 July 2018	1,0917	127,4132	Idamdehe
13	C8	14 July 2018	1,0931	127,4134	Idamdehe
14	C9	14 July 2018	1,0895	127,4123	Idamdehe
15	C10	14 July 2018	1,0874	127,4108	Idamdehe
16	D1	15 July 2018	1,0608	127,4676	Gufasa
17	D2	15 July 2018	1,0514	127,4243	Payo
18	D3	15 July 2018	1,0638	127,4066	Bobo-Idamdehe
19	D4	15 July 2018	1,0803	127,4052	P. Damar Tour
20	D5	15 July 2018	1,1066	127,4300	Marinbati
21	D6	15 July 2018	1,1087	127,4541	Taboso
22	D7	15 July 2018	1,0879	127,4729	Loku Kie
23	D8	15 July 2018	1,0699	127,4719	Gamlamo
24	D9	15 July 2018	1,0240	127,5208	Sidangoli-Jailolo street

25	D10	15 July 2018	1,0478	127,5165	Sidangoli-Jailolo street
26	D11	15 July 2018	1,0633	127,5005	Bukumatiti
27	E1	16 July 2018	1,0668	127,4662	Jalan Baru
28	E2	16 July 2018	1,0725	127,4674	Sokonara
29	E3	16 July 2018	1,0830	127,4722	Hatebicara
30	E4	16 July 2018	1,0560	127,4992	Todowangi
31	E5	16 July 2018	1,0455	127,5091	Tuada
32	E6	16 July 2018	1,0203	127,4970	Motui
33	E7	16 July 2018	0,9741	127,5139	Buku Bualawa
34	E8	16 July 2018	0,9841	127,5063	Buku Maadu
35	E9	16 July 2018	0,9910	127,4980	Tauro
36	F1	17 July 2018	0,9546	127,5156	Tataleka
37	F2	17 July 2018	0,9025	127,5089	Sidangoli Gam
38	F3	17 July 2018	0,8836	127,6422	Tetewang
39	F4	17 July 2018	0,9706	127,6249	Akelamo Kao
40	F5	17 July 2018	0,8643	127,6872	Pasir Putih
41	F6	17 July 2018	0,8030	127,6837	Braha
42	F7	17 July 2018	0,9276	127,5804	
43	G1	18 July 2018	1,1006	127,5036	Porniti
44	G2	18 July 2018	1,2014	127,5119	
45	G3	18 July 2018	1,2951	127,5395	Ibu selatan
46	G4	18 July 2018	1,3664	127,4901	Ibu Utara
47	G5	18 July 2018	1,5495	127,6117	Ibu Utara
48	G6	18 July 2018	1,4963	127,6832	Ibu Utara
49	G7	18 July 2018	1,5002	127,5534	Ibu
50	H1	19 July 2018	1,4296	127,5247	Ibu
51	H2	19 July 2018	1,1338	127,4863	Sahu
52	H3	19 July 2018	1,0669	127,4596	Gufasa
53	H4	19 July 2018	1,0608	127,4556	Galala
54	H5	19 July 2018	1,0419	127,4073	
55	H6	19 July 2018	1,0527	127,4438	
56	H7	19 July 2018	1,1909	127,4116	Todahe
57	H8	19 July 2018	1,1519	127,4345	Sangaji (Sahu)

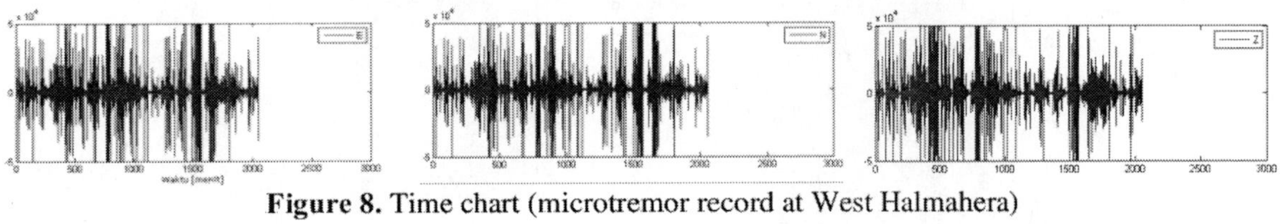

Figure 8. Time chart (microtremor record at West Halmahera)

5. Result

Based on the results of the observations of the ambient waves at 57 points, the three components of the wave component direction Nort – South (N-S), East – West (E-W) and vertical (Z) are obtained. The results of the analysis using the HVSR method will produce a graph of the relationship between H / V and frequency (Figure 9). The value of amplification (A_o) and the dominant frequency (f_{dom}) obtained from the graph will be interpolated so that it can be used as a contour of the values of Amplification (A_o), Frequency (f_{dom}) and Dominant Period (T_{dom}) in the study area.

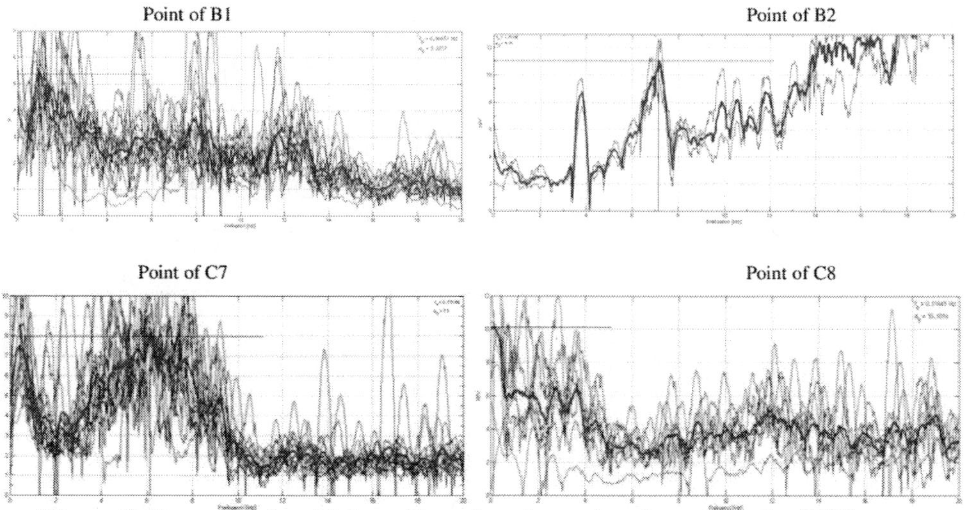

Figure 9. Example of multiple and broad peak graphs of spectra ratio (H/V) curves

The graph of the relationship between H/V and frequency indicates that there is a significant lateral variation in the thickness of the underground or sedimentary structures. Contour maps of dominant frequency values (f_0) in the West Halmahera area vary from 0.0 - 4.9 Hz (Figure 10). From the observation point, around 65% of the regions have a low value of f_0 or less than 1 Hz. The region with the lowest dominant frequency value is the South Mother district, this indicates that the area has a large sediment thickness and is a high-risk earthquake area. Areas with the highest frequency of the dominant frequency will be lower if in areas with deeper basement depth. The dominant frequency becomes higher when in area with shallower basement depths [13].

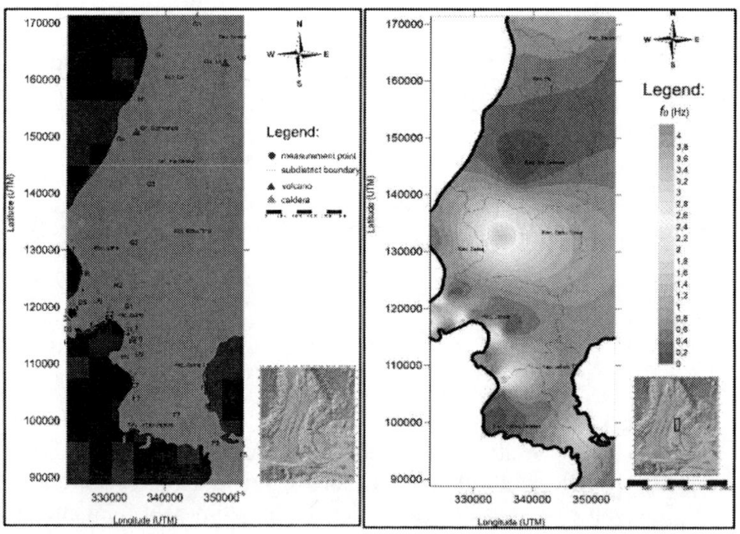

Figure 10. Contour map of dominant frequency (*fo*) in West Halmahera

The amplification value is affected by the wave velocity, the greater the wave velocity, the smaller the amplification value, and vice versa. The value of amplification is closely related to the level of rock density, increasing the value of amplification, the rock will be reduced in density or soft. Contour maps for amplitude H/V (A_0) ratios in West Halmahera (Figure 11) show varying values around 4.7 - 51. The highest amplification area is located at point D10, Jailolo sub-district and is an area with high damage risk. If shaken by an earthquake.

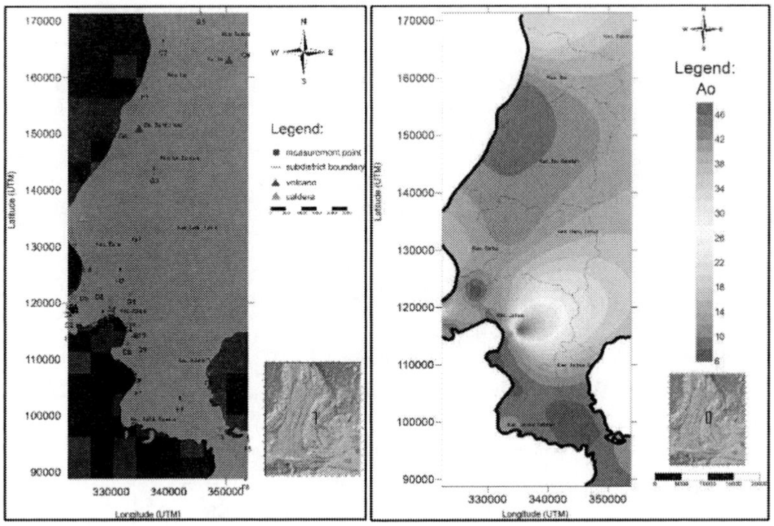

Figure 11. Contour Map of amplitude ratio (*Ao*) in West Halmahera

The influence of topography is very significant with the incoming earthquake waves so that there are some buildings that are damaged and not damaged even though in the same area or still in one village (Table 2).

Table 2. Comparison of HVSR charts on damaged and undamaged village buildings

Point	Damaged		Point	Not Damaged	
	Graph	Building		Graph	Building
B1	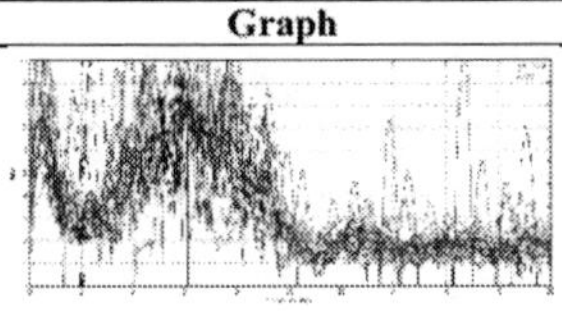	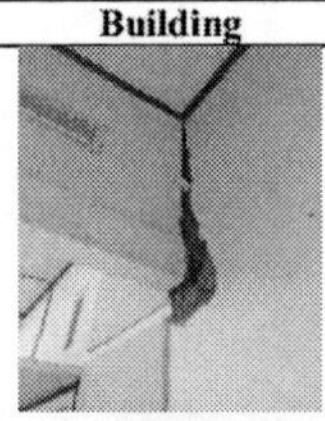	B4	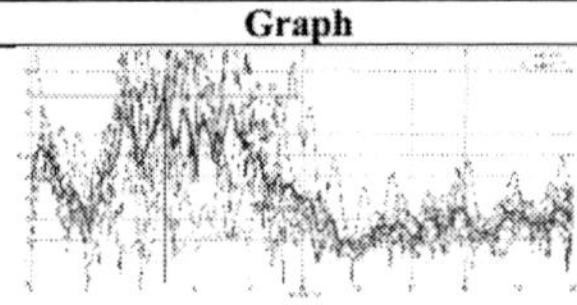	
C2	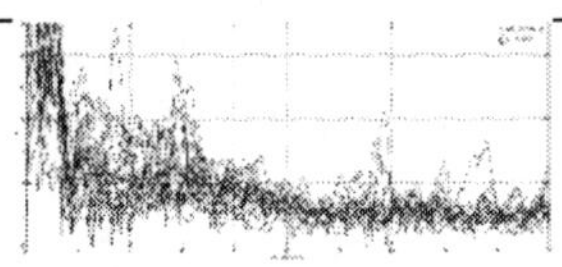		B3	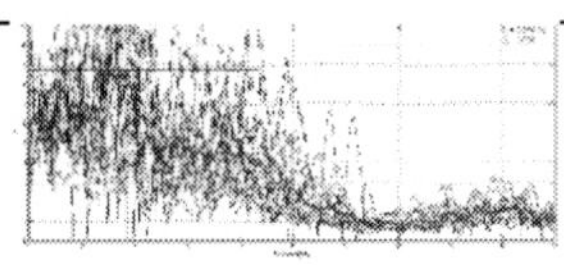	

6. Conclusions

1. The dominant frequency value (f_0) in the West Halmahera area varies from 0.0 - 4.9 Hz. Regions with the lowest dominant frequency value are South Ibu sub-District, this shows that the area has a large sediment thickness and is a vulnerable area of high.

2. The amplitude ratio of H / V (A0) in the West Halmahera area shows varying values around 4.7 - 51. The highest amplification value area is Jailolo sub-district and is an area that has a high risk of damage if shaken by an earthquake.

References

[1] Mogi K 1963 *Bulletin of the Earthquake. Research Institute* **41** pp. 615-658.
[2] Nugroho S P 2016 *Evaluasi Penanggulangan Bencana 2015 Dan Prediksi Bencana 2016*. Kepala Pusat Data Informasi dan Humas BNPB Jakarta.
[3] Passarelli L, Heryandoko N, Cesca S, Rivalta E, Rasmid, Rohadi S, Dahm T, Milkereit C 2018 *Frontiers in Earth Science* **6** Article 79.
[4] Venzke E 2013 *Global Volcanism Program, 2013*. Volcanoes of the World, v.4.5.4. Smithson. The institution, Natl. Museum Nat. Hist.
[5] Gunawan E, Kholil M, Meilano I 2016 *Physics of the Earth and Planetary Interiors* **259** 29-33.
[6] Lee and Lawver 1995 *Tectonophysics* **251** 85-138.
[7] Hall R and Wilson M E J 2000 *Journal of Asian Earth Science* **18** 781-808.
[8] Waltham D, Hall R, Smyth H R, Ebinger C J 2008 *The Geological Society of America Special Paper* 436.
[9] Verstappen H T 1964 *K. Ned. Aardrijkskd. Genoot* **81** 297–316.
[10] Gunawan E, Ghozalba F, Widiastomo Y, Meilano I, Hanifa N R, Hidayati S 2017 *Geotechnical and Geological Engineering* **35** (1) 425-432.
[11] White S M, Crisp J A, Spera F J 2006 *Geochem. Geophys. Geosys* **7** Q03010. doi: 10.1029/2005GC001002.
[12] Passarelli L and Brodsky E E 2012 *Geophys J Int* **188** 1025–1045. doi: 10.1111/j.1365-246X.2011.05298.x.
[13] Parolai S, Bormann P, Milkereit C 2001 *J Earthquake Eng* **5** 541– 564.

Acknowledgments

We would like to thank Seismology Laboratory UGM and Environment Laboratory Universitas Khairun for permits to use and revise the template provided by the laboratory.

IOP Conf. Series: Earth and Environmental Science **279** (2019) 012042 doi:10.1088/1755-1315/279/1/012042

Interpretation of One Dimensional Schlumberger Curve Resistivity Data using "Least Square" Inversion

***Harmita Lestari[1], Muhammad Nasri[1], Andi Zulkifli[1], Sabrianto Aswad[1], Syamsuddin[1]**

Geophysics Departement, Hasanuddin University, South Sulawesi. Indonesia

Email address: HarmitaLestaryMVD@gmail.com

Abstract. The geoelectric method using Schlumberger configuration is one of the most widely used geophysical methods. In this study, a one-dimensional Schlumberger configuration will be applied to modeling the subsurface structure using two types of data, namely synthetic data and acquisition data in the field. The purpose of this study is to calculate resistivity (ρ_i) and determine the thickness (d_i) of the subsurface layer. In order for the model to be obtained to have a high level of accuracy, inversion least square is used. The value of resistivity and layer thickness obtained are very close to the synthetic model used. Comparison of calculation pseudo resistivity curves and observation pseudo resistivity has a small RMS error value which supports the accuracy of the study. This is obtained by giving initial guesses to the resistivity value of the acquisition data. The final results obtained from this study are in the form of five layers with different resistivity values and with a relatively small RMS error value of 1.15.

Keywords: Geoelectric Method, Two-dimensional Schlumberger Configuration, Inverse Modeling, Resistivity.

1. Introduction

In surveying geoelectric methods there are configurations that are often used, including Schlumberger, Wenner, and Dipole-dipole configurations. Schlumberger configurations that have the deepest range compared to other configurations where the potential electrode distance is fixed, but the distance between the current electrodes is changed.

In geophysics generally known models and model parameters are used to characterize a subsurface geological condition. The process of estimating models and model parameters based on the observed data on the surface of the earth to produce a response that matches the observational data or field data is called modeling.

1.1 Electric Potential on the Earth

If a continuous current is channeled into an isotropic homogeneous medium, if ∂A is an element of surface area and J is the electric current density (A / m2), then the magnitude of the current element passing through the surface element [1] is:

$$\partial I = \vec{J}.\partial\vec{A} \quad (1)$$

1.2 Potential by Two Sources

In exploration using geoelectric methods, two pairs of electrodes are commonly used, one pair of current electrodes and one pair of potential electrodes as shown by [2]:

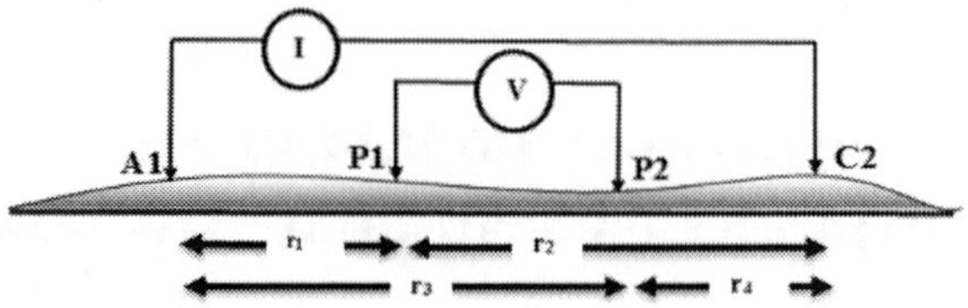

Figure 1. The arrangement of two pairs of current and potential electrodes on the surface of the earth.

The potential difference between points P1 and P2 is given by [3]:

$$\Delta V = \frac{\rho I}{2\pi}\left[\left(\frac{1}{r_1}-\frac{1}{r_2}\right)-\left(\frac{1}{r_3}-\frac{1}{r_4}\right)\right] \tag{2}$$

So, obtained the following equation [1]:

$$\rho = \frac{2\pi}{\left[\left(\frac{1}{r_1}-\frac{1}{r_2}\right)-\left(\frac{1}{r_3}-\frac{1}{r_4}\right)\right]}\frac{\Delta V}{I} \tag{3}$$

1.3 Potential for Layered Earth Models

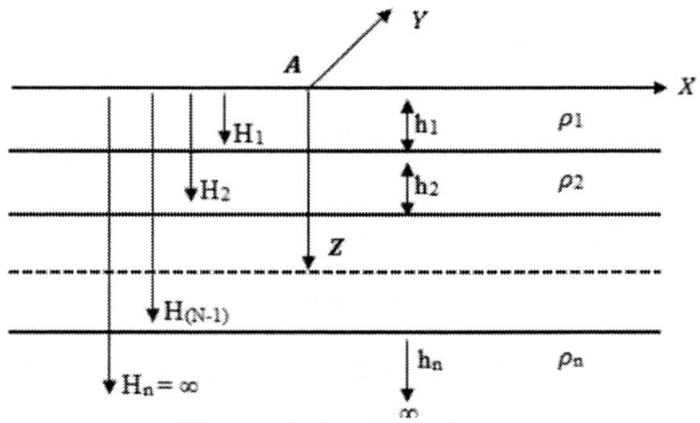

Figure 2. Model n Layer

For a layered earth model consisting of horizontal layer N (Figure 2) that each layer is homogeneous and isotropic, the separation between one layer and another layer is the boundary plane between two different resistivity media [1].

It is known that if the earth is homogeneous and isotropic then the electrical potential generated by a source point on the surface is [1]:

$$V_0 = \frac{I\rho}{2\pi}\frac{1}{R} = \frac{I\rho}{2\pi}\frac{1}{(r^2+z^2)^{\frac{1}{2}}} \tag{4}$$

1.4 Kernel Functions and Resistivity Transformation Functions

The initial stage of interpretation of resistivity measurements of the earth's layers is the analytic search for Kernel functions (K ((λ)), produced [4]:

$$K_i(\lambda) = \left[\frac{K_{i+1}(\lambda)+p_i\,Tanh\,(\lambda d_i)}{p_i+K_{i+1}(\lambda)Tanh\,(\lambda d_i)}\right] \tag{5}$$

1.5 Forward Modelling and Inverse Modelling

Forward modeling states the process of calculating "data" which is theoretically observed on the surface of the earth if it is known the price of certain subsurface model parameters. In modeling geophysical data, a model is sought that produces a suitable response or fit with observational data or field data [5]. In Inverse modeling, the model parameters are obtained directly from the data. [6] defines inversion theory as a unity of mathematical and statistical techniques or methods to obtain useful information about a physical system based on observations of the system [5].

IOP Conf. Series: Earth and Environmental Science **279** (2019) 012042 doi:10.1088/1755-1315/279/1/012042

Geoelectric data inversion is a complex problem. Therefore, a small change in the data can bring a big change to the model, so it needs to be given the first initial (initial model) to approach like the actual model. This problem can be solved by introducing a damping factor (damping factor) into the equation system [7]. This parameter is added to the diagonal a^Ta which helps to increase the level of direct current of eigenvalues so that no eigenvalues can be zero [8]. The damped least squares solution is given by the following equation:

$$\Delta m = (a^T a + \varepsilon^2 I)^{-1} a^T \Delta d \tag{6}$$

Where Δm is a correction vector parameter, Δd is the data difference vector data, 'a" is the Jacobian matrix. "I" is an identity matrix and "ε" is called "a" muffled factor (damping factor) [9].

Singular Value Decomposition (SVD) can be easily applied to small-scale geophysical problems. It is mathematically strong and numerically stable and also provides other important information about the state of the model and data so as to enable the resolution of the model and the study of covariance [10]. Equation (6) can be solved using the SVD scheme in the inversion scheme. A matrix of NxN or NxP matrix a can be factored into products from the other 3 matrices as follows [9].

$$a = USV^T \tag{7}$$

Where for data N and P are measured parameters. U (NxP) and V (PxP) is respectively data spaces and the eigenvector and S parameters are PxP diagonal matrices which consist of non-zero r of a, with r≤m. This diagonal is included in S (λ_1, λ_2 ... λ_p) which is called the singular value of a. This factorization is known as SVD from a [9]. If SVD is used in the least square solution damped. Equation (27) is obtained [9].

$$\Delta m = (VS^2V^T + \varepsilon^2 I)^{-1} VSU^T \Delta d \tag{8}$$

By entering the damping factor into diagonal elements, then inverse it is obtained [9]:

$$\left(Vdiag(\lambda_j^{\,2} + \varepsilon^2)V^T\right)^{-1} = Vdiag\left\{\frac{1}{\lambda_j^{\,2} + \varepsilon^2}\right\}V^T \tag{9}$$

Substitution of equation (8) to equation (9) so that it is obtained:

$$\Delta m = Vdiag\left\{\frac{1}{\lambda_j^{\,2} + \varepsilon^2}\right\}V^T VSU^T \Delta d \tag{10}$$

And vector correction parameters can be expressed as:

$$\Delta m = Vdiag\left\{\frac{\lambda_j}{\lambda_j^{\,2} + \varepsilon^2}\right\}U^T \Delta d \tag{11}$$

Equation (11) provides a damped least square solution with SVD.

2. Result and Discussion

Results of Synthetic Data Models and Inversion Results

(a) Model 1: Two-Layer Earth Model with $\rho_1 > \rho_2$

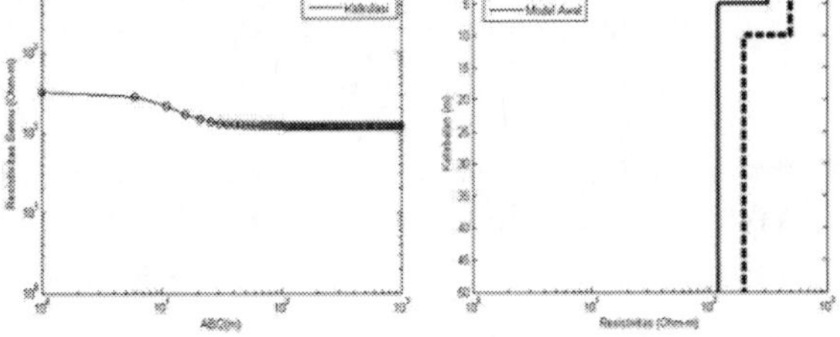

Figure 3. Comparison of observation resistivity curve model and synthetic layer thickness with calculation resistivity curve and inversion layer thickness for the two-layer earth model with $\rho_1 > \rho_2$.

Table 1. Comparison of apparent resistivity values and thickness of synthetic layers with apparent resistivity values and inversion yield layer thicknesses for two-layer earth models with $\rho_1 > \rho_2$.

Number of layer	ρ (Ω.m)			d (m)			Error
	Synthetic	Inversion	Difference	Synthetic	Inversion	Difference	
1	320	320	0.00	5	5	0.00	0.00
2	120	120	0.00	-	-	-	

Model 2 : Three-Layer Earth Model with $\boldsymbol{\rho_1 > \rho_2 > \rho_3}$

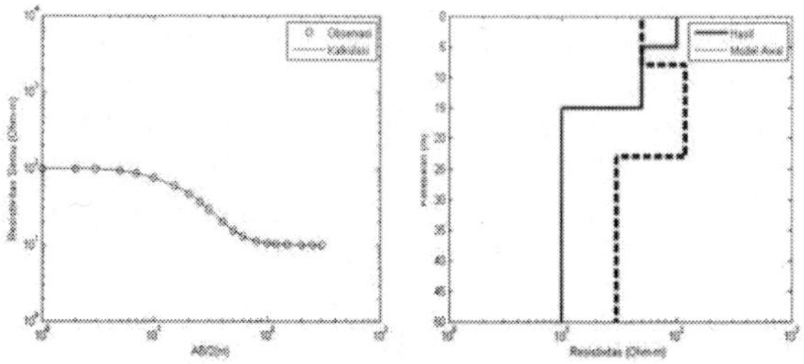

Figure 4. Comparison of observation resistivity curve model and synthetic layer thickness with calculation resistivity curve and inversion yield layer thickness for the three-layer earth model with $\rho_1 > \rho_2 > \rho_3$.

Table 2. Comparison of apparent resistivity values and thickness of synthetic layers with apparent resistivity values and inversion yield layer thicknesses for three-layer earth models with $\rho_1 > \rho_2 > \rho_3$.

Number of Layer	ρ (Ω.m)			d (m)			Error
	Synthetic	Inversion	Difference	Synthetic	Inversion	Difference	
1	100	100	0.00	5	4.99	±0.01	
2	50	50.07	±0.07	10	9.99	±0.01	0.01
3	10	10.00	0.00	-	-	-	

Model 3 : Four-Layer Earth Model with $\boldsymbol{\rho_1 > \rho_2 > \rho_3 > \rho_4}$

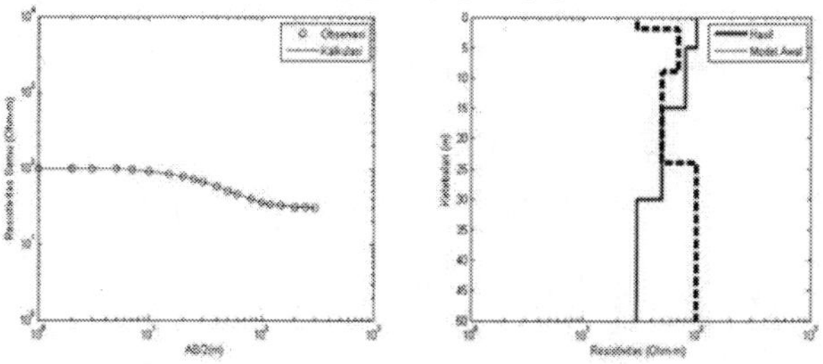

Figure 5. Comparison of observation resistivity curve models and synthetic layer thickness with calculated resistivity curves and inversion yield layer thicknesses for the four-layer earth models with $\rho_1 > \rho_2 > \rho_3 > \rho_4$.

Table 3. Comparison of apparent resistivity values and thickness of synthetic layers with apparent resistivity values and inversion yield layer thicknesses for four-layer earth with $\rho_1 > \rho_2 > \rho_3 > \rho_4$.

Number of Layer	ρ (Ω.m)			d (m)			Error
	Synthetic	Inversion	Difference	Synthetic	Inversion	Difference	
1	100	100	0.00	5	5.00	0.00	
2	80	80.01	±0.01	10	9.99	±0.01	0.01
3	50	50.01	±0.01	15	15.00	0.00	
4	30	30.00	0.00	-	-	-	

Model 4 : Five-Layer Earth Model with $\rho_1 > \rho_2 > \rho_3 > \rho_4 > \rho_5$

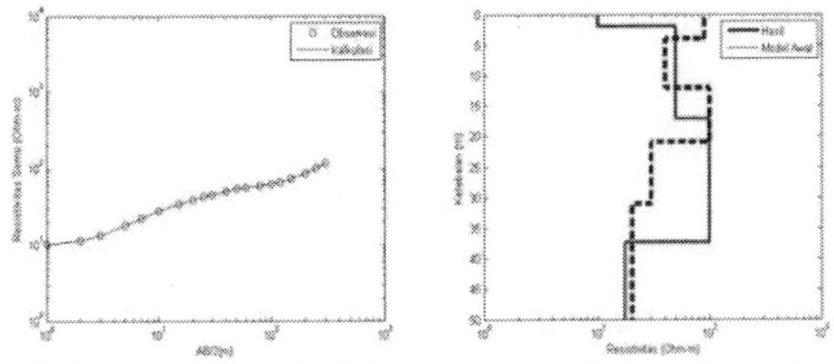

Figure 6. Comparison of observation resistivity curve models and synthetic layer thickness with calculated resistivity curves and inversion yield layer thickness for the five-layer earth models with $\rho_1 > \rho_2 > \rho_3 > \rho_4 > \rho_5$.

Table 4. Comparison of apparent resistivity values and thickness of synthetic layers with apparent resistivity values and inversion yield layer thickness for five-layer earth with $\rho_1 > \rho_2 > \rho_3 > \rho_4 > \rho_5$.

Number of Layer	ρ (Ω.m)			d (m)			Error
	Synthetic	Inversion	Difference	Synthetic	Inversion	Difference	
1	10	10	0.00	2	2	0.00	0.01
2	50	50.08	±0.08	15	15.15	±0.15	
3	100	100.91	±0.91	20	20.13	±0.13	
4	20	17.45	±2.55	25	21.74	±3.26	
5	400	398.86	±1.14	-	-	-	

2.1 Results of Field Data and Inversion Results

Inversion program that has been made in this study was tested using the field data. Located in Bonto Padang Village, Kahu Subdistrict, Bone District, South Sulawesi. From the field data after inverting the data using the program that has been made, the comparison of the results of the observation resistivity curve and the thickness of the layer with the apparent resistivity curve and the thickness of the inversion layer. (Figure 7) are as follows:

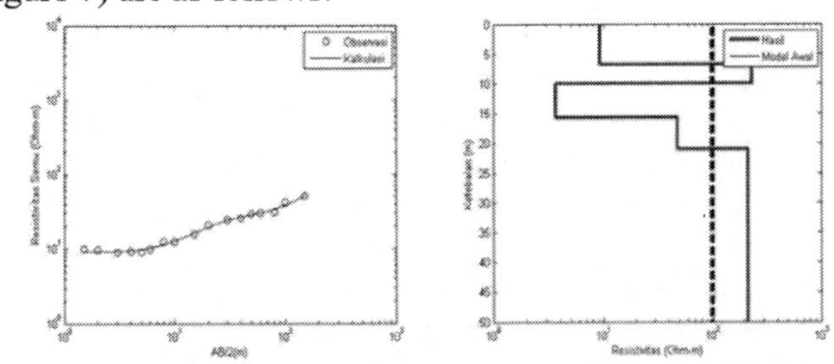

Figure 7. Comparison of observation resistivity curve model and field data layer thickness with calculation resistivity curve and inversion result layer thickness.

For field data that is processed using the program that has been made obtained the comparison of the apparent resistivity value of the observation and the thickness of the field data layer with the apparent resistivity value of the calculation and the thickness of the inversion result layer. (Table 5) as follows:

Table 5. Comparison of apparent resistivity values and thickness of field data layers with apparent resistivity values and inversion result layer thickness.

Number of Layer	ρ (Ω.m)	d (m)	Error
1	9.16	6.87	1.15
2	232.91	3.22	
3	3.66	5.60	

3. Conclusion

The conclusion that can be drawn from this study is that Schlumberger curve interpretation uses least square inversion for some synthetic data models such as a two-layer, three-layer, four-layer, and five-layer models, resulting in actual resistivity and thickness values that are close to synthetic data models.. This can be noted from the results of the apparent resistivity curve model and the resistivity synthetic data calculation approaching for the two-layer, three-layer, four-layer and five-layer models with relatively small RMS error values even though by giving an initial guess for the resistivity and thickness values.

This result is quite satisfactory, it can be seen from the matched curve produced between the apparent resistivity curve of the observation and the apparent resistivity of the calculation with a relatively small RMS error value of 1.15.

REFERENCES

[1] Hendrajaya L 1990 *Resistivity Geoelectric Method* Earth Physics Laboratory. ITB: Bandung.

[2] Syamsuddin and Lantu 2009 *Textbook Report on SCL-Based Learning Module Geoelectric and Geoelectromagnetic Methods* Hasanuddin University.

[3] Telford W M, Geldart L P, Sheriff R E 1990 *Applied Geophysics* Cambridge University Press Cambridge, United Kingdom.

[4] Koefoed O 1979 *Geosounding Principles, 1.* Elsevier Sci. Pub. Co. New York.

[5] Grandis H 2009 *Introduction to Geophysical Inversion Modeling* Indonesian Geophysicists Association.

[6] Menke W 1984 *Geophysical Data Analysis: Discrete Inverse Theory* Academics Press.

[7] Roy I G 1999 *An Efficient Non-Linier Least-Squares 1D Inversion Scheme for Resistivity and IP Sounding data.* Geophys. Prospect.

[8] Raju D C V 2003 *LIMAT: A Computer Program for Least Squares Inversion of Magnetic Anomalies over Long Tabular Bodies* Comput. Geosciens.

[9] Ekinci Y L and Alper D 2008 *A Damped Least-Squares Inversion Program for the Interpretation of Schlumberger Sounding Curves.* Department of Geophysical Engineering, Canakkale Onsekiz Mart University, Turkey.

[10] Meju M A 1994 *Geophysical Data Analysis: Understanding Invers Problem Theory and Practice.* Society of Exploration Geophysics Course Notes Series, No. 6 1st ed., SEG Publisher, Tulsa Oklahoma.

Acknowledgments

We would like to thank to Geophysics Department, Hasanuddin University for all their assistance and advise on this paper. And also thank you for all the people who support this paper so we can finished this so well.

IOP Conf. Series: Earth and Environmental Science **279** (2019) 012043 doi:10.1088/1755-1315/279/1/012043

Groundwater Identification at East Sentani Auwena Formation (Tema) Based on Geoelectric

Virman[1], Muhammad Altin Massinai[2], Muhammad Fawzy Ismullah Massinai[2]

[1]Physics Education Study Program, Faculty of Teacher Training and Education – Cenderawasih University, Jayapura
[2]Geophysics Dept., Hasanuddin University

*altin@science.unhas.ac.id

Abstract. Jayapura City, Papua Province today is preparing to host the National Sport Week (Pekan Olahraga Nasional – PON in Indonesian) 2020 to be held here. For this event, on East Sentani District will build international swimming pool, football stadium and other facilities by government or private. This public facility needs fresh water source to come in useful as maximally. Fresh water from water local company is limited so the best solution is using groundwater. Groundwater is renewable resources, its volume and quality is influenced by vegetation, rainfall, slope, rock surface and others. This existence on subsurface can different based on geological conditions. For this issue, research did to identify groundwater potential on this area. This research used geoelectric 1D resistivity method, Schlumberger configuration. The acquisition held at three places: Kampung Kemiri, Kampung Harapan and Kampung Ifar Gunung. Data processing used IPI2win software, result is resistivity which is explaining lithology and depth. Based on resistivity distribution that groundwater located at Kampung Kemiri on depth 28.6 m with resistivity 81.1 ohm.m and at Kampung Harapan on depth 18.9 m with resistivity 8.9 ohm.m.

1. Introduction

Fresh water is most needed for human that must be met. However, in fact, there are still many people who have difficulty getting it. According to [1] in Indonesia's society freshwater access is still around 67.1%. The lack of fresh water facilities has caused several problems, among others, high public expenditure and high rate of diseases experienced by the community such as diarrhea. Freshwater issue is responsibility of the government and must be immediately addressed by developing infrastructure including freshwater. According to [2] infrastructure development by the government must be a solution for the needs of various segments of society, because it will have an impact on economic growth both directly and indirectly.

East Sentani is one of location in Jayapura regency which is growing quite rapidly in terms of population, as well as physical and non-physical development. Physical construction of stadiums and swimming pools is a facility prepared for sports parties (Pekan Olahraga Nasional – PON in Indonesian, 2020), these facilities definitely require large amounts of water. So far, in East Sentani daily needs of water for community activities rely on surface water. However, with the passage of time the surface water managed by the Regional Water Supply Company (PDAM) is no longer sufficient. This condition is an impact of the inadequate carrying capacity of the Cycloop Mountain due to deforestation. This environmental impact raises a new phenomenon in the rainy season where floods occur and in the dry season the rivers experience drought.

Groundwater is the best alternative if there is insufficient surface water. Groundwater is widely used to meet human freshwater needs for several reasons, namely having relatively good quality and relatively more difficult to experience pollution compared to surface water. Subsurface groundwater is found in geological formations such as sand, gravel or a mixture of sand and gravel that act as aquifers. Aquifers have the nature of storing and passing groundwater in significant quantities to wells or springs. According to [3] groundwater is found in a translucent geological formation known as aquifer, the distribution of groundwater in an area is not the same.

According to [3] the presence of subsurface groundwater is not evenly distributed, this is controlled by, among others, vegetation, rainfall, surface rock structures and so on. In order to obtain groundwater, an investigation of the composition of the earth layer must be carried out so that it can be known whether or not there is a water carrier layer (aquifer). Several methods of surface investigation can be carried out, including geological methods, gravity methods, magnetic methods, seismic methods and geoelectric methods. From these methods, the geoelectric method is a method that is used a lot and the results are quite good. This method studies the nature of the flow of electricity in the earth and how to detect it in the earth's surface [4]. According to [5] geoelectric methods are widely used in hydrogeology because it has advantages in terms of accurate, fast, affordable prices. This study aims to identify aquifer layers based on the value of the type of resistance using geoelectric resistivity method.

Geoelectric estimation is intended to obtain an overview of the subsurface soil and the possibility of groundwater at a certain depth. The geoelectric method is one of the predictive geophysical methods but is recognized as having an important role in the search for natural resources including groundwater. The nature of this prediction is partly due to the interpretation of geophysical data known as ambiguity.

Resistivity geoelectric method uses the assumption that the earth is homogeneous isotropic. Through this assumption the measured type of resistance is actually not dependent on the electrode spacing. In fact, the earth consists of layers with different ρ. So that the measured potential is the influence of these layers. Therefore value of the measured type of prisoner is not just a type of resistance for one layer, this is mainly for wide electrode spacing. The measured resistance type is called apparent resistivity. Pseudo prisoners are formulated as [6];

$$\rho_a = K \frac{\Delta V}{I} \; ; \; R = \frac{\Delta V}{I} \tag{1}$$

Where ρ_a is an apparent resistivity, K is a geometry factor, ΔV is the potential difference between the two potential electrodes and I is the current strength that is injected.

Based on equation (1) it can be seen that the apparent resistivity value depends on the geometry of the electrode configuration used. The type of resistivity geoelectric method has several configurations that can be used including Schlumberger configuration, Wenner configuration, dipole-dipole configuration etc. In this study, Schlumberger configuration was used. From equation (1) the value of geometry factor (K) for Schlumberger configuration is as follows:

$$\rho_s = K_s \frac{\Delta V}{I} \text{ , with } K_s = \frac{\pi \left(L^2 - l^2\right)}{2l} \tag{2}$$

Where L is the length of the current electrode (AB/2) while l is the potential electrode length (MN/ 2).

Generally, this Schlumberger configuration is carried out with the current electrode distance (C1, C2) made 10 times or more potential electrode distance (P1, P2). Each configuration has an unequal depth of penetration, so that in measuring depth penetration is one of the factors that are taken into consideration in the selection of electrode configurations, other factors are the type of structure, the sensitivity of the device, the level of noise available.

Several factors influence the price of this resistivity, including:

- The water content, the medium containing the resistivity water is getting smaller.
- Porosity, which is the ratio between the volume of cavities (pores) to the volume of the rock itself. Porosity is expressed in % volume. The large volume of rock pores will provide more fluid content so that the resistivity price will be smaller. In connection with this, there is an empirical formula referred to as Archie's Law, as follows:

$$\rho = a\rho_w \; \emptyset^n S^m \tag{3}$$

where ρ_w = resistivity of rock containing liquid (ohm.m), a and m are constants (0.5 <a <2.5; 1.3 <m <2.5), $\rho_{(w)}$ = water resistivity (ohm.m), $\emptyset$ = porosity, n = 2 and S are part of the pores of the rock containing fluid.

- Temperature, the resistivity of a rock is inversely proportional to its temperature, the temperature relationship with resistivity is shown by the following equation:

$$\rho_w = k\, e^{-0{,}00821\, t} \tag{4}$$

Where $\rho_{(w)}$ is the fluid resistivity (ohm.m), k = constant, depending on the electrolyte concentration in the fluid and t = temperature. Based on this equation it can be seen that if the temperature rises, the resistivity will decrease exponentially.

2. Methodology

2.1. Research Location

The research location is East Sentani District, Jayapura Regency, which is administratively included in the Sentani District area. To reach the work site from Jayapura City can be reached by land by using two-wheeled vehicles or four-wheeled vehicles with very easy access of approximately 30 minutes. Geographically, Jayapura Regency is located at coordinates 02°32'45"S to 02°36'07"S and 140°32'22"E to 140°37'10"E with an area of 17,516.6 Km^2.

2.2. Research Procedure

Before to the measurement, a preliminary survey was conducted to map the measurement points and the direction of the refractive geoelectric stretch. Based on these considerations, the total geoelectric stretch is 4 sounding points which include two located in the altitude area and 2 (two) in the flat area. The four sounding points are determined based on the results of the initial survey, considered representative and have a relatively long stretch. Field measurements are carried out by measuring the potential difference and current strength generated at each change in AB and MN electrodes. In this measurement the Schlumberger configuration is used, this configuration is good for mapping and sounding. For sounding resistivity, the spacing of the electrode is changed in graduation to a very large point.

Field measurements are carried out by measuring the potential difference and current strength generated at each change in AB and MN electrodes. In this measurement the Schlumberger configuration is used, this configuration is good for mapping and sounding. For sounding resistivity, the spacing of the electrode is changed in graduation to a very large point. In order to Schlumberger electrode rule, spaced current electrodes are much larger than the spacing of potential electrodes. The stages of research activities that will be carried out in detail are data collection, processing, and data analysis.

2.3. Data Acquisition

To support the smoothness of data collection, a number of things must be prepared including the making of a permit, preparing equipment and a literature review. The equipment used includes geological maps, hydrogeological maps, GPS, one unit of geoelectric equipment and several other supporting tools. The types of activities carried out in detail at this stage are as follows:

i. Geoelectric measurement is preparing equipment, setting work functions for students involved in this research activity and preparing a permit. The next activity after arriving at the location is setting up geoelectric equipment components. In setting the mounting components all four electrodes and connectors can be controlled via a geoelectric monitor. After all equipment is properly connected, current injection can be carried out, followed by recording or filling in the data tables that have been prepared including current (mA), potential (volts) and length of the stretch.

ii. The measurement position can use several ways, but in this study the location of each region was chosen based on the displayed expanse. This method is carried out considering the groundwater in the aquifer layer spread laterally is relatively broad.

2.4. Processing and analysis of data

The data obtained will be processed to get apparent resistivity, the important thing to remember in processing or calculating the type of resistance is to do unit conversion. The output from this processing will be obtained in the form of a graph between the distance (AB/2) vs type resistance. The graph obtained is true resistivity, which is a graph that illustrates subsurface lithology, including resistance to type, thickness and depth of the layer. Data analysis is done by comparing the results of processing with some supporting data such as type resistance tables and some other supporting data. Type resistance table is a physical property that explains the relationship between electric current and several types of rock. The relationship is quantitatively as follows: igneous rock has a higher resistance price than compact sedimentary and sedimentary rocks. Loose sediments will have a relatively lower resistivity compared to compact sedimentary rocks. Measured type prisoners will be smaller in value if the rock contains hard water or salt.

3. Results and Discussion

Table 1. Actual resistivity values for three sounding points

Sounding Point	Layers	Interpretation Result			Estimated Lithology
		Depth (m)	Thickness (m)	Resistivity (ohm.m)	
Buffer-Waena	1	0.284	0.284	0.006	Top soil (Humus)
	2	7.66	7.37	457	Coarse fine sand
	3	15.5	7.89	46.6	Coarse fine sand
	4	126	111	1328	Basic rock
	5	>126	~	10.8	Coarse fine sand
Kampung Harapan	1	2.075	2.08	269	Top soil (gravel, sand)
	2	10.53	8.46	122	Coarse fine sand
	3	21.4	10.9	8.9	Coarse fine grain size
	4	42.39	21	104	Sand fine grain size
	5	>42.39	~	0.14	Coarse fine sand
Kampung Kemiri	1	0.32	0.32	176	Top soil (gravel, sand)
	2	5.38	5.7	2783	Sedimentary rocks, breccia
	3	77.43	71.1	858	Clay, Marl
	4	>77.43	~	22.7	Sand
Kampung Ifar Gunung	1	1.36	1.37	433	Top soil (Gravel, Sand)
	2	10.36	8.99	10.1	Coarse fine sand
	3	35.11	24.8	756	Compact sediment rock
	4	>35.11	~	3275	Basic rock

Geoelectric measurement points are spread over 4 (four) locations, locations selected based on growth index, industry, and so on. The four locations include Buffer Waena, Kampung Harapan, Kapiri Kemiri, and Ifar Gunung. Geoelectric data are different in shape and are used to calculate. The right types are the models used by each type and type of stone for each measurement. To find out the types of rocks that exist at each point with large numbers used and geological data in the study area. Data on the type of detainees that have gone through further sorting in Table 1. The results of the analysis of the models that produce the structure below are as follows:

Measurement location of Buffer Waena morphology is not flat (wavy), measurements using Schlumberger configuration geoelectric with a total stretch of 600 meters or AB/2 are 300 meters, the direction of the north-south stretch. Based on the results of data processing (Table 1), it was found that in general the existing rock layers had contrasting physical properties found in layer 4 (four). The type of resistivity value is 1328 ohm.m and is at a depth of 126 meters, the value indicates as a layer that is impermeable (water-resistant) the structure can be either bedrock or clay rock. This layer includes quite thick, which is 111 meters, this is reinforced by a topography that is not flat (bumpy). Layer 5 (five) which is at a depth of below 126 meters of resistivity type 10.8 ohm.m, this layer of structure is generally conductive or in the form of sediment deposits. Contrast types between layers 4 (four) and layer 5 (five) showed that in layer five was categorized as a layer of aquifers which were confined aquifers, this indication was partly strengthened by the condition of the research area where there were Sentani Lakes.

Measurements in Kampung Harapan location, based on Table 1, the aquifer layer is found in the type of resistivity 8.9 ohm.m and 0.14 ohm.m. Between the two layers there is a layer of 4 (four) with a type of resistivity of 104 ohm.m, this layer can act as an impermeable layer with a layering structure in the form of clay. The unconfined aquifer in layer 3 (three) is at a depth of 10.53 meters and has a thickness of 10.9 meters. The second aquifer layer was found in layer 5 (five) with a type of resistivity 0.14 ohm.m at a depth below 42.39 meters. At this measurement location it is about 100 meters from Sentani Lake, and is in the lowlands so that the quality of groundwater is not only affected by the structure of the land but also the water of Sentani Lake. The unconfined aquifer in the 3 (three) layer of material is an alluvial layer or dynamic sand and groundwater flowing from north to south. The Cycloop Mountains with existing vegetation function as catchment areas which drain groundwater graphically.

Measurements in Kampung Kemiri, based on Table 1 below, are high resistivity in layers 2 (two) and layers 3 (three). These two layers have a depth of up to 71.7 meters, presumably as an impermeable layer in the form of clay. The presence of the imperbeable layer causes the conductive layer (layer 4) (22.7 ohm.m) to be classified as confined aquifer. Measuring locations that are not flat, and relatively close to the Cycloop Mountains (at altitude) cause unconfined groundwater not found.

Measurements in Ifar Gunung, based on the distribution of resistivity values (Table 1) consist of 4 (four) layers. Layer 2 (two) is a layer with relatively low (conductive) type resistance which is 10.1 ohm.m usually associated with groundwater. However, this type of low resistance occurs due to non-compact soil structure and is controlled by the activity of organisms and microflora and physical and chemical processes. The type of rock is a mixture of weathering results from native rocks and organic substances from older decomposition. The absence of groundwater is reinforced by a resistive structure with 3275 ohms and this is associated with bedrock.

4. Conclusions

Based on the height difference of the aquifer in this study, it can be said that groundwater is dynamic and the flow direction from north to south. The northern part of Sentani is formed by a Cycloop mountain range, this is a rechager area where groundwater flows towards the south as the output area. Ifar Gunung and Buffer Waena have the same layering structure, but groundwater at Ifar gunung is not obtained so that the possibility of groundwater at Buffer Waena is a catchment from Sentani Lake. This condition is reinforced by the location of the Buau Waena measurement area which is about 100 meters from Sentani Lake and continues to the measurement area in Kampung Harapan. Groundwater from the Cycloop Mountains is more dominant in the measurement area of Buena Waena and the Kampung Kemiri area. Whereas in the Kampung Harapan area the groundwater is more predominantly influenced by Sentani Lake water.

References

[1] Anandini F *Jurnal Perencanaan Wilayah dan Kota* vol 23 no 2.
[2] Alfiah R, Ari I R D and Hariyani S 2017 *Jurnal Rekayasa Sipil* vol 11 no 3.
[3] Mareta N and Hidayat E 2016 *Jurnal Natural B* vol 3 no 4.
[4] Massinai M A, Lantu, Latuconsina H and M Ismullah M F 2018 *IOP Conf. Series: Earth and Environmental Science* vol 132/ 012017.
[5] Todd D K 1980 *Groundwater Hydrogeology* (John Wiley & Sons, New York)
[6] Telford W M, Geldart L P and Sherif R E 1990 *Applied Geophysics* (Cambridge University Press, New York)

IOP Conf. Series: Earth and Environmental Science **279** (2019) 012044 doi:10.1088/1755-1315/279/1/012044

Detection and Delineation of Shallow Channel Morphology Using 3D Seismic Data in the K Field, Malay Basin, Malaysia.

***Nur Hazira Hamsan, Md Yazid Mansor and Liaw Kim Kiat**

Department of Geosciences, Universiti Teknologi PETRONAS, Malaysia.
Email address: nurhazira.hamsan@gmail.com

Abstract. The Malay Basin early 2D seismic exploration programs were designed to test simple structural traps. Large relief folds are easily located and subsequently drilled. Many major oil and gas fields were discovered since the 1970s in middle-late Miocene siliciclastic reservoirs, and eventually produced successfully. However, these easy hydrocarbon accumulations in the Malay Basin are fast depleting due to its long production history. Hence, new petroleum plays are urgently needed to sustain the production. The utilization of high-resolution 3D seismic data in the early 2000's has greatly increased the chance of drilling success, including high-risk prospects such as stratigraphic plays. This paper discusses the subtle channel morphology in the shallow section of the K Field. The 3D seismic data shows multiple-stacked channel geometry within the Pliocene (5.5-5.2 Ma) strata. The seismo-stratigraphic interpretation is aided by attributes analysis including horizon probe to detect and delineate the internal geometries. The results are correlated with modern analogues to predict the probable depositional elements within these fluvio-marine systems. These shallow channel features were previously avoided as hazardous drillings. Instead, until they are tested, these could be considered as good porosity clastic reservoirs filled by medium to heavy density oil.

Keywords: Malay Basin, Pliocene, stratigraphic play, 3D seismic data

1. Introduction

Stratigraphic traps in the Malay Basin are often considered as one of the riskiest exploration targets to be tested. The critical factors to be an excellent stratigraphic exploration play are the presence of a fully charged petroleum system with good seal integrity. Three-dimensional (3D) visualization techniques provide an alternative, interactive means of viewing amplitude and attribute volumes that facilitate the extraction of meaningful information and improves the interpretation accuracy and efficiency [1]. In early years, seismic interpretation was solely based on 2D interpretation workflows. This includes faults and horizons picking on the dip (inline) and strike lines (crossline), to generate time-structure maps through gridding and contouring. It is a common practice that most exploration prospects drilled within the easy structural traps such as folds, fault-blocks, well defined reefs and seismic anomalies with clear direct hydrocarbon indications (DHI's). Since 1984, the advent of 3D seismic technology has allowed better subsurface resolution for solving various problems of geological interpretation [2]. To this day, without support from DHIS's, any subtle stratigraphic traps will be almost impossible to receive approval to drill.

Seismic data 3D visualization helps in displaying the stratigraphic hydrocarbon traps in their true three-dimensional perspective, allowing interpreters to visualize the complex geomorphology of stratigraphic plays and better reservoir characterization. There are many astonishing multi-stacked

IOP Conf. Series: Earth and Environmental Science **279** (2019) 012044 doi:10.1088/1755-1315/279/1/012044

channel features presence within the target area, observed within 1 km from the seabed, which could not otherwise be easily enhanced with 2D seismic data. The application of seismic attribute technique helps to reveal characteristics that is masked by using amplitude data themselves. It allows better geological interpretation of formation, particularly in channels and thin bed reservoir environments. Seismic attributes are defined as "any or all observations" extracted from the 3D seismic data which are directly or indirectly helps in hydrocarbon exploration [3]. Seismic attributes can be clustered into two main categories, which are structural attributes and stratigraphic attributes, which emphasize different properties of the seismic signal [4, 5]. Selection on the types of attributes used are very crucial depending on what analysis we want to develop for our study. It is one of the main steps in recognizing stratigraphic events and to overcome pitfalls in the interpretation.

2. Study area

The study area is located at the K field, in north-central Malay Basin (Figure 1). The field was discovered in 1970, at an average water depth of 65 m. It is a faulted, domal anticline structure with a closure >300 km^2 [6]. Previous explorations were designed to assess the oil-filled potential within this structure at intervals of Groups D and E reservoirs but with no success. Multi-stacked gas-bearing reservoirs were found in Groups B, D and E. As the major reservoirs were the main interest for this exploration, limited reservoirs data were gathered from the Group B reservoirs during exploration period. Other than mud logs and conventional well logs, subsurface data were rather limited. Pressure data and conventional core are absent in this area. Thus, it was expected that mapping and assessment of the reservoir would be very challenging particularly with respect to establishing good understanding of reservoir quality and producibility. This anticline structure is believed to be formed during the extensional and compressional phases of structural development of the Malay Basin during the Oligocene, within an overall dextral shear regime. It is thought to form as deep syncline within half-graben which was inverted during the middle-late Miocene, caused by changing to sinistral shear regime [6]. The structure was started as an elongated E-W ridge which due to changing stress direction, become more circular.

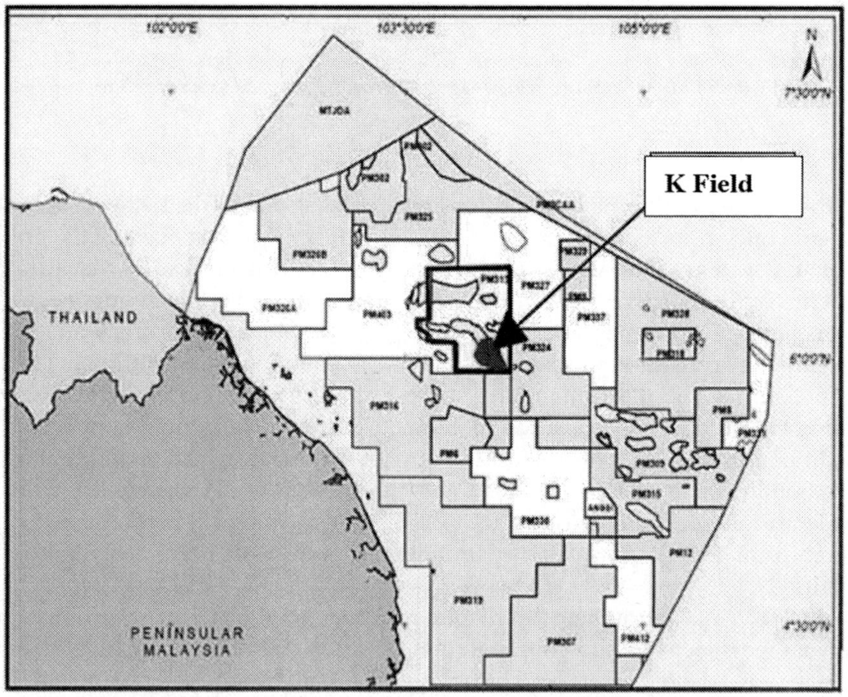

Figure 1. Location map of the K Field (Modified from [8])

IOP Conf. Series: Earth and Environmental Science **279** (2019) 012044 doi:10.1088/1755-1315/279/1/012044

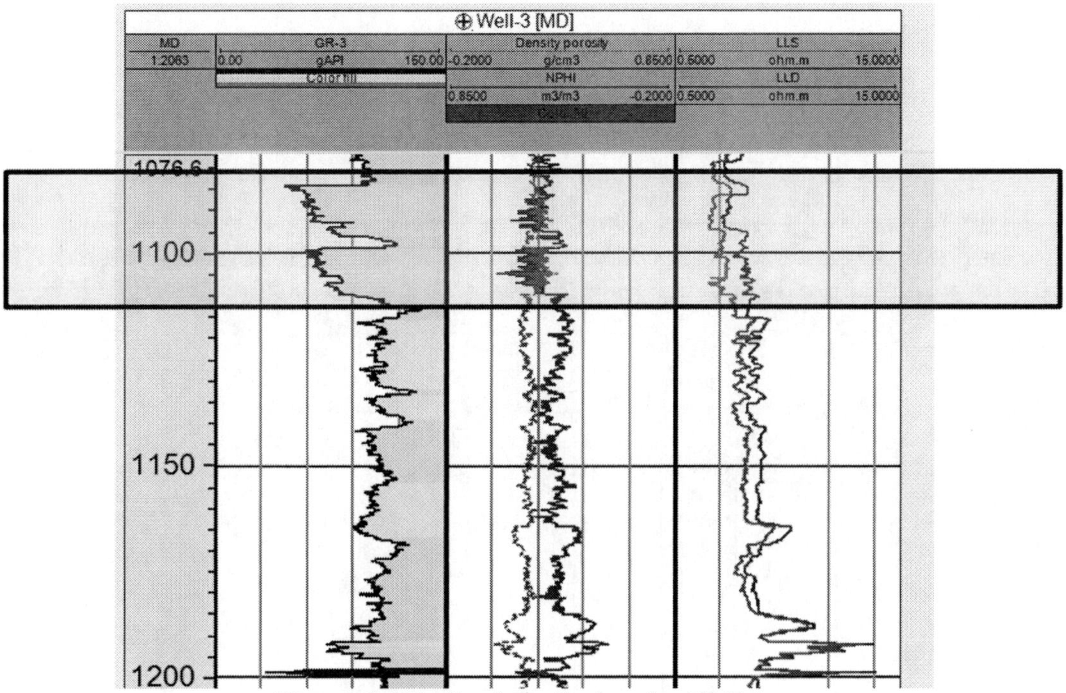

Figure 2. Wireline logs displayed at Well-3.

2.1 Group B reservoir description

The Group B reservoirs were deposited in the Early Pliocene (about 5.5 ma) with an average thickness of 1090 m. This phase was dominated by thermally-induced subsidence which continued in the late post-rift section [7, 8]. Deposition during this phase was mainly within lower coastal plain and shallow marine environment (fluvial deltaic tidal dominated sedimentation) [9]. Group B can be sub-divided into Upper and Lower Group B formation. The Upper section predominantly consists of laterally restricted stacked channels deposited in lower coastal plain environment. Lower Group B, in the other hand, were better developed with thicker in net sand and better reservoir quality than Upper Group B. The reservoirs are interpreted to be incised-valley deposits made up of discrete channels [9]. Early exploration penetrated this formation Group B uneventfully. They penetrate the shallow gas anomaly within Group B reservoirs, it turned out to be 1- 20 m thick of interbedded silty sandstones and shale with low pressure gas. Only mud logs and gas chromatographs were recorded across the shallow intervals, without any testing made in the shallow zones being of no commercial value back then in the 70s and 80s.

3. Dataset and methodology

This study is based on a 2004 vintage 3D seismic data, ca. 335 km^2 in the north-central Malay Basin. The digital data will be loaded onto the 2D/3D seismic interpretation platform, namely Petrel software. Total of four wells used to conduct this study (well-2, well-3, well-4 and well-5). The seismic imaging can be considered as poor to fair as a result of shallow gas imaging effects. The data in the central part of the K field is much affected by gas pockets and poor fault imaging. This brings in challenges during the horizon interpretation. However, most of the reflectors outside the gas cloud area are well-imaged. Previous exploration well reports in this field were collected and analyzed to get better understanding of the subsurface formations and reservoir properties. Wildcat wells' LAS format files were then collected, inclusive of all available wireline logs data. This includes: (i) electrical logs (lateralog deep resistivity, lateralog shallow resistivity and microSFL resistivity log), (ii) lithology logs (gamma ray, spectral gamma ray and corrected gamma ray) and (iii) porosity logs (density, neutron porosity, sonic logs). Well deviation trajectories and lithology information from mud logs are obtained as well.

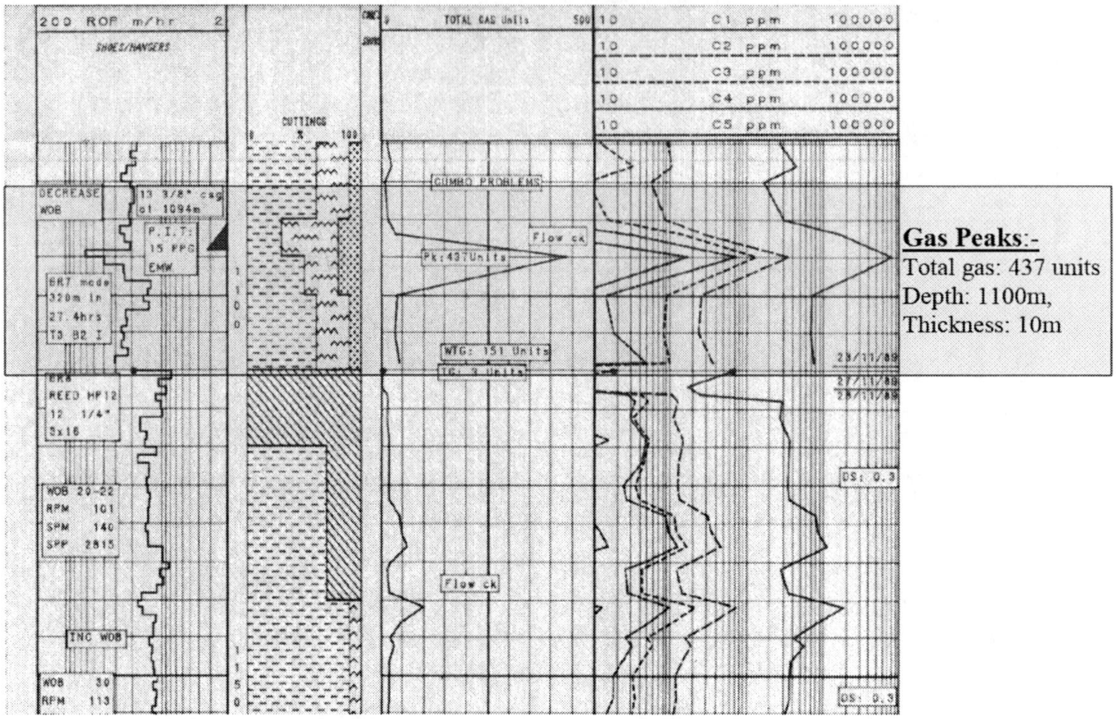

Figure 3. Mud log reports indicate gas peaks at the target reservoir interval. (Modified from [6]).

3.1. *Facies classification and correlation of exploration wells*

The incorporation of multi-well logs and lithological mud logs report helps in the facies classifications and reservoir prediction. Logs such as gamma ray, porosity, density and sonic are used to analyse the lithology of the formations. According to the exploration well reports, Group B are made up of mainly shale and siltstone with thinly sandstone layers. Therefore, the facies classification is filtered accordingly. These facies log then were created for each well to distinguish it reservoir lithology. Well to well correlation is then carried out for all the wells to analyse the lateral continuity of the lithological occurrence, away from the well control.

3.2. *Seismic volumes attributes and direct hydrocarbon indicator (DHI)*

Volume attributes computation allows us to generate new seismic volumes to enhance the interpretation and visualization purposes [10]. Out of the many volume attributes available, only few are proven affective for this study. Variance attribute, or better known as edge detection attribute highlights geologic features that have abrupt boundaries. It uses the local variance as a measure of signal unconformity, associate with the dip corrected slices used to emphasis on the structural features [11]. Although this attribute can easily detect the fault and channel edges, they cannot indicate the thickness of the channel's particularly a very thin bed channel. This workflow is further enhanced by using amplitude attributes such as RMS amplitude and sweetness attribute to analyse its reservoir quality and DHI study. It made use of envelope and instantaneous frequency to highlight bright spots. The basis of this attributes study is based on the quick look interpretations on the 3D seismic attributes response. Interestingly, most of the bright response presence in this study area is still untested. Based on the quick rock physics study on the lithological sequence from previous findings associated with seismic anomaly, we can justify the similar anomaly for our predicted seismic response, away from the well control.

IOP Conf. Series: Earth and Environmental Science **279** (2019) 012044 doi:10.1088/1755-1315/279/1/012044

3.3. Fault and horizon interpretation

This study area consists of numerous north-south trending faults. Due to the surface channel gas pockets in the central part of the structure, it is difficult to map and delineate the fault. However, most of the major faults located outside the 'wipe-out' zone are well-imaged. Fault interpretation is first analysed and mapped before horizon interpretation is carried out. The application of structural attribute (variance) helps to detect the discontinuities or phase changes that occur across the fault zones. This is followed by the stratigraphic interpretation of seismic horizons. Bright reflectance and widespread lateral continuity are the basis for the horizon interpretation. Seeded 3D auto tracking is chosen for conducting the horizon interpretation. This method is not applicable inside the gas cloud within the central structure, due to the poor signal and chaotic response, therefore guided auto tracking is selected to interpret the sections. A total of two horizons were mapped throughout the 3D seismic sections to resolve the thickness as well as lateral extent of any geobody.

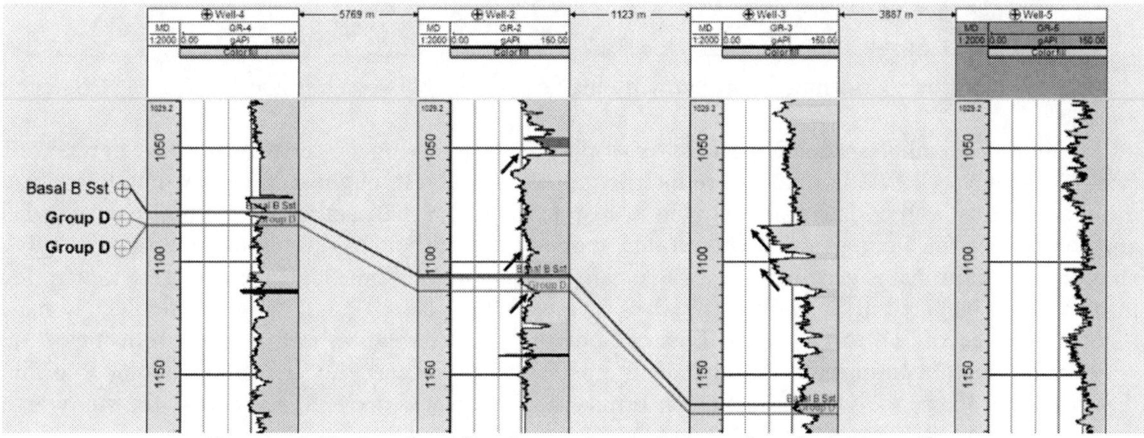

Figure 4. Facies classification and correlation of exploration wells

4. Results and discussion

Figure 2 showing the wireline logs displayed at Well-3. The first track displaying the gamma ray log, showing lower number of counts of radioactive minerals on sand facies, indicating of sandstone lithology. Second track displaying the common industrial convention, neutron-density log in defined scale and polarity. This combination of log response indicates a quick-look interpretation of hydrocarbon bearing zone. The presence of hydrocarbon filled, porous sandstone reservoirs are determined using the crossover between porosity and density log. This is also referring to the terms of 'Butterfly effect'. Gas bearing zone will resulting in bigger separation of crossover, while oil just give a small separation between logs. Following the abnormally low resistivity values (around 2 to 6 ohm-m), despite the indications of neutron-density cross over with the presence of gas peaks from mud log response in Figure 3 at interval 1100 m. The significant increase in total gas units are associated with a zone of increased in porosity and permeability. The gas chromatograph in mudlogging sensor is used for more detailed hydrocarbon analysis. It helps to separate the gas streams into different types of alkane group, namely methane [CH_4] – denoted as C_1 – as well as the following constituents: ethane [C_2H_6] or C_2, propane [C_3H_8] or C_3, the normal and isopolymers of butane [C_4H_{10}] or nC_4 and iC_4 and pentane [C_5H_{12}] or nC_5 and iC_5. The results obtained are displayed in Figure 3. At the interval of 1100 m, it shows that the gas chromatograph detects the alkane group from methane till pentane.

In Figure 4 the gamma ray log displayed at each well to determine the facies, linked with the geological mud log data. It is showing that in Well-2 and Well-3, the sands are better developed compared to Well-4 and Well-5. The bell-shaped log motif within Well-2 indicate the fining upwards succession marked with the presence of channel sands. Funnel-shaped log motif at Well-3 indicates presence of coarsening upward barrier bars. Plus, the "butterfly effect" along with the gas peaks at mud log data also validated the existence of HC-bearing channel sands in Lower Group B.

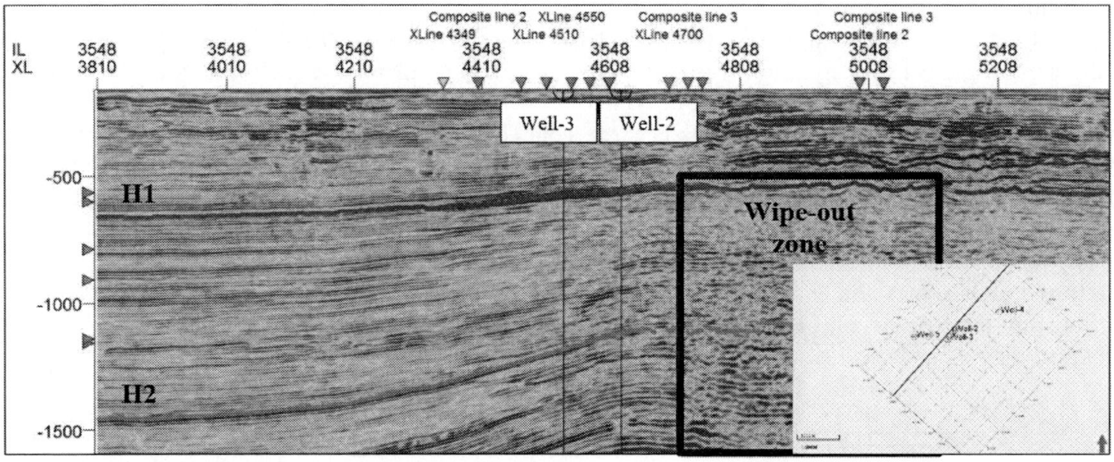

Figure 5. Seismic inline showing the target zone between horizon 1 and 2.

In Figure 5, an inline section is selected for display as it is showing the correlation between two wells (well-2 and well-3). Two horizons were interpreted and mapped throughout the study area to evaluate the thickness as well as lateral extent. These mapped horizons will act as a structural control when conducting horizon slicing from the generated geobody. It will help to reveal any potential but subtle stratigraphic traps that is usually not easily visualized using normal time slice method. Due to very few information's gathered in Group B during the previous drilling campaign, it causes difficulties to make a correlation according to well tops. Basis of horizon 1 interpretation is the bright reflectance and widespread lateral continuity. Horizon 2 is interpreted based on the Pliocene–Miocene unconformity that presence across the wells. This unconformity was correlated and mapped across the study area, marks the boundary between Groups B and D.

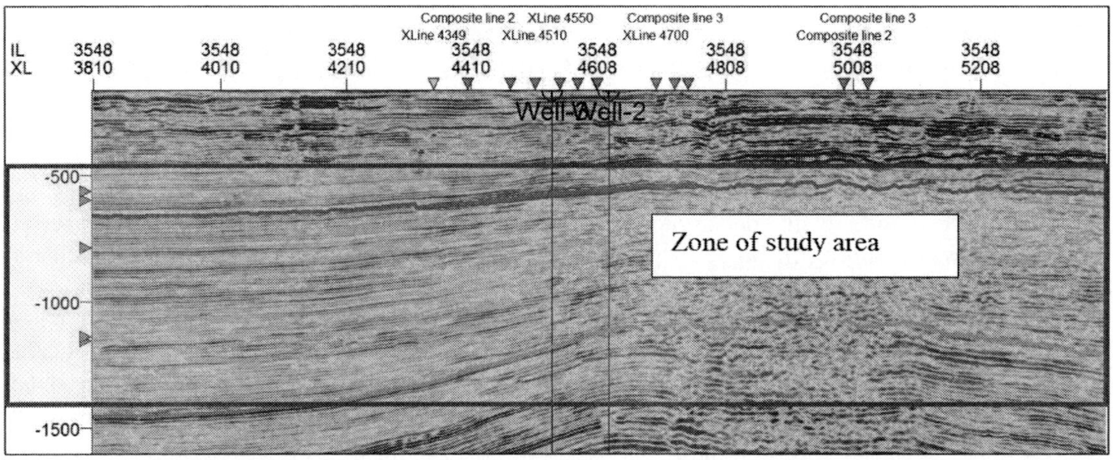

Figure 6. Inline section at interval 3548, marked with the zone of study area.

IOP Conf. Series: Earth and Environmental Science **279** (2019) 012044 doi:10.1088/1755-1315/279/1/012044

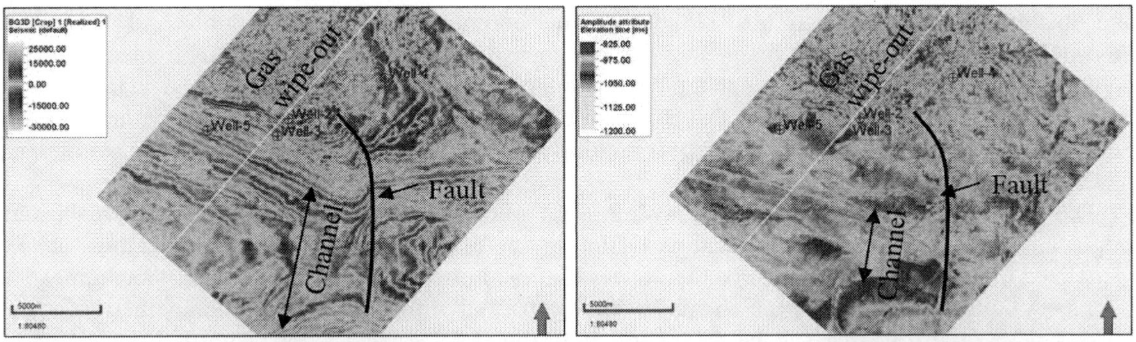

Figure 6a. Original seismic amplitude displaying the time-slice (right) and the horizon slice (left)

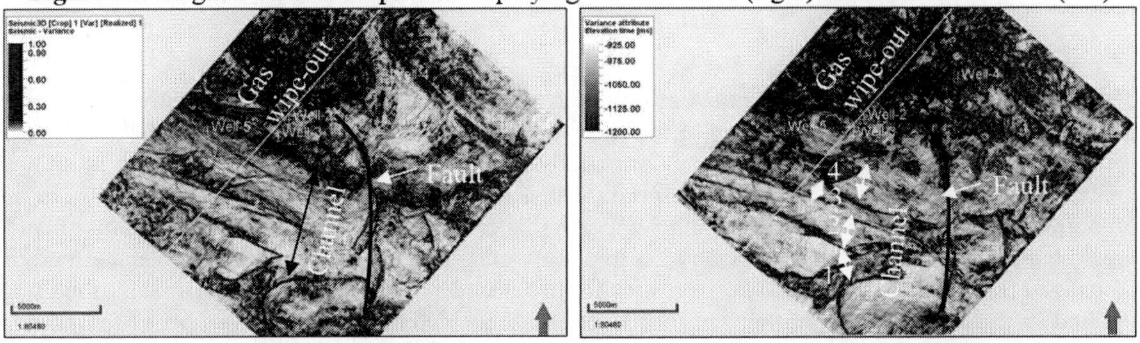

Figure 6b. Variance attribute displaying the time-slice (right) and the horizon slice (left)

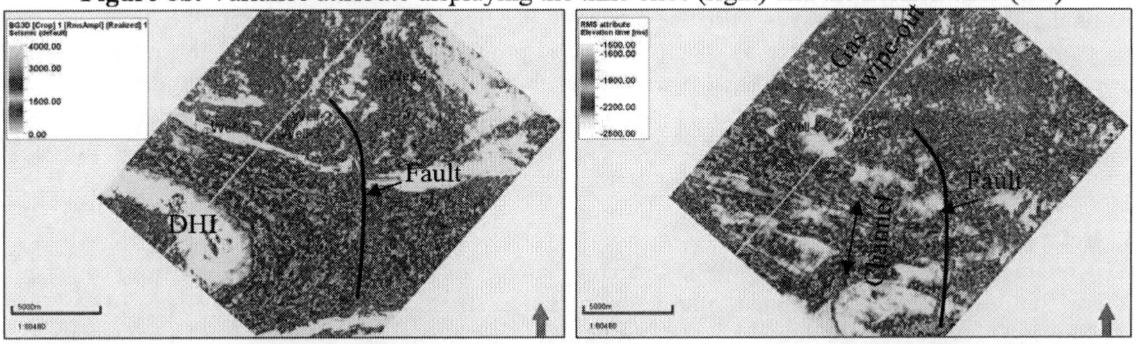

Figure 6c. RMS amplitude attribute displaying the time-slice (right) and the horizon slice (left)

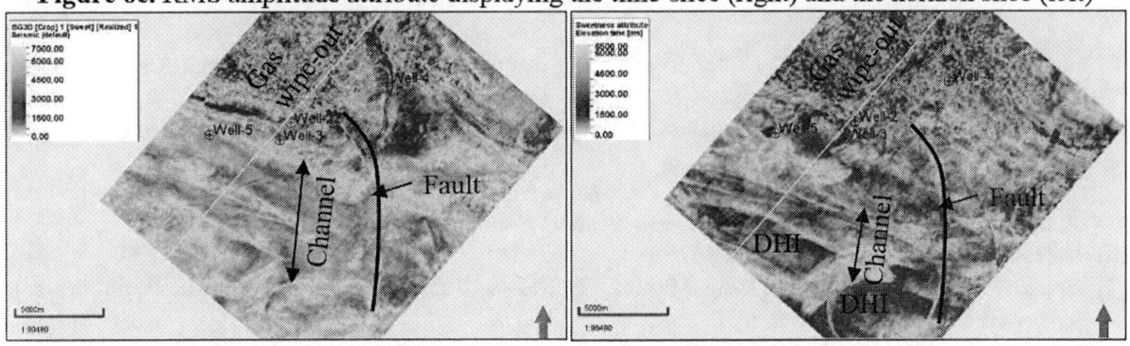

Figure 6d. RMS amplitude attribute displaying the time-slice (right) and the horizon slice (left)

The shallow geomorphology study within the zone of study area between H1 and H2 reveals several channels feature, marked with clear termination pattern (figure 6). The presentation of these channel bodies is further enhanced by applying seismic attributes at the 3D seismic volumes. These time-slices are displayed at depth – 1400 ms TWT and the horizon slices are – 180 ms below the H1. Figures 6a

IOP Conf. Series: Earth and Environmental Science **279** (2019) 012044 doi:10.1088/1755-1315/279/1/012044

and 6b show the same indication of broader channel body at the time-slice sections and more detail morphology of the individual channel at the horizon slice. Variance attributes in figure 6b further enhance the discontinuity events for the stratigraphic interpretation. It is interpreted that the broad channel is an incised valley deposit that made up of multiple-flow channel directions within the same interval. These channels are believed to be migrating towards the NE direction based on the overlapping patterns.

The DHI study in this channel body, using RMS amplitude and sweetness attribute reveals some bright amplitude responses, interpreted as DHI (Figures 6c and 6d). Some of these features are not visualized by using normal time-slice methods. From the horizon geobody generated, there appears to be a sand build-up in between the channel 1 and 2, marked with bright amplitude anomaly. Channel levee sands are also believed to be developed, with most of the sand being deposited outside of the channel bodies.

5. Conclusion

The detection and delineation of channel morphology within Group B formation is greatly enhanced by the application of seismic attributes in 3D seismic data. Several channel patterns were detected on the well-3 log motif, indicated that the sand formation is more compartmentalized. This log motif was analysed to find similar potential for reservoir filled sand away from the well control. There are numerous stacking and overlapping channel patterns that can be observed within the boundary of study areas, which were not penetrated by any exploration wells. The lateral extent of the channel body is visualized by using time-slice and horizon slice for better understanding. Various seismic attributes are applied to the seismic data, further enhanced and helps for detection and delineation of the channel geomorphology. The channel is interpreted to be an incised valley, made up of overlapping discrete channels. With the application of variance attribute, it helps to separate the individual channels that overlapped within the same interval. The presence of probable sand build-up and levee deposits area revealed by the application of RMS and sweetness attributes, marked with bright amplitude event, which is believed to be a DHI indicator. The individual channel may not bring in any values for the hydrocarbon exploration, however if we can cluster all this compartmentalized sand bodies, it will bring in improved economic value for tie-in to existing production facility.

6. References

[1] Chopra, S., & Marfurt, K. J. (2009). Detecting stratigraphic features via crossplotting of seismic discontinuity attributes and their volume visualization. *The Leading Edge, 28*(12), pp. 1422-1426.

[2] Ghosh, D., Sajid, M., Ibrahim, N. A., & Viratno, B. (2014). Seismic attributes add a new dimension to prospect evaluation and geomorphology offshore Malaysia. *The Leading Edge, 33*(5), pp. 536-545.

[3] Taner, M. T., Schuelke, J. S., O'Doherty, R., & Baysal, E. (1994). Seismic attributes revisited. In *SEG Technical Program Expanded Abstracts 1994,* pp.1104-1106.

[4] Robertson, J. D., & Fisher, D. A. (1988). Complex seismic trace attributes. *The Leading Edge, 7*(6), pp. 22-26.

[5] Chopra, S. and Marfurt, K.J. (2005). Seismic attributes for prospect identification and reservoir characterization: Oklahoma: *SEG Geophysical Development Series No. 11*, pp.464.

[6] Well-3, Appraisal well (n.d.)

[7] Mansor, M. Y., Rahman, A. H. A., Menier, D., & Pubellier, M. (2014). Structural evolution of the Malay Basin, its link to Sunda Block tectonics. *Marine and Petroleum Geology, 58*, 736-748.

[8] Madon, M., Abolins, P., Hoesni, M.J., & Ahmad, M., (1999). Malay Basin, in: The Petroleum Geology and Resources of Malaysia (Ed. by R. Selley & L.K. Meng), Petronas, Kuala Lumpur, pp. 173-217.

[9] Nasir, M., Azlin, M., Shamsuddin, S. A., & Bakri, M. S. R. (2008). From Shallow Hazard to Productive Opportunity: Opening a New Gas Play in the North Malay Basin. In *International Petroleum Technology Conference*. International Petroleum Technology Conference.

[10] Chopra, S., & Marfurt, K. J. (2007). *Seismic attributes for prospect identification and reservoir characterization*. Society of Exploration Geophysicists and European Association of Geoscientists and Engineers.
[11] Randen, T., Monsesn, E., Signer, C., Abrahamsen, A., Hansen, J.O., Seater, T., Schlaf, J. & Sonneland, L. (2000). Three-dimensional texture attributes for seismic data analysis: SEG International Meeting.

Acknowledgement
I am are very thankful to PETRONAS for providing the data to conduct this MSc research study. Special thanks to my supervisor, Dr Liaw Kim Kiat, and my co-supervisor, Mr Md Yazid Mansor, for their continuous support and assistance throughout my studies.

Geochemical Fluids Assessment of Gunung Endut Geothermal Area, Banten Province, Indonesia

C. Risman[1]*, D. N. Sahdarani[1], F. M. H. Sihombing[1], Supriyanto[1,2]

[1]Geology, Faculty of Mathematics and Natural Sciences, Universitas Indonesia
[2]Geophysics, Faculty of Mathematics and Natural Sciences, Universitas Indonesia

*Corresponding author: choirul.risman17@gmail.com

Abstract. Geochemical assessment was completed in Gunung Endut geothermal area to determine fluid characteristics discharging from their natural surface manifestations. Gunung Endut geothermal area is centralised in the western part of the region. Three surface manifestations located in Desa Handeuleum, Cikawah and Gajrug appears along the apparent structural (fault) line. Diluted chloride- carbonate water belongs to partially equilibrated group may indicate mixing between chloride water and groundwater. Quartz and Na-K geothermometer indicated reservoir temperature of 107 – 127 °C and 111 – 132 °C respectively. Geoindicators of Cl/B indicated that there are two separate reservoirs in the sub-surface. Geoindicators of Na/K and Na/Ca revealed that all the manifestations lie in the outflow zone of Gunung Endut geothermal area.

Keywords: geothermal, geochemical assessment, Gunung Endut, Banten

1. Introduction

Gunung Endut geothermal area is located in Lebak region, Banten province (Figure 1) and is currently stated as geothermal working area by the government [1]. Surface manifestations in a form of warm springs are mostly primitive and unutilized by the local residents. Previous geochemical assessment was reported by the Ministry of Energy and Mineral Resources suggesting that the speculative energy can reach up to 61 MWe [1]. This research focuses on geochemical fluid assessment in Gunung Endut geothermal area to investigate fluid type, reservoir temperature using geothermometry, common reservoir and the relationship between apparent geological structure as the main control for surface manifestations. A collaborative geological mapping in the area surrounding Gunung Endut, where most of its lithologies consists of quaternary volcanic deposits and limestone sediment deposits [2]. Previous geophysical studies in Gunung Endut area also indicated that faults and volcanic activities were the main control of surface manifestations and alterations [3].

IOP Conf. Series: Earth and Environmental Science **279** (2019) 012045 doi:10.1088/1755-1315/279/1/012045

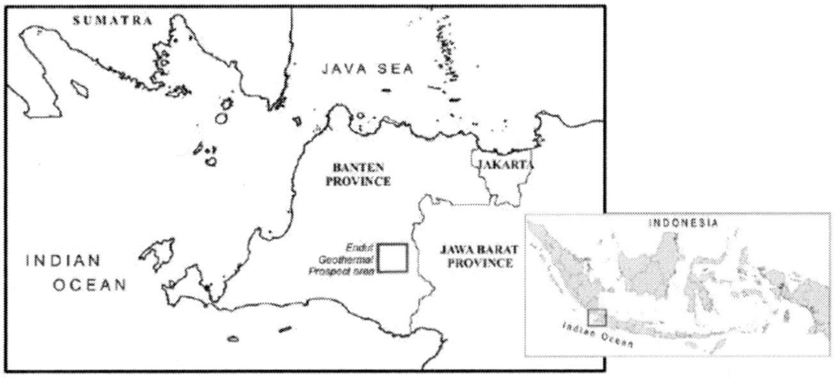

Figure 1. Gunung Endut geothermal area located in Banten Province, Indonesia [3].

2. Geological Settings

A collaborative geological mapping in the area surrounding Gunung Endut, where most of its lithologies consists of quaternary volcanic deposits and limestone sediment deposits [2]. Previous geophysical studies in Gunung Endut area also indicated that faults and volcanic activities were the main control of surface manifestations and alterations [3].

An updated geological mapping revealed that the lithologies of the area consists of Andesitic Lava (Qpal), Tuff Kasar (Qptk), Tuff Halus (Qpth), Volcanic Breccia (Qpbv) and Limestones (Tmg) (Figure 2). Two structural features of normal fault with strike/dip value of N45E/46° and horizontal fault with strike/dip value of N160E/32° were found in Desa Sobang and Desa Handeuleum respectively. Three surface manifestations were found in Cikawah and Handeuleum 10 km northwest from Gunung Endut and Gajrug 15 km north from Gunung Endut. Surface alterations were also obtained from the proximity of the surface manifestation. Altered rock analysed by XRD method and show hydrothermal minerals such as montmorillonite, calcite, and secondary quartz. Sample alteration taken around from Cikawah hot springs and indicated argillic alteration with temperature < 200°C [4].

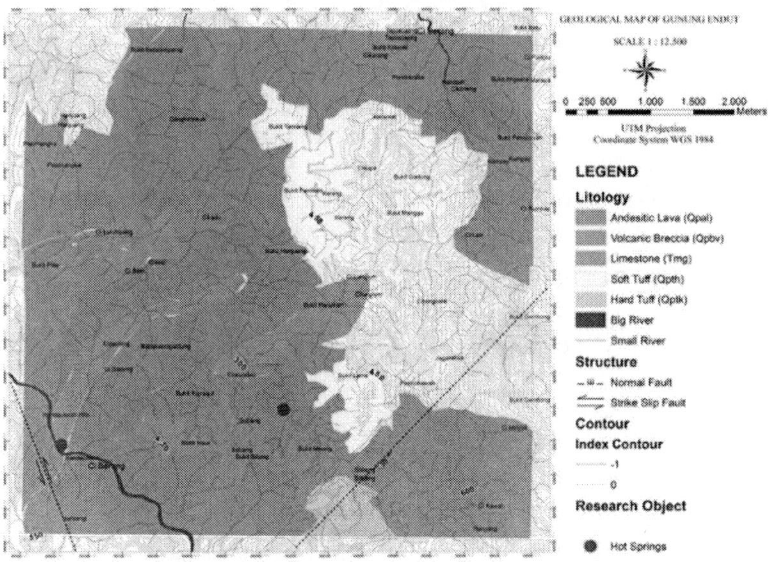

Figure 2. Geological map of Gunung Endut

3. Fluid Chemistry

Five samples from three different manifestations (CKW1, CKW2, HDL, GJG 1, GJG 2) were obtained and sampled according to the procedure [5]. The samples were collected both untreated and with treatment (the addition of HNO_3) [5]. Surface manifestation temperatures were recorded at 73 – 76 °C and the pH of the water is almost neutral (between 6.8 to 7.3). Laboratory analysis of ion balance shows charge balance between -4 - 1% allowing all the samples to be used in the calculations [5]. Total dissolved solids ranges from two manifestations (Cikawah and Handeuleum) were reported at 160 – 358 mg/L indicating high content of Si, Na, K, and Ca. As for manifestation in Gajrug, total dissolved solids were reported at 350 – 529 mg/L. The waters in Gunung Endut geothermal area can be classified as diluted chloride-carbonate waters (Figure 3) with HCO_3^- species as the dominant anions. Bicarbonate waters can form due to heating condensation reaction with near surface ground water.

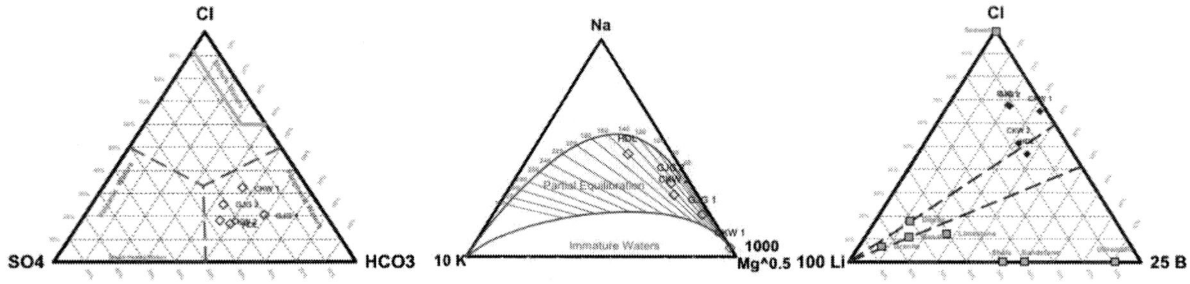

Figure 3. Ternary diagram of fluid samples obtained from Gunung Endut area. (A) CSH (B) NKM, (C) CLB

Samples from Cikawah and Gajrug (CKW 1, CKW 2 and GJG 1) shows an anomalous value of Mg indicating that the fluid has undergone a series of mixing with ground water. This also support the high content of HCO_3 in the fluid as the counter-ions for Mg possibly derived from the limestone lithologies.

4. Geothermometer

Reservoir temperature can be estimated using Na-K-Mg diagram [6] by plotting chemical compositions of Na, K and Mg. Quartz geothermometer is also applied to the data. Four samples indicate a partially equilibrated fluid but one fluid indicating an immature water meaning the interaction between host rock and fluid are not fully completed. These data are then calculated into each respective geothermometer.

Silica geothermometer predicted that the reservoir temperature lies between 100 – 127 °C, whereas Na/K geothermometer predicted reservoir temperature between 56 – 132 °C. Na/K geothermometer [7] was applied due to its compatibility with the fluid chemistry that contains high concentration of bicarbonate anion [5]. A mixing of groundwater can also affect the concentration of Mg and consequently affecting Na-K-Mg geothermometer calculation.

Table 1. Geothermometer calculation

Sample Name	$T_{Na-K-Mg}$ (Giggenbach, 1988)	Quartz	$T_{Na/K}$ (Fournier, 1979)	$T_{Na-K-Ca}$
CKW1	60	117	111	87
CKW2	150	100	56	8
HDL	160	127	132	79
GJG1	110	147	140	95
GJG2	130	107	91	57

5. Geoindicator

Fluid chemistry can also be applied as geoindicators [5]. Compositions of Na, K, Ca, and Mg can indicate whether the manifestation is an upflow or an outflow. High ratio of Na/K and low ratio of Na/Ca indicate an outflow zone. It is likely that all samples are an outflow zone of Gunung Endut geothermal area.

Common reservoir can also be determine based on Cl-Li-B ternary diagram and Cl/B ratio. Chloride and Boron are conservative ions that makes a reliable geoindicator [5]. Surface manifestations that lies in Cikawah and Handeulum shows similar value whereas manifestation in Gajrug shows a significant difference. It is likely that these 2 separates manifestations originate from two different reservoir, one that lies to the northern part and one that lies to the northwestern part of Gunung Endut respectively. Higher Cl content when compared to Li and B, indicates that the manifestation originate from an older hydrothermal system and the fluid were able to migrate through the basement rock [5].

Table 2. Geoindicator calculations

Sample Name	Na/K	Na/Ca	Mg/Ca	Cl/B
CKW1	163.8	6.4	0.112	50.0
CKW2	33.1	13.3	0.005	40.0
HDL	29.1	15.4	0.001	31.2
GJG1	72.7	13.0	0.021	83.3
GJG2	48.1	22.1	0.009	80.0

6. Geothermal system in Gunung Endut geothermal area

Fluid chemistry analysis revealed that all the manifestations found around Gunung Endut area are indicator of outflow zones. Outflow zone with dilute-bicarbonate chloride water is a typical manifestation for geothermal system in an eroded volcanic terrain [8]. Both normal and horizontal faults found near the manifestations are predicted to be the main control of the surface manifestations. Highly altered rocks appear as a result of the interaction between geothermal fluid and the surrounding rocks. A geophysical study of magneto-telluric found some low resistivity pattern proximate to the manifestations and to the peak of Gunung Endut [3]. However, the heat source of the geothermal system is yet to be determined.

An upflow zone-type manifestations can likely be found at the central/peak of Gunung Endut. Surface alteration and structural pattern in the field may aid to the direction of the upflow zone. Further field mapping and alteration studies can help create a better conceptual model in Gunung Endut geothermal area.

7. Conclusion

Geochemical assessment of Gunung Endut area revealed that three surface manifestations are indicators of outflow zones bases on the geoindicator analysis. All fluid type has diluted-bicarbonate water with near-neutral pH with high concentration of Total Dissolved Solids. There are two apparent reservoir overlying to the north and northwest of Gunung Endut geothermal area. The distribution of surface manifestations is thought to be controlled by faults (normal and horizontal) based on field observation and the findings of altered rocks along the structural line. Reservoir temperatures were obtained using geothermometry and yield the temperature ranging from 107 – 127 °C using quartz geothermometer and 111 – 132 °C using Na/K geothermometer.

Acknowledgements

The authors would like to acknowledge the Faculty of Mathermatics and Natural Sciences for granting the research fund through PITTA scheme (Publikasi Internasional Terindeks untuk Tugas Akhir Mahasiswa) 2018, PUPT and PSDG, BG, ESDM for conducting fluid analysis.

References

[1] Kementerian Energi dan Sumber Daya Mineral. 2017. Buku potensi panas bumi jilid 1.
[2] Sujatmiko dan S. Santosa. 1992. Peta Geologi Lembar Leuwidamar 1109-03, Peta Geologi Bersistem Jawa skala 1:100.000. Direktorat Geologi Indonesia
[3] Supriyanto et al. 2018. The Effect of Gravity Measurement Distribution Points on Interpretation of Gravity Data in The Gunung Endut Geothermal Prospect Area, Indonesia. International Journal of GEOMATE, Jan, 2018 Vol.14, Issue 41, pp.60-67
[4] Wohletz, Kenneth, and Grant Heiken. 1992. Volcanology and Geothermal Energy. Berkeley: University of California Press.
[5] Nicholson, Keith. 1993. Geothermal Fluids Chemistry and Exploration Techniques. Berlin. Springer-Verlag
[6] Giggenbach, W. F., Geothermal solute equilibria. Derivation of Na–K–Mg–Ca geoindicators; Geochim. Cosmochim. Acta 52, 1988, pp 2749–2765.
[7] Fournier, R.O., Truesdell, A.H. 1973. An empirical Na/K/Ca geothermometer for natural waters. Geochimica et Cosmochimica Acta, 1972, Vol.37, Issue 5, pp.1255-1275.
[8] Hochstein, M.P., Browne, P.R.L. 2000, Surface Manifestation of Geothermal Systems With Volcanic Heat Sources. Editors: Haraldur Sigurdsson, Encyclopedia of Volcanoes, Academic Press

IOP Conf. Series: Earth and Environmental Science **279** (2019) 012046 doi:10.1088/1755-1315/279/1/012046

Identify the Distribution of Galena using Induced Polarization and Resistivity Methods in central of Lombok, West Nusa Tenggara

Asnur Azis[1], Muh. Adimaher Zamhuri[1], Muh. Iqbal Rais[1], Syamsuddin [1], Sabrianto Aswad[1], Obed Patiung[2], Yanto Sudianto[3]
[1]Geophysics Department, Hasanuddin University, South Sulawesi. Indonesia
[2]Geological Engineering Department, Hasanuddin University, South Sulawesi. Indonesia
[3]Agency for the Assessment and Application of Technology - Center for Mineral Resource Technology

asnurazis10@gmail.com

Abstract. This study aims to identify the distribution of galena in central of Lombok, West Nusa Tenggara. On the surface of the area Silicified Rock shows the shape of quartz veins, which are formed as fillers in volcanic breccia, tuff and igneous andesitic rocks. Quartz veins generally have a relative direction following the main stress, with a vein thickness of 0.5 cm - 3 cm (stockwork of vein), while the main vein has a thickness ranging from 22 cm - 2.5 m. Mineralization conditions are very strong in quartz vein outcrops indicated by the presence of pyrite mineralization, chalcopyrite and galena. One of the geophysical exploration methods used to detect the presence of galena is the geoelectric method. The geoelectric method used in this study is the resistivity method and the induced polarization method (time domain). The resistivity method is used to detect the distribution of rocks associated with galena, while the induced polarization method detects the effect of the capacitor on subsurface metal minerals. This research has 3 line, the length of each line is 235 meters and one line is used as a reference for measuring 24 meters in length using the Wenner schlumberger configuration. Interpretation results obtained galena distribution is in all line with chargeability value 35.0-45.7msec

Keywords: Resistivity, Induced Polarization, Wenner Schlumberger, Galena

1. Introduction

Along with good industrial activities, both at home and abroad there will be a need for raw materials or black wood (PbS), especially China. So many activities are carried out to find areas that have the potential to contain black ingredients. Correspondingly, the government has included the private sector in an effort to find and develop the potential for lead (Rosana, 2007).

From several previous investigations, especially metal minerals, it is known on the island of Lombok that there are sulphide minerals in the rocks of Pengulung Formation (Gunradi, 2005). Preliminary exploration carried out by the Directorate of Mineral Resources (1997) in the regions of West Lombok and Central Lombok, shows that there are several prospect areas of metal mineralization in this area (Gunradi, 2005).

One of the appropriate methods to detect the distribution of the presence of ore deposits below the surface is to use geoelectric methods. The geoelectric method itself is defined as a geophysical method that studies the nature of electricity in the earth. A good geoelectric method used for metal mineral exploration is a method of Induction Polarization. (Telford, et. al , 1990).

Induced polarization Method (IP) is a relatively new method of geophysics and has been used in the field of basic metal exploration . This method is widely used in exploration of base metals because of the polarization phenomenon that occurs to a rock medium . The polarization phenomenon indicates the presence of metal below the surface that is not detected properly if it only uses the resistivity geoelectric method (Telford , et . al , 1990).

2. Resistivity Method

The flow of electric current in rocks and minerals can occur if rocks or minerals have many free electrons so that the electric current is flowed in rocks or minerals by the free electrons. This electric flow is also influenced by the nature or characteristics of each rock that passes through it. One of the characteristics or characteristics of the rock is resistivity. Resistivity or commonly called types resistance is a parameter that shows the inhibitory power of a medium in flowing electric current.

In the isotropic homogeneous earth model (figure 1), an electrode C inside the earth is coupled with other electrodes on the surface at a considerable distance. Electrode C can be seen as a source point that emits electrical currents in all directions within the earth with type resistance. When the distance between 2 current electrodes is determined , the magnitude of the potential for point P_1 is due to current sources C_1 and C_2 (Telford , et . al , 1990):

$$V_1 = \frac{I\rho}{2\pi}\left(\frac{1}{r_1} - \frac{1}{r_2}\right) \tag{1}$$

With the same rules, the potential for point P_2 due to current sources C_1 and C_2 is :

$$V_2 = \frac{I\rho}{2\pi}\left(\frac{1}{r_3} - \frac{1}{r_4}\right) \tag{2}$$

So that the potential difference between points P_1 and P_2 is as follows :

$$\Delta V = \frac{I\rho}{2\pi}\left[\left(\frac{1}{r_1} - \frac{1}{r_2}\right) - \left(\frac{1}{r_3} - \frac{1}{r_4}\right)\right] \tag{3}$$

ΔV = potential difference between P_1 and P_2 (Volt),
I = current strength through current C_1 C_2 (Ampere) electrode,
r_1 = distance between C_1 and P_1 (meters),
r_2 = distance between C_2 and P_1 (meters),
r_3 = distance between C_1 and P_2 (meters),
r_4 = distance between C_2 and P_2 (meters)

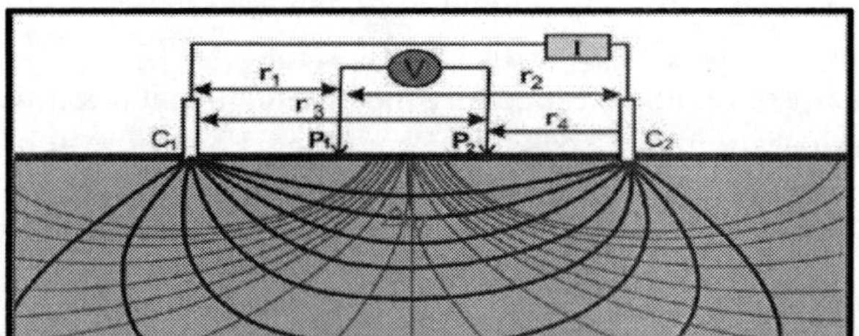

Figure 1. Electricity Potential by Two Current Sources in the Earth (Kearaey, et., al, 2002).

Based on the arrangement of measurement electrode placement there are various types of configurations including Wenner Alpha configuration , Wenner Beta , Wenner Gamma , Pole-Pole , Dipole-Dipole , Pole-Dipole , Wenner-Schlumberger , and Equatorial Dipole-Dipole .

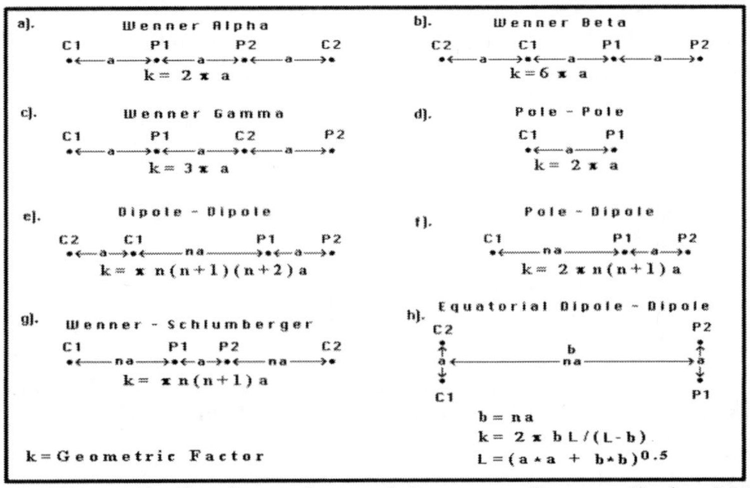

Figure 2. Electrode Configuration and Geometry Factors (Loke, 2004).

Figure 2 shows each configuration has its own characteristics, so that each configuration has advantages and disadvantages. Each configuration produces different geometry factors. Where the geometry factor is a quantity that changes into the distance apart (space) of the electrode and depends on the arrangement of the electrodes that is formulated (Loke , 2004) .

$$K = \frac{2\pi}{\frac{1}{r_1}-\frac{1}{r_2}-\frac{1}{r_3}+\frac{1}{r_4}} \quad (4)$$

K is a geometry factor that depends on the arrangement of 4 electrodes . Substituting the geometry factor of all resistivity equations can be simplified to :

$$\rho = K \, \frac{\Delta V}{I} \quad (5)$$

ρ is an apparent resistivity with Ohm . Meter

A problem is better done with one type of electrode configuration, but it is not certain that the problem can be solved if another type of configuration is used. Therefore, before making a measurement, the goal must be clearly known so that it can choose which type of configuration to use (William, 2003).

3. Induced Polarization Method

The induced polarization method is part of the geophysical method that utilizes polarization properties due to the presence of electric currents passed in the medium. This method arises because of the limitations of the Geoelectric method, one of the techniques for measuring this method is in the time zone, therefore it is called Time Domain Induced Polarization (TDIP) (Yatini, et, al., 2014).

The effect of induced polarization (IP) can be illustrated by using a four-electrode configuration in a resistivity measurement as in figure 1. When the current flow at the current electrode are stopped , the potential difference value of the two potential electrodes is not directly zero but decreases slowly until worth zero .

The same phenomenon occurs when an electric current is turned on. In the initial state, the value of the potential difference increases slowly over a certain time interval to a constant value. The medium that experiences the effect is called a medium that can be polarized. The IP effect is shown during the time interval decrease in potential difference as shown in figure 3.

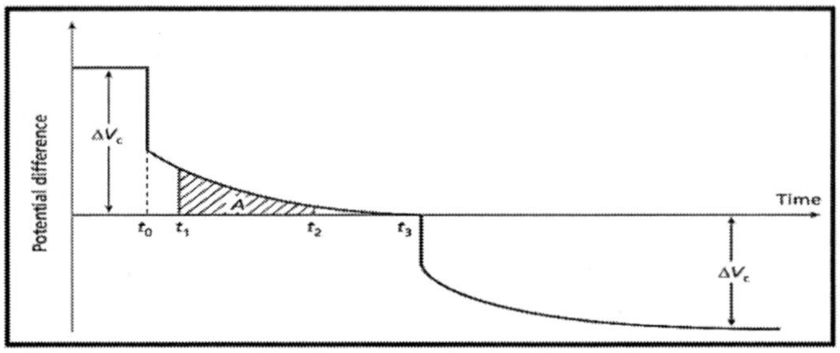

Figure 3. Phenomenon of Induced polarization (Kearey, et.al, 2002).

3.1 Measurement Priciples

Induced polarization method uses the Wenner Schlumberger configuration when taking measurements of the field as illustrated with Figure 1. Data acquisition method of polarization induction can be done in two ways, time domain measurement and frequency domain measurement.

Measurements of the frequency domain are carried out by flowing electric current into the ground in two different frequencies of a certain time interval. The use of these two different frequencies is because each material has a different response for certain types of currents with the same frequency.

The measurement of the IP time domain is used to measure the potential difference after the current is stopped. One measurement parameter is chargeability M, which is defined as an area of an under a decrease curve over a certain time interval (t_1-t_2) which is normalized by a potential difference ΔV_C. Chargeability is measured from a certain time interval when the current is stopped (Kearaey, et., al, 2002).

$$M = \frac{1}{\Delta V_C} \int_{t_1}^{t_2} v(t)dt \quad (6)$$

M has a time dimension with second or millisecond units

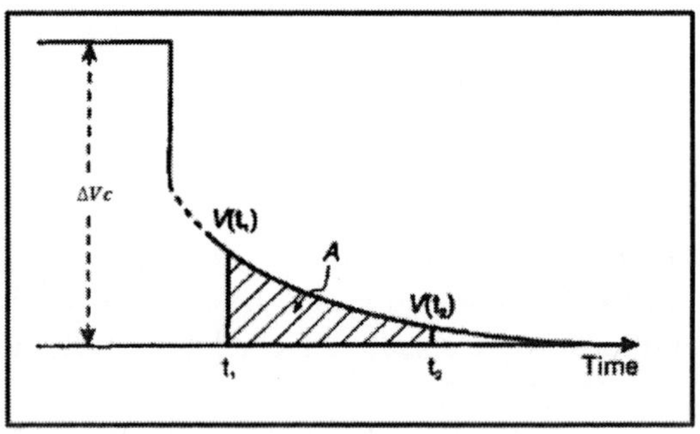

Figure 4. Residual potential for time (Kearaey, et., al, 2002).

4. Result and Discussion

In this study the interpretation is based on several things where each parameter used is the result of literature studies and field parameters that are used as reference measurements. Where in this reference has a quartz vein whose relative direction follows the main stress with a vein thickness of 0.5 cm - 3 cm (stock work of vein), while the main vein have a thickness ranging from 22 cm - 2.5 m . Mineralization conditions are very strong in quartz vein outcrops indicated by the presence of pyrite mineralization, chalcopyrite and galena. Galena are found to be associated with tuff (figure 5).

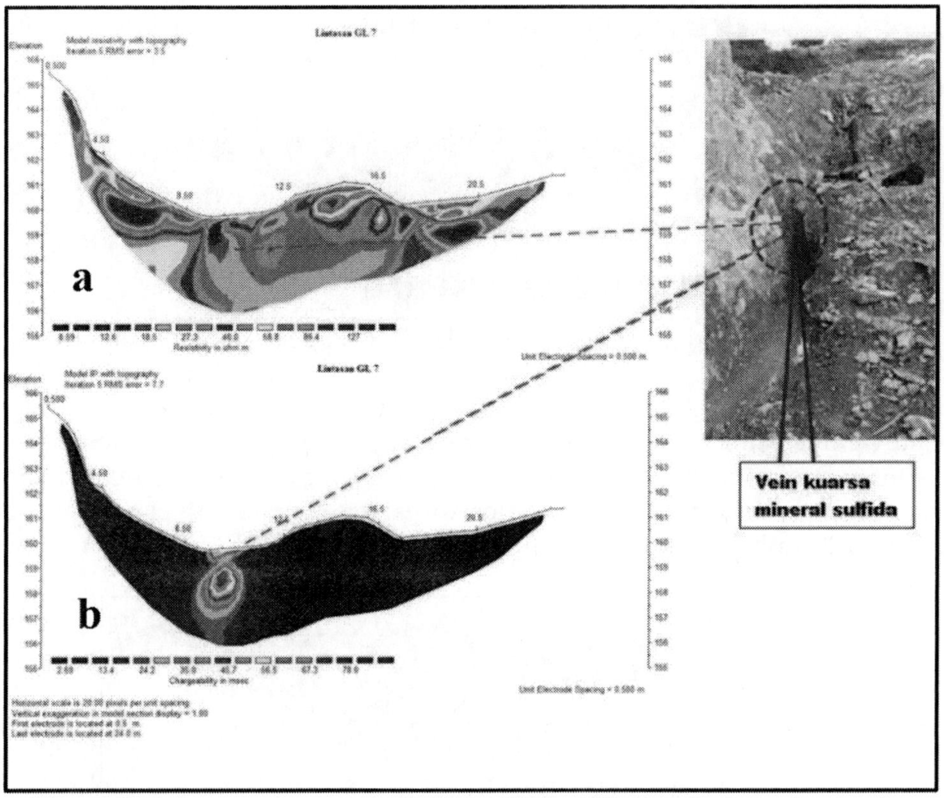

Figure 5. Reference measurements in the field. a). Resistivity inversion results, b). Chargeability inversion results

The inversion result from figure 5 shows the range of resistivity values between 8.59 - 152.93Ωm and the chargeability value between 2.69-83.41msec. Galena has a chargeability response > 35.0 - 45.7msec with a low resistivity value. The reference distance parameter aims to obtain the model parameter values to obtain theoretical data that matches the observation data. If there is a response to a model that matches the observation data, the model used to obtain the response can be considered to represent the subsurface conditions where the data is measured.

4.1 Analysis of Measurement Results

The processed data generates resistivity values and chargeability values then are converted in Res2Dinv which is then displayed in the Surfer section of each path to facilitate interpretation.

Line_1

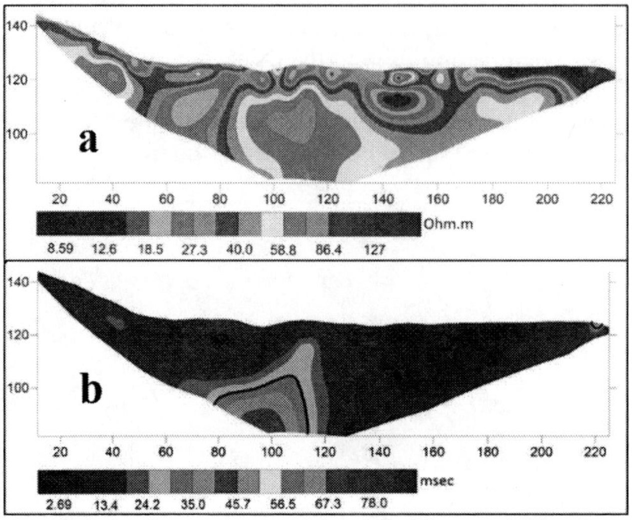

Figure 6. a). Resistivity inversion results, b). Chargeability inversion results

Figure 6 Shows the resistivity response to a high resistance anomaly with a resistivity value of 58.8 - 86.4 μm which is thought to be the origin rock. Chargeability response to a value of 35.0 - 45.7msec shows the mineral prospects of galena which are locally related, indicated by a black line indicated where the accumulation of galena is, with a thickness of ± 20 m at a distance of 80-120 with a depth of 20-45m.

Line_2

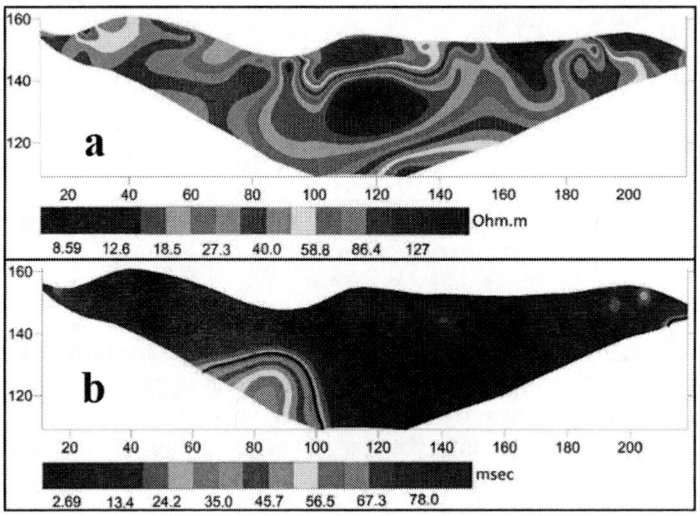

Figure 7. a). Resistivity inversion results, b). Chargeability inversion results

Figure 7. Chargeability response to a value of 35.0 - 45.7msec shows the prospect of galena marked with a black line indicated where the mineral accumulation of galena is accumulated, with a thickness of ± 20m at a distance of 60-100m with a depth of 20-45m. Resistivity response shows the dominance of low in medium resistivity values with a range of values of 27.3-40 Ωm, but at a distance of 80-100m there is a weak zone anomaly that has a lower resistivity value than its surroundings.

5. Discussion

Interpretation is done by combining resistivity and chargeability sections. This is to strengthen the assumption of the results of the analysis in determining the distribution of rocks containing galena from each measuring path. From the resistivity section and the chargeability section of the inversion results which are associated with the geological conditions of the study area, obtained estimates of the chargeability and resistivity values of rock containing galena based on observations can be seen in Table 1.

Table 1. Resistivity value based on rock lithology

Resistivity (Ωm)	Rock Lithology
40.0-127	Breccia rock is composed of polymict fragments consisting of andesite , diabase and basalt
8.59-27.3	Tuffaceous rocks have experienced physical weathering due to exogenous and endogenous influences (hydrothermal processes).

Galena have a low chargeability value <3.7 msec (Kearey, 2002). The presence of galena associated with the surrounding environment causes the presence of galena will be indicated by a greater chargeability value, which is between 35.0 - 83.41 msec.

Line_1

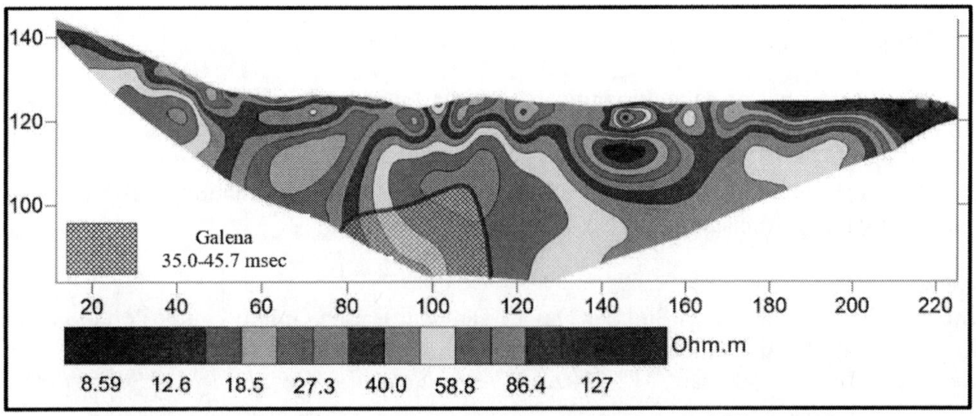

Figure 8. Overlay resistivity section and chargeability section

The data processing results of this path shows the chargeability response which is locality with a chargeability value of ≥35.0 msec, can be seen in Figure 8 is the result of overlaying resistivity section and chargeability section can be seen shaded zones at distances between 80-120m, in this zone the prospects of mineralization galena is quite large with a thickness of ± 20m, a high resistivity value ≥58.6 Ωm looks like a large chunk. This can be interpreted as bedrock associated with galena mineral, then linked to the geological conditions of the study area, then this value is considered as breccia intrusion rock beneath the surface.

Line_2

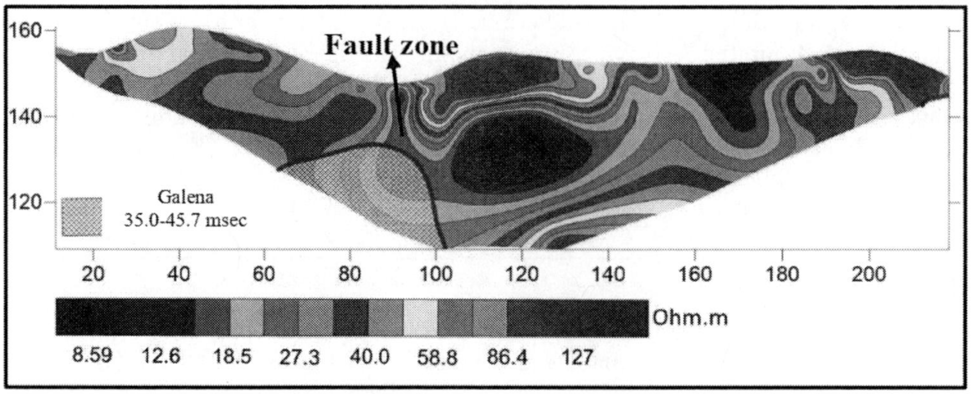

Figure 9. Overlay resistivity section and chargeability section

The data processing results in this path show the chargeability response in the similarity of the model on the line_1 path (figure 7.b), It is assumed that the continuity of galena on the line_1 and line_2 paths. Figure 9 is the result of overlaying the resistivity section and chargeability section can be seen shaded zones at a distance between 60-100m is the mineral prospect of galena. At this layer it is assumed that there is a weak zone at a distance of 80-100m which has a lower resistivity value than its surroundings which is thought to be composed of alteration rocks consisting of sericite rocks (interlaced tuffs) where these rock units have high porosity and permeability.

6. Conclusion

On each line, the prospect of galena is indicated on the line 1 and 2, line 1 is associated with breccia rock, while line 2 is associated with sericite rocks (altered tuffs).

Excellent prospects for mineralization are on lines 1 and 2. The continuation of rocks in the form of breccia and tuff which is the accumulation of galena.

References

[1] Gunradi, R. 2005. Evaluasi Sumber Daya/Cadangan Bahan Galian Untuk Pertambangan Sekala Kecil, Daerah Pulau Lombok, Provinsi Nusa Tenggara Barat.

[2] Kearey, P.,Michael, B., Ian., H . 2002. *An Introduction to Geophysical Exploration*. Blackwell Science Ltd. London.

[3] Loke, M.H. 2004. Rapid 2D Resistivitas & IP Inversion using the least-square method, Geotomo Software. Malaysia.

[4] Rosana F,M. 2007. Keterdapatan Bahan Galian Galena di daerah Ceigemblong, Kabupaten Lebak, Propinsi Banten. Jurusan Geologi. FMIPA. Universitas Padjadjaran.

[5] Telford, W.M., Geldart, L.P., Sheriff, R.E. 1990. *Applied Geophysics*. Cambridge Univ. Press. Cambridge.

[6] William J.J. 2003 Application Geoelectric Resistivitas method for detectionof underground Mine Working, Geophysical Technology for detecting underground Coal.

[7] Yatini, S. D., Laesanpura, A., 2013, Physical Modelling studies of the Time Domain Induced Polarization (TDIP) Response, Case : Homogeneous Isotropic Medium, Proceedings, 3rd Annual Basic Scienc.

IOP Conf. Series: Earth and Environmental Science **279** (2019) 012047 doi:10.1088/1755-1315/279/1/012047

Analysis of salt pond cluster in Arungkeke sub-district Jeneponto Regency

Sakka[1,*], Baharuddin[2], W Samad[3]

[1] Geophysics Study Program, Hasanuddin University
[2] Marine Science Study Program, Lambung Mangkurat University
[3] Marine Science Study Program, Hasanuddin University
[*] Center for Environmental Research and Development, Hasanuddin University
E-mail : sakka.fisika@yahoo.com

Abstract. Jeneponto Regency is one of the largest national salt production locations in South Sulawesi. One of them is located in Arungkeke Sub-district with a land area of 328.61 ha (21.62%), but salt pond activities in this area are still carried out traditionally, so that not all locations can produce and some produce less than 71 tons / ha with NaCl levels <90% and not meeting industry salt standards (NaCl> 97%). This research was conducted to analyse the area of salt ponds based on clusters on the similarities they have in managing salt ponds. The results showed that the total area mapped by 243.9 ha could be divided into 14 clusters / groups with a wide range of 10-30 ha / cluster. Each cluster is divided into bozem ponds, weeds and salt tables with a ratio of 5: 3: 2. The channel network system to be developed is divided into two primary channels, namely SPPA1, and SPPA2 and two secondary supply channels, namely SSPA1.1 and SSPA1.2. The filling of water into the ponds of each cluster can only be done using a pump with a water debit requirement ranging from 0.116 - 0.347 m3 / s. To achieve good salt quality, increasing the density value is carried out in stages, namely in bozem, hatchery and salt table are 4 - 11 °Be, 11-22 °Be and 22-30 °Be, respectively.

1. Introduction

Jeneponto Regency is one of the central regions of many salt producers in Indonesia. The regency which is located 90 km south of Makassar City has long been known as one of the areas with extensive salt ponds and is spread in four sub-districts namely Bangkala Barat, Bangkala, Tamalatea and Arungkeke sub-districts. Based on the Quickbird satellite image analysis (2014), shows the pond area in Jeneponto Regency is 1,519.89 ha, where Bangkala District has the largest pond area of 479.43 ha (31.54%), Tamalatea 454.86 ha (29.93 %), Arungkeke Subdistrict amounted to 328.61 ha (21.62%) and West Bangkala District 256.98 ha (16.91 ha). However, not all ponds in Jeneponto Regency can produce well. This is due to various factors, such as the influence of the season on the volume of availability of sea water to irrigate salt ponds, the activity of salt ponds is still traditional, and namely the salt table is still grounded, so that salt is still fused with the soil when harvested. Besides for this reason, the development of salt pond land must be based on the technical potential of the development area. Efendy *et al.* (2014) suggested that the technical potential in question is a variety of suitability of producing land such as: (1) availability of land in coastal areas to facilitate obtaining raw water for sea water; (2) flat slope topography for tidal mechanisms when entering sea water; (3) non phorus soil types to guarantee the availability of old water and perfect deposition; (4) low rainfall to increase evaporation of young sea water; (5) the river flow that empties into the area is small so that the process of dilution of sea water as raw material does not significantly affect salinity and (6) the availability of land and sea transportation routes to support marketing.

Salt production in Jeneponto Regency in 2012 decreased compared to 2011. Salt production in 2011 amounted to 41,304 tons while in 2012 only produced salt of 37,877 tons. In other words, there was a decrease in salt production by 3,427 tons. This is caused by natural factors and the absence of technology application to overcome this. In the span of 5 years, salt production in Jeneponto Regency continues to fluctuate (Arzal, 2014). The Ministry of Maritime Affairs and Fisheries (KKP) assessed that the quality of salt produced by Indonesia has not met the quality standards of export salt. So that in the near future, in addition to targeting salt self-sufficiency this year, the government will also improve the quality of people's salt according to industry standards with 97 NaCl levels up to 98 percent (CNN Indonesia 2015). In an effort to encourage increased capacity for domestic salt production to meet national salt needs, at the end of 2009 the Government began launching the National Salt Self-Sufficiency Program. The declaration of the program was carried out in the framework of implementing Presidential Decree No. 28 of 2008 concerning National Industrial Policy. The Presidential Decree states that the people's salt industry is included in the national priority industry cluster, which is a national industrial cluster that has high prospects to be developed based on its ability to compete in international markets, and industries that have sufficient and competitive production factors in Indonesia.

This research was conducted to map salt pond land in Arungkeke District based on cluster analysis for organizing a collection of patterns into clusters (groups) for the similarities they have in managing salt ponds with variables such as: pond productivity (tons / ha / season), management land (ha / person), farmer production capacity (ton / person), harvest period from Crystal table (days), average length of dry season in the last 10 years (daily), proportion of quality of salt production (%), condition of access to major road and primary channel conditions.

2. Method

This research was carried out in the salt pond area of Arungkeke District, Jeneponto Regency (Figure 1) in September - December 2015. The data collected included primary data, namely topographic survey, salt quality, tides, pond channels, soil test, infrastructure and secondary data, namely physical condition of the region, socio-cultural conditions, economic conditions, condition of facilities and infrastructure and other related plans / studies.

Survey and data collection activities and Focus Group Discussion (FGD) activities in capturing the aspirations of all components of society are included in the spatial database.

Cluster analysis for organizing a collection of patterns into clusters (groups) based on the similarities they have in managing salt ponds with variables including: farm productivity (ton / ha / season), land management (ha / person), farmer production capability (ton / person), harvest period from Crystal table (days), average length of dry season in the last 10 years (daily), proportion of quality of salt production (%), condition of access to major roads and condition of primary channels.

From the results of the analysis of socio-economic and technical studies, then the determination of the salt pond cluster is presented in an area with a minimum size of 5 - 50 ha.

Calculation of water volume requirements is carried out to determine salt production in each cluster. Pond water needs are calculated by taking into account technical factors, among others; desired productivity, water replacement needs, water loss (evaporation and seepage). The thing that is done is analyzing the depth of water to determine the volume of water needed and the elevation of the water level in each pond (salt table, hatchery, bozem) in order to achieve salt production of at least 120 tons / ha / year. The calculations are carried out for the most downstream ponds, namely the salt table, followed by the pool of interest, and finally for bozem. Water requirements in each pond are calculated based on the planned water depth constant at the salt table and the water thickness (Be) setting in each pond. Before analyzing the volumes of the needs of each pond, then an analysis of the water requirements for the pond channels is done first for both the primary and secondary channels.

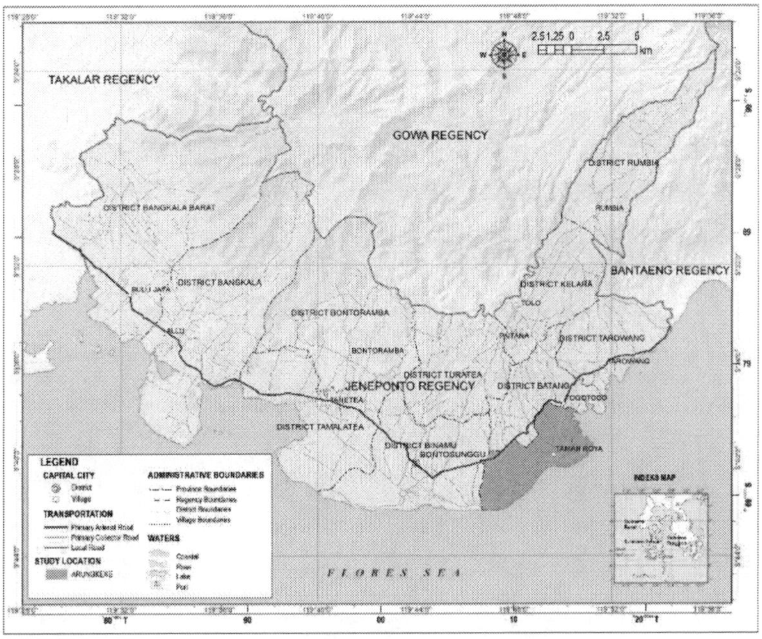

Figure 1. Location map of research

3. Result and Discussion

The process of making salt requires land that is close to the sea, has low soil porosity or the soil is not sandy. The source of sea water used must be clean / not contaminated with municipal wastewater. Besides that topography and soil physical properties and climate are very influential in the process of making evaporation salts.

The shape and size of the pond plot must be designed in accordance with the local physical condition, so that the land use and processing of the pond unit are more efficient. In general, the shape of the pond map is rectangular, square or multi-faceted. Of course each form has its own advantages and disadvantages. (Afrianto *and* Liviawati, 1991).

3.1 Distribution of Salt Pond Clusters

Based on the results of the survey and discussion (FGD) with the owner of the pond, the salt pond cluster in the study location can be divided into 14 groups / clusters. Each cluster is divided into bozem ponds, weeds and salt tables. The division of clusters for each location is shown in Figure 2.

Cluster distribution is done by considering:

1. The area of each cluster is 5 to 50 ha.
2. Cluster boundaries follow the existing pond boundaries (not dismantling community pond embankments).
3. The farmer groups that have been formed by the community are kept in one cluster.
4. Each cluster has at least 5 landowners.
5. The limit of bozem ponds, ponds and salt tables is attempted to follow the existing ponds.

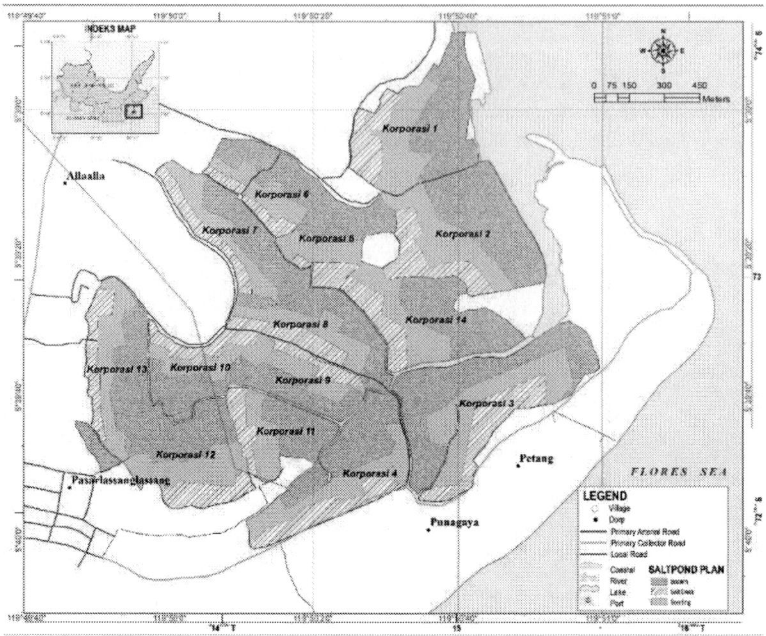

Figure 2. Distribution of Salt Pond Clusters, Arungkeke sub-district

Based on the above considerations, the form of each cluster in Arungkeke Sub-district is irregular and has a different area (Table 1). Determination of the location of bozem, seeding and salt table on each cluster is done by considering the topography and location of the existing channel / irrigation. The location of the salt table is placed in an area that has a lower elevation than the seeding, while the bozem location is attempted to be adjacent to the existing channel. Based on the Marine and Fisheries Research Agency (BRKP, 2006), this construction includes the tafel complex is salting construction in which a large saline complex (natural) is made irregular (natural) as a collective interest group, the concentrated water (old water) then flowed to a table for crystallization.

Table 1. Land area of each Cluster in Arungkeke sub-district

Cluster/ Corporation	Land area (ha)			
	Bozem	Weed ponds	Salt table	Total
Corporation 1	10,7461	6,8895	4,4279	22,0636
Corporation 2	13,2966	7,3087	5,4912	26,0966
Corporation 3	14,7421	9,6601	6,4452	30,8474
Corporation 4	8,8679	6,7753	4,9465	20,5896
Corporation 5	6,0266	3,3908	2,0188	11,4361
Corporation 6	5,9649	3,8993	2,1550	12,0192
Corporation 7	7,3614	4,6786	3,3251	15,3651
Corporation 8	7,2955	4,9192	3,6568	15,8715
Corporation 9	4,8424	3,0285	2,1082	9,9790
Corporation 10	5,6478	4,3559	3,0261	13,0298
Corporation 11	5,2898	3,6309	2,3665	11,2872
Corporation 12	9,2551	6,4857	3,4770	19,2177
Corporation 13	7,4896	5,8487	3,3865	16,7248
Corporation 14	9,7114	5,7286	3,9639	19,4040
Total	**116,5373**	**76,5999**	**50,7946**	**243,9318**

3.2 Calculation of Water Needs

The water requirements in each cluster are very much determined by the area of the salt table in each cluster and the magnitude of evaporation and the rate of change in density (°Be). Water volume and water level in each pond are determined by the depth of water in the salt table. The calculation of water requirements in each cluster is done by using the data from the measurement of the relationship of the rate of change in the density of sea water with changes in the volume of sea water due to evaporation. Changes in water density were observed from 4 °Be to 30 °Be by observing changes in water depth every time there was a change in density, as shown in Table 2 and Figure 3.

According to BRKP (2006) in the reservoir, salinity is expected to be more than 35 ppt (3.5 °Be), in this basin it is expected that precipitation of mud particles in the seawater is expected to occur at the same time evaporation occurs, resulting in an increase in salinity ranging from 50 ppt (5 °Be) - 100 ppt (10 ° Be) which is then transferred to the drainage pond I. In this reservoir the Guluk (K / G) first work is done after table water reaches salinity 40 ppt - 60 ppt (4 °Be - 6 °Be).

Table 2. Results of measurement of water density in the experimental pond

Date	Concentration (°Be)	Depth (m)	Volume (m^3)	Evaporation (m^3)	Cumulative Evaporation (m^3)
28-11-2015	4	0,1	1,000	0	0
29-11-2015	8	0,082	0,820	0,180	0,18
30-11-2015	11	0,058	0,580	0,240	0,42
01-12-2015	15	0,047	0,470	0,110	0,53
02-12-2015	18	0,036	0,360	0,110	0,64
03-12-2015	22	0,025	0,250	0,110	0,75
04-12-2015	26	0,021	0,210	0,040	0,79
05-12-2015	28	0,012	0,120	0,090	0,88
06-12-2015	29	0,007	0,070	0,050	0,93
07-12-2015	30	0,006	0,060	0,010	0,94

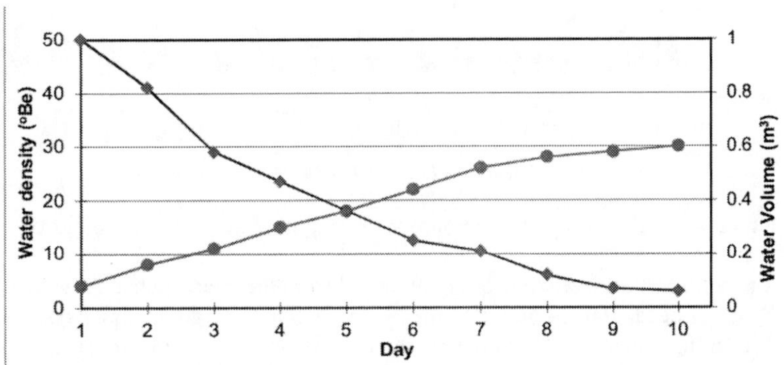

Figure 3. Relationship between changes in the density and the water volume

Calculation of water volume and water level elevation in each pond (salt table, hatchery, bozem), is carried out using the data in Table 2. At first the calculation of volume and water elevation is carried out on the salt table pool, then in the drainage pond, and last in the bozem pool. The density of water during the stay in each pool is set at 4 - 11 ° Be in bozem, 11 - 22 ° Be in flushing, and 22 - 30 ° Be on the salt table. The water depth at the salt table at a concentration of 22 °Be was also set at 0.10 m. From

IOP Conf. Series: Earth and Environmental Science **279** (2019) 012047 doi:10.1088/1755-1315/279/1/012047

the depth of the water in the salt table, the water depth in the seeding and bozem is calculated based on the rate of change in water volume and the rate of change in density.
The results of the calculation of water volume and water level elevation in salt table ponds, seeding and bozem in each corporation in Arungkeke District are shown in Figure 4-5.

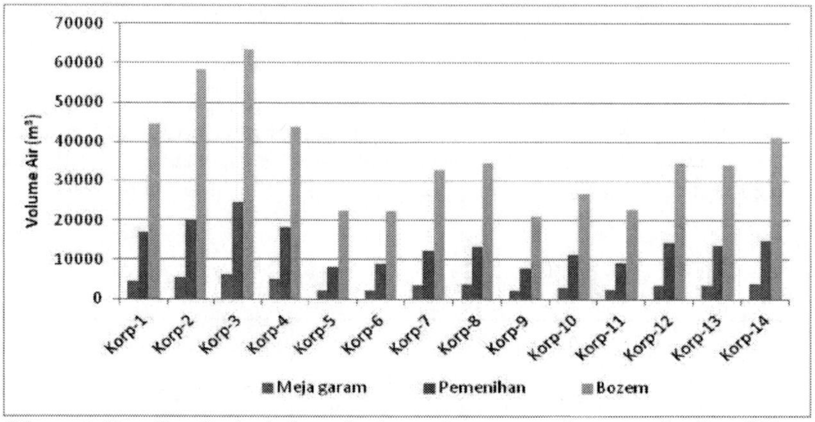

Figure 4. Results of calculation of salt pond water requirements for first filling

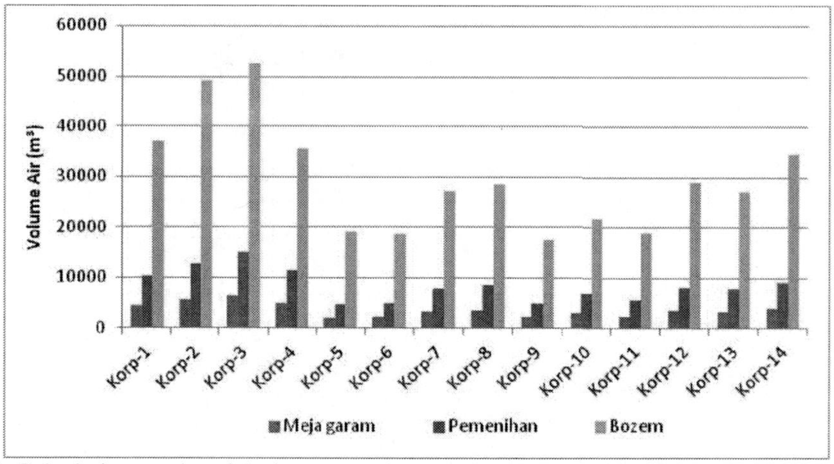

Figure 5. Calculation results of the image pond water requirements for the second filling and so on

Water needs in the pond for salting activities are obtained from the nearest channel with bozem location. Thus, to meet the water needs in the pond, it is necessary to plan the right channel system. The planning of the salt pond irrigation canal system is determined by considering the location of the fish farm and the ways of the salt farmers. Based on these things the following limits are set:

- Utilize as many channels as possible
- Every corporate area will get irrigation canal services to be able to enter water directly into bozem.

Based on the above constraints, the channel network system that will be developed in Arungkeke Subdistrict is divided into two primary channels, namely SPPA1, and SPPA2 and two secondary supply channels, namely SSPA1.1 and SSPA1.2 (Figure 4).

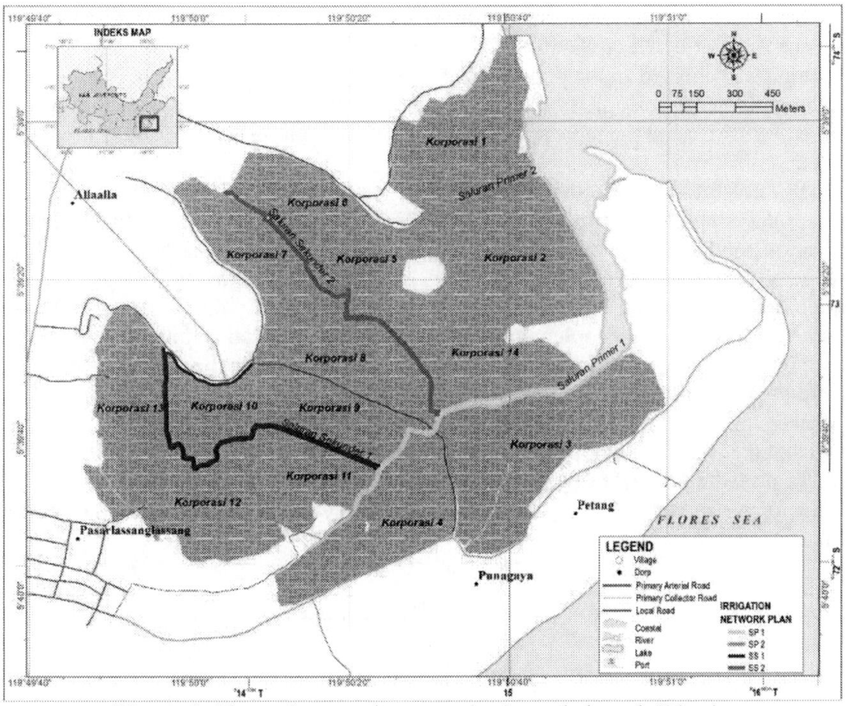

Figure 6. Channel network system in Arungkeke sub-District

Based on the results of the existing study of water governance carried out by salt farmer farmers in the current study location, it can be explained as follows: Salt farm farmers enter water from the primary channel to the dominant salt pond (bozem) using a pump machine. The water supply from the channel to the first pond (bozem) is carried out until the height of the water in the bozem is at least 15 cm from the bottom of the pond. At this time the community of pond farmers in the district. Arungkeke also did not take into account the concentration (oBe) in the drainage of water from bozem to harvesting and from seeding to salt table. Farmers keep water for 5-10 days in the first pond (bozem), then the water is channeled to the second pond (seeding) until it reaches a minimum height of 10 cm and is stored for 4 - 7 days. Water is flowed to the third pond (table salt) until it reaches a height of about 5 cm, and is stored for 5 days in a table of salt until salt crystals are formed and ready to be harvested.

To improve production and quality, it is necessary to improve water governance when salting activities are carried out. Basically the process of salt formation is achieved by evaporation of sea water so that an increase in the density value of 4 oBe becomes 30 oBe. The increase in density value is carried out in stages, namely in the 4 - 11 oBe bozem pond, where in this basin it is expected that the mud particles will be deposited again and some Magnesium, Calcium and Sulfate compounds settle at this stage. According to the results of BRKP's research (2006), in bozem (hatchery tank I) began to be added Oxalate or Carbonate to form Magnesium Oxalate / Magnesium Carbonate compounds, Calcium Oxalate / Calcium Carbonate so that Magnesium and Calcium settle. In addition to the addition of Oxalate or Carbonate in this tub can also be stocked (inoculated) Artemia which can help reduce compounds that are not needed in the making.

In this first seeding was expected to increase salinity to 150 ppt (15 ° Be). Then the water is channeled to the second water bath, in the second stage precipitation also occurs so that the sea water (brine) that will enter the crystallization pond I is already high with the NaCl compound which will become salt. In crystallization pond II where salinity has reached> 150 ppt (15 ° Be) the work of Kesap Guluk II (K

II) was carried out after table water reached salinity 180 ppt - 220 ppt (18 ° Be - 22 ° Be) and table / the above pool is carried out by Kesap Guluk II (K / G II) with the same treatment.
The second stage is channeled in the pond so that it reaches 11-22 oBe, in the process of increasing the density reaching above 15 ° Be it is better to do the Guluk II (K / G II) work done after table water reaches 18 ° Be - 22 ° Be and the table / pool above is carried out by Kesap Guluk II (K / G II) with the same treatment.
The third stage or at the salt table the concentrations range from 22 ° Be - 30 oBe. In this pond it is expected that NaCl contained in seawater is> 98%. According to the results of BRKP's research (2006) the crystallization pond or precipitated salt table of NaCl> 72% and the remaining water known as bittern water can be used for health. Bittern water contains a lot of Magnesium compounds that can be used as health supplements, so far water bittern has always been discarded by the farmer, with the introduction of bittern as a health / medicine supplement and the presence of shelter in bittern marketing is expected to be a source of side income from the farmer besides salt. In the crystallization process, it is necessary to pay attention to the maintenance of the salt table and to do *aflak* (leveling of the salt base surface).
Thus the transfer of water from one pond to another (bozem-seeding-salt table), requires the right timeliness so that the water transfer process needs to be controlled. The transfer of water from bozem to seeding and from seeding to table salt must be in accordance with the concentration level that has been designed. The process of water drainage from the bozem pond to the harvesting and from the seeding to the salt table is shown in Figure 6.

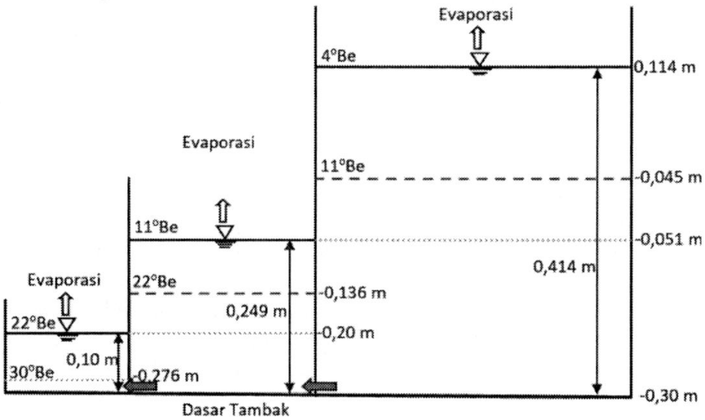

Figure 7. High water profile in bozem, seeding and salt table

4. Conclusion

To increase the production and quality of salt in Arungkeke Subdistrict, Jeneponto Regency can be done with a cluster system. From the mapped land of 243.9 ha, it can be divided into 14 clusters / groups with a wide range of 10-30 ha / cluster. Each cluster is divided into bozem ponds, weeds and salt tables with a ratio of 5: 3: 2. The channel network system to be developed is divided into two primary channels, two secondary channels. Tidal conditions with small elevations cannot be carried out by gravity, so filling the water into the ponds of each cluster can only be done using a pump with a discharge ranging from 0.116 - 0.347 m3 / s. In order to achieve good salt quality, the increase in density value is carried out in stages, namely in bozem ponds, hatchery and salt table are 4 - 11 °Be, 11-22 °Be and 22-30 °Be, respectively.

References

[1] Adiyoso, R., dkk. 2014. Perencanaan Sistem Jaringan Tambak Garam Di Pemongkong Kabupaten Lombok Timur. Jurnal Karya Teknik Sipil. Universitas Diponegoro, Semarang.
[2] Arzal, S. 2014. Analisis Tingkat Pendapatan Petani Garam di Kabupaten Jeneponto, Sulawesi

Selatan. Skripsi Jurusan Ilmu Ekonomi Fakultas Ekonomi Dan Bisnis, Universitas Hasanuddin, Makassar.
[3] Efendy, M. dkk. 2014. Pemetaan Potensi Pengembangan Lahan Tambak Garam Di Pesisir Utara Kabupaten Pamekasan. Jurnal Kelautan Universitas Trunojoyo, Madura
[4] Afrianto, Eddy dan Evi Liviawaty. 1991. Teknik Pembuatan Tambak Udang. Bandung: Kanisius.
[5] [BRKP] Badan Riset Kelautan dan Perikanan, 2006. Buku Panduan Pengembangan Usaha Terpadu Garam dan Artemia. Departemen Kelautan dan Perikanan, Jakarta.
[6] CNN Indonesia. 2015. Kualitas Garam Indonesia Tak Layak Ekspor [Internet]. 7 Januari 2015. Tersedia di www.cnnindonesia.com/ekonomi/...92.../kualitas-garam-indonesia-tak-layak-ekspor/.
[7] Purbani, Dini. 2011. Proses Pembentukan Kristalisasi Garam. Pusat Riset Wilayah Laut dan Sumberdaya Nonhayati Badan Riset Kelautan dan Perikanan Departemen Kelautan dan Perikanan.

IOP Conf. Series: Earth and Environmental Science **279** (2019) 012048 doi:10.1088/1755-1315/279/1/012048

Phase Changes in the Atmospheric Hydrology Cycle

Aqramunnisah, D ASuriamihardja, A H Assegaf and B A Samad
Department of Physics
Faculty of Mathematics and Natural Sciences, Hasanuddin University, Makassar

E-mail: akramunnisaliazminah@gmail.com

Abstract. A natural distillation process as an atmospheric machine in producing fresh water from seawater is of importance to be relearned. This study aims to scrutinize a thermodynamic cycle involving evaporation on sea surface and condensation at a possible height to occur in lower temperature and pressure by comparing reversible and irreversible process. The main stratification of pressure and temperature are obtained from local radiosonde record. The result of this study will give an explanation about how the diabatic process restores the energy to maintain the system equilibrium.

1. Introduction

The global climate results from the interaction between the atmosphere and the ocean with warming from solar radiation, the gravitational force of the earth and the dynamics of the Coriolis force originating from the earth's rotation.[1]Heating occurs in annual and daily cycles, dependence on latitude, and increases by reflecting on the surface of the earth and clouds.[2]The formation of convective clouds in the tropics results in high rainfall in the Indonesian Maritime Continent.[3]These convective clouds in the upper troposphere can release latent heat, which will cause interference with the surrounding atmosphere.

When the moist air rises and the value of the environmental lapse rate is in the range between the adiabatic lapse rate and saturated adiabatic lapse rate, will make the unstable condition of the atmosphere.[4]So the stability of the air is determined by the condition of humidity. Therefore, the amount of annual rainfall, intensity, duration, frequency and distribution of space and time varies greatly. [5]

In daily hydrologic cycles, several evaporation processes, the growth of atmospheric boundary layers, convection and precipitation can be viewed in a thermodynamic framework. [6]In the morning to the afternoon, the height of boundary layer has increased influencing stability on the atmosphere.[7]Several studies have also discussed atmospheric circulation by thermodynamic approaches.[8][9]This makes it possible to associate heat and humidity transport through atmospheric movements assuming maximum thermodynamic efficiency. [10]In addition, to analyze the classification of rain, a formulation of the thermodynamic parameters of the climate is needed.

In the condition of convective and stratiform condensation occurs, parameterization schemes have also been carried out which are used as modeling in predicting numerical fine mesh weather. [11]Parameterization of other atmospheric models has also been carried out in analyzing the properties of single cloud scattering and scaling of optical thickness under conditions of partial blurring.[12]The determination of rainfall also can be found by using ENSO and IOD parameters. [13]But the formulation of the thermodynamic physical parameters regarding the transformation of dry air and water vapor with its interaction with the condensation cells to form clouds is still not much discussed.[14]Therefore, it is necessary to conduct a study on the formulation of climate thermodynamic parameters in the process of rain formation to better understand the physical mechanism of the hydrologic cycle. So, the idealized hydrologic cycle described here is to scrutinize a thermodynamic cycle involving evaporation on the sea surface and condensation at a possible height to occur in lower temperature and pressure.

2. Hydrologic Cycle

A working engine system (ΔW) uses heat from the heat reservoir (ΔQ_{in}) and releases the heat to the cold reservoir (ΔQ_{out}). In 2012, Konings et al. [6]examined the hydrologic cycle between two hot and cold reservoirs. Hot reservoirs are at sea level with temperature $T_H$doing evaporation process, and cold reservoirs at a certain height with temperature T_C doing the condensation process.

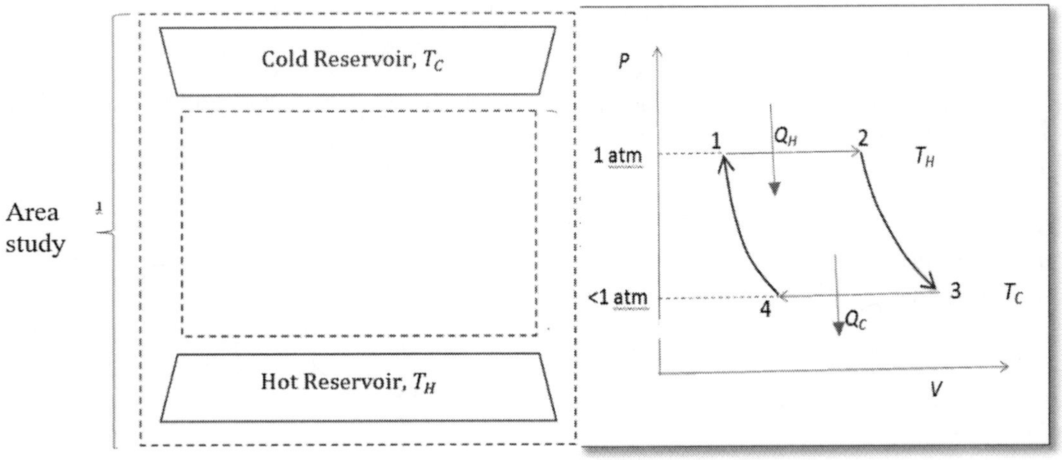

(a) (b)

Figure 1. (a) Scoping of the study area. In hot reservoir, the evaporation process takes place with surface temperature T_H, while in the cold reservoir there is a process of condensation with upper temperature over T_C. This process will cause air circulation by involving a phase change from moist air to liquid air. (b)P-V diagram on the quasi-Carnot cycle following by 4 processes.Evaporation, adiabatic expansion, condensation, and compression. On P-V diagram above, the compression occurs diabatically. [15]

2.1. Evaporation

Underwater dome of the p-V diagram on **Figure 1,** transition from state1 to state 2 occurs at sea surface with temperature T_H as the process of isotherm-isobaric evaporation. A parcel of liquid water mass m_l is transformed into water vapor mass by keeping that $m_l = m_g$. This water vapor will joint with m'_g as a residual water vapor mass come downed from upper layer. The quality of mixing is expressed as:

$$x = m'_g/m_t \tag{01}$$

Where m_t is a total of liquid water and water vapor masses which is expressed by$m_t = m_l + m'_g$. Because ΔW=$m_t R T_H$, accordingly $m_l/m_t = m_g/m_t = 1 - x$.

In this process, we also assumed that the volume of liquid air is V_l evaporated to become V_{g2} and joining with residual vapor volume from diabatic process V'_{g1} to V'_{g2}, so the two processes can be written as $V_{t1} \rightarrow V_{t2}orV_{t1} = V_{l1} + V'_{g1} \rightarrow V_{t2} = V_{g2} + V'_{g2}$.

Change of volume due to isotherm isobaric process:

$$\Delta V_{12} = (V_{g2} - V_{l1}) + (V'_{g2} - V'_{g1}) \tag{02}$$

IOP Conf. Series: Earth and Environmental Science **279** (2019) 012048 doi:10.1088/1755-1315/279/1/012048

Because of: $V_{l1} \ll V_{g2}$ simply that:

$$\Delta V_{12} = V_{g2} + \left(V'_{g2} - V'_{g1}\right) \tag{03}$$

By dividing equation (03) with m_t and considering the ongoing process will also isochoric, the specific volume of the process can be expressed,

$$\Delta v_{12} = (1 - x)v_{g2} \tag{04}$$

Based on the process of phase change above, we found that specific work exerted by the system to the environment is,

$$\Delta w_{12} = (1 - x)RT_H \tag{05}$$

Due to the isotherm process, the change of internal energy is$\Delta u_{12} = 0$. While the latent heat for evaporation is L_H, hence the heat required for evaporation is,

$$\Delta q_{12} = (1 - x)L_H \tag{06}$$

2.2. Adiabatic Expansion

Transition from state-2 to state-3 is in the process of adiabatic expansion. A parcel of water vapor mass m_g is rising and joining with a residual vapor m'_g come downing from upper layer at the sea surface just after evaporation. Total water vapor mass is $m_t = m_g + m'_g$to the level where a process of condensation will be possible.The respective total volume will be$V_{t2} = V_{g2} + V'_{g2}$and $V_{t3} = V_{g3} + V'_{g3}$, the change of estimated volume is,

$$\Delta V_{23} = V_{t3} - V_{t2} \tag{07}$$

By dividing equation (07) by m_t, we find the specific volume is,

$$\Delta v_{23} = (1 - x)\left(v_{g3} - v_{g2}\right) + x(v'_{g3} - v'_{g4}) \tag{08}$$

In the right hand side of equation (08) is actually no need to distinguish, the phase is in vapor phase. Along the process which is occurred adiabatically, the specific work is,

$$\Delta w_{23} = -c_v(T_C - T_H) \tag{09}$$

Based on the first law of thermodynamics $\Delta w_{23} = \Delta u_{23}$. Although this process occurs adiabatically, in reality there is still a heat needed by the system to change its potential thermodynamic ($\mu = T \ln \mathcal{H}$,$\mathcal{H}$ is the humidity) from the sea surface to the height of possible condensation. Therefore, this process is also referred as a pseudo-adiabatic process with heat input is added by potential thermodynamics in the last section,

$$\Delta q_{23} = R(T_H \ln \mathcal{H}_H - T_C \ln \mathcal{H}_C) \tag{08}$$

2.3. Condensation

Transition from state-3 to state-4 occurs at the level of condensation with temperature T_C as the process of isotherm-isobaric condensation. A parcel of water vapor mass m_t is partly transformed into liquid water mass m_l, and partly remained as a residual water vapor mass m'_g, by keeping that the total water vapor remains as $m_t = m_l + m'_g$.

The volume $V_{t3} \rightarrow V_{t4}$, partly condensate, and is expressed as$V_{t3} = V_{g3} + V'_{g3} \rightarrow V_{t4} = V_{l4} + V'_{g4}$.
The change of volume due to condensation, $\Delta V_{34} = V_{t4} - V_{t3}$

$$\Delta V_{34} = \left(V_{l4} - V_{g3}\right) + \left(V'_{g4} - V'_{g3}\right) \tag{09}$$

Because of $V_{l4} \ll V_{g3}$, and vapor component is undergoing in isochoric process, simply the equation (09) will be,

$$\Delta V_{34} = -V_{g3} \tag{10}$$

The work performed on the system is given by:

$$\Delta w_{34} = -(1-x)RT_C \tag{11}$$

Due to the isotherm change, the internal energy $\Delta u_{34} = 0$, and the latent heat for condensation is L_C, the heat deliberated during condensation is,

$$\Delta q_{34} = -(1-x)L_C \tag{12}$$

2.4. Diabatic compression

Transition from state-4 to state-1 is in the process of diabatic compression. There are two parcels water masses come down to the sea surface. The first is liquid water mass m_l as a precipitation, and the second is water vapor mass m_g' experiencing diabatic compression. The compression of volume is $V_{t4} \to V_{t1}$, where $V_{t4} = V_{l4} + V_{g4}'$ and $V_{t1} = V_{l1} + V_{g1}'$. Due to the incompressibility of liquid water, $V_{l4} = V_{l1}$, and the change of volume due to diabatic compression will be:

$$\Delta V_{41} = V_{t4} - V_{t1}$$

$$\Delta V_{41} = (V_{l1} - V_{l4}) + (V_{g1}' - V_{g4}') \tag{13}$$

Simply it is rewritten as:

$$\Delta V_{41} = (V_{g1}' - V_{g4}') \tag{14}$$

Change of specific volume due to diabatic compression will be:

$$\Delta v_{41} = x(v_{g1}' - v_{g4}') \tag{15}$$

Following Konings et al. (2015) the specific work that is done to the system and is related to the change in internal energy,

$$\Delta w_{41} = x\left(L_H \ln\left(\frac{T_H}{T_C}\right) - R(T_H - T_C)\right) \tag{16}$$

Because the change of internal energyis$\Delta u_{41} = -c_v(T_C - T_H)$the specific heat will be,

$$\Delta q_{41} = -c_v(T_C - T_H) + x\left(L_H \ln\left(\frac{T_H}{T_C}\right) - R(T_H - T_C)\right) - xR(T_H \ln \mathcal{H}_H - T_C \ln \mathcal{H}_C) \tag{17}$$

The third term in the right hand side of equation (17) is due to the change in potential thermodynamics.

3. Irreversible System of Thermodynamic Processes

Following standard book of thermodynamics,[15] a sum of change of the system and reservoir entropy expresses the change of universe entropy,

$$dS_{\text{system}} + dS_{\text{reservoir}} > 0 \tag{18}$$

There is a flow of heat $đQ$ from the reservoir:

IOP Conf. Series: Earth and Environmental Science 279 (2019) 012048 doi:10.1088/1755-1315/279/1/012048

$$dS_{\text{reservoir}} = -\frac{dQ}{T} = -\frac{[illegible]}{T} \tag{19}$$

Substitute (19) to (18), we have:

$$dS_{\text{system}} - \frac{dQ}{T} > 0 \quad \text{or} \quad dQ - TdS_{\text{system}} < 0 \tag{20}$$

Using the first law of thermodynamics$dQ = dU + dW = dU + pdV$, equation (18) becomes:

$$dU + pdV - TdS < 0 \text{or} d(U + pV - TS) < 0 \text{ or } dG < 0 \tag{21}$$

$$TdS > d(U + pV) \tag{22}$$
$$dQ > d(U + pV) \tag{23}$$

Equation (21) explains that Gibbs function during the irreversible process at temperature and pressure constant will remain to decrease until a final equilibrium condition is achieved. Equation (22) and (23) informs that latent heat absorbed and deliberated by the system is larger than its enthalpy $H = U + pV$.If the thermodynamic potential of liquid water μ_l and thermodynamic potential of gas is μ_g, the general expression for Gibbs function will be,

$$dG = -SdT + Vdp + \mu_l dm_l + \mu_g dm_g \tag{24}$$

In the isotherm-isobaric process, Gibbs function will be,

$$dG = \mu_l dm_l + \mu_g dm_g \tag{25}$$

Specific Gibbs function,

$$dg = \frac{dG}{m_t} = \mu_l d\left(\frac{m_l}{m_t}\right) + \mu_g d\left(\frac{m_g}{m_t}\right) = -\mu_l dx + \mu_g dx = (\mu_g - \mu_l)dx \tag{26}$$

In the process of evaporation and condensation (isotherm-isobaric), specific Gibbs function which is previously written as $G = H - TS = U + pV - TS$, its change during the process is,

$$dg = pdV - Tds \tag{27}$$

Substitute (25) into (26) will be found that,

$$Tds = pdv - (\mu_g - \mu_l)dx \tag{28}$$

During evaporation at the sea surface, the mass of liquid water phase m_l is transformed into a mass of vapor water phasem_g, hence $m_t = m'_g + m_g$. Following Pauluis(2010), if $\mathcal{H}_H$ and $\mathcal{H}_C$ is humidity in the surface and upper air, Gibbs function at the sea surface where evaporation takes place:

$$G_H = \mu_l m_l = m_l RT_H \ln \mathcal{H}_H ;\mu_l = \mu_H = RT_H \ln \mathcal{H}_H \tag{29}$$

Specific Gibbs function at the sea surface:

$$g_H = (1 - x)\mu_l = (1 - x)RT_H \ln \mathcal{H}_H \tag{30}$$

Gibbs function at the height where condensation takes place, but the vapor mass (m_t) is partly (m_l) condensed attemperature $T_C$and constant pressure $p(T_C)$ with the remaining mass of water vapor of the amountm_g:

$$G_C = \mu_g m_l = m_l RT_C \ln \mathcal{H}_C;\ \mu_g = \mu_C = RT_C \ln \mathcal{H}_C \tag{31}$$

IOP Conf. Series: Earth and Environmental Science **279** (2019) 012048 doi:10.1088/1755-1315/279/1/012048

Specific Gibbs function at the condensation level:

$$g_C = (1-x)\mu_g = (1-x)RT_C \ln \mathcal{H}_C \tag{32}$$

All processes of the quasi-Carnot cycle are listed following the first law of thermodynamics as depicted in **Table 1**.

Table 1. Thermodynamic Process of Evaporation and Condensation

Trans	Δu	Δw	$\Delta \mu$	Δq
1 → 2	0	$(1-x)RT_H$	0	$(1-x)L_H$
2 → 3	$C_v^g(T_C-T_H)$	$-C_v^g(T_C-T_H)$	$-R(T_C\ln\mathcal{H}_C - T_H\ln\mathcal{H}_H)$	$R(T_H\ln\mathcal{H}_H - T_C\ln\mathcal{H}_C)$
3 → 4	0	$-(1-x)RT_C$	0	$\Delta q_{34} = -(1-x)L_C$
4 → 1	$C_v^g(T_H-T_C)$	$\Delta w_{41} = x\left(L_H\ln\left(\frac{T_H}{T_C}\right) - R(T_H - T_C)\right)$	$-xR(T_H\ln\mathcal{H}_H - T_C\ln\mathcal{H}_C)$	$C_v^g(T_H-T_C) + x\left(L_H\ln\left(\frac{T_H}{T_C}\right) - R(T_H-T_C)\right) - xR(T_H\ln\mathcal{H}_H - T_C\ln\mathcal{H}_C)$
TOTAL	$\oint du = 0$ Where: $C_v^g \approx C_v^l$	$\oint dw = C_p^g(T_H-T_C) + x\left(L_H\ln\left(\frac{T_H}{T_C}\right) - 2R(T_H - T_C)\right)$ Where: $C_p^g = R + C_v^g$	$\oint d\mu = (1-x)R(T_H\ln\mathcal{H}_H - T_C\ln\mathcal{H}_C) = (1-x)R\Delta\mathcal{H}$	$\oint dq = (1-x)(L_H-L_C) + C_v^g(T_H-T_C) + x\left(L_H\ln\left(\frac{T_H}{T_C}\right) - R(T_H - T_C)\right)(1-x)R(T_H\ln\mathcal{H}_H - T_C\ln\mathcal{H}_C)$

If we assumed that $\Delta T = T_H - T_C$, and $\Delta q_{in}$is total heat inputted by the system for $q > 0$in each process. By using formulas in **Table 1**, the efficiency of the Carnot cycle will be as follows:

$$\eta = \frac{\oint dw}{\Delta q_{in}} \tag{33}$$

Because there is a thermodynamic potential effect on the process that takes place, so the calculation of efficiency can also be done using a comparison between the total heat in each process and the heat entering the system. In other hand, efficiency can be written as follows,

$$\eta = \frac{\oint q_{net}}{\Delta q_{in}} \tag{34}$$

4. Role of Diabatic Process

In equation (24) we defined that during the irreversible process, latent heat absorbed and deliberated by the system dQ is larger than its enthalpy $H = U + pV$. If assumed that Q_{in} in the heat flux input from evaporation and diabatic compression process, then Q_{out} is heat flux output from the condensation process, it's shown from Table 4 that estimations of Q_{out} is always larger than Q_{in} because the latent heat of evaporation is always lower than the latent heat of condensation.

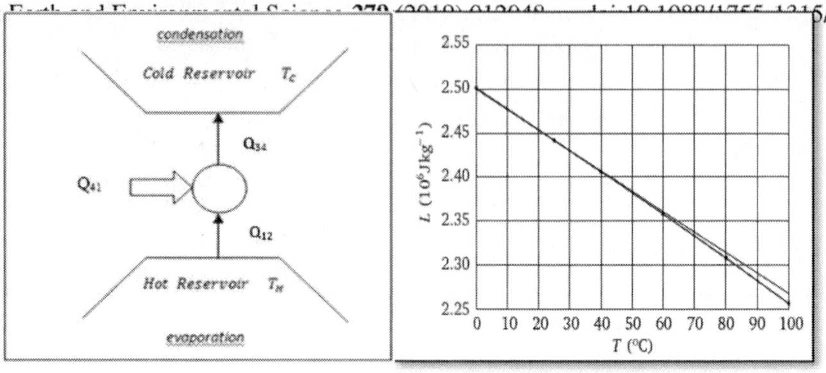

(a) **(b)**

Figure 2(a) shows a representation of the irreversible process on evaporation and condensation. The process occurs between the hot reservoir at sea level and cold reservoir in the atmospheric boundary layer. (b) The variations of Latent heat through the atmosphere temperature. [16]

In order to bring the condition to the initial process, it is need to restore work from diabatic process. To recognize this condition, the condensation process in the hydrological cycle is examined as part of the diabatic process rather than the adiabatic process.

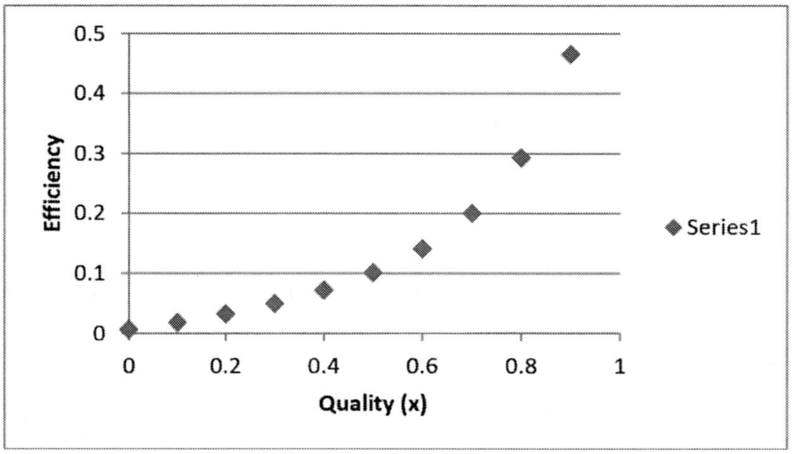

Figure 3. Efficiency of the system through the quality (x) assumed that temperature of sea surface and lifting condensation level were 300 K and 273 K. The C_v^g, C_v^g, R are assuming constant.

The efficiency through the quality system *(x)* is ranged between 1 <x<0. This situation fulfills the process of rain formation which can be viewed in the quasi-Carnot cycle which is not all the mass of air that has been condensed falls in the form as rain m_l. There is always residual vapor remaining at m_g. When the $x = 0$, then all mass of m_g that has been condensed to m_l falls in the form of rain. As for $x = 1$, there is no rain because the entire air parcel that has been condensed has re-evaporated as m_g. The higher the efficiency, the smaller the amount of water vapor converted to precipitation. On the contrary, the lower the efficiency, the higher water vapor converted to precipitation. So, the efficiency and quality of the system that shows the potential for rain to occur are inversely proportional.

Table 2. The energy estimations due to reversible processes, assumed that temperature of sea surface and lifting condensation level were$\mathcal{H}_H = 0.9$ and $\mathcal{H}_C = 1$. The c_v^g, c_v^g, R are assumed to be constant.

X	Qin	Qnet	Efficiency
0	16846.74	-74034.8	-0.02976

0.1	42065.21	-42935.6	-0.01894
0.2	67283.69	-11836.4	-0.00579
0.3	92502.16	19262.78	0.010557
0.4	117720.6	50361.98	0.031408
0.5	142939.1	81461.18	0.058929
0.6	168157.6	112560.4	0.09693
0.7	193376.1	143659.6	0.152807
0.8	218594.5	174758.8	0.24305
0.9	243813	205858	0.413442
1	269031.5	236957.2	0.85606

In the **Table 2**, the energy flux enters the system and the amount of energy produced continues to increase as the value of quality increases. In one complete cycle, not all air masses are transformed into liquid in the form of precipitation. But there is a vapor phase of the overall air mass undergoes evaporation before reaching the surface.

5. Conclusion

This paper analyzed the evaporation and condensation in the hydrologic cycle by using irreversible process. Due to the irreversible process, latent heat absorbed and deliberated by the system $đQ$ is larger than its enthalpy $H = U + pV$. In order to bring the condition to the initial process, it is need to restore work from diabatic process. To recognize this condition, the condensation process in the hydrological cycle is examined as part of the diabatic process rather than the adiabatic process. The efficiency of the system is proportional to the quality. The higher the efficiency, the smaller the amount of water vapor converted to precipitation

Acknowledgement
We are indebted to the Center of Environmental Research and Development insupporting the completion of this study.

References

[1]Dijkstra H. Ocean atmosphere interaction and the tropical climatology , part II: Why the Pacific Cold Tongue Is in The East. J. Climate. 1995: p 1343-1359
[2]Yamanaka M. Physical Climatology of Indonesian Maritime Continent; An Outline to comprehend Observational Studies. 2016: p 178-179.
[3]TjasyonoB.Aktivitas Konvektif di Atas Benua Maritim Indonesia dan Keterkaitannya dengan Variabilitas Iklim Regional dan Lokal. Bandung: Institut Teknologi Bandung. 2004
[4]Ahrens DC. Meteorology Today:An Introduction to Weather, Climate, and the Environment.Canada:Cengange Learning. 2009
[5]Pabalik I, Ihsan N, and ArsyadM. Analisis Fenomena Perubahan Iklim dan Karakteristik Curah Hujan Ekstrim di Kota Makassar. Jurnal Sains dan Pendidikan Fisika. 2015
[6]KoningsAG, Feng X, Molini A, and Manzoni S. Thermodynamics of an Idealized Hydrologic. Water Resources Research. 2012:Vol 48
[7]AssegafAH, Samad W, Sakka. Some Characteristics Of Atmospheric Boundary Layer Over Makassar. Spermonde. 2018:p.1-6
[8]LorenzR, Lunine JI, Withers PG, McKay CP, and Titan. Mars and Earth: Entropy Production by Latitudinal Heat Transport. Geophys. Res. Lett. 2001: p.415-418.
[9]KleidonA, FraedrichK, Kunz T, and LunkeitF. The Atmospheric Circulation and States of Maximum Entropy Production. Geophys Res Lett. 2003: p.30.
[10]KleidonA, and Renner M. Thermodynamics Limits of Hydrologic Cycling Within the Earth System. Hydrol. Earth Syst. Sci. 2013: p.2873-2892
[11]SundqvistH, Berge E, and KristjanssonJE. Condensation and Cloud Parameterization Studies with a Mesoscale Numerical Weather Prediction Model. Monthly Weather Review.1988: Vol117

[12]Chou M, & Suarez MJ. Parameterization for Cloud Overlapping and Shortwave Single-Scattering Properties for Use in General Circulation and Cloud Ensemble Models. Journal of Climate. 1997:Vol 11. p. 551-513.
[13]Bannu, Suriamiharja DA, Takeuchi N, Kuze H. Impacts of the ENSO and IOD phenomena: Long-term analysis in Indonesia region. 2003
[14]Caesar M. Fundamental on Thermodynamics Processes Behind Clouds' and Rainfalls' Formation. Atmospheric and Climate Sciences. 2015: 257-265.
[15]ZemanskyM,and Dittman RH. Heat and Thermodynamics. 7th edition. The McGraw-Hill Companies. 1957
[16]Pauluis, O. Water Vapor and Mechanical Work: A Comparison of Carnot and Steam CyclesVAmerican Meteorology Society. 2010: Vol 68
[17]Ambaum MHP. Thermal Physics of the Atmosphere. Wiley-Blackwell. 2010: p

IOP Conf. Series: Earth and Environmental Science **279** (2019) 012049 doi:10.1088/1755-1315/279/1/012049

Hydrologic cycle of moist air in the lower atmosphere

D Rahmawati, D A Suriamihardja, A H Assegaf and B A Samad
Department of Physics, Faculty of Mathematics and Science, Hasanuddin University, Makassar 90245, Indonesia.

Email: derahmawati17@gmail.com

Abstract. A hydrological cycle in the atmosphere plays an important role in the Earth system. [1] The hydrological cycle in the atmosphere is a source of fresh water for life on earth. Starting from the density of heavy water, it is converted into a mild vapor density which allows an increase in water vapor from the sea surface to the dew point where the process of changing water vapor into liquid phase can take place. An atmospheric machine is represented by a modified Carnot cycle below water dome containing mixed water vapor and liquid in an atmospheric layer. This study aims to learn changes in thermodynamic parameters, such as heat and work during the moist air experiencing thermodynamic processes and the correlation of temperature and humidity to rainfall. Basic information recorded by radiosonde is used as mixed water vapor and liquid parameters representing an initial atmospheric condition in producing fresh water. The results will provide information required to figure out the process of evaporation and condensation in the troposphere and the relationship between temperature and humidity to rainfall.

1. Introduction

A hydrological cycle in the atmosphere plays an important role in the Earth system. [1] The hydrological cycle in the atmosphere is a source of fresh water for life on earth. Starting from the density of heavy water, it is converted into a mild vapor density which allows an increase in water vapor from the sea surface to the dew point where the process of changing water vapor into liquid phase can take place. Water in the atmosphere can be in three phases, liquid, vapor, and ice. A change in a liquid phase into a vapor phase is called evaporation and its reverse is called condensation. While a change in a liquid phase into a solid phase is called freezing and its reverse is called liquefaction. A change in a solid phase into a vapor phase is called sublimation and its reverse is called a deposition. When water vapor which is part of natural air turns into liquid or solid particles, these particles become foreign matter in the atmosphere and cause clouds, fog, rain, snow, moisture or hailstone.

Formation of water in a vapor phase in the air is very important in determining weather conditions. [2] Atmospheric stability has to do with the weather conditions of an area, one of which can be seen by understanding the characteristics of the atmospheric boundary layer. In a study stated, the comparison between day and night tends to show the atmosphere during the day is not stable compared to the night. [3] So it is quite important to conduct a study related to the hydrological cycle during the day. In the atmospheric hydrological cycle, the evaporation process or the occurrence of condensation can be described in the thermodynamic framework. [4] Several other studies also discuss a thermodynamic approach to the hydrological cycle in the atmosphere. [5][6] The thermodynamic framework referred to this study relates to a Carnot cycle in terms of each process to obtain efficiency values and the relation of temperature and humidity to rainfall.

Our study will be focused on the atmosphere above the Indonesian Maritime Continent (IMC). In a study stated, one of the classifications of climate areas in Indonesia can be determined by looking at rainfall patterns. [7] In general, the presentation of rainfall depends on the evaporation process of sea levels but some global factors such as the presence of ENSO or IOD events, resulting in drought despite evaporation. [8]

However, the focus of this study is to understand the hydrological cycle in the atmosphere locally in terms of the thermodynamic framework. Consequently, it is interesting to study the hydrological cycle above the sea surface of the equatorial region discussing thermodynamic parameters on the evaporation and condensation processes. This study aims to learn changes in thermodynamic parameters, such as heat and work during the moist air experiencing thermodynamic processes and the correlation of temperature and humidity to rainfall.

2. Modified Carnot Cycle

Carnot cycle consists of taking a working substance in a system through the four processes that together constitute a reversible and cyclic transformation. [9] In a case of an ideal Carnot cycle, process 1 to 2 is isotherm, process 2 to 3 is an adiabatic expansion, process 3 to 4 is isotherm and process 4 to 1 is adiabatic compression. In this study, the Carnot cycle is modified to slightly different processes underwater dome, namely process 1 to 2 is isotherm and isobaric (evaporation), process 2 to 3 is adiabatic expansion, process 3 to 4 is isotherm and isobaric (condensation) and process 4 to 1 is adiabatic compression which is depicted in **Figure 1**. The equation of state of water vapor as an ideal gas can be written using subscripts v as follows

$$PV = mR_vT;\ R_v = 461.5\ \mathrm{Jkg^{-1}K^{-1}} \tag{01}$$

$$P = \frac{m}{V}R_vT = \rho R_vTP\alpha = R_vT;\ \alpha = \frac{1}{\rho} \tag{02}$$

Where P, V, m, T, R_v, ρ and α stand for respective pressure, volume, mass, temperature, gas constant for water vapor, density, and specific density. [10]

The relative humidity is the ratio of the actual amount of water vapor to the saturation amount at a certain temperature, and expressed by the following equation: [11] At the ocean surface where the process of evaporation takes place, relative humidity will be: [12]

$$\mathcal{H}_H = \frac{e_H}{e_s(T_H)} \tag{03}$$

At the top in the process of condensation is:

$$\mathcal{H}_C = \frac{e_C}{e_s(T_C)} \tag{04}$$

Where $\mathcal{H}_H, \mathcal{H}_C, e_H, e_{dry}, e_s(T_H)$ and $e_s(T_C)$ stand for respective humidity at the ocean surface and at the top, hot and cold vapor pressure and saturated vapor pressure at the given hot and cold temperature.

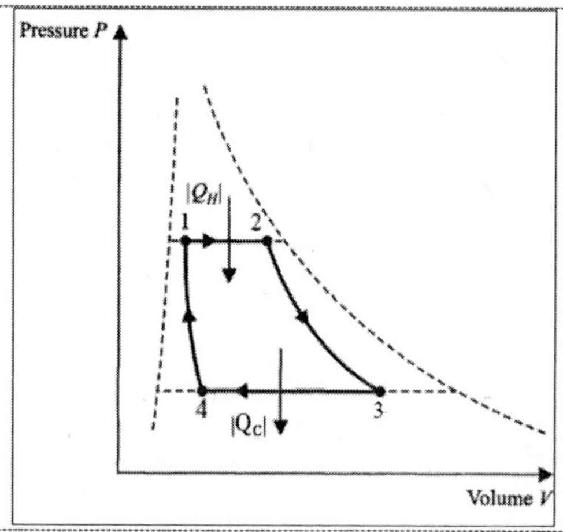

Figure 1. *A modified Carnot cycle (Zemansky and Dittman, 1997).*

IOP Conf. Series: Earth and Environmental Science **279** (2019) 012049 doi:10.1088/1755-1315/279/1/012049

3. Heterogeneous system

Heterogeneous systems are systems of two phases, the liquid phase and the vapor phase. Heterogeneous systems have temperatures and pressures that vary with height. Gibbs G free energy, expressed in equations involving U, T, S, V, p which stands for internal energy, temperature, entropy, volume and pressure. [14] Gibbs free energy function:

$$G = U + pV - TS \quad (05)$$

$$dG = dU + pdV + Vdp - SdT - TdS \quad (06)$$

$$dG = -SdT + Vdp + (dU + pdV - TdS); \quad (07)$$

Because of $TdS = dU + pdV$, hence:

$$dG = -SdT + Vdp \quad (08)$$

The system is containing liquid water mass (m_l) and water vapor mass (m_g). Gibbs free energy function has additional terms of two phases of water.

$$dG = -SdT + Vdp + \mu_l dm_l + \mu_g dm_g$$

$$dU + pdV - TdS = \mu_l dm_l + \mu_g dm_g$$

$$dU + pdV - \mu_l dm_l - \mu_g dm_g = TdS$$

$$dU + đW - d(m_l\mu_l + m_g\mu_g) = TdS \quad (09)$$

Expression in specific terms:

$$du + đw - d\mu = đq \quad (10)$$

The transition from state-1 to state-2 occurs at sea surface with temperature T_H as the process of isotherm-isobaric evaporation. A parcel of liquid water mass m_l is transformed into water vapor mass by keeping that $m_l = m_g$. This water vapor will joint with m_g' as a residual water vapor mass come down from the upper layer. The quality of mixing is expressed as $x = m_g'/m_t$. Where m_t is a total of liquid water and water vapor masses which is expressed by $m_t = m_l + m_g'$.

Accordingly, $m_l/m_t = m_g/m_t = 1 - x$. Thermodynamic potential (specific Gibbs free energy) during the evaporation is zero implied that $\mu_g = \mu_l$. This is supported by the fact that during a change of phase the Gibbs free energy remains the same value, nevertheless the first differential of Gibbs free energy to the temperature having different values.

For the change of volume and specific volume due to the isotherm isobaric process (evaporation):

$$\Delta V_{12} = (V_{g2} - V_{l1}) + (V_{g2}' - V_{g1}') = \Delta v_{12} = (1 - x)(v_{g2} - v_{l1}) + x(v_{g2}' - v_{g1}')$$

Because of: $V_{l1} \ll V_{g2}$ and $v_{l1} \ll v_{g2}$, the change of volume:

$$\Delta V_{12} = V_{g2} + (V_{g2}' - V_{g1}') = \Delta v_{12} = (1 - x)v_{g2} + x(v_{g2}' - v_{g1}') \quad (11)$$

This transition is isotherm-isobaric; consequently, the second term in the right-hand side will be isochoric which is having a zero value. Simply the change of specific volume is $\Delta v_{12} = (1 - x)v_{g2}$. During an isotherm process, there will be no change in internal energy, $\Delta U_{12} = 0$. Heat and specific heat required to evaporate are:

$$\Delta q_{12} = (1 - x)R(T_H) \quad (12)$$

Since the change of phase form liquid to vapor involves expansive volume, the system will exert the work to the environment, an important note $\frac{RT_H}{P(T_H)} \gg v_l$

$$\Delta W_{12} = m_l p(T_H)\left(\frac{RT_H}{p(T_H)} - v_l\right) = (1-x)R(T_H) \tag{13}$$

The transition from state-2 to state-3 is in the process of adiabatic expansion. A parcel of water vapor mass m_g is rising and joining with a residual vapor m_g' come downing from the upper layer at the sea surface just after evaporation. Total water vapor mass is $m_t = m_g + m_g'$ to the level where a process of condensation will be possible. During this transition, there will be a change in thermodynamic potential at the sea surface to its value at the possible level of condensation,

$$\Delta m_t \mu = m_t \mu_C - m_t \mu_H$$

$$\Delta \mu = R(T_C \ln \mathcal{H}_C - T_H \ln \mathcal{H}_H) \tag{14}$$

Because there will be a change in thermodynamic potential at the sea surface, this process called pseudo-adiabatic. So, $\Delta q_{23} \neq 0$. Heat specific for the process:

$$\Delta q_{23} = R(T_H \ln \mathcal{H}_H - T_C \ln \mathcal{H}_C) \tag{15}$$

The work is exerted by the system to the environment:

$$\Delta W_{23} = m_t \frac{c_v^g}{R}(RT_H - RT_C) = m_t C_v^g (T_H - T_C) > 0 \tag{16}$$

Due to the work exerted to the environment, the internal energy decreases:

$$\Delta U_{23} = -\Delta W_{23} = m_t C_v^g (T_C - T_H) \rightarrow \frac{\Delta W_{23}}{m_t} = \Delta w_{23} = -C_v^g (T_C - T_H) \tag{17}$$

For the change of Volume and specific volume due to the adiabatic expansion process:

$$\Delta V_{23} = V_{t3} - V_{t2} = \Delta v_{23} = (1-x)(v_{g3} - v_{g2}) + x(v_{g3}' - v_{g4}') \tag{18}$$

The transition from state-3 to state-4 occurs at the possible level of condensation with temperature T_C as the process of isotherm-isobaric condensation. A parcel of water vapor mass m_t is partly transformed into liquid water mass m_l, and parts remained as a residual water vapor mass m_g', by keeping that the total water vapor remains as $m_t = m_l + m_g'$.

Thermodynamic potential (specific Gibbs free energy) during the condensation is partly zero implied that $m_g \mu_g = m_g \mu_l$. The other part is also zero due to the isotherm-isobaric will be also isochoric, and its thermodynamic potential is also zero.

During an isotherm process, there will be no change in internal energy, $\Delta U_{34} = 0$. Heat is released out of the system:

$$\Delta q_{34} = -(1-x)R(T_C) \tag{19}$$

The work performed on the system is given by:

$$\frac{\Delta W_{34}}{m_t} = \Delta w_{34} = -(1-x)RT_c \tag{20}$$

Change of volume and specific volume due to condensation:

$$\Delta V_{34} = (V_{l4} - V_{g3}) + (V_{g4}' - V_{g3}') = \Delta v_{34} = (1-x)(v_{l4} - v_{g3}) + x(v_{g4}' - v_{g3}')$$

Because of: $V_{l4} \ll V_{g3}$ and $v_{l4} \ll v_{g3}$, the changes of volume are :

$$\Delta V_{34} = -V_{g3} + (V_{g4}' - V_{g3}') = \Delta v_{34} = -(1-x)v_{g3} + x(v_{g4}' - v_{g3}') \tag{21}$$

IOP Conf. Series: Earth and Environmental Science **279** (2019) 012049 doi:10.1088/1755-1315/279/1/012049

By the same reasoning the change in specific volume will be $\Delta v_{34} = -(1-x)v_{g3}$

The transition from state-4 to state-1 is in the process of adiabatic compression. Through this transition, some of the remaining vapor and precipitation in liquid form is brought to the surface and reheated to T_H. The specific work performed to the system and related to the change of internal energy:

$$\Delta w_{41} = -C_v^g(T_H - T_C) \tag{22}$$

There are two parcels water masses come down to the sea surface. The first is liquid water mass m_l as precipitation, and the second is water vapor mass m_g' experiencing adiabatic compression. During this transition there will be a change in thermodynamic potential from the level of condensation to the sea surface:

$$\Delta m_g \mu = m_g \mu_C - m_g \mu_H$$
$$x\Delta\mu = xR(T_C \ln \mathcal{H}_C - T_H \ln \mathcal{H}_H) \tag{23}$$

For the change of volume and specific volume due to adiabatic compression will be:

$$\Delta V_{41} = \left(V_{g1}' - V_{g4}'\right) = \Delta v_{41} = x\left(v_{g1}' - v_{g4}'\right) \tag{24}$$

Table 1. Recapitulation of all process in reversible Carnot cycle

Processes	Δu	Δw	$\Delta\mu$	Δq
$1 \to 2$	0	$(1-x)RT_H$	0	$(1-x)RT_H$
$2 \to 3$	$-C_v^g(T_H - T_C)$	$C_v^g(T_H - T_C)$	$-R(T_C \ln \mathcal{H}_C - T_H \ln \mathcal{H}_H)$	$R(T_H \ln \mathcal{H}_H - T_C \ln \mathcal{H}_C)$
$3 \to 4$	0	$-(1-x)RT_C$	0	$-(1-x)RT_C$
$4 \to 1$	$xC_v^g(T_H - T_C) + (1-x)C_v^l(T_H - T_C) = C_v^g(T_H - T_C$	$-C_v^g(T_H - T_C)$	$-xR(T_H \ln \mathcal{H}_H - T_C \ln \mathcal{H}_C)$	$-xR(T_H \ln \mathcal{H}_H - T_C \ln \mathcal{H}_C)$
Total	$\oint du = -(1-x)C_v^g(T_H - T_C) + (1-x)C_v^l(T_H - T_C) = (1-x)(C_v^l - C_v^g)\cdot(T_H - T_C) \approx 0$ Where: $C_v^l \approx C_v^g$ is used.	$\oint dw = (1-x)R(T_H - T_C) + (1-x)C_v^g(T_H - T_C) = (1-x)C_p^g(T_H - T_C)$ Where: $C_v^g + R = C_p^g$	$\oint d\mu = (1-x)R(T_H \ln \mathcal{H}_H - T_C \ln \mathcal{H}_C) = (1-x)R\Delta\mathcal{H}$	$\oint dq = (1-x)R(T_H - T_C) + (1-x)R\Delta\mathcal{H} \approx (1-x)R(T_H - T_C) + (1-x)R\Delta\mathcal{H} \approx (1-x)\{R(T_H - T_C) + R\Delta\mathcal{H}\}$

Based on **Table. 1**, we can find out the total heat dq of each process, namely:

$$\oint dq = \oint dw - \oint d\mu + \oint du \approx (1-x)\{R(T_H - T_C) + R\Delta\mathcal{H}\} \tag{25}$$

Where to find the value of $\Delta\mathcal{H}$, that is $\Delta\mathcal{H} = T_H \ln \mathcal{H}_H - T_C \ln \mathcal{H}_C$

So, we can calculate efficiency as follows

$$\eta = \frac{\oint dq}{\Delta q_{in}} = \frac{(1-x)\{R(T_H-T_C)+R\Delta\mathcal{H}\}}{(1-x)RT_H+R\Delta\mathcal{H}} \approx \frac{(1-x)[R(T_H-T_C)+R\Delta\mathcal{H}]}{(1-x)RT_H+R\Delta\mathcal{H}} \approx \frac{(1-x)[(T_H-T_C)+\Delta\mathcal{H}]}{(1-x)T_H+\Delta\mathcal{H}} \tag{26}$$

For $x = 1 \to \eta = 0$

$$\text{For } x = 0 \to \eta = \frac{[R(T_H-T_C)+R\Delta\mathcal{H}]}{RT_H+R\Delta\mathcal{H}} = \frac{(T_H-T_C)+(T_H \ln \mathcal{H}_H - T_C \ln \mathcal{H}_C)}{T_H + T_H \ln \mathcal{H}_H - T_C \ln \mathcal{H}_C} = \frac{(T_H + T_H \ln \mathcal{H}_H)-(T_C + T_C \ln \mathcal{H}_C)}{T_H+(T_H \ln \mathcal{H}_H - T_C \ln \mathcal{H}_C)} \tag{27}$$

Efficiency Carnot:

$$\eta_C = 1 - \frac{T_C}{T_H} \tag{28}$$

For the T_H value is 300, T_C is 273, R is 461.5, $\mathcal{H}_H$ is 0.82, $\mathcal{H}_C$ is 0.78 and x is from 0 to 1 with interval 0.1.

4. Application

The data to be applied in this study is obtained from Sultan Hasanuddin Airport of Makassar. Software used in this study for processing data in Microsoft Excel 2007. The data collected is in the form of upper air data for 1 year in 2010, such as temperature, relative humidity and rainfall. The focus of this study does not involve phase changes, it does not involve latent heat but discusses about the cycle of humidity, illustrated in **Figure 2**.

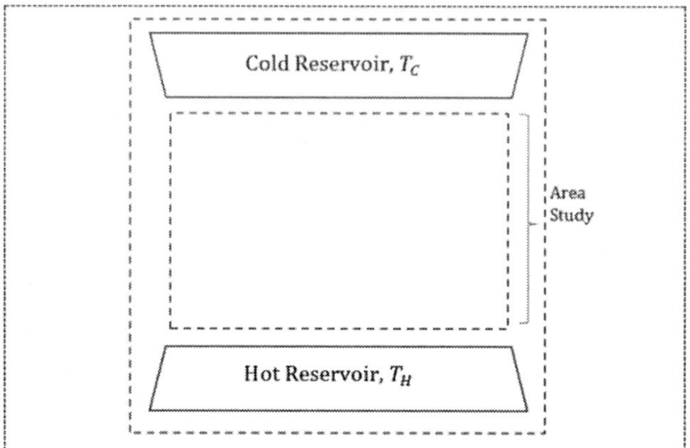

Figure 2. Focus of the area study

5. Result

Based on the calculated data, we have the curves q value versus x as depicted in **Figure 3**, and efficiency with x as depicted in **Figure 4**,

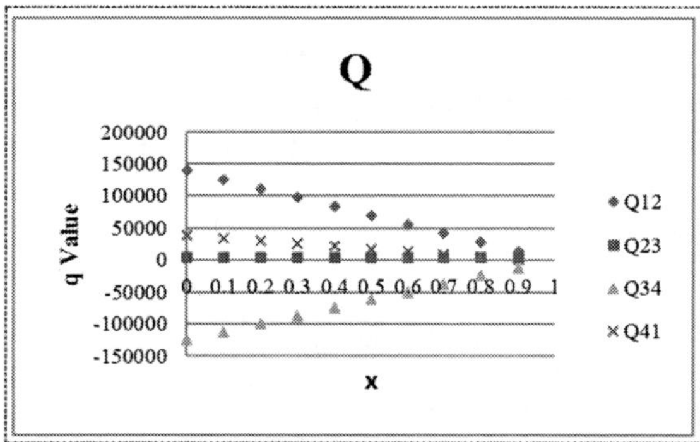

Figure 3. Graph q in each process against x

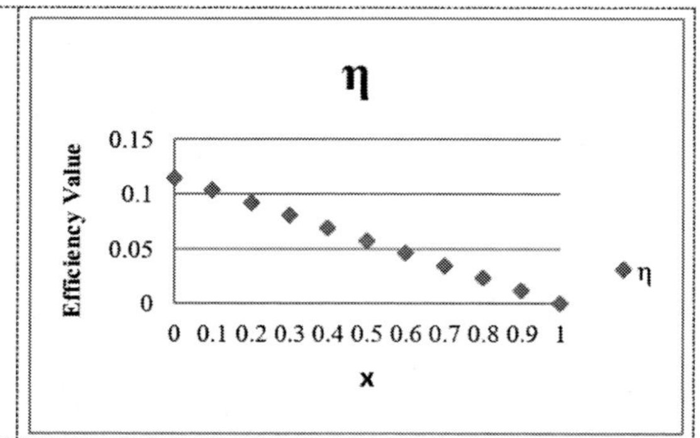

Figure 4. The graph of the efficiency value with x

The result provides information concerning the process of evaporation and condensation in the troposphere. From each process it can be explained that when evaporation, between specific heat and mixing quantity (x) is inversely proportional to the greater the mixing quantity, the smaller the specific heat. In this process, heat is needed which means heat enters the system. Conversely, when the opposite condensation of the evaporation process with the high value of x, the specific heat is also high. In this process, the heat is released which means the heat is released out of the system.

Based on existing data, to see the relationship rainfall to temperature and humidity can be illustrated in **Figure 5**, by displaying a 1-year graph of data collected every 3 months.

IOP Conf. Series: Earth and Environmental Science **279** (2019) 012049 doi:10.1088/1755-1315/279/1/012049

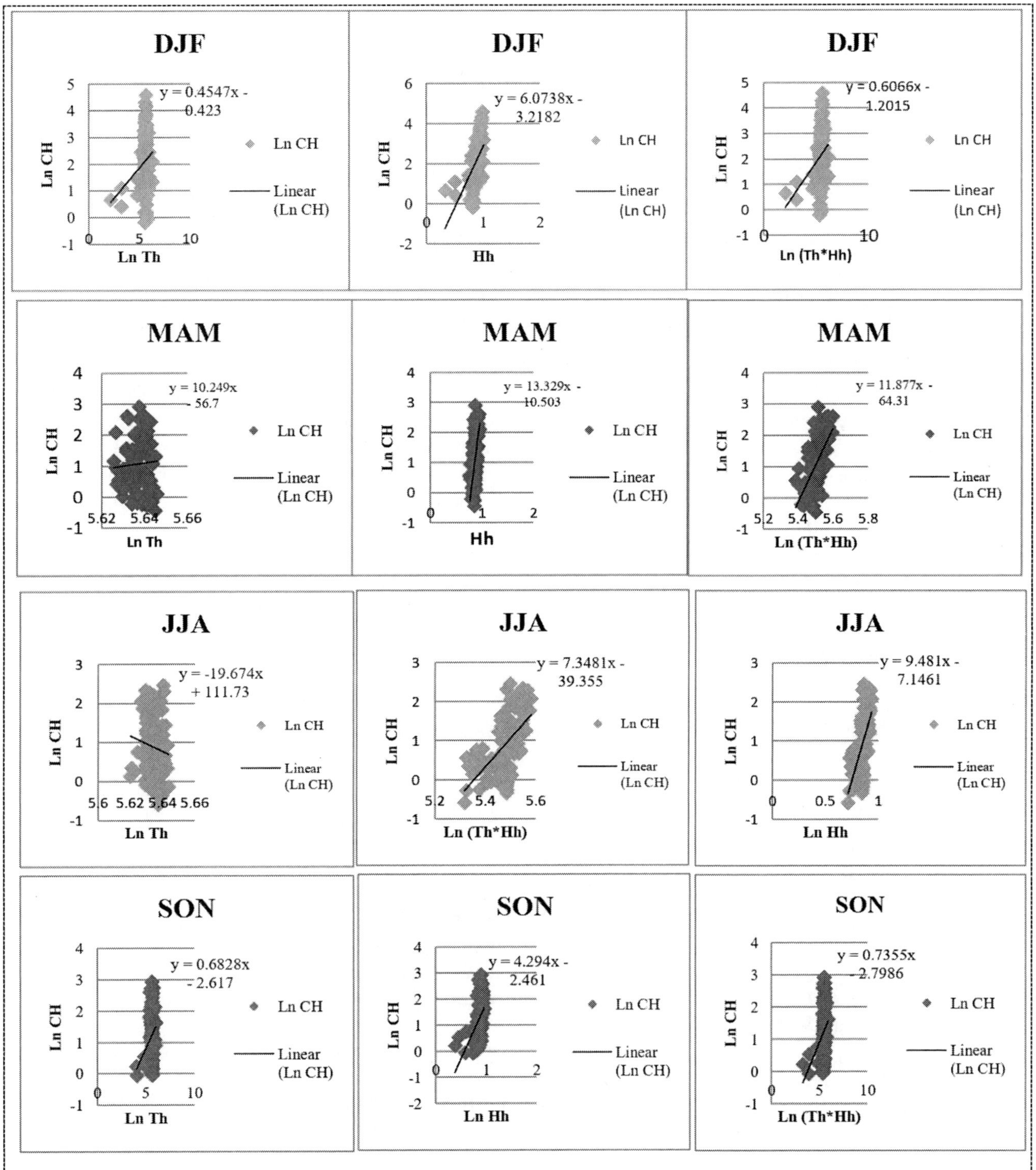

Figure 5. Graph of the relationship of rainfall to temperature and humidity in 1 year for 3 month

The results from the graph above explain the correlation between temperature and humidity to rainfall. In theory, December, January and February (DJF) the real motion of the sun is in the southern hemisphere so that the wind blowing from North to South is called the West Monsoon, during the third month, the rainy season generally occurs. Six months later, in June, July and August (JJA) precisely, the movement of air masses from south to north is called the East Monsoon, the dry season occurs.

In this study, one-month representative data was presented from DJF, MAM, JJA and SON. In order to be able to compare the differences between the two seasons for one year, the rainy season is represented in January and the dry season is represented in August. January is the month that has high rainfall and almost every day there is rain. As for August, is the month with the lowest rainfall and relatively little amount of rain. It can be seen in the graph above, the higher the humidity, the higher the amount of rainfall applies to each month. Likewise with the temperature affects the quantity of rainfall, it's just different for JJA because at 3 months the rainfall is very minimal.

6. Conclusion

In the troposphere, it is worthwhile to construct a thermodynamic point of view to the hydrologic cycle comprising of a mixture water vapor mass and liquid water mass under the curve of water dome. The simplest cycle is modified Carnot cycle that is enough to figure out the process of evaporation and condensation. From this cycle comes out the efficiency that depends on the difference of temperature and relative humidity between the levels of evaporation and condensation. This result is a preliminary study on the application of thermodynamics in the hydrological cycle. The correlation between temperature and humidity to rainfall is the higher the temperature and humidity, the greater the quantity of rainfall. Needless to say that larger and deeper development is of importance to be conducted further.

7. Acknowledgements

We would like to express our appreciation to all parties who have helped the completion of this study, especially to the Center of Environmental Research and Development, Institute for Research and Services, Hasanuddin University for supporting facilities during the study.

Reference

[1] Chahine, Moustafa T. "The Hydrological Cycle and its Influence on Climate." *Nature* 359 (Oktober 1992): 373-379.

[2] HK, Bayong Tjasyono. *Meteorologi Indonesia Volume I.* Bandung: Badan Meteorologi Klimatologi dan Geofisika, 2006.

[3] Assegaf, Alimudin Hamzah, Wasir Samad, and Sakka. "Some Characteristics of Atmospheric Boundary Layer Over Makassar." *Spermonde*, 2018: 1-6.

[4] Konings, Alexandra G., Xue Feng, Annalisa Molini, Stefano Manzoni, Giulia Vico, and Amilcare Porporato. "Thermodynamics of an idealized hydrologic cycle." *WATER RESOURCES RESEARCH,* 48 (Mei 2012): 1-13.

[5] Vehkamaki, Hanna, and Ilona Riipinenb. "Thermodynamics and kinetics of atmospheric aerosol particle formation and growth." *Chem Soc Rev*, 2012: 5160–5173.

[6] Kleidon, A., and M. Renner. "Thermodynamic limits of hydrologic cycling within the Earth system: concepts, estimates and implications." *Hydrology and Earth System Sciences* (Copernicus Publications on behalf of the European Geosciences Union), July 2013: 2873-2892.

[7] Liong, The Houw, Bannu Bannu, and P M Siregar. "Peranan Pengelompokan Samar dalam Prediksi Kekeringan di Indonesia Berkaitan dengan ENSO dan IOD." *Jurnal Matematika dan Sains*, 2009: 57-61.

[8] Bannu, Bannu, Dadang A Suriamiharja, N Takeuchi, and H Kuze. "Impacts of the ENSO and IOD phenomena: Long-term analysis in Indonesia region." *36th COSPAR Scientific Assembly.* Beijing: The ADS, 2006. 2888.
[9] Lynch, Peter. *Physical Meteorology.* Belfield: Mathematical Computation Laboratory Dept. of Maths. Physics, UCD, Belfield., 2015.
[10] Bechtold, Peter. *Atmospheric Thermodynamics.* England: ECMWF, 2015.
[11] Stull, Roland. *Meteorology for Scientists and Engineers.* Canada: Warren Faidley, 2015.
[12] Ambaum, Maarten H.P. *Thermal Physics of the Atmosphere.* United Kingdom: WILEY-BLACKWELL, 2010.
[13] Pauluis, Olivier. "Water Vapor and Mechanical Work: A Comparison of Carnot and Steam Cycles." *Journal of The Atmospheric Sciences* 68 (Agustus 2010): 91-102.
[14] Mark W.Zemansky, Richard H.Dittman. *Heat and Thermodynamics.* New York: McGraw-Hill, 1997.

Development of the Total Ozone Column Model Prediction based on Stratospheric Zonal Wind and Temperature over the Maritime Equatorial Region

Putri Wulandari*, H Halide and Paharuddin

Physics Departement, Geophysics Study Program, Hasanuddin University, Jl. Perintis Kemerdekaan Km. 10, Makassar 90245, Indonesia

*wulandarip14h@student.unhas.ac.id

Abstract. As one of the constituent gases which plays an important role in protecting the earth from UV radiation, the characteristics of ozone important to be studied and even modeled. This study mainly concerned to the temporal analysis of the ozone concentration which is total ozone column (TOZ) in the atmosphere by using zonal wind and temperature in the stratospheric layer. Spectral techniques applying to find out the predominant peak oscillation and investigate anomalous at each layer that varies from 100 – 10 hPa layers. The result showed that TOZ anomaly over the maritime equatorial region has a similar pattern with the zonal wind which related to the Quasi-Biennial Oscillation (QBO) phenomena and stratospheric temperature anomaly, especially at 30 hPa layer. Then, we developed a model prediction for TOZ by using multiple linear regression and validate for 7 lead ahead using statistical approach; Pearson correlation, RMSE, and Euclidean distance. Moreover, our developed model then compared with the persistence model. Finally, we found a significant result with statistical correlation, r = 0.92 at lead 1.
Keywords: Equatorial TOZ, Model Prediction, Stratospheric Temperature, Zonal Wind

1. Introduction

In the recent decade, at least there have been two main atmospheric phenomena impacted the general public life, well-known as climate change and ozone depletion. Consequently, massive destruction of the catalytic ozone cycle by chlorine and bromine that resulting from the photodissociation of chlorofluorocarbon (CFC) compounds emitted at the surface [1]. About 90% of ozone concentration located at the altitude of 15 to 30 km in the stratospheric layer [2]. The higher the latitude, the lower down the ozone. Near the equatorial region, the highest concentration lies within an altitude of 25 km [2]. The total ozone column (TOZ) which referred to column density mostly used to specify the amount of ozone in the atmosphere above specific region. This quantity expressed in Dobson unit.

Generally, ozone (O_3) produced by the photochemical processes that occur in the tropical and subtropical stratosphere, where the sunshine almost intense throughout the year. At the first, ozone formation obtained from photolysis of molecular oxygen (O_2) at wavelengths less than 240 nm, which produces two oxygen atoms in their electronic ground states (3P). This atom then recombines with another molecular oxygen to form ozone by interfering of molecule M, mostly nitrogen or oxygen. At the same time of the ozone-forming, conversely, it is destroyed again by the photolysis of O_3 at wavelengths less than 300 nm and by reaction with an oxygen atom [2].

Stratospheric ozone is an essential parameter that controls UV radiation on the surface since ozone will be the only filter remain when the cloudiness and atmospheric conditions are clean. The ozone in the stratospheric layer that so-called ozone layer filters a large portion of UV radiation emitted by the sun (about 100 - 380 nm). Among three types of UV radiation, the emission of solar UV-B radiation extremely danger for plants, animals, and humans. Accordingly, increases in UV-B radiation particularly impacting on photosynthesis, skin cancer, and weaken the immune system [2]. Therefore, the role of ozone being necessary for protecting life on the earth.

Chemical processes in the stratosphere affect the thickness of the ozone layer, but it is important to note that physical and dynamic processes also play a significant role. As one of the atmospheric trace gases, ozone transported over long distances by the stratospheric wind system, which affects global ozone distribution [1]. Furthermore, the characteristic of ozone that absorbs solar UV radiation causes the temperature of stratosphere increases with altitude. Consequently, affect the vertical distribution of the stratospheric temperature [1]. Many chemical reactions in the atmosphere depend on the prevailing temperature. The main ozone depletion cycles in the middle and upper stratosphere (brought about by the catalysts HOx, NOx, ClOy, and BrOy) slow down with decreasing temperature, which results in higher ozone concentrations [3]. Fluctuations of the ozone thickness both in the short and long-term possibly explained by the interaction of the entire process. This knowledge is necessary to comprehend the future estimate of the ozone layer. In this present study, we mainly concerned to develop a model prediction of TOZ by including stratospheric wind parameter that represented by zonal wind and temperature in the stratospheric layer, especially at 30 hPa.

2. Data and Method

2.1 Data

To study ozone variations, we used a monthly total ozone columns (TOZ) dataset of a MERRA-2 satellite from 1988 to 2018, obtained from https://disc.gsfc.nasa.gov/datasets/ [4]. Whereas, the zonal wind derived from QBO Singapore's data, available since January 1979 until present. This dataset observed by regularly radiosonde measurement of daily vertical wind profiles at the levels 100 to 10 hPa. Moreover, stratospheric temperature data from NOAA NCEP-NCAR CDAS-1 reanalyses also used. This dataset accessed through the IRI operational website, online data repository and analysis tools. Both zonal wind and stratospheric temperature data on a month-by-month basis.

2.2 Method

In this present study, we apply spectral analysis techniques, hereafter referred to power spectral density and wavelet. Time series data decomposed into time-frequency space, then determine variability dominant modes and variations modes in time as well. Power spectral density function shows the strength each variation (energy) as the function of frequency, mostly used to maintain the magnitude of frequency variations. This method useful to determine a level in which stratospheric zonal wind has similar characteristics with ozone. From here, then we cross-correlate between TOZ dataset with zonal wind and temperature at the stratospheric layer, respectively. By combining significant lags as the result of cross-correlation, we specify a set of predictors in modeling the ozone.

The ozone model was developed to predict the future monthly TOZ by using zonal wind and stratospheric temperature data, especially at 30 hPa. In this case, total ozone column is a single layer property, where an extracted data typically represent the equatorial region in 0° latitude and 100°E longitude. TOZ anomaly obtained by subtracting monthly dataset with the average of TOZ from January 1988 to March 2018 for each month. Regarding the TOZ model with two parameters as mentioned before, the total ozone column model also formed from time series dataset itself. Then, these two models were compared and validate up to 7 lead ahead. Finally, the best model selected based on a statistical approach, which is deterministic skill score.

To asses the prediction skill of an ozone model, we used three skill measures. Two commonly used in a statistical score, which are the Pearson correlation and the RMSE (root-mean-square error). Another skill measure is *Euclidean* distance, which calculated after performing a 45° rotation for each observation – prediction pair relative to the line of perfect forecast [5]. These measures of skill are defined as follows [6]:

$$r = \frac{\sum_{m=1}^{n}(p_m - \bar{p})(o_m - \bar{o})}{\left[\sum_{m=1}^{n}(p_m - \bar{p})^2\right]^{1/2}\left[\sum_{m=1}^{n}(p_m - \bar{p})^2\right]^{1/2}} \tag{1}$$

where r is a Pearson correlation coefficient, p_m and o_m are the mth prediction and observed value, respectively (m = 1, 2, ..., n). while $\bar{p}$ and $\bar{o}$ are the mean values of the prediction and observation. Therefore root-mean-squared error and Euclidean distance are defined as:

$$\mathrm{RMSE} = \left[\frac{1}{n}\left(\sum_{m=1}^{n}(p_m - o_m)^2\right)\right]^{1/2} \quad (2)$$

$$\text{Euclidean distance} = |Y'| = \left(\sum_{i=1}^{n} Y_i'^2\right)^{1/2} \quad (3)$$

3. Result and Discussion

3.1 The characteristics of Ozone, Zonal Wind and Temperature

Figure 1 is showing the PSD of monthly zonal wind data in the stratospheric layer at 100 to 10 hPa for the period of January 1988 – March 2018. From this figure, clearly shown that pre-dominant peak oscillation is about 28 months with the highest spectral energy around 12 x 10^8 $m^2/s^2/Hz$.

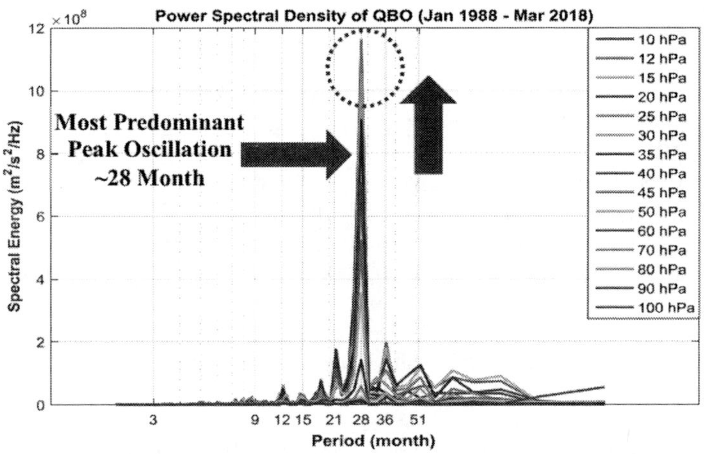

Figure 1. The Power Spectral Density (PSD) of zonal wind for a period of January 1988 - March 2018

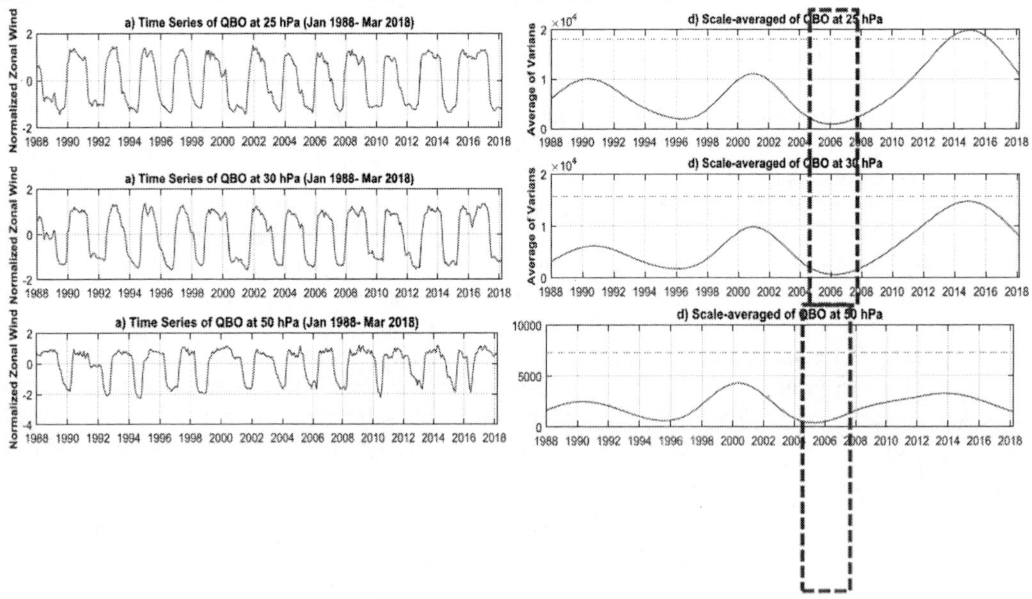

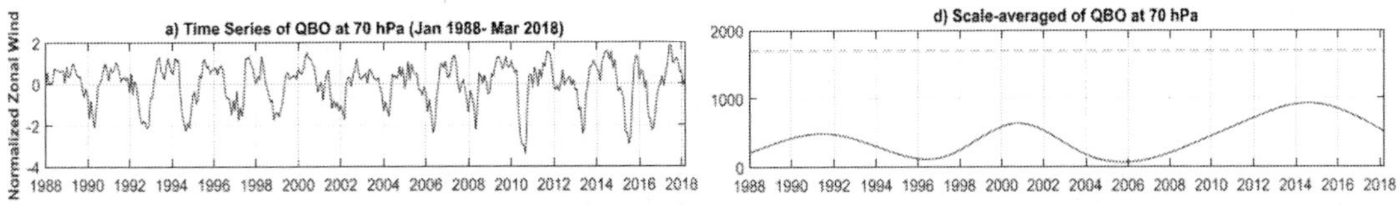

Figure 2. Time series of zonal wind (left panel) and the average of variance (right panel) at selected layers 25, 30, 50, and 70 hPa for a period of January 1988 - March 2018

Wavelet analysis applied to zonal wind data at all levels of the stratospheric layer. But, only selected layers presented. The results show that zonal wind is an atmospheric phenomenon that oscillates about 28 months. This phenomenon contains a periodically of easterly and westerly phase, which vary from cycle to cycle. In the left panel in Figure 2 above, we can see characteristics of zonal wind oscillation at 25, 30, 50, and 70 hPa since January 1988 - March 2018. A positive value indicates westerly wind, conversely in the negative value. Along with decreasing in altitude, these two phases of wind are quite hard to distinguish caused by atmospheric disturbance. The inconsistency turns up at 50 and 70 hPa which showed an average of the variance in the right panel.

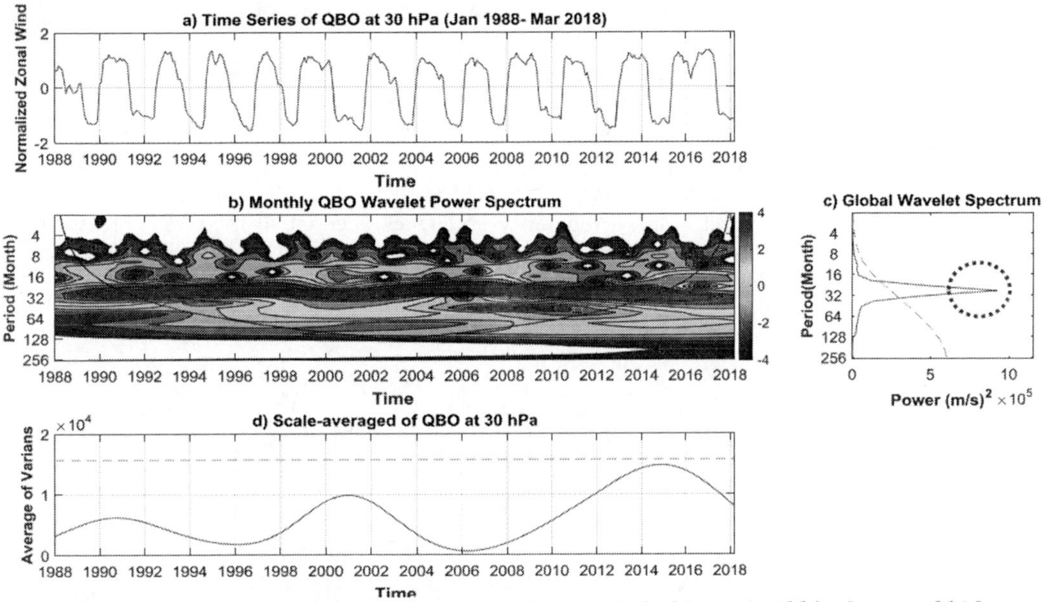

Figure 3. Wavelet of zonal wind at 30 hPa for a period of January 1988 - January 2018

Since our analysis only focused on zonal wind data at 30 hPa, hereafter the complete wavelet results presented. As explained before, the top panel in Figure 3 (a) illustrates the time series data resulting from the normalization of zonal wind data against its standard deviation, while the bottom panel (d) is the average of zonal wind variance after scaling. The wavelet power spectrum of zonal wind (b) displayed in the middle panel. A red color bar indicates the westerly phase of wind, and conversely for blue. From the global wavelet spectrum (c), clearly show that zonal wind still has a periodic oscillation in 28 months, as same as the power spectral density result.

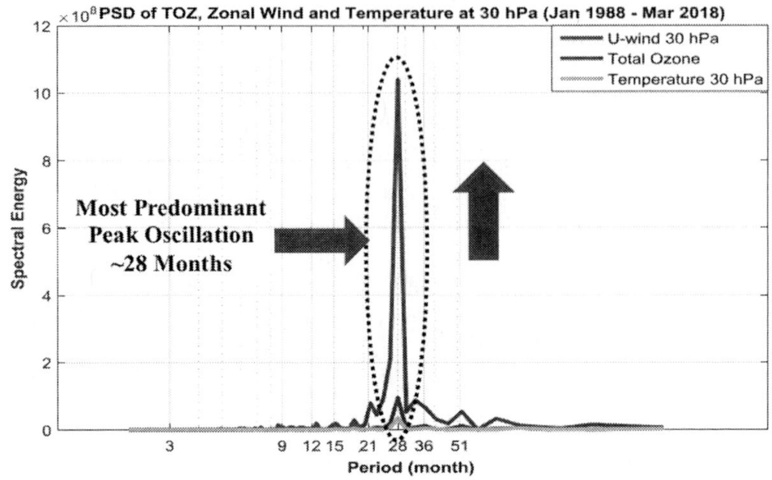

Figure 4. PSD of TOZ, zonal wind, and temperature at 30 hPa from January 1988 - March 2018. Blue, red and orange line represent zonal wind at 30hPa, TOZ and temperature at 30 hPa respectively

Peak oscillation of TOZ, zonal wind, and temperature at 30 hPa determined after processing the data into anomalies in each month. As shown in Figure 4, our three dataset evidence with the highest spectral energy in 28 months period. This result supported by the plot of TOZ time series on zonal wind and temperature at 30 hPa, respectively (Figure 5). In general, both parameters almost similar to each other, while the negative anomalies in TOZ related to the negative anomalies in zonal wind and temperature at 30 hPa. However, some irregularity still takes account in this case. Please note the similar result also obtained by Hermawan [7] while comparing the QBO and ozone anomalies from AIRS satellite over Kototabang, West Sumatera, Indonesia.

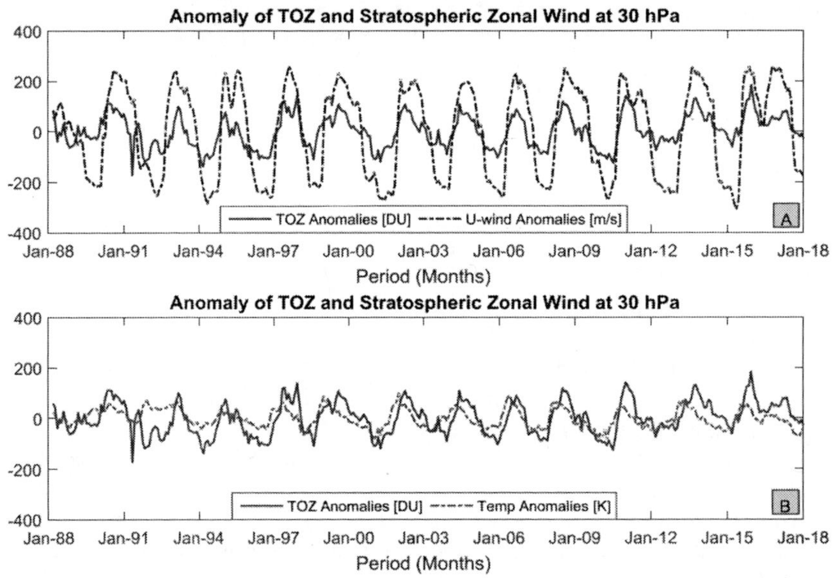

Figure 5. Time series of TOZ, zonal wind, and temperature anomaly for January 1988 - March 2018

3.2 Development of TOZ Model Prediction

Model prediction of total ozone determines by the significance lag of zonal wind and temperature at 30 hPa layer. Selected monthly TOZ predictions for lead 1, 3, 5 and 7 of MR (purple line) and persistence (yellow line) model presented in Figure 6. Its clearly showed that the longer the lead times,

IOP Conf. Series: Earth and Environmental Science **279** (2019) 012050 doi:10.1088/1755-1315/279/1/012050

the smaller the amplitude and shifted at a later time. In other words, lead time influence on expected forecast skill.

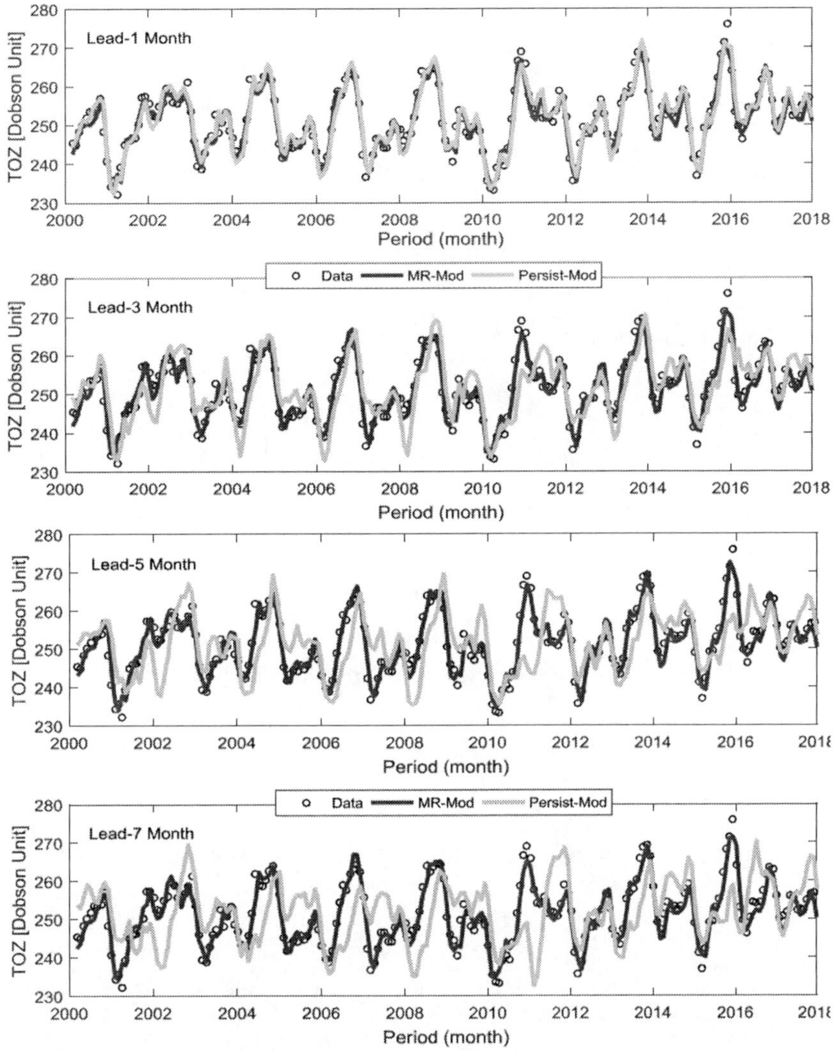

Figure 6. Monthly total ozone predictions for lead 1 to 7 of two ozone model (MR-Mod and Persist-Mod) against data from January 2000 - March 2018

In general, our MR model prediction of ozone was successful in predicting the moderate peak and amplitude of TOZ in October 2010, August – September 2013, and August 2015. However, its lack to predict high value on October 2015. If we compare it with a simple persistence model, both of them still have a good result until lead 3. But, the persistence model fails to predict the peak and trough level due to shift time. Therefore, only MR model still be able to follow the TOZ pattern up to lead 7.

3.3 Skill Measures

The scatterplot of both MR and persistence model presented by the figure below. This graph illustrates the distribution of data toward the line of perfect forecast (red-dotted line) only for selected lead 1, 3, 5 and 7.

IOP Conf. Series: Earth and Environmental Science **279** (2019) 012050 doi:10.1088/1755-1315/279/1/012050

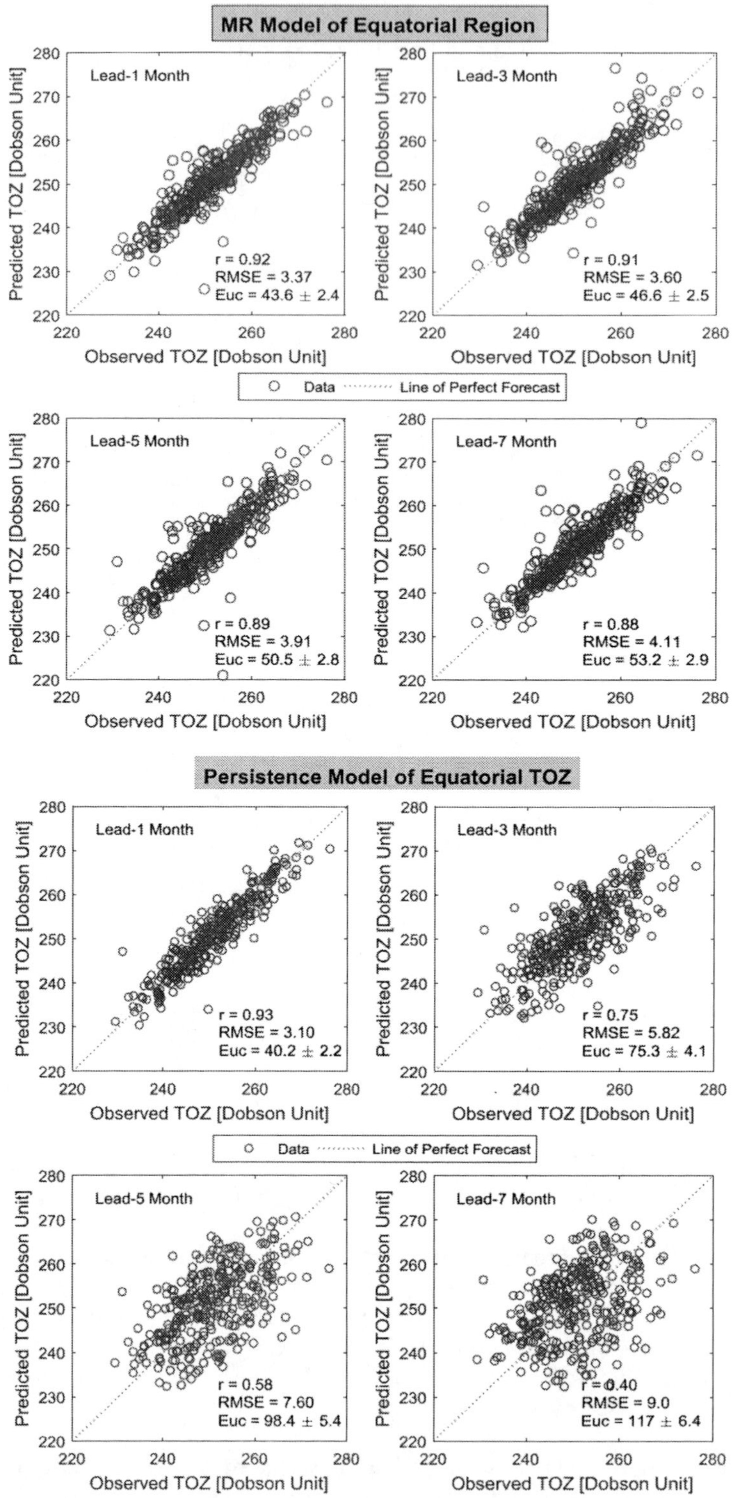

Figure 7. Scatterplot of ozone model prediction for lead 1, 3, 5, and 7

As mentioned before that we used a statistical approach, which is deterministic score to assess prediction skill of the model. This method has been used by Halide [6] in predicting ENSO using the SSTA Nino3.4. Figure 8 below present the Pearson correlation score, RMSE, and Euclidean distance of each model, respectively from the left to the right side. Based on this result, we can see that our model developed much better than the comparison model.

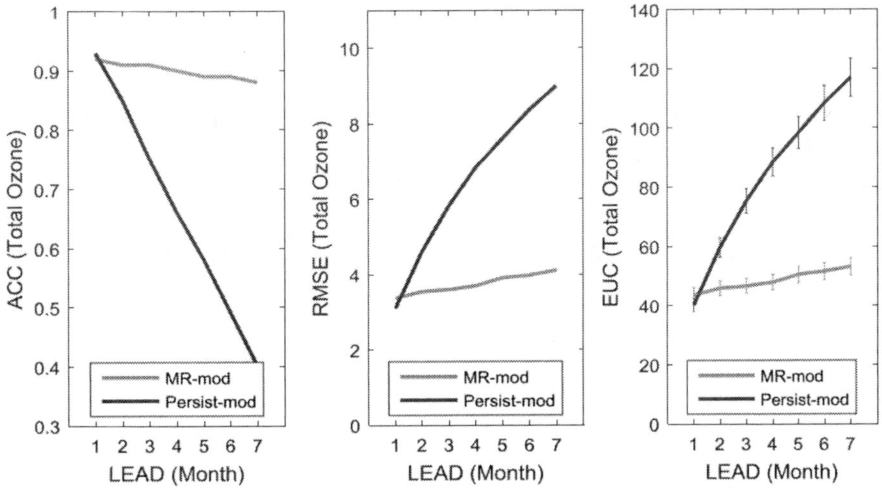

Figure 8. Skill measure metrics; Pearson correlation, RMSE, and Euclidean distance of the two ozone models. Green line for MR model, whereas Persistence model represented by red line

4. The Future Works of Ozone Model Prediction

It showed that our model prediction of TOZ produces useful up to 7 monthly leads, with the latter correlation value about 0.88. As the spectral analysis result, we know that TOZ has peak periodic oscillation in 28 months. Therefore, we have questioned the model performed in the long-term trends. Validate for 28 months ahead of expected to fulfill this gap. However, the comparison of our model developed with powerful model either statistical and dynamical probably reliable to predict the future model of total ozone.

5. Conclusion

As one of the atmospheric gases, ozone protecting the life on earth from harmful UV-B radiation. Ozone concentration in the atmosphere influenced by chemical, physical and dynamic processes that interact with each other. Since the equatorial TOZ related well to zonal wind and stratospheric temperature, in this present study we developed the ozone model prediction. The spectral analysis results of PSD and wavelet showed that the predominant peak oscillation of total ozone anomalies, zonal wind and stratospheric temperature at an altitude of 30 hPa was 28 months. Both parameters have almost similar characteristics on time series. Therefore, the development of the TOZ prediction model has been done using these two parameters, where the prediction model equation in lead 1 month is Yt+1 = -0.1754X1t + 0.0431 X1t-1 + 0.0607 X1t-2 - 0.0368X2t + 0.7880X2t-1 - 0.4838X2t-2 - 0.0096X2t-3 - 0.0430X2t-4 - 0.0029X2t-5 - 0.0464Yt-1 + 0.7623Yt. Up to 7 lead months, our MR model still has the best fit on observed data than the persistence ones.

References

[1] Dameris M 2010 *Angew. Chem. Int. Ed* **49** 8092-8102
[2] Boucher O 2010 *Weather* **65** 105-110
[3] J Haigh and Pyle J 1982 *J R Meteorol. Soc* **108** 551-574
[4] Global Modeling and Assimilation Office (GMAO) 2015 MERRA-2 instM_2d_asm_Nx: 2d, Monthly mean, Single-Level Assimilation, Single Level-Diagnostics V5.12.4, Greenbelt, MD, USA, Goddard Earth Sciences Data and Information Services Center (GES DISC),

Accessed: [***August 23, 2018***], 10.5067/5ESKGQTZG7FO
[5] Halmar Halide 2017 *AIP Conf. Proc* **1801** 070001
[6] Halide H and P Ridd 2008 *Int. J. Climatol* **28** 219-233
[7] Eddy Hermawan 2018 *IOP Conf. Ser.: Earth Environ. Sci* **149** 012064

Acknowledgments
We wish to thank the NASA GSFC, NOAA NCEP-NCAR, and Institut für Meteorologie Freie Universität Berlin, who have provide the TOZ of MERRA-2 Satellite, temperature, and QBO Singapore's data available to access online. We also express our gratitude to Hasanuddin University for supporting this study.

IOP Conf. Series: Earth and Environmental Science **279** (2019) 012051 doi:10.1088/1755-1315/279/1/012051

Cooling water dispersion delivered from Jeneponto Steam Power Plant

Bahrul Ulum[1*], Alimuddin Hamzah Assegaf[1], Sakka[1], Wasir Samad[2]

[1]Study Program of Geophysics, Department of Physics, Faculty of sciences, Hasanuddin University

[2]Department of Marine Sciences, Faculty of Marine and Fisheries, Hasanuddin University.

bahrululum1808@gmail.com

Abstract. Generally, a steam power plant is built near the shore in order to obtain water in large volume as a main material in cooling process. It is the case to Jeneponto power plant, where the cooling water is redistributed from its source with different temperature condition. This temperature will disperse in coastal water out of power plant which is driven by tidal current. The aims of this study is to model a hydrodynamic pattern around steam power plant and temperature distribution pattern of cooling water in spring and neap tidal phases. The method used in this study is a 2D-mathematical model of hydrodynamics and heat transport. The result showed the tidal current tends northward during high tide and southward to during low tide. The temperature tends to disperse northward during the high tide and southward during low tide. The southward disperse is more dominant than northward disperse.
Keywords: temperature dispersion, tidal currents direction, and steam power plant

1. Introduction

In an energy sector, especially the electricity subsector is one of the main drivers of economic development in all countries including Indonesia. The need for electrical energy in Indonesia has continued to increase in the last few decades, to meet the electricity needs, a number of power plants were built. The types of generators built generally are thermal power plants / steam power [1]. In the South Sulawesi region there is a Jeneponto steam power plant (PLTU) with a capacity of 2x125 MW which is built on an area of approximately 100 hectares to meet the electricity needs of the people of South Sulawesi [2].

The operation of a power plant installation generally uses sea water as a cooler. The sea water that has been used as a cooler will be discharged back to the sea at different temperatures, but to reduce the temperature before being discharged back into the sea, the cooling water is first flowed through a cooling channel. Cooling water that re-enters the sea still have temperatures above the ambient temperature of sea water [3]. If cooling water discharged into the sea does not match the normal temperature of the sea, then the life of the marine biota around the disposal of cooling water will be disrupted [4]. Some results of the study suggest that only fish, crustaceans and mollusc can withstand high temperatures. The maximum temperature that can be tolerated by fish is 38.1^{0}C, crustaceans 37.9 ^{0}C and molluscs 36.7^{0}C [3]. Heat distribution are a diffusion process, where the distribution occurs to

IOP Conf. Series: Earth and Environmental Science **279** (2019) 012051 doi:10.1088/1755-1315/279/1/012051

the difference in gradient temperature, so the temperature moves from high temperature to low temperature [5].

To find out the distribution of cooling water from waste products, a mathematical model of hydrodynamic model and heat transport simulation is used [6]. Water temperature greatly determines the quality of water because the temperature changes affect the physical, chemical and biological processes of water bodies [7]. In a study conducted by Hasriyani (2014), a prediction simulation of heat distribution plan for PLTU phase II units 3 and 4 with an increase in temperature of 2.4^0C-3.8^0C to the waters of Punagaya from ambient temperature 32^0C [8]. Therefore, this research was carried out in order to model the distribution of cooling water from the drainage channel of the PLTU Jeneponto to the surrounding waters when the PLTU was operating, by creating a distribution scenario based on the temperature of the cooling water coming out of the disposal outlet.

2. Steam Power Plant

The steam power plant (PLTU) is a power plant that uses steam power as the main driver of the turbine, to produce electricity. The PLTU has a by-product in the form of hot water where the temperature is higher than the temperature of the water before being used for cooling. The amount of cooling water needs depends on the maximum capacity of the PLTU units, in general the use of cooling water at full load for each megawatt is 45-55 m^3/second. Cooling water with a relatively high temperature, with a large volume, and continuously discharged into the water will affect the rise in water temperature from the initial conditions (ambient temperature) [4]. Based on Regulation of the Minister of Environment number 8 of 2009 concerning waste water quality standards for businesses and / or activities of thermal power plants, the maximum temperature of the cooling source (heat water) is 40^0C and salinity at a radius of 30 m from the disposal location of waste water to the sea , the salinity level of wastewater must be the same as the natural salinity [9].

3. Impact of Hot Water Waste on Marine Biota

Temperature changes unnaturally indirectly affect biota, where the carrying capacity of the habitat is lost. For example in coral reef habitats, with changes in temperature, the solubility of oxygen and calcium carbonate (calcite or aragonite) in water will change. Furthermore, it will affect the solubility of metal (metal) contamination and other toxic materials which are assimilated by the physiological process of biota. Hot water waste from a power plant installation is usually discharged directly into the sea, increasing the water temperature and causing thermal pollution. A temperature increase of 10 degrees can accelerate the metabolic activity of aquatic biota to twice the usual. Because each type of aquatic biota has a different metabolic speed, the aquatic biota can only live at a different range of temperatures for each group of biota [10].

4. Continuity and Momentum (2DH)

Integration of continuity and momentum equations to look for 2DH equations, used assumptions and simplifications as follows [11]:

1. The average depth value is considered representative enough to represent values of magnitudes that change along the flow depth
2. The speed and acceleration of the vertical direction are considered very small, so they are ignored
3. The hydrostatic pressure distribution applies throughout the depth
4. Base slope of both small horizontal directions

With this simplification, the continuity and momentum equations for the 2DH model are as follows [12]:

Continuity equation

$$\frac{\partial h}{\partial t}+\frac{\partial uh}{\partial x}+\frac{\partial vh}{\partial y}=0 \tag{1}$$

Equation Momentum

On the x axis:

$$\frac{\partial hu}{\partial t}+\frac{\partial hu^2}{\partial x}+\frac{\partial huv}{\partial y}=-gh\frac{\partial \eta}{\partial x}-\frac{h}{\rho_0}\frac{\partial p_a}{\partial x}-\frac{gh^2}{2\rho_0}\frac{\partial \rho}{\partial x}+\frac{\tau_{sx}}{\rho_0}-\frac{\tau_{bx}}{\rho_0}-\frac{\partial}{\partial x}[hT_{xx}]+\frac{\partial}{\partial y}[hT_{xy}] \quad (2)$$

On the y axis:

$$\frac{\partial hv}{\partial t}+\frac{\partial huv}{\partial x}+\frac{\partial hv^2}{\partial y}=-gh\frac{\partial \eta}{\partial y}-\frac{h}{\rho_0}\frac{\partial p_a}{\partial y}-\frac{gh^2}{2\rho_0}\frac{\partial \rho}{\partial y}+\frac{\tau_{sy}}{\rho_0}-\frac{\tau_{by}}{\rho_0}-\frac{\partial}{\partial x}[hT_{xy}]+\frac{\partial}{\partial y}[hT_{yy}] \quad (3)$$

Where :

H (x,y,t) : The depth of water varies with time (m)
η (x,y,t) : Elevations of Sea level (m)
u (x,y,t) : Average speed of depth on the x axis (m/s)
v (x,y,t) : Average speed of depth on the y axis (m/s)
g : Gravitational acceleration (m^2/s)
ρ : Sea water density (kg/m^3)
ρ_0 : Reference water density (kg/m^3)
p_a : Surface pressure
τ_{sx} : Direction surface tension of x axis
τ_{sy} : Direction surface tension of y axis
τ_{bx} : Basic voltage of the x axis
τ_{by} : Basic voltage of the y axis
T_{xx} : Shear stress of x axis direction
T_{xy} : Normal voltage y direction towards x axis
T_{yy} : Shear stress of y axis direction

5. Transport Equations

The distribution of water quality which is a substance in the form of solution and particles can be known by the transport module approach. Some dynamic model approaches used to describe water quality refer to the development of a 2-D model based on momentum equations and continuity equations by considering the depth where h = η + d is [12]:

$$\frac{\partial h}{\partial t}+\frac{\partial h\bar{u}}{\partial x}+\frac{\partial h\bar{v}}{\partial y}=hS \quad (4)$$

Where:

h : Total depth water (m)
η (x,y,t) : Sea level elevation (m)
d : Depth Water (m)
S : Magnitude of discharge in this case channel discharge (m^3/s)
u(x,y,t) : Average speed of depth on the x axis (m/s)
v(x,y,t) : Average speed of depth on the y axis (m/s)

Transport of a component in the waters depends on the current, where in the waters, the dominant currents are generated by tides. For the temperature distribution the equation used in modelling is the following 2D heat transport equation:

$$\frac{\partial h\bar{T}}{\partial t}+\frac{\partial h\bar{u}\,\bar{T}}{\partial x}+\frac{\partial h\bar{v}\bar{T}}{\partial y}=hF_T+hT_sS \quad (5)$$

Where :

h : Total water depth (m)

$\bar{T}$: Average temperature to depth (^{0}C)

T_S : Temperature at source */source* (^{0}C)

S : Magnitude of discharge in this case channel discharge (m^3/s)

u (x,y,t) : Average speed of depth on the x axis (m/s)

v (x,y,t) : Average speed of depth on the y axis (m/s)

F_T : Horizontal diffusion temperature

For the decay process in the heat transport module the following equation is used:

$$F = 0.2388/(\rho.Cp.h).[(4.6\ 0.09(Tref + Tr) + 4.06w)\ \exp\ (0.033\ (Tref + Tr))] \quad (6)$$

Where :

Tref : Temperature references (^{0}C)

Tr : Temperature increase (^{0}C)

ρ. : Water density (kg/m^3)

Cp : Specific heat (cal/kg ^{0}C)

W : Wind velocity (m/s)

6. Method

Research area is administratively located on coordinates of 5°37'16,095" SL and 119 ° 32'41,827" EL. Location of research in particular is disposal region of wastewater/cooling water of PLTU of Jeneponto, Punagaya village,Bangkala sub-district, Regency of Jeneponto which get influences from tidal waves of the sea. The method used in this study is 2D hydrodynamic models and heat transport.

7. Conclusion

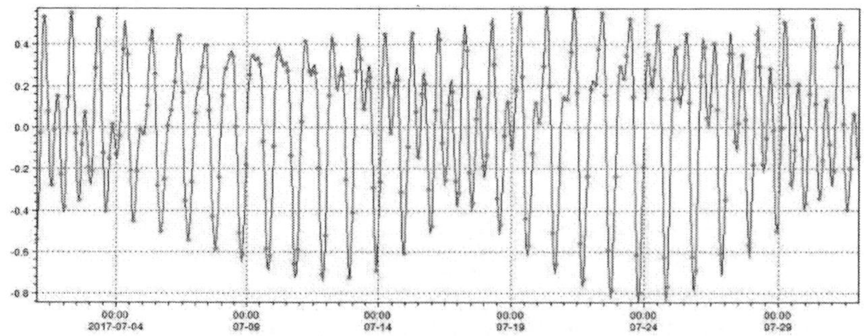

Figure 1 Tidal Domain Research

Tidal

From the results of this research, it is obtained that type of mixed tide mainly diurnal in which within a single day one pair of high tide and low tide occur for once each with different height of elevation. but sometimes, two pairs of high tide and low tide occur within one day with a massive differences on elevation within certain time . Numbers of highest high tide in 0.573 m and numbers of lowest low tide is -0.844 m.

Table I. Determination of high and low tides

Charcteristic	Formula	Level (cm)	Tidal Range (cm)

IOP Conf. Series: Earth and Environmental Science **279** (2019) 012051 doi:10.1088/1755-1315/279/1/012051

HAT	LAT + 2(K1+O1+S2+M2)	172.251	72.825	
MHHWS	LAT + 2*(S2+M2) + (K1+O1)	128.603	29.177	
MHHWN	LAT + O1 + K1+ 2*M2	122.869	23.443	
MSL		99.42	0	Neap tide = 46.8 Spring tide =58.34
MLLWN	LAT + K1 + O1+ 2*S2	75.982	-23.443	
MLLWS	LAT + K1 + O1	70.248	-29.177	
LAT	MSL-K1-O1-S2-M2	26.600	-72.825	

Where :
HAT : *Highest Astronomical Tide*
LAT : *Lowest Astronomical Tide*
MHHWS : *Mean Higher High Water Spring*
MLLWS : *Mean Lower Low Water Spring*
MHHWN : *Mean Higher High Water Neaps*
MLLWN : *Mean Lower Low Water Neaps*
MSL : *Mean Sea Level*
From the table above to show that value ride install it amounting to 145.65 cm.

Ocean current
Current velocity in the surrounding area of Jeneponto PLTU could be seen on the following models:

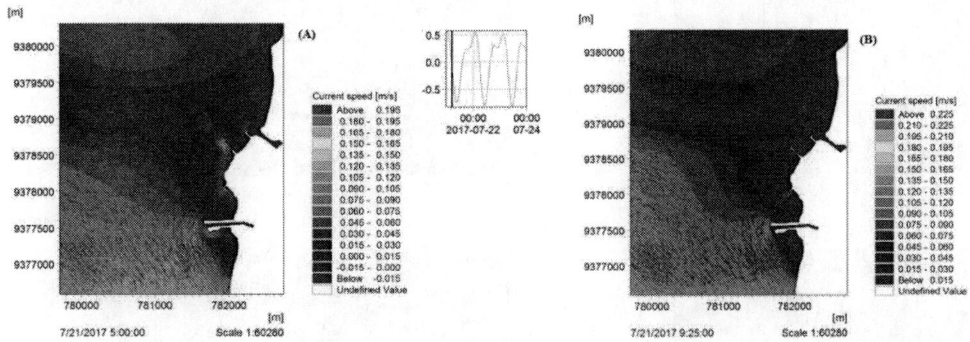

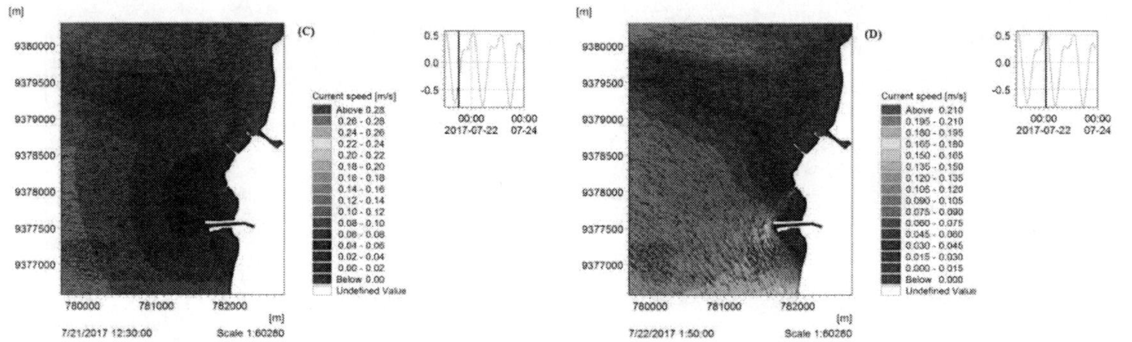

Figure 2 Velocity of Ocean Current

Figure **A** portray current velocity of phase heading to low tide condition, figure **B** portray current's velocity of lowest low tide phase, figure **C** portray current's velocity of phase heading to high tide condition and figure **D** portray current's velocity of high tide phase. On condition heading to low tide, velocity of the current's is 0.03-0.12 m/s with the current moves southward, on the lowest low tide velocity of the current is 0.01-0.03 m/s with current begin turning around and moving northward . On condition heading to the high tide velocity of the current is 0.02-0.1 m/s with current moves northward, on high conditions of high velocity of the current is 0.01-0.04 m/s with current already move southward.

Heat Dispersion

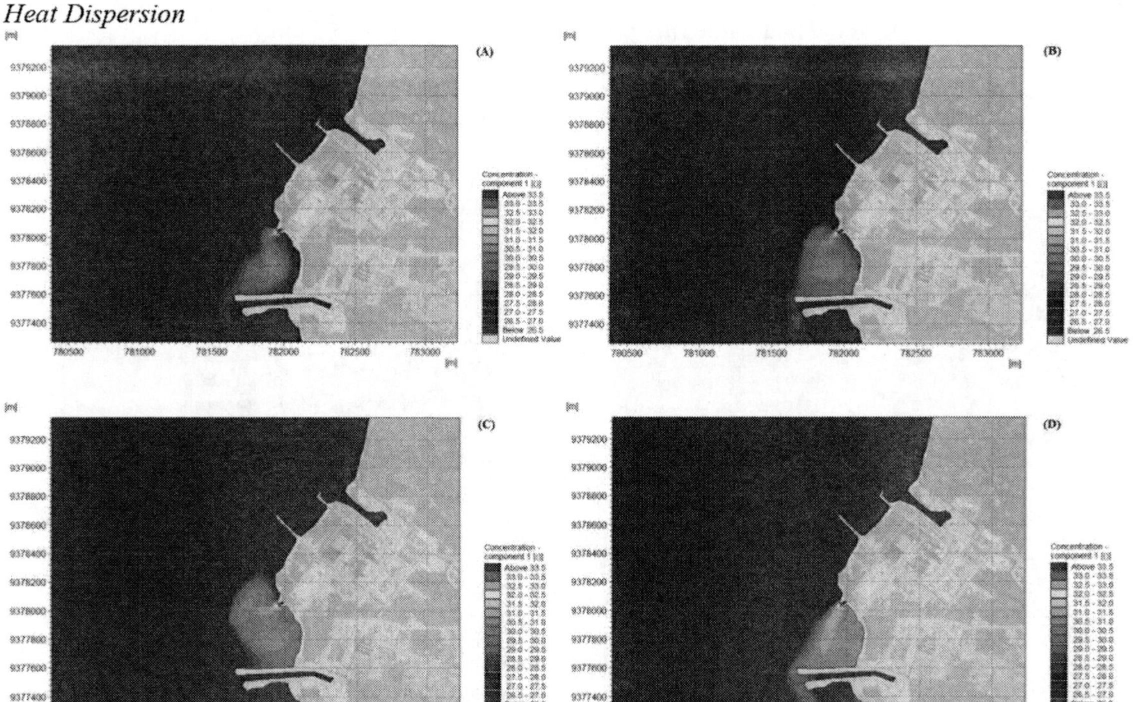

Figure 3 Heat Dispersion

Figure **A** portray heat dispersion of phase heading to low tide condition, figure **B** portray heat dispersion of lowest low tide phase, figure **C** portray heat dispersion of phase heading to high tide condition and figure **D** portray heat dispersion of high tide phase. From heat distributions simulations,

It can be seen that farthest distance of heat dispersion not exceeding numbers of 500 m already changes due to ambient temperature / temperature of environment , escalation of highest temperature dominantly happen Southern part owing to the current that moves affect the process of heat dispersion on waters. From those models statistic test was then conducted by utilizing linear regression method by comparing the data of field and model measurement. Result of statistic test shows value of model's accuracy at 1.54891, this value increasingly approach number of 0 (zero) means that it is increasingly accurate. With that much of value then it can be stated that accurate constructed model is at 98.45109% with significances value is at 0.027.

References

[1] Susila I M, Faridha, et al 2011 *Ketenagalistrikan Dan Energi Terbarukan* **10** (1), 35 – 50. ISSN 1978 – 2365.
[2] Tetty, Baso A, Syamsuddin R 2014 *Dampak Sosial Ekonomi Pembangkit Listrik Tenaga Uap Punagaya Terhadap Budidaya Rumput Laut Di Kabupaten Jeneponto*.
[3] Cahyana C 2011 *Model Sebaran Panas Air Kanal Pendingin Instalasi Pembangkit Listrik Ke Badan Air Laut* Thesis Universitas Indonesia : Jakarta
[4] Fudlailah P, Mukhtasor, Zikra M 2012 *Pemodelan Penyebaran Limbah Panas di Wilayah Pesisir (Studi Kasus Outfall PLTU Paiton)*.
[5] Wibowo N T, Sugianto D N, Indrayanti E 2012 *Journal of Oceanography* **1** (1) 102-110.
[6] Susiati H and Mellawati J 2012 *Jurnal Pengembangan Energi Nuklir* **14** (2).
[7] Efendi H 2003 *Telaah Kualitas Air Bagi Pengelolaan Sumberdaya dan Lingkungan Perairan* PT.Kanisius 257.
[8] Hafid H 2014 *Prediksi Sebaran Suhu Air Panas PLTU dan Pengaruhnya Terhadap Ekosistem Perairan (Studi Kasus : Rencana Pembangunan PLTU Jeneponto Tahap II unit 3 dan 4 di Desa Punagaya, Kec.Bangkala, Kab.Jeneponto)* Thesis Universitas Hasanudddin.
[9] Peraturan Menteri Negara Lingkungan Hidup Nomor 08 Tahun 2009 Tentang Baku Mutu Air Limbah Bagi Usaha Dan/Atau Kegiatan Pembangkit Listrik Tenaga Termal
[10] Mukhtasor 2007 *Pencemaran Pesisir dan Laut.* Jakarta: PT. Pradnya Paramita.
[11] Istiarto 2011 CFD di Bidang Hidraulika Saluran Terbuka, JTS FT UGM. Yogyakarta
[12] Mike by DHI 2012 *Mike 21 & Mike 3 Flow Model FM, Hydrodynamic and Transport Modul, Scientific Documentation*, DHI Software

Study of Climate Determination Analysis Based On Pallontara/Papananrang and Rainfall Opportunities in Sidrap

Amir Yassi, Kaimuddin, Abdul Haris Bahrun, Asmiaty Sahur*

Agrotechnology Study Program, Faculty of Agriculture, Hasanuddin University

*asmiatyasmiaty@gmail.com

Abstract. Climate change is shown by the increasing imbalance in the amount of water in the dry season and rainy season. Thus, people experience water shortages in the dry season and floods in the rainy season, thus disrupting all schedules and cropping patterns which result in crop productivity decreasing until the harvest. Often the results of observations via satellite or meteorological instruments by the Agency for Meteorology and Geophysics are opposite in issuing a forecast of seasons with pallontara and papananrang in determining the weather / climate conditions that will occur. So that the results of these different forecasts can cause panic in planning agricultural development programs in the following year, no wonder the planning of agricultural development programs at the central level is often also out of sync especially in the timing (season) Estimates of the season that will occur in a year can be determined from the results of the scavenger gathering (meetings of civil servants, stake holders, researchers and universities). Specific objectives of the study to obtain a description of next year's climate forecast indicators are determined by several variables used pallontara and as a comparison analysis rainfall from several years. The results of the study that some indicators used by Pallontara in predicting climate are based on Islamic calendar, astronomical factors, behavior of organisms, observation of vegetation and observations of the moon. Whereas as a comparison used monthly rainfall data from several Sidrap Regency stations.

Keyword. Pallontara / Pananrang, climate

1. Introduction

Drought and flooding are two forms of natural disasters that quite often hit the agricultural sector in Indonesia. In extreme climatic conditions (years of El-Nino and La-Nina), the extent and intensity of agricultural land affected by the disaster increased sharply. Observations of El-Nino in 1994 and 1997 showed that the accumulation of rice fields that experienced drought from May to August exceeded 400 thousand ha while in normal years and La-Nina was less than 75 thousand ha. Furthermore, in La-Nina in 1995, accumulation of widespread flooding from October to December reached 250 thousand ha while in normal years and years El-Nino was generally less than 100 thousand ha [1]. Many climate experts look down on traditional climate or seasonal predictions, this is because traditional predictions do not use scientific methods and do not produce scientific reports. In addition, it does not make measurements with tools that are carried out regularly so that they can be repeated by others [2]. This happens because the observations they make are only based on animal behavior, the state of plants and using astrological observations.

The forecasting or forecasting practices of the season have long been carried out such as Pranata Mangsa on Java Island. Porhalaan in the Batak area, Lontara in South Sulawesi, Wariga in Bali and Nyali in East Flores. It is practiced by implicating the existence of public trust in the truth of the statements of smart people at that time. Observation methods do not use analytical methods so that they do not report in writing. However, their knowledge is collected from a hereditary experience, which is certainly a result of hereditary observations.

There is often a difference between the papayarang / pallontara and BMG in the estimated season. So that traditional forecasting (pallontara / papananrang) has the advantage that the forecasting model of human findings does not have an analytical and scientific method. Excellence is obtained from the ability of plants or animals to integrate all information obtained from the environment, into an information about the future conditions.

Climate forecasting in South Sulawesi has long been carried out since the government program namely Bimas on a large scale on a large scale to plant rice using fertilizing technology and superior seeds. With the support of the pallontara a "sipulung tudang" is held as a meeting place in determining the time to get down the rice fields and the planting schedule aims to avoid damage caused by plant pests and diseases, avoid excess water or lack of water so that rice production increases.

Ranging from descending to rice fields, plowing, until the time for harvest is over. There is an appalili ceremony before the land piracy. There is Appatinro pare or appabenni ase before rice seedlings are sown. This ritual is also commonly carried out when storing rice seedlings in a balla position, a special place located in the center of the house aimed at keeping no animal passing on it. Then the ritual was assembled with massureq, reading meong palo karallae, one of Lagaligo's epics about rice. And when the harvest arrived, a bokko coup was held, a harvest ritual which was usually accompanied by kelong pare. After going through a series of rituals, Mapadendang was implemented. In Sidrap and surrounding areas this ritual is known as appadekko, which means adengka ase lolo, the activity of pounding young rice. Appadekko and Mappadendang are said to have originated from this activity.

Papananrang from the Jeneponto area uses indicators about the layout and brightness of star sightings or star clusters. By looking at the culmination of two star clusters, namely lambarua and Mangiwanga, as an indicator of meeting two ocean currents that can cause large waves. It was different from what was done in Wajo District besides using 9 star clusters, also paying attention to the brightness of its.

1.1 Formulation of the Problem
The older age of Pallontara / Papananrang which is one of the local wisdom that must be preserved, so that the extent of the results of observations and years of experience from pallontara / papananrang can be utilized and a comparison of the results of analysis of rainfall opportunities in determining the condition of rain that will occur. Scheduled patterns every year are carried out through scavenger gatherings by bringing together technicians, practitioners, researchers, BMG and universities to discuss season forecasts that will occur. Another section of your paper.

1.2 Research purposes
This study aims to (1) increase the enthusiasm and participation of the pallontara / papananrang in determining season forecasts / predictions, (2) increase and carry out annual records of the occurrence of natural signs so that the quality of plants / trees in determining the forecast for the season will be more precisely so that it can be studied scientifically, (3) printing a new generation of young pallontara / papananrang that is more dynamic and able to combine traditional forecasting with modern, (4) determine the object of forecasting in a manner that is so that it is sharper in forecasting.

1.3 Expected output
(1) Obtained the results of forecasting methods according to the conditions of the local area.
(2) A forecasting method that is more specific and sharp so that it can be used as input for the agro-complex sector.

(3) The passion for work is higher and directed so that regional assets can be preserved as a culture of South Sulawesi.
(4) New findings from the combination of traditional forecasting methods and modern forecasting methods based on scientific foundations.

2. Methods

2.1 Place and time

The survey activity was carried out in Sidrap Regency, South Sulawesi Province. The location of the activity is the area where the success of the rice harvest in Sidenreng Rappang was due to the firmness of 'Nene Mallomo' in carrying out the law, this was seen in the culture of the local community in determining the planting period through TUDANG SIPULUNG (Tudang = Sit, Sipulung = Gathering or can be translated as a Great Deliberation) which was attended by Pallontara '(experts on Lontara books)), papananrang (experts in determining the season) and indigenous community leaders. The region has two peak rainfalls which are equatorial in December / January and April / May. Time of activity for 7 (seven) months or one planting season.

The materials used in this activity are: cameras, recording devices, computers as analysis tools using Excel programs and writing instruments.

The activity was carried out in the form of lectures, training and FGDs and evaluation of the climate prediction results of pontontara / papananrang. While as a comparison, secondary data was obtained from various Sidrap Meteorology / Climatology Stations, Maros Climatology Station and Tiroang TPH Station and data from PU Irrigation / PSDA. Secondary data in the form of rainfall data that occurs over 10 years, data is sampled from stations representing several sub-districts.

2.2 Extension method

This counseling activity was carried out before tillage by involving members of the New TaniWala group as partners which included farmer owners and cultivators by including family members, stake holders and related agencies. The extension materials include:

a. How the participants can know that the plants are not independent of the influence of climate and the impact that will be caused.
b. How to use observational instruments for weather / climate elements and utilize these data as a comparison of traditional climate forecasting
c. How to avoid the effects of damage caused by the influence of extreme climates (floods, droughts and pests).
d. How to increase human resources (farmers) by changing the mindset of farmers from monoculture farming or just by using one variety every season to polyculture farming or the pattern of rotation of varieties, so that income from one commodity can be added from the crops.
e. Creating regeneration of pallontara / pananrang that is able to combine traditional climate predictions with modern climate predictions through the use of weather / climate data from the results of statistical analysis.

3. Results and Discussion

3.1 Traditional season forecast

Climate forecasting / weather has actually been started since humans were able to see natural phenomena and connect them with animal behavior with certain weather events empirically. Since centuries ago people had assumed that if male wolves and donkeys often howled and shook their ears, goats and rams gore at each other, pigs snored and fidgeted all this was a sign of bad weather such as heavy rain accompanied by Guntur [3].

In the past three decades the climate in Indonesia has experienced a dynamic change. One of the conditions that can be felt is the increasing temperature and the increasing diversity of current climate patterns. Higher temperatures influence the continued increase in evaporation and evapotranspiration which leads to increasingly depleted water availability [4]. Climate change is also indicated by the increasing imbalance in the amount of water in the dry season and the rainy season. Thus, people experience water shortages in the dry season and floods in the rainy season.

The occurrence of thousands of hectares of agricultural land due to the dry season. Paradoxically, on the affected land in December, January, February, March, there will be an overflow of water that is difficult to control. In fact, these two phenomena are natural phenomena at a reasonable level that come every year. If later irregularities arise, solely because of errors in handling, the greed of a few people in utilizing natural resources, mistakes using agricultural technology, and lack of knowledge [5].

Often the results of observations via satellite or meteorological instruments by the Agency for Meteorology and Geophysics are opposite in issuing a forecast of seasons with pallontara and pananrang in determining the weather / climate conditions that will occur. So that the results of these different forecasts can cause panic in planning agricultural development programs in the following year, no wonder the planning of agricultural development programs at the central level is often also out of sync especially in the timing (season) [2]. Estimated seasons in Sulawesi The South is determined by the results of the scavenger workshop (meetings of civil servants, stake holders, researchers and universities).

It must be admitted that our ancestors did not have analytical methods or written reports about the results of observations carried out every day. However, their knowledge is gathered from hereditary experiences and this is where the first step of knowledge at any time can be used as science to meet the tastes of the times by putting it on a solid foundation in the form of a critical and analytical approach.

In Lontara, many state indicators are taken from the location and brightness of star appearances or star clusters. For example papananrang from Jeneponto uses star clusters as sources of predictions such as lambarua, mangngiwanga, purung-purunga, pa'jekoa, janganga and balla keppanga.

The entire star cluster is a sign of carrying rainfall throughout the year, but is bound to the area where the star clusters are in their respective circular circles. The effect of rainfall can be seen when allimbang (crossing the kulaninasi point). Circulation of the star clusters together with the sun's path to the earth. Indicators of high rainfall that will occur at the time of allimbang and concurrent occurrence of Sa'ra (sinking of star clusters along with solar radiation in the west). The star groups that have the most rainfall are purung-purunga, pa'jekoa and janganga [6].

In Sidrap, a prominent Bugis intellectual figure lived during the Addatuang Sidenreng period and Addatuang Rappang (Addatuang = a kind of district government in the past) called 'Nene Mallomo'. He is not from the royal family, but his intelligence in state law and government makes his name quite famous. A legal order that is still perpetuated in Sidenreng, namely: Naiya Ade'e De'nakkeambo, de'to nakke'ana. (Translation: in fact the custom does not know the father and does not know the child). The wise word was issued by 'Nene Mallomo' when summoned by the King to decide the sentence to Nene Mallomo's son who stole the plow's neighbor's plow equipment. In Lontara 'La Toa, ene Nene Mallomo' is in line with other Bugis-Makassar figures, such as I Lagaligo, Puang Rimaggalatung, Kajao Laliddo, and so on. The rice harvest success in Sidenreng was due to the firmness of 'Nene Mallomo' in carrying out the law, this was seen in the culture of the local community in determining the planting period through deliberations called TUDANG SIPULUNG (Tudang = Seated, Sipulung = Gathering or can be translated as a Great Deliberation) by the Pallontara '(experts on Lontara books)), papananrang (experts in determining the season) and indigenous community leaders [7].

3.2 Rain pattern analysis

Based on Figure 1, the predicted rainfall that occurs from the results of statistical analysis shows that during the planting season October-March the peak of rain and its number experience a shift, namely in the last 5 and 7 years relatively the same, whereas the last 17 years there has been a decrease of only 100 mm per month. Conversely, in the April-September planting season all three did not change even though in the last 5 years the rainfall in June had decreased by 105 mm. Forecasts made by the pallontara said that the gadu planting season (April-September) in Dua PituE Subdistrict tends to be wetter so that pest attacks increase in intensity, so farmers are expected to prepare anticipation of control by planting resistant varieties, the timing of planting must be appropriate.

IOP Conf. Series: Earth and Environmental Science **279** (2019) 012052 doi:10.1088/1755-1315/279/1/012052

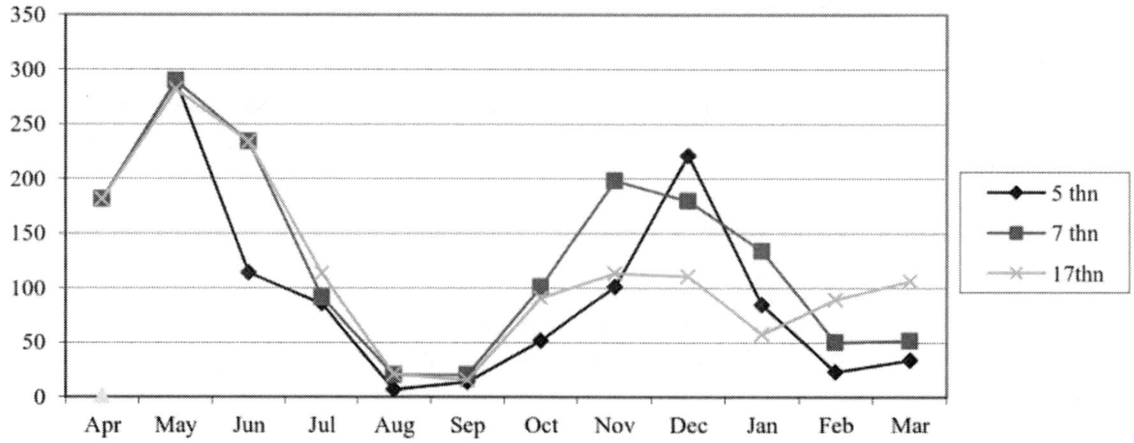

Figure 1. Rainfall distribution from TanruTedong Station at 5, 7 and 17 years Representing Two PituE Districts.

Based on the results of rainfall analysis in Figure 2, the pattern of rainfall in Baranti sub-district changes in patterns in the region in the last 5 years of the planting season from October to March the amount of rain is higher than the April-Sep planting season, whereas in the last 12 years relatively constant despite the amount of rain which occurs on average below 200 mm, so that conditions must be assisted by irrigation water to meet the water needs of rice plantations. According to the pallontara this region has changed due to the influence of Pinrang regency which blows east monsoon with relatively little water vapour.

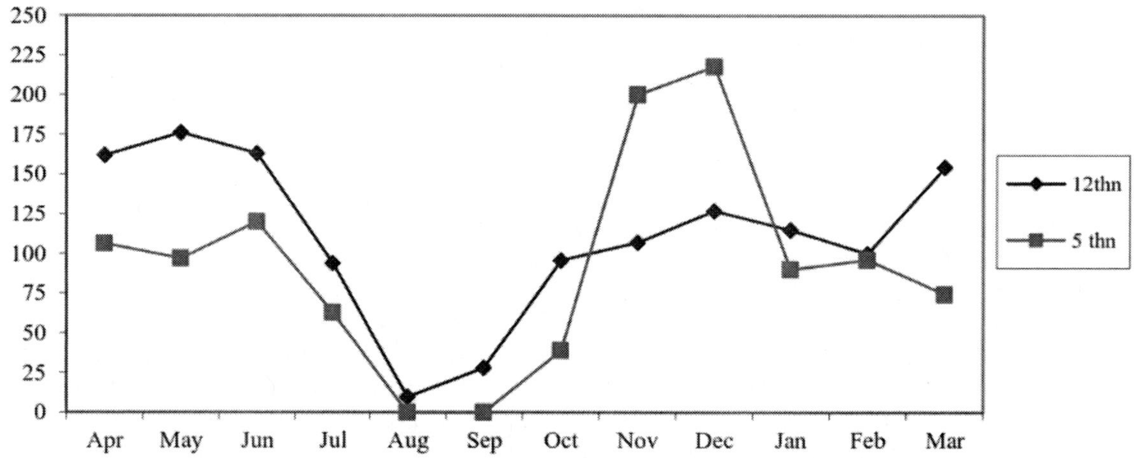

Figure 2. Rainfall distribution that occurs from Baranti 5 and 12 years Station Represent Baranti District.

The results of the rainfall analysis in the last 5 and 12 years in Figure 3 show that the area is relatively unchanged in the second planting season, except the peak of rain during the October-March planting season for the last 5 years the highest rainfall is reached at the end of the month and the dry season is more long from mid-July to mid-September and the rainfall distribution is relatively lower, while the last 12 years occur at the end of March and the lowest rainfall occurs only in September. Pallontara's prediction said that in this area it was relatively wet because it was influenced by Enrekang district which had a lot of Orographic rain.

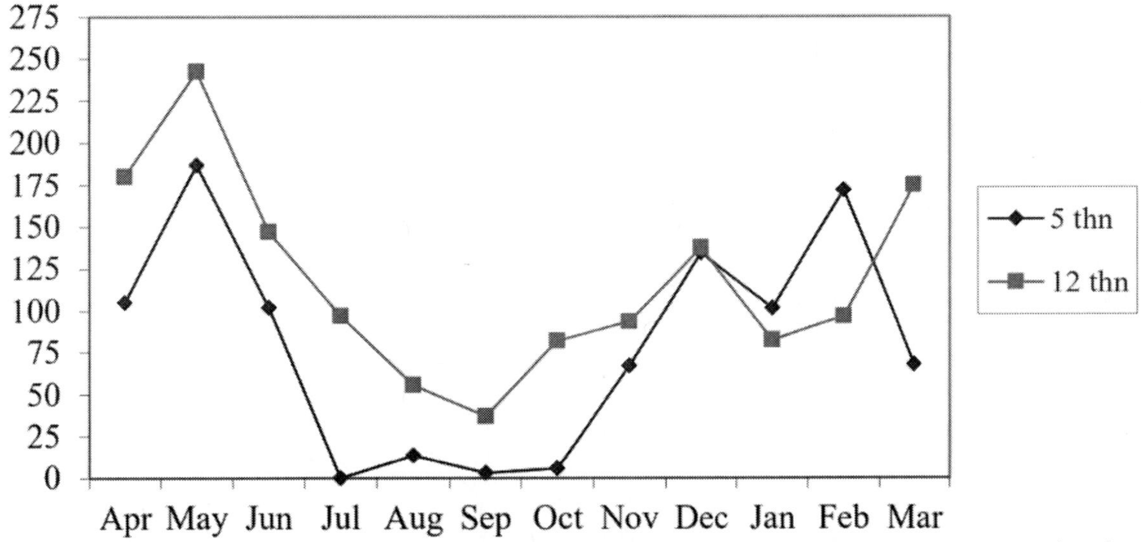

Figure 3. Rainfall distribution that occurs from Rappang Station 5 and 12 years Representing the Panca Rijang District.

The results of the analysis in Figure 4 show that in the MaritengngaE sub-district located in the center / district center the pattern of rainfall that occurred in the two planting seasons was relatively the same in the last 5 years and 12 years, except for the relatively higher rainfall intensity in the last 12 years but almost all months range below 200 mm per month. Based on the predictions of Pallontara.

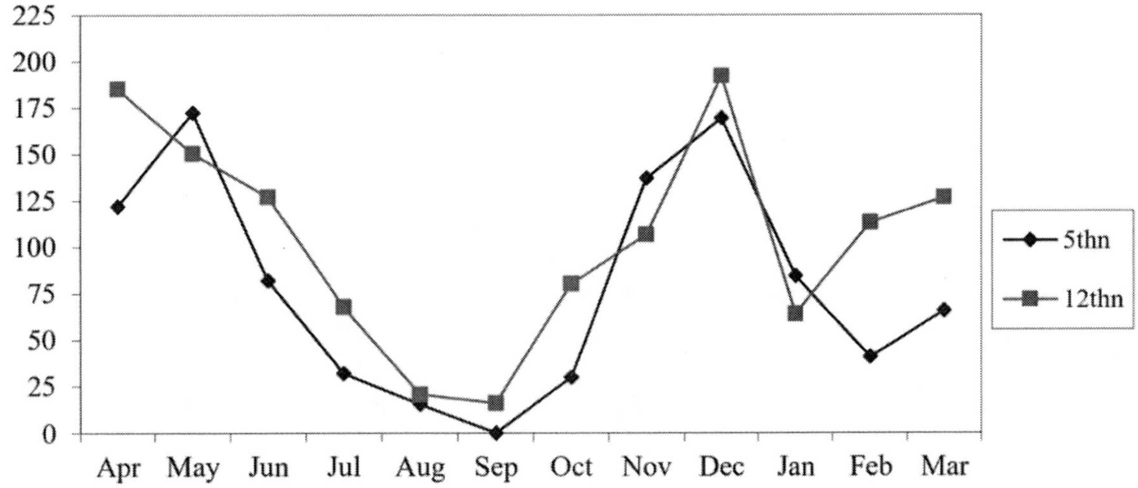

Figure 4. Distribution of rain that occurs from Sereang Station 5 and 12 years representing the MaritengngaE District.

The results of the analysis of rain patterns from the Panca Lautang sub-district are shown in Figure 5, that the two planting seasons April-September and October-March are relatively the same except that the rainfall intensity is relatively high in the last 12 years, even though the last 5-year rain peak occurred in March with 250 mm. In general, it shows that the rain that occurs on average is below 160 mm per month so it is expected that this region should be suitable for development only for crops. Based on the pallontara prediction, the occurrence of rain followed the influence of Soppeng Regency where the rain that occurred was limited to soaking the soil.

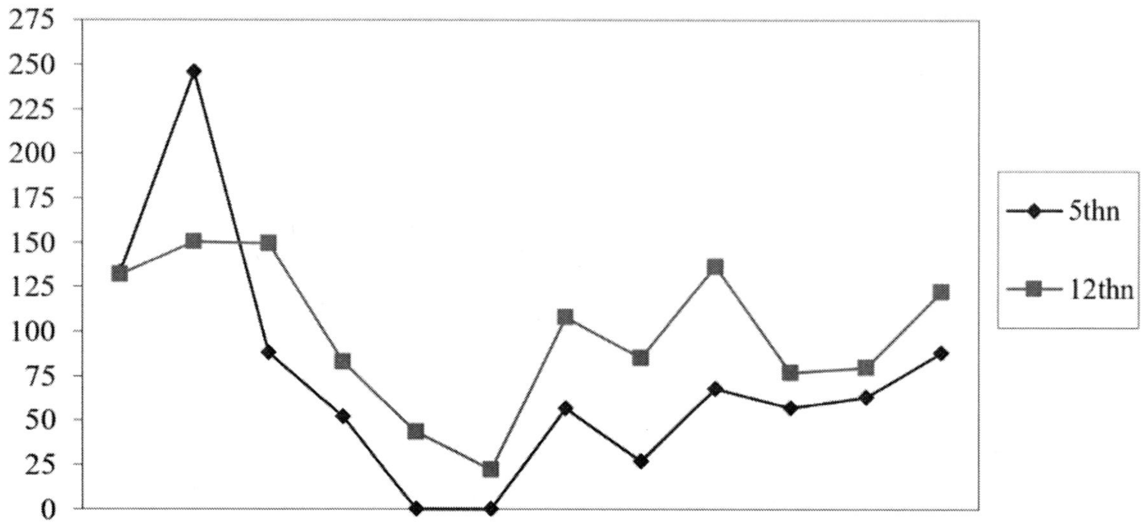

Figure 5. Distribution of rain that occurs from Bilokka Station 5 and 12 years.

Representing the Panca Lautang District stated that the rainfall pattern in this area was influenced by four adjacent territories, namely Pare-pare village, Soppeng regency, Wajo and Enrekang regencies, so that they received rain shipment from these regions.

Based on the results of the analysis of rain patterns in Figure 6, it shows that in this region the amount of rain that occurs in the planting season of October-March is relatively higher than the planting season from April to September. From July to October. According to Pallontara this region is strongly influenced by southeast Passat winds originating from Pare-pare.

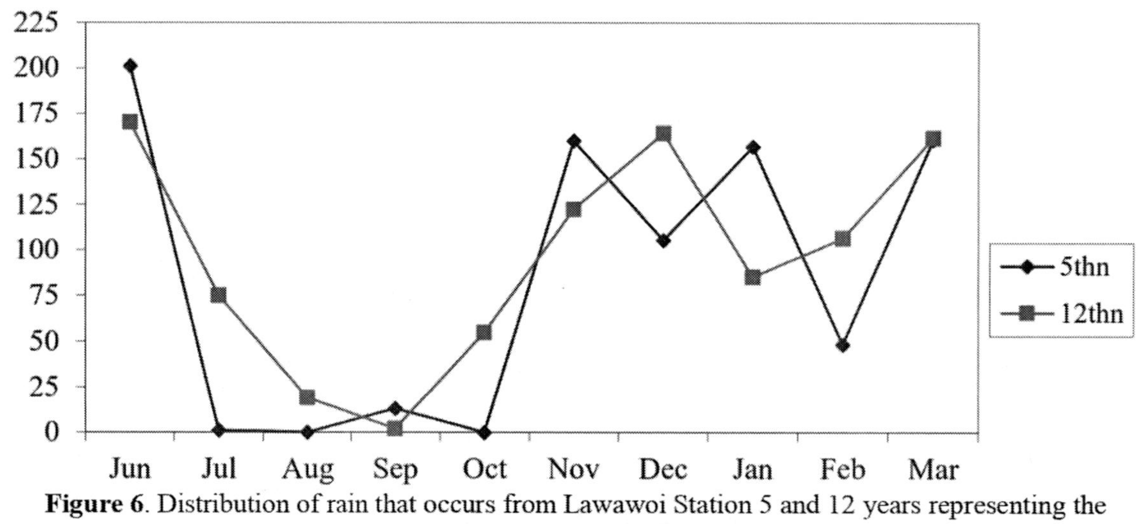

Figure 6. Distribution of rain that occurs from Lawawoi Station 5 and 12 years representing the Watang Pulu District.

4. Conclusions and Recommendations

4.1 Conclusion

a. Climate forecasting based on pallontara using the eight-year cycle indicator made every year with Islamic calendar (hijriah) with the formula of the hijri year reduced by two then divided by eight, the remainder of the year will be determined based on years starting from Alif to Dal.
b. Climate / rain forecasts with dates when the days of Muharram appear and natural conditions that occur up to three Muharram.
c. Climate / rain forecasts with indicators of insect behavior and plants are a sign of the arrival of the rainy season or dry season.
d. The results of rainfall analysis in Sidrap Regency in seven sub-districts, which experienced changes in the 2014 planting season occurred in the MaritengngaE sub-district, Panca Lautang and Lawawoi.

4.2 Suggestions

a. Research should be continued to obtain both qualitative and quantitative data using weather / climate observations, so that more scientific data can be obtained.
b. Research is needed on planting rice and secondary crops according to the planting time obtained both from pallontara and from the results of statistical analysis of rain distribution

References

[1] Boer R., Notodiputro K A, and Las I 2000 Prediction of daily rainfall characteristics from monthly climate indices. *Proceeding of "The Second International Conference on Science and Technology for the Assessment of Global Climate Change and Its Impacts on Indonesian Maritime Continent*, 29 November-01 December 1999.

[2] Blantran de Rozari 1994 *Bulletin Meteorologi Pertanian Indonesia* **II**, No. 1 dan 2.

[3] Baharsyah J S, Pawitan H, Las I, Bey A, Boer R, Handoko, Winarso P A, Suharsono H 1994 *Buletin PERHIMPI* **II**. No. 1 dan 2. Hal. 18-30. Tahun 1994. Bogor.

[4] Koesmaryono Y 2004 *Dampak Pola Iklim Terhadap Sumber Daya Air di Indonesia.Suara Pembaruan. Simposim Perhimpunan Meteorologi Pertanian dan Kongres ke-5 Perhimpi di Jakarta.*

[5] Karama A S 2004 *Kekeringan dan Banjir, Bom Besar bagi Pertanian Indonesia*. Staf Ahli Menteri Pertanian RI. Last Modified 16 Sep 2004.

[6] Toto P R 1997 *Pokok-Pokok Pikiran "Pallontara Turatea" Dalam Ramalan Cuaca. Hasil Tudang Sipulung Di Kabupaten Jeneponto.*

[7] Badaruddin B A 2011 *Tudang Sipulung Langkah Awal Mappalili. Disalin dari buku Lontaraq H. A. Badaruddin* Kab. Sidrap.

IOP Conf. Series: Earth and Environmental Science **279** (2019) 012053 doi:10.1088/1755-1315/279/1/012053

Distribution of Soft Coral *Octocorallia (Alcyonacea)* in coastal waters of Gonda, Polewali-Mandar, West Sulawesi

Ayub Wirabuana, Magdalena Litaay*, Willem Moka, Dody Priosambodo

Lab Environmental and Marine Sciences, Department of Biology
Faculty of Mathematic and Natural Sciences Hasanuddin University, Makassar, Indonesia 90245

*mlitaay@fmipa.unhas.ac.id

Abstract. The research on "Distribution of Soft Coral *Octocorallia: Alcyonacea* population in coastal waters of Gonda, Polewali-Mandar, West Sulawesi" has been done from August to December 2017. This study aims to determine the soft coral distribution and diversity at different depths in Coastal waters of Gonda, Polewali-Mandar, West Sulawesi. The data collection was performed by using the *"Line Intercept Transect (LIT)"* method with 50 meters transect at a depth of 3 meters and 10 meters at each station. The transect line was following the contour lines. Sample identification was performed by means of video recording. Environmental parameters including temperature, salinity, dissolved oxygen, current, brightness, turbidity, and pH was measured during field campaign, respectively. Data analysis was done by using the *Shannon-Wiener* Diversity Index, Uniformity index, Dominance index and Morisita Index. The results showed that 16 species from four families of Soft Coral has been found in Gonda Beach, where the number of species found at 10m depth are higher than 3m depth, respectively. The average value of the diversity index at 3m depth (1.62) is higher than at 10m depth (1.34), but both are still classified in the moderate category. The pattern of population distribution in the overall site at this location tends to be random.

1. Introduction

The Gonda coastal area is one of travel destination in West Sulawesi, this location placed in the middle of Coral triangle which streched out in Indonesia-Filipina-Papua New Guinea coastal water. Gonda coastal water not have been explored yet. This area has a big biodiversity of marine biota, one of the biota that can be found in this area is soft corals.

Soft corals (*Octocoraliia:Alcyonacea*) belongs to Coelenterata [1], which distributed along a shallow water in tropical and subtropical regions. Nowadays, many soft corals have known to have some important use in variuos filed such as ecology, pharmachy as economic perspective. Some member of genus *Sarcophyton* are notified as current indicator for surrounding environment. Beside, many soft coral also are known for their potency in producing bioactive compound that basic material for drugs [2]. Soft coral from South west Sulawesi province is far less documented. Most of soft coral from Indonesia was recorded from Sumatera, Jawa, Borneo, South Sulawesi Province and eastern part of Indonesia [3][4]. There are 96 genera from 4 families that found in Indo-Pasific, while 28 genera of it found in Indonesia [5].Study done by Manuputty in 2002, recorded approximately 219 species of soft coral are distributed in Indonesia [5]. This present study was first attempt to record a distribution of soft coral from Gonda waters in South West Sulawesi province of Indonesia.

IOP Conf. Series: Earth and Environmental Science **279** (2019) 012053 doi:10.1088/1755-1315/279/1/012053

2. Material and method

2.1. Sampling site

Sampling site was determine according visual habitat characteristic. Data collection was done at three different location as follows: 3°29'91"S, 119°07'62"E (Site I), closed to unexist mangrove forest, 3°30'50"S 119°07'33"E (Site II), closed to mangrove forest, and 3°30'82"S,119°07'25"E (Site III), closed to inhabited area.

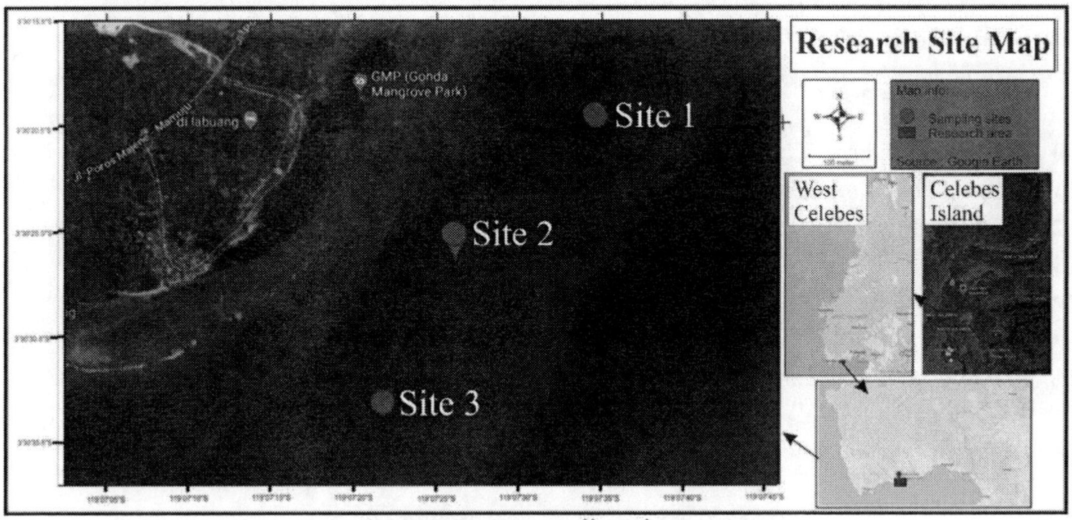

Figure 1. Sampling site

2.2. Sampling method.

Data on soft coral was collected by using visual census method in combination with Belt transect [6]. A 50 m belt transect was applied at three sites, in which observation on soft coral was done at area of 2.5 m at left and right side of the transect. All soft coral (solitary or colony) inside observation area was photograph, counted and sub samples was collected for further identification. Sample photograph was done by help underwater camera. Species identification are based on species characteristic according to Tropical Pacific Invertebrates [7],and Soft Corals and Sea Fans A Comprehensive Guide to the Tropical Shallow Water Genera of the Central-West Pacific, the Indian Ocean and the Red Sea [8].

3. Result and Discussion

3.1. Result

3.1.1. Species composition

The result of species identification indicates that there are sixteen species belongs to four families of soft coral present in Gonda waters (**Table 1**). As can be seen from **Table 1**, Family Alcyoniidae consist of 9 species; Family Nephtheidae (5 species); Family Tubiporidae (1 species) and Family Xeniidae (1 species). Some example of soft coral, most common ones, are shown in **Figures 2,3** and **4**.

Table 1. Soft coral species composition at different sites in Gonda coastal waters

No	Species	Site I (m)		Site II (m)		Site III (m)	
		3	10	3	10	3	10
	Alcyonidae						
1	*Cladiella sp.* Gray, 1869	√		√	√	√	

IOP Conf. Series: Earth and Environmental Science **279** (2019) 012053 doi:10.1088/1755-1315/279/1/012053

2	*Klyxum sp.* Alderslade, 2000				√		
3	*Lobophytum sp.* Marenzeller, 1886	√		√	√	√	√
4	*Lobophytum compactum* Tixier-Durivault, 1956	√			√		
5	*Lobophytum crassum* von Morenzeller, 1886				√		
6	*Sarcophyton crassocaule* Moser, 1919	√			√		√
7	*Sarcophyton elegans* Moser, 1919	√		√	√	√	√
8	*Sinularia sp.* May, 1898	√		√	√	√	
9	*Sinularia flexibilis* Quoy & Gaimard, 1833			√	√		√
	Nephtheidae						
10	*Paralemnalia sp.* Kükenthal, 1913				√		
11	*Paralemnalia thyrsoides* Ehrenberg, 1834	√			√	√	√
12	*Litophyton sp.* Forskål, 1775				√	√	√
13	*Scleronephthya sp.* Studer,1887		√		√		√
14	*Stereonephtya sp.* Kükenthal, 1905				√		
	Tubiporidae						
15	*Tubipora musica* Linneaus, 1758	√					
	Xeniidae						
16	*Heteroxenia sp.* Koelliker, 1874				√		

Figure 2. *Sinularia flexibilis* Quoy & Gaimard, 1833

Figure 3. *Paralemnalia sp.* Kükenthal, 1913

Figure 4.*Sarcophyton elegans* Moser, 1919

3.1.2. Species Diversity and distribution

The result of analysis on species diversity as indicated by Shannon-Wienner diversity index is given in **Table 2,** while soft coral distribution pattern is shown in **Table 3**. As described in **Table 2**, diversity index range from 0 (site I 10m) to 23.6 (site II 10m), this means that most of value reflect as moderate category except low category was found at site I at 10 m depth, recpectively.Value of Morisita dispersion indexranges from ~ to 0.52 (**Table 3**), indicating that soft coral in the study area are mostly random distributed.

Table 2. Shanon-Wienner Diversity Index

Site	Depth (m)	Diversity	Category
I	3	1,96	Moderate
	10	0	Low
II	3	1,30	Moderate
	10	2,36	Moderate
III	3	1,60	Moderate
	10	1,66	Moderate

Table 3. Value of Morisita Index

Site	Depth (m)	Morisita Index	Category
I	3	-0,06	Random
	10	~	-
II	3	0,52	Random
	10	-0,15	Random
III	3	0	Evenly
	10	0,03	Random

Data on environmental parameters is shown in **Table 4**. As seen the Table, each parameter are not significantly differ amongst sites except for current at site II (0.4±0,30 km/h), lower than at site I (1.03 ± 0.09 km/h) and site III (1.3 ± 0.25 km/h).

Tabel 4. Environmental parameters at different sites at Gonda coastal waters (mean ± sd)

Site	Temperature (°C)	Salinity (‰)	DO (mg/L)	Current (km/h)	Brightness (m)	Turbidity (%Transmitant)	pH
I	27 ± 1.00	26± 1.00	8.3 ± 0.25	1.03 ± 0.09	6.5 ± 0,29	89 ± 0.00	8.4 ± 0.06
II	27± 1.15	25± 1.3	9 ± 0.30	0.4 ± 0.30	6 ± 0,00	98 ± 0.01	8.4 ± 0.00
III	28± 1.00	26 ± 0.0	9.2 ± 0.06	1.3 ± 0.25	6.5 ± 0.32	98 ± 0.01	8.5 ± 0.20

3.2 Discussion

The result of present study indicates that number of soft coral are less compared to one reported from Biak waters that of 18 species, those belongs to 5 families [4]. However, number of species of soft coral found in Gonda coastal water is higher than one found by Goh [9] that recoded 8 species in Singapore waters.

The composition of species found in Gonda coastal waters are generally dominating by *Lobophytum sp* and *Sarcophyton elegans* (Figure 4), *counted for* 83.33% in total population. It is known that these two species can almost be found in all observed sites at different depths, except at site I at 10 m depth. This result is very reasonable considering that site I with a depth of 10 m has a substrate that is entirely sand. This shows that both species have a fairly good level of adaptation. On the other hand, the six species shows low species composition in which 16.67% is the lowest species composition value among other species.

The existence of soft coral in waters is strongly influenced by natural factors that exist in the region. Some soft coral species also have different substrates to grow, but majority of soft corals grow on hard substrates such as rocks or hard corals that have died. The depth level also affects soft coral growth. Depth is one of factors that influence the amount of light penetration that reaches the bottom substrate. In addition, the level of turbidity also affects the amount of light penetration that entering a body of water.

Different soft coral also has a tolerance limit for different environmental factors. For example as stated by Fabricus and Adersladde 2001 [8], vapor salinity that supports growth of soft coral range from 25°/₀₀ to 45°/₀₀. This level of tolerance can be seen from the distribution of soft coral growth.
Direct observations made in the field reveals that there was damage to coral reefs in several different places. At these three observation sites, we observed almost the same cause of coral reefs damage, which is due to the use of cyanide poison. This can be seen from the pattern of coral damage that follows the direction of sea water flows, with the lime frame still intact. In some observed places, we found rubble which became a marker of bombings or anchoring activities in the region. Damage that occurs at the site also occurs due to natural factors, such as waves that can break several types of coral. At some point, there are also dead coral that covered by sediments from land. This dead coral organism will gradually become a growing place for some soft coral species.

4. Conclusion

The present study reveals that 16 species of soft coral belongs to 4 families is recorded in Gonda waters, in which more species was found in 10 m compared to 3 m. Environmental parameters are in vapour to support soft coral in this area. In addition, diversity index shows species diversity is moderate and soft coral tends to randomly distributed in the area.

References

[1] World Registes of Marine Science 2018 *Alcyonacea* http://www.marinespecies.org/aphia.php?p=taxdetails&id=1365 Access on 2018-05-16.
[2] Soedharma D Kawaroe M Haris A 2005 *J Ilmu-Ilmu Perairan dan Perikanan Indonesia* **12** 2:121-128.
[3] Manuputty A E W 2010. *J Oseanologi and Limnologi in Indonesia*. **36**:211-225

[4] Manuputty A E W 2016 *J Oseanologi and Limnologi in Indonesia* **1**:47-59.
[5] Manuputty A E W 2012 Soft Coral. *Perairan Indonesia*. LIPI. Jakarta
[6] Brower J E, Zar JH and Ende C 1989 *Field and Laboratory Methods of General Ecology*. McGraw Hill.
[7] Collin P and Arneson C 1995 *Tropical Pacific Invertebrates*. Coral Reef Press. California
[8] Fabricius K and AldersladeP 2001 *Soft Corals and Sea Fans A Comprehensive Guide to the Tropical Shallow Water Genera of the Central-West Pacific, the Indian Ocean and the Red Sea*. Australian Institute of Marine Science. Townsville.
[9] Goh PLH, TanG E and Tan LT 2009 *Journal of Coastal Development*. **12**(2) 89-98

Acknowledgment
The Author would like to thank Mr. Philip Alderslade for helping in species identification. Thank you very much to Mudatsir Zainuddin, Hasnawi, Andy Nugraha and Bachtiar Anas for assisting data collection.

IOP Conf. Series: Earth and Environmental Science **279** (2019) 012054 doi:10.1088/1755-1315/279/1/012054

The dynamic system analysis of lemuru fishery in Bali strait by using biological parameter yield of some surplus production models.

Georgina Maria Tinungki, Nasrah Sirajang

Department of Mathematics, Faculty of Mathematics and Natural Science, Hasanuddin University, Jl Perintis Kemerdekaan KM 10, Tamalanrea, Makassar, Indonesia

Abstract. Stock assessment in any fisheries scientific study is an effort to determine the productivity of fishery resource, the effect of catch on resources and changes in catch patterns, for example due to implementation of management and development policies. It involves 3 (three) main aspects: the first, determine total biomass by using surplus production model; the second, determine recruitment (the number of fish that has reached the age during one recruitment season) can use Ricker model; the third, determine the equation of growth in the fishery can use Von Bertalanffy model. Mathematical models are mostly used to describe the dynamics of fish populations. In the use of appropriate models in stock assessment will result in a more appropriate basis in selecting fishing methods and advanced analysis of catches, hence required a plan in the sustainable management and utilization of fish resources. The study of sustainable dynamic of lemuru fisheries management in Bali strait is the dynamics of biomass of lemuru fishery by using the biological parameters yield of some surplus production models. By obtain the biological parameter values r, q and K, we obtain the dynamic trajectory between biomass and time, indicates that at the first year of observation, the biomass level is relatively high, but when several years later the biomass tends to decrease until it reaches the stable of biomass obtained for about 30 years and so on ($t > 30$).

1. Introduction

The main objective of this study was to analyze the dynamics of lemuru fishery biomass in Bali strait by using the biological parameter yield of some surplus production models [1]. Generally, the Surplus Production Model is used in fish stock assessments because this model group can be estimated only by using catch data and available efforts in common [2]. It ignores the complicated biological processes in a fish stock by assuming that the stock can be treated as aggregate biomass [3]. When all other factors remain constant, aggregate biomass from a fish stock will decrease as pressure is made on the resource by increases catch effort [4]. The surplus production model is a popular model in fisheries literature and has been used for over forty years [5]. This is due to not only is the model relatively simple to calculate, but it also requires only catch data and catch efforts time series as it available in most fishing centers [6].

In fact, according to [7], as the Graham-Schaefer model that not all biomass follows logistical growth. In this matter, the decrease of catch per unit effort ($CPUE_t$) yield against catch effort (E_t) follows the linear regression pattern, and the relationship between the catch yield and the symmetric parabolic-shaped biomass with the peak point (maximum) at the biomass level of $K/2$ [8]. So, the Fox Model [9] has several different characteristics from the Graham-Schaefer model that the growth of biomass follows the Gomperz growth model and the decrease of *CPUE* against the catch effort [10].

[11] argues that the surplus production model as constructed by Graham (1935) and completed by Schaefer (1954) called as the Graham-Schaefer Method (1954) or logistic production models play an important role in the dynamics of fishery population. This model still has a disadvantage in which a model with only two parameters cannot predict three biological parameters so that there are several other surplus production models that can predict these three parameters [12]. [13] argues that *CPUE* is not only determined by the current catch effort (as in the Graham-Schaefer model) but also by the previous catch effort. Gulland assumes that there is a relationship between abundance and previous efforts, when recruitment and natural mortality is constant.

[14] completely abandoned the special properties possessed by the Gaham-Schaefer model, that by inserting a variable m into the Graham-Schaefer model, it would mean that *MSY* can be generated from various biomass sizes varying from 0 to K, where $X_{opt} = K/2$. [1] model is known as a different model because the use of a simple differential equation from the Schaefer model. The difference between the Walters and Hilborn models and the Schaefer is the Walters and Hilborn models using discrete versions of the biological model. In fact, this model can also show what is known in mathematical form as a *chaotic* behavior for a large value of r [15]. Moreover, the behavior of model is as true as the Schaefer model, especially for the low rates of r, q and E [16].

[15] provides a dynamic time model ("*discrete in time*"), and stochastic in which it contrary to the Graham-Schaefer model that is static time and deterministic and continuous. Thus, [15] reconstituted the surplus production model in the form of dynamic, time-specific and stochastic models, which made it superior to the traditional Graham-Schaefer model [17]. The Clarke Yoshimoto Pooly (CYP) (1992) model estimates the biological parameters of r (natural growth rate), q (catch capability coefficient), and K (environmental carrying capacity). Reliable and realistic model that can describe *Maximum Carrying Capacity* (MCC), and *Maximum Sustainable Yield* (MSY) and other variables related only Cushing's model, where CPUE is obtained from CPUE of sustainable surplus production model [12].

2. Materials and Methods

This study is a desk study using primary and secondary data. Primary data were obtained from direct interviews with local fisheries,in Bali on May 2013. While, secondary data were obtained from Central Bureau of Statistics, and other related institutions. The method used is dynamic system approach, which is able to analyze a system dynamically and change according to time.

2.1. The Aplication of Fishery Population Dynamic Equation

Following [11] in analyzing dynamic systems is not necessarily quantitative analysis because by conducting qualitative analysis, the nature of dynamic equations and its stable can also be studied with graphic assistance that will be very helpful in analyzing the dynamic model qualitatively.

For example, the dynamics function of the fish resources stock is often written as a model called the *lumped* parameter model in which fish stock (X) is often written as:

$$\frac{dX}{dt} = \dot{X} = F(X) \tag{1}$$

The most common example of this model is logistic model or Verhulst, where the above model is written as:

$$\dot{X} = rX\left(1 - \frac{X}{K}\right) \tag{2}$$

Where :

r = growth rate (net) and

K = carrying capacity

The above equation can be described as follows:

As figure 1, it appears that stock (X) will grow ($X > 0$)

- On interval $0 < X < K$. This is indicated by the arrow to the downward right of the $F(X)$ curve.

- On the contrary, in region where $X > K$, function X will decrease so that the stock will also decrease. This is indicated by the arrow of upward left of the $F(X)$ curve.
- Point $X = K$ is called the equilibrium point. In this case the function $F(X)$ will also be equilibrium at $X = 0$. From these two equilibrium points, only at $X = K$ the equilibrium will be stable (indicated by the opposite arrow counter) [11].

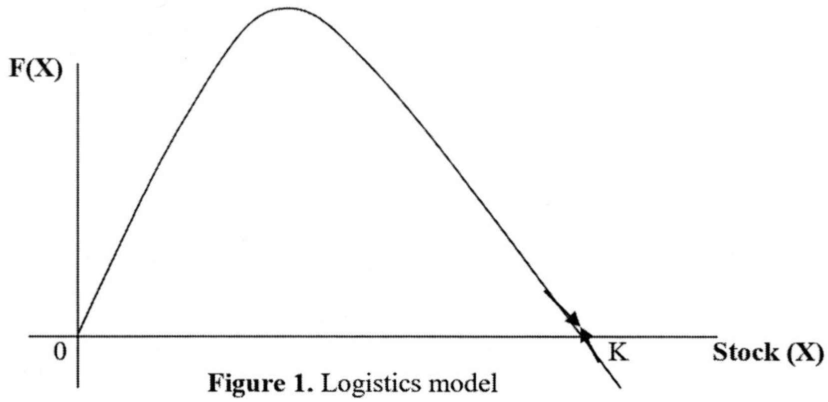

Figure 1. Logistics model

An example of a logistic curve with the equation : $\frac{dX}{dt} = aX - bX^2 - y$ can be depicted in Figure 1

In figure 1, the logistic curve with the equation : $\frac{dX}{dt} = aX - bX^2 - y$,

where :

$$0 < X < X_1 \quad : f(X) < y \quad \rightarrow \quad dX/dt \downarrow \;\Rightarrow\; X \uparrow$$

$$X_1 < X < X_2 \quad : f(X) > y \quad \rightarrow \quad dX/dt \uparrow \;\Rightarrow\; X \uparrow$$

$$X_1 > X_2 \quad : f(X) > y \quad \rightarrow \quad dX/dt \downarrow \;\Rightarrow\; X \downarrow \qquad (3)$$

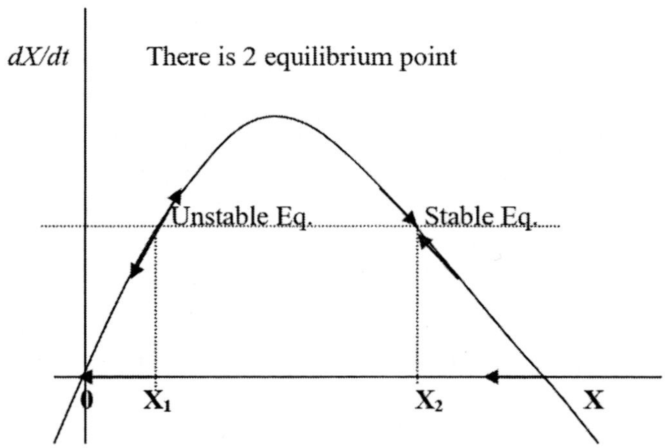

Figure 2. Logistic curves with there are two equilibrium points

2.2. Surplus Production Model

The surplus production model is a method that can be used in estimates fish stock, by using catch yield data, which require simple data containing one variable that is Y_t in ton/year as independent variable,

IOP Conf. Series: Earth and Environmental Science **279** (2019) 012054 doi:10.1088/1755-1315/279/1/012054

and E_t in days/year as dependent variable. In addition, there are three parameters of the natural growth rate r, the carrying capacity K and the catch capability coefficient q.

Also, this model is used in estimating the size of population, based on the large of catch yield for a particular catch effort in water. It can depict the presence of fish stocks in the past and can predict future yield based on catch per unit effort ($CPUE_t$) data. In determining the optimum level of effort, efforts that can produce a maximum sustainable yield catch without affecting long-term stock productivity. The surplus production model to be examined in this case is an overall stock, total effort and total catch yield obtained from the stock, without including details of such things as growth and mortality parameters or the effect of mesh size on aged fish captured other.

The surplus production model allows for an analysis where little information, especially on yields, stock abundances, and the amount of fishing available. Unfortunately, these models cannot be adapted to detailed biological information about the fish in question, or provide detailed advice on fishing patterns. The estimation of optimum catch effort (E_{opt}) and maximum sustainable catch (MSY) is approximated by the Surplus Production Model, as it is known that between catch per unit effort ($CPUE_t$) and effort can be either linear or exponential (Gulland, 1983).

The Surplus Production Model consists of two basic models: the Schaefer Model (linear relationship) and the Gompertz Model as developed by Fox (exponential relationship) (Gulland, 1971). With parameters and variables as defined previously, the procedures for estimating the parameters as described by Schaefer, Fox, Gulland, Pella & Tomlinson, Walter & Hilborn, Schnute and Clarke, Yoshimoto and Pooley are as follows :

Schaefer : $U_t = a - bE_t$

Where $a = qK$ and $b = q^2K/r$

Fox : $\ln U_t = a - bE_t$

Where $a = \ln(qK)$ and $b = q/r$

Gulland : $U_t = a + b\overline{E}_t$

Where $a = qK$ and $b = q^2Kr$

Pella & Tomlinson: $U_t = a - bE^{m-1}$,

Where $a = qK$, $b = q^mK / r^{m-1}$

Walter & Hilborn : $\frac{U_{t+1}}{U_t} - 1 = r - \frac{r}{Kq}U_t - qE_t$

Where $a = r$ and $b = r/qK$

Schnute : $\ln (U_{t+1}/U_t) = a + b[(U_t+U_{t+1})/2] - c[(E_t+E_{t+1})/2]$

Where $a = r$, $b = rqK$ and $c = q$

CYP : $\ln(U_{t+1}) = a \ln(qK) + b \ln(U_t) - c(E_t + E_{t+1})$

Where $a = 2r/(2+r)$, $b = (2+r)$ and $c = q/(2+r)$

Cushing : $\ln (S_{t+1}+Y_t) = \ln a + b \ln S_t$;

Where $a > 1$ and $0 < b < 1$ (4)

3. Results and Discussions

The Dynamic System Analysis of Lemuru Fishery in Bali Strait

Fishery resources are dynamic natural resources as explored in the preceding section, resulting in the dynamics of biomass from lemuru fisheries in Bali strait by using the biological parameters yield of some surplus production models as follows:

1. Schaefer model

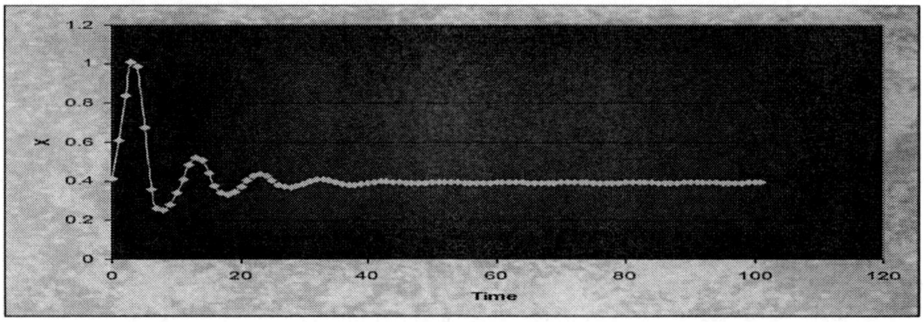

Figure 3. Dynamic trajectory of fish biomass (vertical axis, X_t) (100.000 tons) of lemuru in Bali strait, Year (horizontal axis/time), Schaefer model 2013-2017.

In figure 3, shows equilibrium after almost 30 years, at a production rate less than 40.000 tons, it is obtained by entering the following initial values:

$$X_0 = \frac{h}{qE} = 481.762{,}047 \text{ ton} \quad \rightarrow \text{initial value}$$

$$X_t = (1+r)X_0 - rX_0^{\,2} \quad \rightarrow \text{biomass at time } t$$

2. Fox model

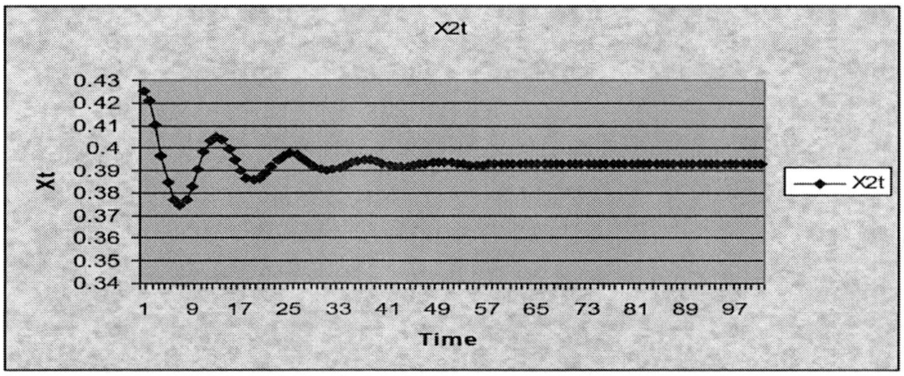

Figure 4. Dynamic trajectory of fish biomass (vertical axis, X_t) (100.000 tons) of lemuru in Bali strait, Year (horizontal axis/time), Fox model 2013-2017.

In figure 4, shows equilibrium after almost 40 years, at a production rate less than 40.000 tons, it is obtained by entering the following initial values:

$$X_0 = \frac{h}{qE} = 31.349{,}894 \text{ ton} \quad \rightarrow \text{initial value}$$

$$X_t = (1+r)X_0 - rX_0^{\,2} \quad \rightarrow \text{biomass at time } t$$

IOP Conf. Series: Earth and Environmental Science **279** (2019) 012054 doi:10.1088/1755-1315/279/1/012054

3. Schnute model

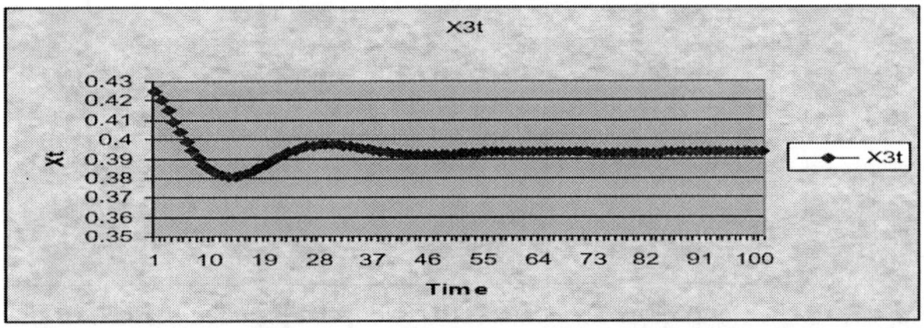

Figure 5 Dynamic trajectory of fish biomass (vertical axis, X_t) (100.000 tons) of lemuru in Bali strait, Year (horizontal axis/time), Schunte model 2013-2017.

In figure 5, shows equilibrium after almost 40 years, at a production rate less than 39.000 tons, it is obtained by entering the following initial values:

$$X_0 = \frac{h}{qE} = 201.415,62 \text{ ton} \quad \rightarrow \text{ initial value}$$

$$X_t = (1+r)X_0 - rX_0^{\ 2} \quad \rightarrow \text{ biomass at time } t$$

4. Walter-Hilborn Model

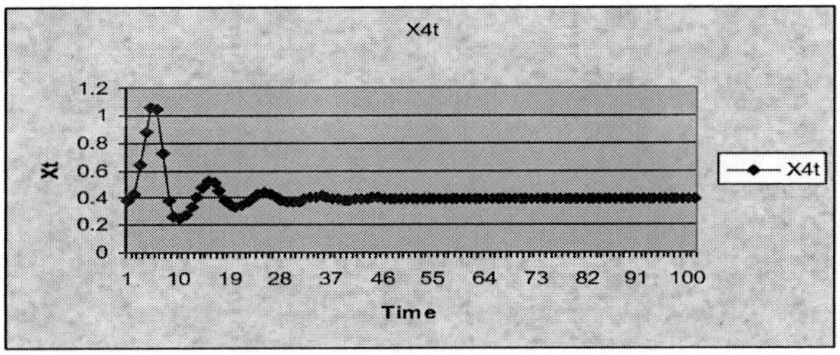

Figure 6. Dynamic trajectory of fish biomass (vertical axis, X_t) (100.000 tons) of lemuru in Bali strait, Year (horizontal axis/time), Walter-Hilborn model 2013-2017

In figure 6, shows equilibrium after almost 30 years, at a production rate less than 40.000 tons, it is obtained by entering the following initial values:

$$X_0 = \frac{h}{qE} = 797.676,16 \text{ ton} \quad \rightarrow \text{ initial value}$$

$$X_t = (1+r)X_0 - rX_0^{\ 2} \quad \rightarrow \text{ biomass at time } t$$

5. Clarke Yoshimoto Pooly Model

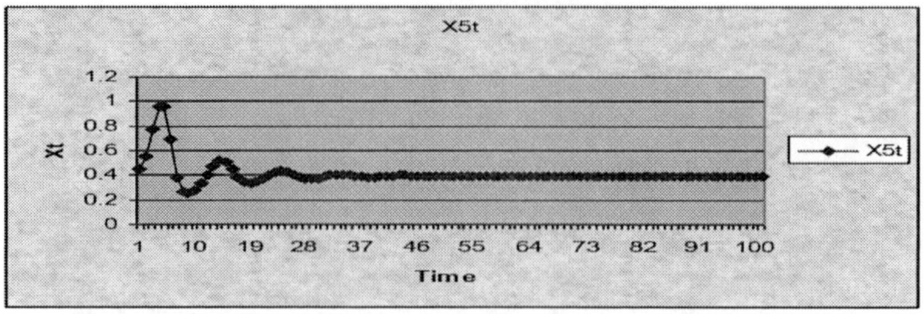

Figure 7. Dynamic trajectory of fish biomass (vertical axis, X_t) (100.000 tons) of lemuru in Bali strait, Clarke Yoshimoto Pooly model 2013-2017.

In figure 7, shows equilibrium after almost 30 years, at a production rate less than 40.000 tons, it is obtained by entering the following initial values:

$$X_0 = \frac{h}{qE} = 76.355,682 \text{ ton} \quad \rightarrow \text{ initial value}$$

$$X_t = (1+r)X_0 - rX_0^{\,2} \quad \rightarrow \text{ biomass at time } t$$

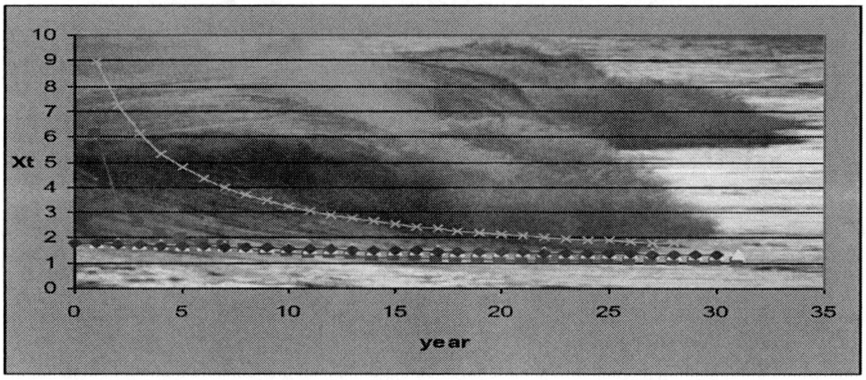

Figure 8. The Graph of dynamic trajectory of Lemuru fish stock in Bali strait, 2013-2017.

The results show that the most statistically appropriate model for estimating biological parameters in Bali strait are the Fox and CYP models. This is in line with previous studies related to fish resources in tropical regions such as Indonesia, the use of Fox and CYP models is considered the most appropriate compared to using other models such as Walters Hilborn and Schnute [3].

4. Conclusion and Recomandation

4.1. Conclusion

In this study, By obtaining the values of biological parameters r, q and K, then the dynamic trajectories between biomass and time are shown in Fig. 9 which is, that at the beginning of the first year of observation, the biomass of 105..000 tons is relatively high, but when several years later biomass appears to decrease until then reach a stable biomass obtained about 30 years and so on (t > 30) at a biomass level of about 27.000 tons/year, can be seen at figure below which shows equilibrium after almost 30 years.

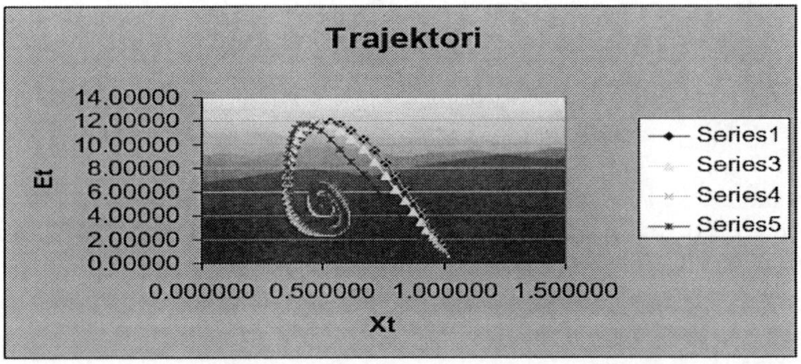

Figure 9. The relationship of fish stock (X_t (100.000 tons)) with catch effort (E_t) of Lemuru fishery in Bali strait, 2013-2017.

In figure 9, it shows that if catch effort is increased, then the trajectory toward balance will be faster (from blue to red). Appear stable focus on the fish stock level of 50.000 tons, while the catch effort is more than 40.000 days.

4.2. Recommendation

This study still needs further research focusing on the the biological parameter dynamics of any fish species present in these for sustainable management. In addition, there should also be an optimal measurement of input and output for each species of fish in the Bali strait , so that the management can be directly conducted on each species of fish to obtain the optimal and sustainable conditions.

REFFERENCES

[1] Wudianto 2001 *Analisis Sebaran dan Kelimpahan Ikan Lemuru di Perairan Selat Bali: Kaitannya Dengan Optimasi Penangkapan*. Dissertation. Program Pascasarjana-IPB. Bogor. [Indonesian]

[2] Saha A K, Setegn S G 2015 *Ecohydrology for sustainability of IWRM: A tropical/subtropical perspective. In: Sustainability of Integrated Water Resources Management* Springer, Switerzerland.

[3] Tinungki G M 2005 *Evaluation of Production Model in Expecting Maximum Sustainable Yield to Support Lemuru Fisheries Management Policy in Bali Strait* Dissertation. Institut Pertanian Bogor, Bogor. [Indonesian].

[4] Anna Z 2017 *BIODIVERSITAS journal* **18** (4) October 2017 p. 1468-1474.

[5] Ghofar A and Mathews C P 2000 The Significance of Environmental Information in The Management of Oil Sardine Fishery in the Bali Strait. *Workshop on Inter-annual Climate Variability and Pelagic Fisheries* Noumea, New Caledonia.

[6] Asep Sholahuddin 2015 *The Application of ANN-Linear Perceptron in the Development of DSS for a Fishery Industry* Journal Elsevier, Available online at www.sciencedirect.com.

[7] Widodo 1986-1987 *Models of Population Dynamics for Fisheries*. A Serial Seminars Published by Oseana. Jakarta.

[8] King R S Tsai C M 2017 *Sustainability* **9** (8) p.1387.

[9] Fox W W 1970 *Trans. Am. Fish. Soc.*, **99** (1) 80-88.

[10] Mace P M 2001 *A New Role for MSY in Single Species and Ecosystem Approaches to Fisheries Stock Assessment and Management*. National Marine Fisheries Service, Woods Hole, MA, USA.

[11] Fauzi A 2001 *An Econometric Analysis of the Surplus Production Function: An Application for Indonesian small pelagic fishery*. IPB Darmaga Bogor.

[12] Tinungki G M, Boer M, Monintja D R, Widodo J, Fauzi A 2004 *Jurnal Ilmu-ilmu Perairan dan Perikanan Indonesia* **11** (2) 135-138. [Indonesian]

[13] Walters C and Hilborn R 1976 *Journal Fish Resource. Board Can* **33** 145-159.

[14] Anderson L G, Seijo J C 2010 *Bioeconomics of Fisheries Management* John Willey and Sons, New York, USA.
[15] Schnute J 1977 *J. Fish. Res. Board Can* **34** 583-663.
[16] Anurani P, Andreas L, Wendel K, Chris F, Peter D, John G, Tom J 2015 *Polar Biol* **38** 651-664.
[17] Anna Z 2016 *Marine and Fisheries Socio-Economi Journal* **11**(2) 161-172. [Indonesian].

IOP Conf. Series: Earth and Environmental Science **279** (2019) 012055 doi:10.1088/1755-1315/279/1/012055

ENSO impacts during west and east monsoon to the ocean current and water mass characteristics in the Makassar Strait

F Jamil[1], Sakka[1] and D Berlianty[2]

[1] Geophysics Study Program, Hasanuddin University, Jalan Perintis Kemerdekaan KM 10, Tamalanrea, Sulawesi Selatan 90245

[2] Institute for Marine Research and Observation, Ministry of Marine Affairs and Fisheries, Jalan Baru Perancak, Negara, Jembrana, Bali 82251

fatmajamilp20@gmail.com

Abstract. Ocean and atmospheric interactions are known by the presence of El-Nino Southern Oscillation (ENSO) in the Pacific Ocean, dipole mode in the Indian Ocean, and Indonesian-Throughflow (ITF). Those interactions have impacts to oceanographic conditions of Indonesian waters, especially Makassar Strait as main route of ITF which carries warm water masses to Indian Ocean due to the El-Nino and La-Nina as part of global scale oceanic belt. The purpose of this research is to understand the ENSO effect on spatio-temporal of current patterns and characteristics of water mass, namely temperature and salinity during west and east monsoon in the Makassar Strait. This study are using the ENSO index data from NOAA between year 2000-2015 and ¼ degrees resolution of temperature, salinity, and currents monthly model from Copernicus Marine Environment and Monitoring System (CMEMS) with focus in February and August 2010 and 2015 to represents respectively for west and east monsoon period during the ENSO year. The results showed that surface currents were stronger during La-Nina with speeds of 0.8 to 1.2 m/s compared to El-Nino event with speeds of 0.6-0.7 m/s and it is higher in east monsoon. During west monsoon the current leads to the east and reversely the current pattern changes to the west during east monsoon. The temperature and salinity when La-Nina events are about 1°C higher and 0.7 psu lower than El-Nino events. Based on the results, in general that the impacts of El-Nino an La-Nina events to the Makassar Strait are clearly shown during east monsoon.
Keywords: ENSO, monsoon, oceanographic conditions, CMEMS, Makassar Strait

1. Introduction

The interaction of ocean and atmospheric can be seen from the Indonesia's climate interaction with ENSO phenomenon in the Pacific Ocean, dipole mode in the Indian Ocean, and the Indonesia throughflow. The dynamics of the ocean and atmosphere in the relation with these large-scale phenomena begin with the formation of potential clouds, cyclones even El Niño Southern Oscillation (ENSO) and Indonesia throughflow (ITF) [1]. The condition of Indonesian Oceanography are mostly affected by monsoon winds and Indonesia throughflow. Monsoon winds, which furthermore cause the monsoon flow caused by different pressure between Asia Tenggara and Australia. West monsoon there are in December-February, East monsoon in June-August, and between of them called transition. Indonesia throughflow is current of warm water mass from Pacific Ocean to Indian Ocean through Indonesian's waters that caused by the different of sea surface height between of these ocean [2]. Characteristics of an ocean can be known both through physical properties and the chemical properties of water such as temperature, salinity, conductivity, dissolved oxygen, and nutrient content. Among these variables, temperature and salinity have an important role to reflecting the sea water mass conditions [3].

Based on that cases, the East Indonesian waters conditions is very interesting to research because the East Indonesian waters are located at the intersection of the water mass movement that come through the Indonesian Monsoonflow and the Indonesian Throughflow. The Makassar Strait is the main traffic lane through of Arlindo, so in this research just limited to the Makassar Strait area to know the effect of strong ENSO that occur during 2010 (La-Nina) and 2015 (El-Nino) to the ocean current patterns and water mass characteristics temperature and salinity.

1.1. El Niño Southern Oscillation (ENSO)

ENSO is a phenomenon of atmospheric oceanic interactions that centered in the Pacific Ocean equatorial [4]. The oscillations of this phenomenon carry two phenomena namely El Niño and La Niña, are warm and cold phases of the Pacific equatorial region. Normally there is a warm pool in north of the Papua island which is a gathering place for surface currents from global oceanic belt. This warm pool also is a circulation place of Walker where there is an air lift (convection center). However, during El Niño, there is a movement warm pool area to the east of the Pacific equator and leaving the subsidence area (contrast to the convection area) in north of Irian. The warm pool area to be cools and with Indonesian throughflow brings the cold currents to the Indonesian waters which also causes drought due to obstruction of evaporation due to the cold of the sea. The contrary indication when La Niña period, occurs more warming in the warm pool area. The accumulation of water masses in the warm pool area causes Indonesian Throughflow are strength [1].

The accumulation in warm pool causes sea surface height in the warm pool area rather than Indian Ocean. The warm accumulation on surface are brings implications for the warm accumulation to the subsurface which thickens the thermocline layer in the warm pool area. When El Niño, occurs the warm water mass movement to the east on the surface and below, so there is a change in the thinning thermocline layer in the western Pacific Ocean and the warm water mass flowing to east. The main cause of the warm water mass flow to the east is currently unknown but the possibility is due to differences in temperature and energy gradients due to energy accumulation in the warm pool area compared to the central Pacific [1].

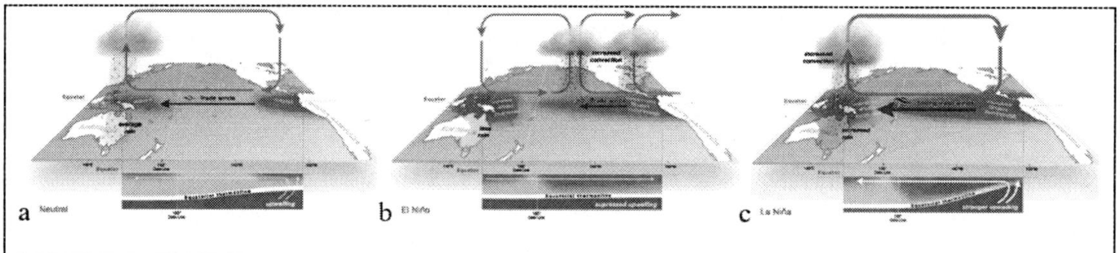

Figure 1. Global climate phenomenon (a) normally, (b) El-Nino, (c) La-Nina (*Source:* Bureau of Meteorology)

1.2. Oceanic Nino Index (ONI)

ONI is the main parameter used by NOAA to monitor, assess and predict ENSO (El Nino and La Nina), a phenomenon of climate anomalies in a global scale. This ONI is based on the changes of SST value from the average Nino 3.4 area. Taken an average of three months. El Niño is marked with a positive ONI greater than or equal to + 0.5°C while La Niña is marked with a negative ONI less than or equal to -0.5°C. ONI graph National Oceanic and Atmospheric Administration (NOAA) shown that from 2000 until 2015 have been occurs many times El Nino and La Nina phenomenon, but very strong occurs on 2010 and 2015. La Nina 2010 phenomenon are known with the NINO 3.4 SST Index value highest -1.6°C. While the El Nino 2015 very strong the highest NINO 3.4 SST Index is +2.6°C.

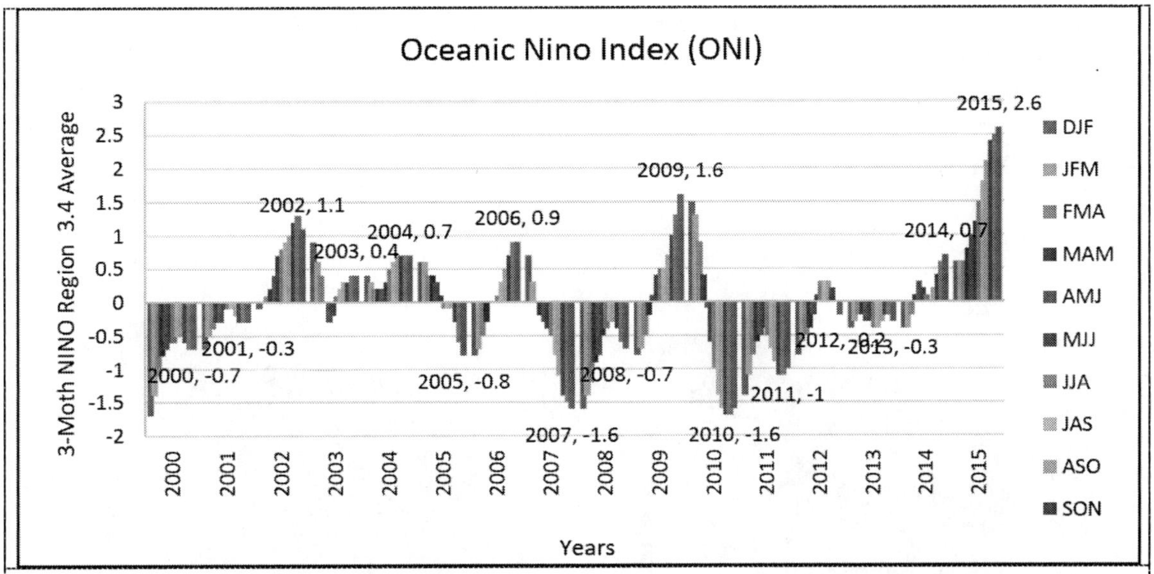

Figure 2. Graph of Oceanic Nino Index (ONI), 2000-2015 (*Source:* NOAA Climate Prediction Center).

2. Methods

The research area is Makassar Strait. The data used in this study include the ONI (Oceanic Nino Index) data 2000-2015 from NOAA http://origin.cpc.ncep.noaa.gov between year 2000-2015 to see the El Nino and La Nina occurrence periods. Temperature, salinity, and currents monthly model ¼ degrees spatial resolution from Copernicus Marine Environment and Monitoring System (CMEMS) http://marine.copernicus.eu. with focus in February and August 2010 and 2015 to represents respectively for west and east monsoon period during the ENSO year. The data thus processing to be spatial and temporal distribution and next to analyzing the conducted between the distribution of temperature, salinity and current patterns when the El Nino and La Nina period.

3. Results and discussions

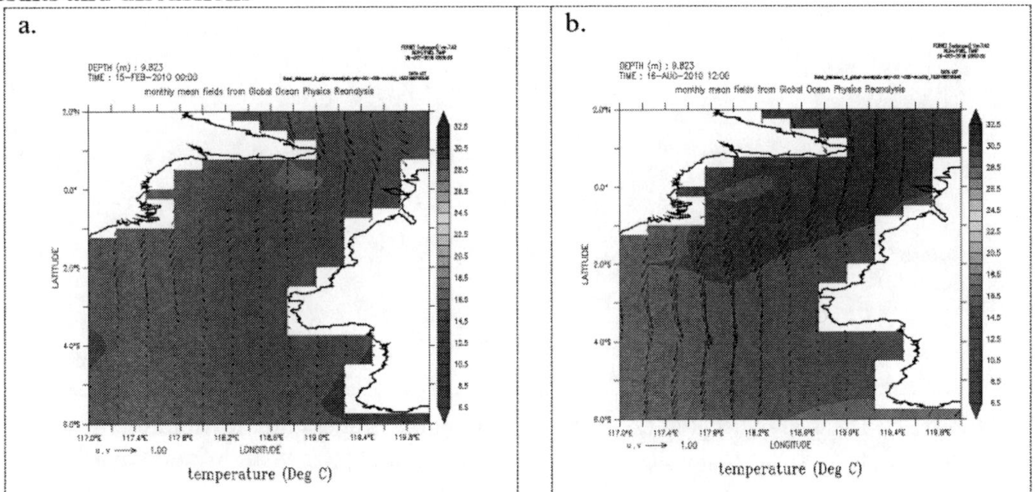

Figure 3. Temperature spatial distribution in February (a), August (b) at a depth of 9,823 m (Mixed Layer) of the Makassar Strait during La Nina 2010 period.

Temperature distribution in February is relatively homogeneous throughout the waters area with a temperature of around 30°C. While in August in east monsoon, the temperature distribution is different 2°C where in the north of waters which start from latitude 2°S-2°N have a greater temperature is 30°C and at south until west part of waters which connection with Java Sea is from 0°-6°LS have a temperature around 28°C.

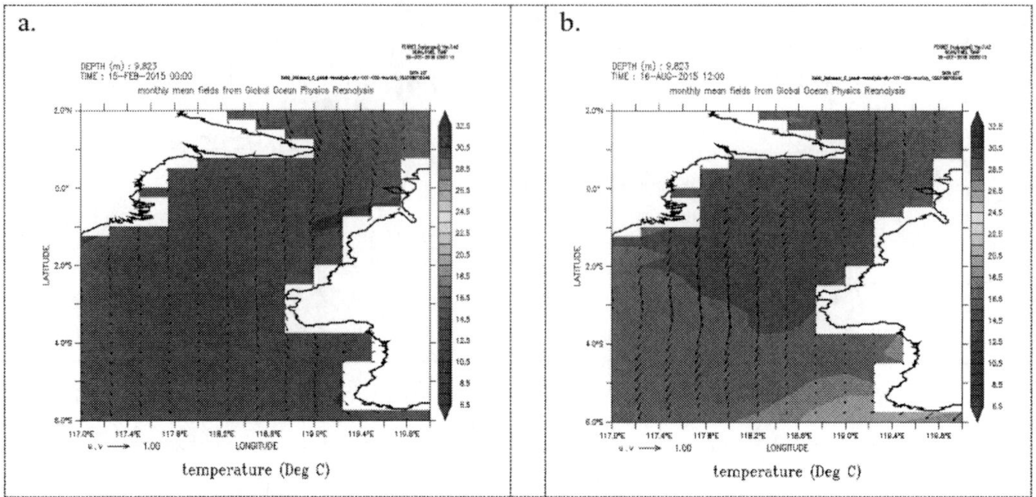

Figure 4. Temperature spatial distribution in February (a), August (b) at a depth of 9,823 m (Mixed Layer) of the Makassar Strait during El Nino 2015 period.

In the El Nino 2015 period, the temperature distribution along the waters there is no significant change from the La Nina 2010 period is in February it has a temperature of 29°C as well as in August that decreased to 28°C in the south part of the strait, just only at the El Nino period, the temperature in the south-east is at longitude 118.5°-120°6-6.5°LS has a cooler temperature until 24°C due the influence of upwelling coming from east waters.

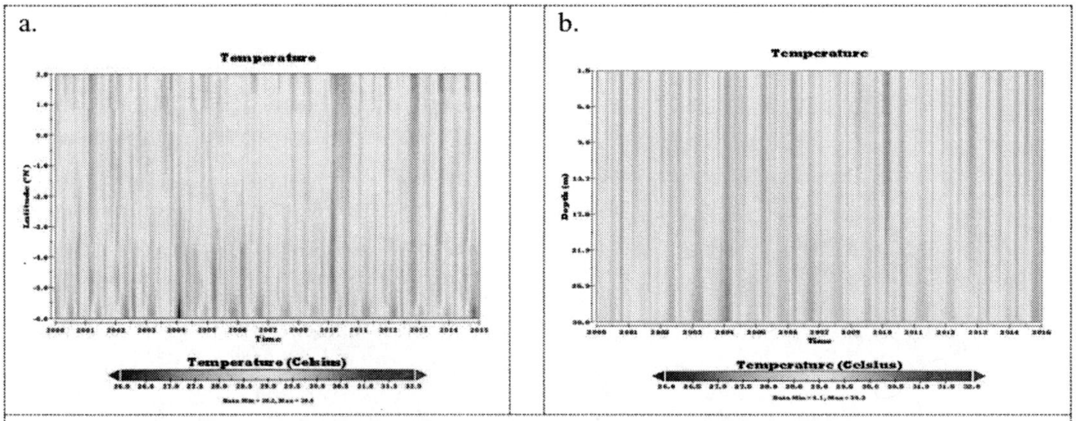

Figure 5. The average spatial (latitude) and temporal temperature distribution (a) and temperature vertical spatial distribution (b) at a depth of 9,8225 m (Mixed Layer) of Makassaar Strait.

The total average spatial distribution of temperatures along latitude 2°LU-6°LS from 2000 to 2015 on dominant shows two different temperature distributions in each year but there was a differences significant temperature anomaly in 2010 with 2015 where throughout 2010 in the figure has a dominant color dark orange which shows a high temperature of around 31°C than in 2015 with a dominant blue color that shows a low temperature of 31°C. Based on the Ocean Nino Index (ONI) data in 2015 there was a strong El-Nino, causing sea level elevation in the western Pacific Ocean to fall compared to the Indian Ocean, so Arlindo will tend to weaken. According to Gordon (2005) that the variability of sea surface temperature in Indonesian waters, especially in the eastern part, is expected to follow Arlindo's fluctuations, when El-Nino sea surface temperature will drop while in the event of sea surface temperatures will rise.

Vertically, the surface temperature distribution is seen to a depth of 28 m. at a surface of 1.5 m to a depth of 28 m in 2010 has a thicker layer (mixed layer) which is still at a depth of 28 m, while in 2015 when the occurrence of El Nino showed no or thinning homogeneous layer that was not visible from the surface. The same thing with the statement (Kunarso, 2010) which says that when La Nina is strong, the upper boundary of the thermocline or the lower layer of the mixed layer decreases so that the homogeneous layer is mixed layer thicker. Whereas at the time of El Nino occurrence the lower limit of the mixed layer layer is thinning. This is due to the occurrence of La Nina when there was a buildup of water masses in the western Pacific and Indonesian waters so that it pressed down the upper limit of the thermocline than when the El Nino occurred.

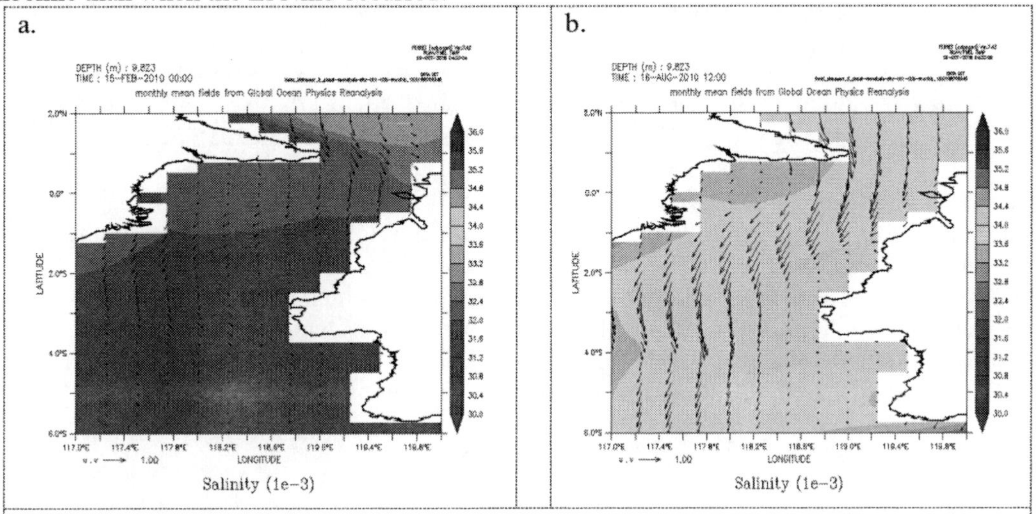

Figure 6. Salinity spatial distribution in February (a), August (b) at a depth of 9,823 m (Mixed Layer) of the Makassar Strait waters in the La Nina 2010 period.

Salinity distribution in this La Nina period showed that the average salinity in the Makassar straits in February was lower than August, which ranged from 30.0 to 33.2 psu in February, and 32.5 to 34.4 psu in August. Salinity variations also occur between the north and south parts, especially in February where the southern part of the Makassar straits has a lower salinity than the northern part. Water mass that comes from pacific which go into Makassar strait in north part has high salinity is 33,2 psu that start decrease until 30,0 psu after the water mass bent through headland. In august, variation of salinity shown in west and north where the north part waters have salinity 34,4 psu and the west part waters is 32,5 psu. Salinity distribution in this month is visible following the flow pattern of water mass that through Makassar straits.

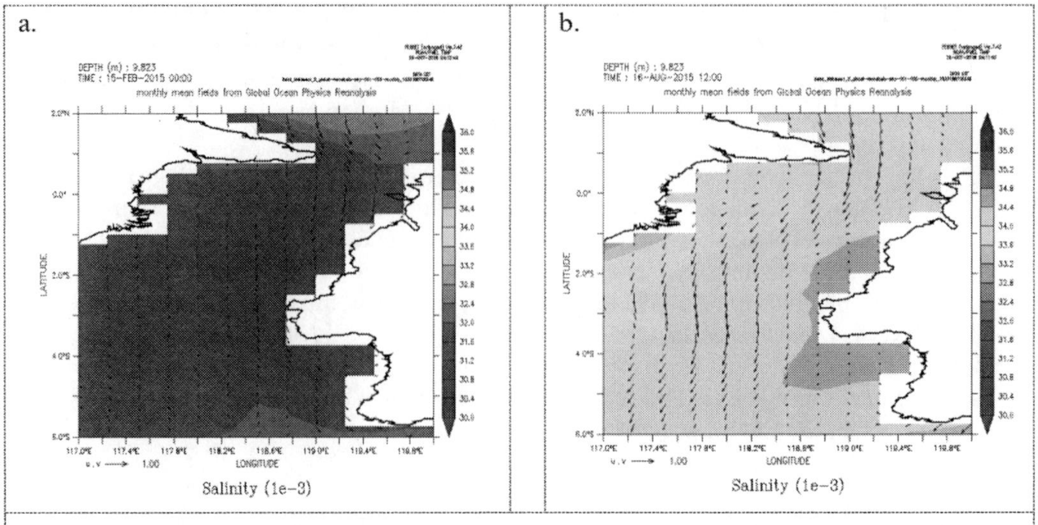

Figure 7. Salinity spatial distribution in February (a), August (b) at a depth of 9,823 m (Mixed Layer) of the Makassar Strait waters in the El-Nino 2015 period.

The spatial distribution of Makassar straits salinity in the El-Nino 2015 event is generally the same as the distribution of salinity in the 2010 La-Nina event which is lower in February and higher in August. However, the incidence of El-Nino which was characterized by a long drought in Indonesia, so that the greater evaporation affected the salinity of the Makassar strait, which ranged from 30.2 to 32.7 psu in February and 32.9 to 34.6 psu in August. In the February, the waters with lower salinity shift to the center at 5 ° LS to 0 ° at 30.2 psu. The highest salinity is in the east to south, which is 34.6 psu and low salinity is in the northern part, starting from the entry of the water mass to the west with a value of about 30.2 psu.

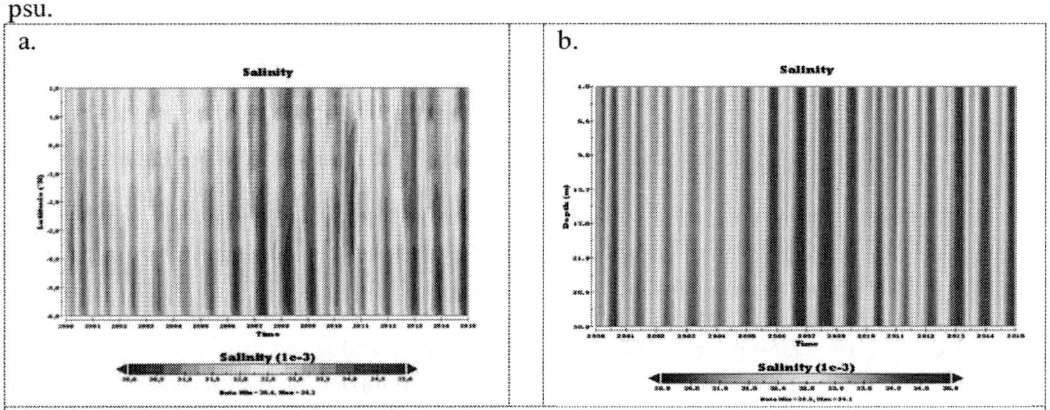

Figure 8. Average spatial total distribution of salinity (latitude) and temporal salinity (a) and salinity vertical spatial distribution, (b) at a depth of 9.8225 m (Mixed Layer) in the Makassar Straits.

The average spatial total distribution of salinity along latitude 2° N-6° S from 2000 to 2015 shows the almost same salinity every year but in 2010 up to early 2011 there was a significant decrease of salinity where in the image has dark blue has low salinity that reaches 30.0 psu. While in 2015 showed high salinity which is around 34.6 psu.

The vertical distribution of salinity that seen at a surface of 1.5 m to 28 m in 2010 is inversely proportional with temperature distribution so that the salinity of the mixed layer is thinner in 2010 when La Nina period than in 2015 when El Nino period.

IOP Conf. Series: Earth and Environmental Science **279** (2019) 012055 doi:10.1088/1755-1315/279/1/012055

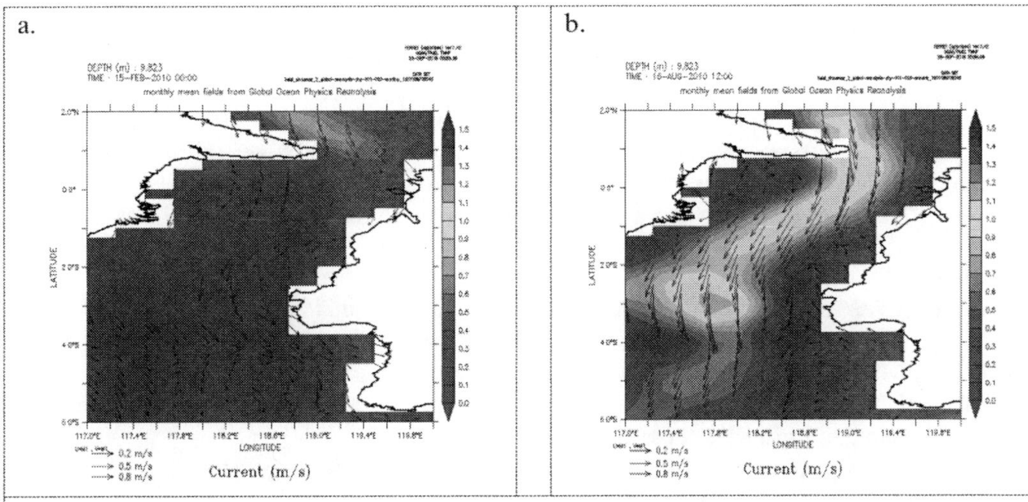

Figure 9. Ocean Current Speeds and Direction in February (a), August (b) at a depth of 9,823 m (Mixed Layer) of the Makassar Strait of La Nina 2010 period

The circulation pattern of surface currents in the February shows the current pattern that through the Makassar Strait to the south and turns to east (Flores Sea). In the north part, the incoming water mass collides with the topography of the sea floor from the headland so that the current turns towards to the west by strengthening the current on the east side as it passes through the headland. While in August the current circulation of the south part of the Makassar straits turns to west (Java Sea) according to the direction of the monsoon wind. In 2010, there was an intensification of high currents until August in the middle of the strait with an average speed around 1 m/s than the current in February that representing the east monsoon which only around 0.1-0.3 m/s.

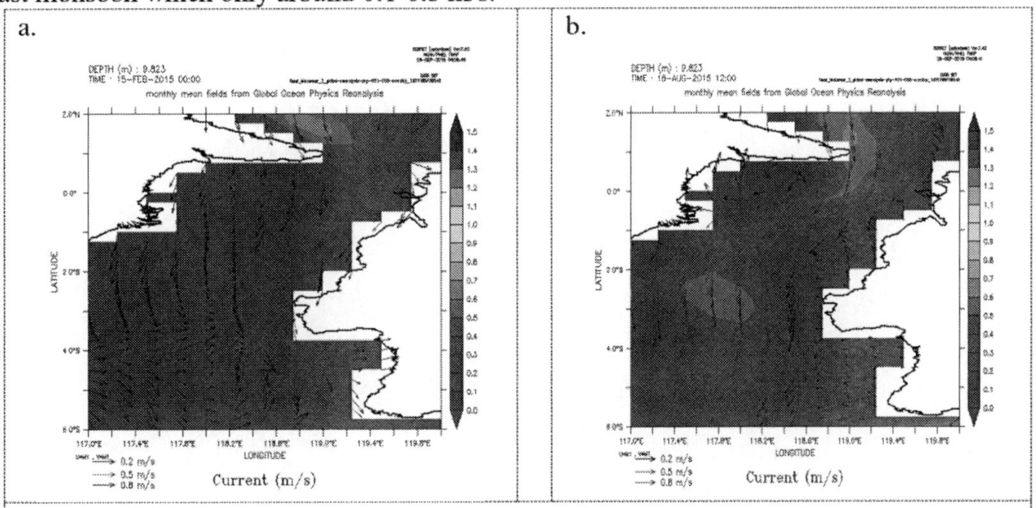

Figure 10. Ocean Current Speeds and Direction in February (a), August (b) at a depth of 9,823 m (Mixed Layer) of the Makassar Strait in El Nino 2015 period

The ocean currents in 9,823 m which there in mixed layer during El Nino 2015 period still show the same current pattern in the La Nina 2010 period. However, in El Nino period, the current speeds in February reached 0.2 m/s and in August the current speeds increased to 0.6 m/s where the current speeds in this year is weaker than in 2010 when La Nina period.

4. Conclusion

The current pattern in mixed layer in the La Nina 2010 period is stronger than the El Nino 2015 period with the highest intensification occurring in East Monsoon (August). The mixed layer temperature distribution during the La Nina 2010 period was higher than the El Nino 2015 period and was more homogeneous in the west monsoon than the east monsoon. The mixed layer salinity distribution when the La Nina 2010 period was lower than the El Nino 2015 period and more varied than the temperature distribution. El Nino and La Nina were more clearly influential on east monsoon (August).

5. References

[1] Aldrian Edvin 2008 *Meteorologi Laut Indonesia* Jakarta: Badan Meteorologi dan Geofisika.
[2] Ilahude A G, Nontji 1999 *Oseanografi Indonesia dan Perubahan Iklim Global (El Nino dan La Nina)* Makalah Lokakarya: Akademi Ilmu Pengetahuan Indonesia, Jakarta.
[3] Haikal M V 2012 *Jurnal Perikanan dan Kelautan* **3** (1) Universitas Padjajaran: Fakultas Perikanan dan Ilmu Kelautan.
[4] Pickard G L and Emery W J 1990 *Descriptive Physical Oceanography: An Introduction* Fifth (Enlarged ed. Oxford: Pergamon Press).
[5] Bureau of Meteorology. http://www.bom.gov.au/
[6] NOAA Climate Prediction Center. http://origin.cpc.ncep.noaa.gov.

Acknowledgements

Thank you to all parties who helped implement this research. This research is part of the Practical Work at the Institute for Marine Research and Observation, Bali.

AUTHOR INDEX

AUTHOR INDEX